Space Shuttle Columbia

Her Missions and Crews

D1290724

Ben Evans

Space Shuttle Columbia

Her Missions and Crews

Published in association with

 Springer   **Praxis Publishing**
Chichester, UK

Ben Evans
Space Writer
Atherstone
Warwickshire
UK

SPRINGER–PRAXIS BOOKS IN SPACE EXPLORATION
SUBJECT *ADVISORY EDITOR*: John Mason B.Sc., M.Sc., Ph.D.

ISBN 0-387-21517-4 Springer Berlin Heidelberg New York

Springer is a part of Springer Science + Business Media (*springeronline.com*)

Library of Congress Control Number: 2005928166

Cover design: Jim Wilkie
Project Copy Editor: Alex Whyte
Typesetting: BookEns Ltd, Royston, Herts., UK

Printed in Germany on acid-free paper

To Michelle – for always being there

Contents

List of illustrations

Preface

On 1 February 2003, high above Texas, the unthinkable happened. With horrifying suddenness, Space Shuttle Columbia disintegrated during her descent from a highly-productive 16-day science mission. Her entire crew of seven, including Israel's first astronaut, perished in the disaster. On that terrible Saturday, a spectre that had haunted NASA for 17 years, since the 1986 loss of Challenger, returned with a vengeance. I was only nine years old when Challenger exploded and remember little of what happened; the destruction of Columbia, by stark contrast, seemed far closer and more personal.

Almost two years earlier, in the spring of 2001, I was fortunate to speak to Rick Husband, who was in command of Columbia on her final mission. I found him to be courteous and warm, with an enthusiasm and openness so typical of many spacefarers. An avid collector of astronaut autographs, I have personalised signed portraits of five of the ill-fated STS-107 crew – Husband, Willie McCool, Dave Brown, Kalpana Chawla and Mike Anderson – and countless others of Columbia veterans. I treasure them all, but those of the STS-107 fliers I now prize particularly highly.

This book is not meant as a study of what happened to Columbia on her last flight, but rather as a celebration of her entire incredible career. Over the past two years, I have spoken to many people who, upon hearing the name of America's first Space Shuttle, have remembered her only as "the one that broke up during re-entry". I feel that this is, to say the very least, an unfair epitaph. Columbia has a long and chequered history and in her 28 missions since April 1981 has scored some remarkable triumphs.

She was the first 'used' manned spacecraft to be blasted into orbit more than once. She supported the debut of the European-built Spacelab research facility and Canada's Remote Manipulator System (RMS), the latter of which is routinely used today to build the International Space Station. She was first to prove the Shuttle's worth as a mini-space station in its own right and as an Earth-circling launch pad from which to boost satellites into geosynchronous and higher orbits to provide communications and reconnaissance, support research in geodetics and electro-dynamics, process semiconductors and peer into the Universe with state-of-the-art X-ray and ultraviolet eyes.

Columbia has also plucked spacecraft *from* orbit in delicate, 28,000 km/h orbital ballets, staged ambitious spacewalks to upgrade the Hubble Space Telescope and holds the current record for having flown the longest Shuttle mission, at almost 18 days. Yet this is scarcely half of her story. Her history is far more than 'just' 28 standalone mission reports: she has countless other tales from the 126 men and women, representing eight different nations, who have spent more than 300 days gazing Earthwards or outwards into the Universe through her windows.

My goal in writing this book was to gather technical esoterica about Columbia and attempt to balance it with 'human' stories from the remarkable pilots, physicists, engineers, doctors and professors who flew on her. I have attempted to weave 'mundane' technical facts, payload details and processing issues with stories of "incredible" rides into orbit, "oh, wow" glimpses of the grandeur of Earth, the unique and highly coveted ability to pull one's trousers onto both legs at the *same time* and a congressman's light-hearted disappointment at being greeted with Californian, rather than Floridian, oranges and grapefruits upon landing at Edwards Air Force Base.

At the time of writing, Space Shuttle Discovery is only weeks away from undertaking the crucial Return To Flight mission and NASA's three surviving orbiters will have their work cut out over the next five years if they are to complete the International Space Station before being retired in 2010. It is with intense sadness, as astronaut Jay Buckey once said, that Columbia will never take pride of place in the Smithsonian; but her mission and legacy are far from over. In fact, her shattered remains are now used by engineers and materials scientists to help to design future hypersonic vehicles.

A fitting tribute, if ever one were needed, for a quite remarkable spacecraft.

Acknowledgements

This book would not have been possible without the assistance of a number of individuals far more knowledgeable than I on the intricacies of the Solid Rocket Boosters, Shuttle main engines, External Tank fuelling procedures, crew training activities and, of course, the operation of the orbiter's multi-million-dollar toilet. I must thank the ever-patient Clive Horwood of Praxis for his support and advice, David Harland for reviewing each chapter, pointing out my mistakes and sharpening up the text, and to the project's copy editor, Alex Whyte.

I owe an immense debt of gratitude to Ed Hengeveld for 'saving my bacon' by kindly giving up his time to supply high-quality illustrations for this book from his own extensive collection. Had it not been for his efforts, and David Harland's tireless work to prepare illustrations for final production, this project would not have reached fruition.

Several astronauts who have flown Columbia have provided fascinating insights into what they all agree was an incredible machine and without doubt the flagship of the Shuttle fleet. Thanks go to Gordon Fullerton and Vance Brand and the late Rick Husband for taking the time to speak with me over the telephone, and to Kacy Carraway, Roberta Ross and Beth Hagenauer of NASA for arranging the interviews.

My interest in space has spanned two decades and might not have lasted so long if not for a remarkable group of friends at the Midlands Spaceflight Society. Andy Salmon has proved to be a goldmine of space information, with the infectious enthusiasm of a true expert in his field, while Rob and Jill Wood have shared more facts about astronauts and cosmonauts with me over the years than I suspect even the astronauts and cosmonauts *themselves* know! None of my work would have seen the light of day were it not for Mike Bryce, editor of the society's newsletter, *Capcom*.

To each of them, I extend my thanks. My family have constantly supported my interest and I must thank my fiancée, Michelle Chawner, for her endless love and enthusiasm; it is to her, for making all this possible, that I would like to dedicate this book. Additional love and thanks go to my parents, Marilyn and Tim Evans, to Sandie Dearn and Ken Jackson, to Malcolm Chawner and Helen Bradford and, of course, to our golden retriever, Rosie.

1

"It will take a hundred flights ..."

THE ASTRONAUTS' ASTRONAUT

John Young was out of this world when he learned that NASA's plans for a reusable manned spacecraft called the Shuttle had finally won Congressional approval.

It was Sunday 23 April 1972 and the 41-year-old Young stood in the sun-drenched desolation of the Cayley plains, participating in the penultimate Apollo lunar-landing mission. A short, dark-haired man with a quiet, country-boy drawl, he had been an astronaut for 10 years and even now, on his fourth spaceflight, showed little desire to do anything else. In fact, when he first set foot on the Moon two days earlier – becoming only the ninth person to do so – his words included the enigmatic phrase, "I'm glad they got ol' Brer Rabbit, here, back in the briar patch where he belongs."

Some Apollo historians have explained the quote by identifying Young himself with Brer Rabbit and the briar patch with his love of space exploration. If this was the case, it could hardly have been more fitting, for until 2002 Young held the record for having been launched into space more times than any other human being, with an impressive six missions under his belt. Even the respected Shuttle astronaut Jerry Ross, who finally broke the record, has pointed to Young as his personal hero. Truly, John Young has become 'The Astronauts' Astronaut'.

He came to NASA in 1962, a little over a year after President John F. Kennedy had promised to land a man on the Moon before the end of the decade. Even at the age of just 32, Young's credentials as a test pilot were already impressive: earlier that same year, he had set world time-to-climb records in the F-4 Phantom fighter. His first four spaceflights were devoted to accomplishing, step-by-step, the complex chain of objectives – engineering tests, rendezvous and docking exercises and risky spacewalks – needed to achieve Kennedy's goal.

His first flight, in March 1965, was a short, five-hour 'hop' on board the first two-man Gemini spacecraft, which did little more than whet his appetite. During the mission, he and Commander Gus Grissom had the task of showing that the Gemini, which would be used to rehearse procedures needed for lunar missions in the relative safety of Earth orbit, was spaceworthy. The flight succeeded and Young returned with a reputation as something of a practical joker, having craftily smuggled a corned beef sandwich – Grissom's favourite fare – on board the Gemini before launch and offered it to his partner while in orbit.

A year later, he was back in space on board another Gemini; this time as Commander, teamed with rookie Mike Collins. Their three-day mission featured rendezvous with two unmanned Agena target rockets, one of which boosted them into a higher orbit, en route to the other. Collins also made a spacewalk to recover micrometeoroid material affixed to one of the Agenas. On Young's third mission, Apollo 10 in May 1969, he and his crewmates conducted a rehearsal of the first lunar landing in orbit around the Moon. This set the stage for Armstrong and Aldrin's "one small step" two months later.

Three years after his return from Apollo 10, and as one of NASA's senior astronauts, Young finally stood in the unrelenting glare of the lunar Sun, in a place where the temperature difference between daytime and nighttime could top 400 Celsius, and joined an elite club of moonwalkers that even today numbers no more than a dozen. In the wake of such a stupendous achievement, one could be forgiven for expecting Young's astronaut career to end after his return to Earth. Surely he could do few other things in his professional life to match or possibly upstage a Moon landing?

Young could not have felt more differently and unlike so many of his fellow astronauts, who left NASA for pastures new, he was eager to tackle America's next challenge in space. His devotion was perfectly epitomised that Sunday in April 1972 when, as he stood amidst the grandeur of the ancient lunar mountains, his breathing harsh and laboured after a long day's work in a bulky spacesuit, he received a call from Mission Control. It was from a rookie astronaut named Tony England, who was acting as the control centre's liaison (nicknamed the 'Capcom') with the men on the Moon.

"This looks like a good time for some good news," England began. "The House passed the space budget yesterday, 277 to 60, which includes the vote for the Shuttle."

Immediately, and in unison, Young and fellow moonwalker Charlie Duke exulted, "Beautiful! Wonderful! Beautiful!" Then Young quietly added, "The country needs that Shuttle mighty bad. You'll see."

Four days later, Young, Duke and the third member of their crew, Ken Mattingly, splashed down in the Pacific Ocean, their 11-day mission over. For Duke, it would be the end of his astronaut career; following a foray into the world of business, he became a born-again Christian and later described his experiences as being so much more fulfilling that walking on the Moon was "the dust of my life" in comparison. Both Young and Mattingly, on the other hand, remained with NASA and each would command two Shuttle missions during the course of the 1980s.

By the mid-1970s, Young had retired from the US Navy with the rank of Captain and was able to concentrate fully on his new duties as chief of NASA's Astronaut Office and immerse himself in the development of the Shuttle. It seemed inevitable, with his breadth of expertise, that he would be a leading contender to command its maiden flight into orbit; by the middle of 1978 it was official and he began training with rookie astronaut Bob Crippen for what would be his fifth mission overall and, in many ways, the most challenging of his entire career.

BIRTH PAINS

With less than eight weeks to go before her first orbital flight, Space Shuttle Columbia finally got the chance to flex her muscles on 20 February 1981. By now, the Shuttle was running three years behind its advertised schedule; its first launch was originally targeted for 1978 and its highest-profile mission – a delicate orbital ballet to reboost America's Skylab space station and prepare it for reoccupation – had been missed. Unexpectedly fierce solar activity in the closing months of the decade caused Earth's atmosphere to inflate, increasing air drag at orbital altitude and Skylab burned up during re-entry in July 1979.

Billions of dollars had been invested in the Shuttle, which was to be the most advanced spacecraft yet to depart Earth. The achievement had not, however, come without problems. Since the original contracts to build the Shuttle had been signed almost a decade earlier, its designers had faced setback after setback: frustrating problems with the development of a patchwork of heat-resistant tiles to shield it during its searing, high-speed re-entry and maddening failures of its throttleable, liquid-fuelled main engines. There was political fallout, too, with the Shuttle's powerful Congressional opponents questioning the need for a multi-billion-dollar reusable manned spaceplane.

For this was another of its advertised qualities: the Shuttle, said NASA, would be the world's first reusable manned spacecraft, capable of flying once every fortnight and carrying commercial satellites, scientific laboratories, space probes, astronomical instruments and – for the first time – *ordinary civilians* into orbit. Plans were already afoot to send teachers, journalists and foreign nationals into space, with up to seven seats available on each flight. The Shuttle, it seemed, was aptly named: it would whisk people into orbit frequently, reliably, relatively cheaply and in conditions a world away from the cramped, one-use-only capsules of the 1960s.

Before such an advanced machine could be declared 'operational', it had to be exhaustively tested. Many of these tests had taken place during and after its construction and a series of high-altitude approach and landing runs were conducted in mid-1977 using a dummy vehicle called Enterprise, taken aloft by an adapted 747 airliner. Fred Haise, one of her pilots, later called it "a magic carpet ride". Although she was never actually capable of flying into space and now sits gathering dust in the Smithsonian, Enterprise demonstrated the Shuttle's aerodynamic performance and ability to make precision landings on predetermined runways.

Ground tests, however, were no substitute for actually flying it in space. Original plans called for six Orbital Flight Tests (OFTs), each carrying two astronauts – a Commander and Pilot – to demonstrate the Shuttle's capabilities, test its manoeuvrability and evaluate its Canadian-built robot arm with different-sized payloads. The number of OFTs was later reduced to four, all of which would be flown by Columbia, the first space-rated Shuttle. Assuming that all four went to plan, the system would be declared 'operational' and become eligible to fly commercial missions on its fifth flight.

Columbia was physically identical to Enterprise, at least at first glance. Both vehicles were not dissimilar in shape and dimensions to the DC-9 airliner, roughly 36 m long with wings spanning 24 m from tip-to-tip, and comprising a two-tier cockpit, cavernous, 18-m-long payload bay with clamshell doors and an aft compartment to house a cluster of three main engines, bulbous Orbital Manoeuvring System (OMS) pods and vertical stabiliser fin.

Unlike Enterprise, however, Columbia would fly further than just the last few minutes from the low atmosphere to the runway. She would, for the first time, undertake the violent climb into space under the combined thrust of her main engines and two behemoth Solid Rocket Boosters (SRBs), withstand the swinging extremes of heat and cold in Earth orbit and bear the full brunt of a fiery descent back through the atmosphere. Moreover, in spite of carrying thousands of items whose failure could doom the Shuttle and kill the crew, Columbia's audacious first launch would be done *with astronauts on board!*

A CALCULATED RISK?

Never in the history of the American space programme had a crew been on board for the first launch of a new spacecraft. The Mercury capsules, which carried Al Shepard and John Glenn on their historic ventures into space, had been extensively tested unmanned, as had Gemini and Apollo. The Soviets, too, had never flown a crew without first testing their spacecraft in an unmanned capacity. The risks were just too high. There can be little doubt, therefore, of the heroism and bravery of the first Shuttle crew: Commander John Young and Pilot Bob Crippen.

One astronaut closely involved in the Shuttle's development was Fred Haise, a veteran of the ill-fated Apollo 13 mission. He had been instrumental not only in its aerodynamic testing in the low atmosphere, but also would probably have commanded one of its early missions had he not resigned from NASA in 1979. He saw an unmanned first flight as potentially much trickier than a manned one. "It would have been very difficult to have devised a scheme, in my view, to have flown [the Shuttle] unmanned," Haise recalled in a 1999 interview.

"I guess you could've used a [communications] link and really had a pilot on a stick on the ground like they have flown some other programmes. But to totally mechanically programme it to do that, and inherent within the vehicle, would have been very difficult. There *was* initially a planned unmanned flight, [but] it was of great complexity and handling the myriad of potential systems problems [made it hard to automate]. With a crew on board [to] be able to handle the multitude of things that you could work around, inherently made the success potential of a flight a lot greater."

Others, including NASA's former director of engineering and development, Henry Pohl, were more sceptical about sending the first Shuttle aloft with a crew on board. "I didn't see any need in risking humans and I didn't think humans would be as proficient as automated equipment," he said. "By that time [the late 1970s], we had the know-how and we could build robots or the automated equipment that can detect things long before a human can detect it, and I thought the vehicle was going to be so difficult to land that we really ought to land it with automated equipment."

THE BUTTERFLY AND THE BULLET

Paradoxically, compared to flights that would follow, the first mission of the Space Transportation System – dubbed 'STS-1' – was relatively straightforward. Its objective was to fly Columbia into space, test her systems and bring her home two days later to a desert landing in California. Yet with so many unknowns and a history of technical problems, it was also the most complicated mission ever attempted. Not only would Columbia herself be tested, but so would the untried boosters and the giant External Tank (ET) that would feed the Shuttle's engines with over 1.9 million litres of liquid propellants.

"Like bolting a butterfly onto a bullet" was how veteran astronaut Story Musgrave, who flew Columbia in late 1996, described the unusual appearance of the combined Shuttle, tank and boosters. It is an appropriate description. The 46.6-m-long ET, reminiscent of an enormous aluminium zeppelin standing on end, is indeed bullet-like, but is actually far more than 'just' a container. Its upper quarter houses liquid oxygen and its lower three-quarters carry liquid hydrogen.

Separating the two sections of the ET is an unpressurised 'intertank', which contains instrumentation and umbilical interfaces to the launch pad's purging and hazardous-gas-detection systems. Above the intertank, the liquid oxygen tank holds up to 542,640 litres of oxidiser and, beneath it, the liquid hydrogen tank holds around 1.4 million litres of fuel. Both are then fed through two 43-cm-wide fuel lines into disconnect valves in the Shuttle's aft compartment, and from thence into the main engines' combustion chambers.

"[The main engine is] very high-performance," said Henry Pohl, "[with a] very high chamber pressure for that day and time [and] very lightweight for the thrust that they were producing. I would say that we came out with that program in the only time in [US history] when it would have been successful. If we had waited another two years before starting development on [the] Shuttle, we probably would not have been able to do it, [because] the people that designed the main engine were the same that designed [previous rocket] engines. That group of people had designed and built seven different engines before they started the Shuttle development. A lot of [them] retired [and so] if we'd waited another two or three years, those people would all been gone and we would have had to learn all over again on the engine development."

Built by Rocketdyne – formerly part of the Shuttle's prime contractor, Rockwell International, but now owned by Boeing – the engines burn for about eight minutes of ascent and are shut down a few seconds before the ET is jettisoned, right on the edge of space. Each engine measures 4.2 m long, weighs 3,400 kg and is 'throttleable' at 1% incremental steps from 65% to 104% rated thrust. This ability, which is controlled by the Shuttle's onboard General Purpose Computers (GPCs), helps to reduce stresses on the vehicle during periods of maximum aerodynamic turbulence.

Despite the immense thrust generated by each engine and the colossal amount of propellant needed to run them for such a short length of time, they in fact provide only 20% of the power needed to get the Shuttle into space. The remainder comes from the two 45.4-m-long SRBs, which are the only solid-fuelled rockets ever used in conjunction with a manned spacecraft. Loaded with a powdery aluminium fuel and an oxidiser of ammonium perchlorate, the boosters, built by Morton-Thiokol in Utah, are mounted like a pair of Roman candles on either side of the ET.

This unusual combination, referred to as 'the stack', is not, in fact, totally reusable and came about following a series of financial and technical compromises dating to the early 1970s. The Shuttle in its present form is designed to fly a hundred times before major modifications become necessary and the SRBs about a quarter of that figure. The ET, on the other hand, is discarded about eight-and-a-half minutes after launch to burn up in the atmosphere over the Indian Ocean. It was considered more costly to modify the tank for reusability than to simply build a new one for each mission.

"The Shuttle is an asymmetric vehicle," said former NASA flight director Neil Hutchinson. "It doesn't look like it ought to launch right, because it's not a pencil! Some of us, in the early days, wondered how that was going to work. In fact, it's [still] a very tricky vehicle to launch. It has to be pointed carefully in the right direction at certain times or you'll tear the wings off or tear it off the External [Tank]. It's *not* a casual launch process."

THE ROCKY ROAD TO STS-1

Preparing for each Shuttle flight takes several years, but the actual bringing together of the components begins with setting up the boosters on a Mobile Launch Platform (MLP) in the gigantic Vehicle Assembly Building (VAB). This 160-m-tall structure – the world's largest scientific building, so vast that *clouds* once formed in its upper reaches before an air-conditioning system was fitted – has dominated the swampy landscape of the Kennedy Space Center (KSC) on Merritt Island in Florida for the best part of four decades. It was used to assemble the massive Saturn V Moon rockets and, since 1980, the Shuttle.

Each SRB comprises six blocks, called 'segments', each of which is hauled with pinpoint precision into place, one on top of the other. To prevent a leakage of searing gases during ascent, a series of rubbery O-rings seal the joints between the segments. After propelling the Shuttle and ET to an altitude of about 45.7 km, explosive rockets at the nose and tail of each booster push them away and parachutes lower them to a gentle splashdown just off Cape Canaveral in the Atlantic Ocean. They are then recovered, refurbished and reused.

When the assembly of the SRBs is complete, the ET is moved into position between them and connected by a series of spindly attachment struts. Following checks of their mechanical and electrical compatibility, the Shuttle itself is moved from the nearby Orbiter Processing Facility (OPF), tilted on its tail and mated to the ET. In the early days of the programme, the Commander and Pilot of the mission boarded the vehicle while it was in a vertical position inside the VAB to rehearse pre-launch procedures. Nowadays, this is done by the entire crew after the Shuttle has been rolled to the pad.

The transfer of the 1.8-million-kg Shuttle stack from the VAB to one of two pads

Attached to her External Tank and Solid Rocket Boosters, and mounted on top of the crawler, Columbia is readied for her first launch.

of Launch Complex 39 – a distance of 5.6 km – takes six hours, with the aptly named 'crawler' inching the MLP and its precious, $2-billion national asset along a track made of specially imported Mississippi river gravels. Once the stack is 'hard-down' on the pad surface, further checks are conducted, payloads installed and the crew participates in a Terminal Countdown Demonstration Test (TCDT), essentially a full dress-rehearsal of the last part of the countdown, followed by a simulated main engine failure and emergency escape procedures.

By 20 February 1981, these preparations in readiness for STS-1 had long been completed; in fact, attached to her tank and boosters, Columbia had sat majestically on Pad 39A since 29 December the previous year. She had been at KSC for even longer. Her construction took almost five years from the start of work to build her cockpit in June 1974 to rollout of the finished article in March 1979. A week later, to the amazement of motorists in the sweltering California heat, Columbia was towed overland from prime contractor Rockwell International's Palmdale plant to Edwards Air Force Base.

By late March, she had been flown 'piggyback' on top of the modified 747 aircraft to KSC and ensconced in one of two bays in the OPF. The latter is positively dwarfed by the immense VAB and is still used to prepare the Shuttle fleet for their missions, to repair and refurbish them and to install and remove their payloads. It is, however, far more than just a spacecraft hangar; due to the extreme volatility of the propellants carried on board the Shuttle, the OPF is fitted with detectors that are so sensitive to explosions that visitors are forbidden from using camera flashes when taking photographs.

After her arrival at KSC, Columbia underwent a protracted period of pre-launch preparations that lasted almost two years. Although she was structurally 'complete', she was far from ready to fly. She had no main engines, her thermal protection system needed attention and her ET and SRB segments were not destined to arrive until the summer of 1979. By the end of the following year, however, significant progress had been made and in November Columbia rolled into the VAB for stacking. Following checks, she moved to the launch pad a few days after Christmas in readiness for launch the following spring.

The thermal protection system – particularly its thousands of tiles, each of which was *individually designed* and not interchangeable – had been the biggest headache during this time because of the sheer novelty of its design. "When it took off on the back of the 747 from Palmdale, a whole bunch of the tiles came off as they went down the runway," former Shuttle manager Arnie Aldrich wryly recalled. "That led to the requirement to have a better understanding of how the tiles were attached and how to know they were well attached and that problem took two years to solve."

Much work still had to be done before Young and Crippen could even board the orbiter, however. One of the most critical exercises was a Wet Countdown Demonstration Test (WCDT), which lasted six days and culminated on 20 February 1981 in a 20-second firing of Columbia's engines. This Flight Readiness Firing (FRF) was necessary to demonstrate their ability to throttle between 94 and 100% thrust, and gimbal just as they would be expected to do in flight. Similar 'wet' – or fully fuelled – tests had been performed before the Saturn V launches, although on those occasions the engines had not been test-fired.

Preparations for the FRF proceeded in a manner not dissimilar to a real countdown: launch controllers started the clock at T − 53 hours when they powered-up the SRBs, ground-support equipment and Columbia herself. Four seconds before the simulated 'liftoff', the Shuttle's engines roared to life at 120-millisecond intervals, reaching 90% rated thrust within three seconds and hitting the 100% mark precisely at T − zero. Three seconds later, engineers simulated retracting the ET umbilical and the SRBs' hold-down posts; a further 15 seconds elapsed before shutdown commands were issued to all three engines. The test was a success and a significant milestone had been cleared.

According to the STS-1 press kit, released around this time, the launch was provisionally booked for "no earlier than" 17 March, but a number of technical issues and a human tragedy conspired to delay Columbia's first flight by several weeks. Following the FRF, engineers had to repair a section of super-light ablator insulation, which had become debonded from the ET during a test of its cryogenic propellants back in January. This pushed the target date for launch back to 5 April, followed by another delay until the 10th, caused by a strike against Boeing by machinists and aerospace workers.

Throughout March, the attention of the world's media focused on Young and Crippen as they maintained their proficiency training, participating in a TCDT and practising how to escape from Columbia in the event of a main engine failure seconds before launch. The biggest fear in such a scenario was the presence of invisible hydrogen flames, through which the pressure-suited astronauts would have to run to reach a slidewire escape basket that would whisk them from the 58-m level of the launch pad down to the ground and a waiting M-113 armoured personnel carrier.

Although Young and Crippen had not been directly involved in the Enterprise approach and landing runs – which had been conducted by two other teams of astronauts, Joe Engle and Dick Truly and Fred Haise and Gordon Fullerton – they nevertheless achieved proficiency in flying the Shuttle Training Aircraft (STA). This Grumman Gulfstream had been modified to fly almost exactly like the Shuttle, approaching the runway at several times the angle of a commercial aircraft and nearly twice the speed. The men also honed their skills in flight software laboratories, visual motion simulators, full-scale Shuttle mockups and on board Columbia herself.

It was just a few days after Young and Crippen had returned to Houston, following their TCDT, when the Shuttle claimed its first two lives. Several technicians working inside Columbia's aft compartment were rendered unconscious by a dangerous build-up of nitrogen gas; and although they were pulled out, one died that same day and another two weeks later. The cause was traced to a breakdown of communications: a warning sign had been mistakenly removed and a supervisor called away. Crippen would later pay a personal tribute to the two dead men, John Bjomstad and Forrest Cole, while in space.

SOFTWARE: THE BIGGEST STUMBLING BLOCK

Meanwhile, despite the setbacks, Columbia was now firmly on schedule to lift off on 10 April 1981, which, as it happened, was two days shy of the 20th anniversary of Yuri Gagarin's pioneering spaceflight. The six-and-a-half-hour launch 'window' – necessitated by a need to have adequate lighting conditions to satisfactorily photograph Columbia's ascent for engineering analysis – opened that day at 11:50 am GMT.* The window also provided for daylight landing opportunities at White Sands Missile Range in New Mexico, should a launch abort require Young and Crippen to perform an emergency return to Earth after one orbit.

Shortly before 9:00 that morning, after a traditional astronauts' breakfast of steak and eggs in the crew quarters, Young and Crippen boarded Columbia for what turned out to be an uneventful countdown – at least, that is, until its final stages. Then, with just nine minutes to go, during a pre-planned hold, a problem cropped up in one of Columbia's five GPCs. It was described as a 'timing skew'; in effect, the backup flight software was unable to synchronise with the primary set.

Unlike previous manned spacecraft, the Shuttle is totally dependent upon its computers to run the main engines, move the elevons, control its heading and operate the thrusters, to name just a few of many thousand different functions. These units are so critical that five GPCs are carried: four primaries, which run the same software and 'vote' before issuing commands, and a backup. If one of the primaries disagrees with the others, it is 'outvoted' and considered faulty. The backup contains its own, different set of flight software, so that if *all four* primaries became corrupted, it can take over control.

The problem that Columbia experienced on 10 April was essentially that the four primary GPCs were not communicating with each other correctly. Taking advantage of the lengthy launch window, the liftoff was rescheduled for 3:20 pm as computer engineers wrestled with the software, but when a solution could not be found it was decided to stand down until 12 April. A disappointed Young and Crippen clambered out of Columbia and would spend the next couple of days maintaining their proficiency flying the STA.

Meanwhile, the GPC problem was isolated late on the 10th and the countdown resumed next day. "The software", remembered Gordon Fullerton, who flew Columbia in March 1982, "became the biggest stumbling block. The software in these computers not only control where you fly and the flight path, but almost *every other* subsystem! Getting the software wrung out and simulators writing the checklists ... we didn't really have it nailed down by STS-1. There were a lot of

---

* All times throughout this book are given as Greenwich Mean Time (GMT). This has been done to avoid possible confusion, as different time zones apply in Florida, California, New Mexico and other locations mentioned in these pages. As a general rule of thumb, Florida is approximately five hours 'behind', New Mexico some seven hours 'behind' and California around eight hours 'behind' GMT.

unknowns [but] you just finally have to set a launch date and say 'We're going to go'. You cannot be 100% sure of everything."

Young and Crippen again departed the crew quarters in the early hours of 12 April and took their seats on Columbia's flight deck. Both were clad in bulky US Air Force high-altitude pressure suits, which afforded them full-body protection and were destined to be worn by the first four OFT Shuttle crews. Since these were considered 'test flights' and were also equipped with ejection seats, the full-pressure garments were mandatory; on later operational missions, when restrictions were relaxed somewhat, it was intended for astronauts to fly in lighter overall-type flight suits and helmets.

If an emergency had necessitated their use, the rocket-propelled ejection seats would have fired Young and Crippen through two overhead hatches, but they could only be used to an altitude of 30.5 km, meaning they would not realistically work during the ascent phase and only at selected intervals during re-entry. Astronaut Jack Lousma, who commanded Columbia's third test flight in March 1982, would later remark that his Shuttle launch was far riskier than his Apollo ascent a decade earlier and his opportunities to escape in the event of an emergency were much reduced.

Whereas most previous manned spacecraft had taken the form of ballistic capsules attached to the top of expendable boosters – so that, in the event of problems, an escape rocket could lift them several thousand metres into the air and parachute them a couple of kilometres out to sea – the Shuttle did not offer that option. An on-the-pad emergency would have precluded the use of the ejection seats, because the astronauts would have hit the ground before their parachutes had opened; also, ejections during the first couple of minutes of ascent would have sent them straight into the SRBs' roiling exhaust plumes.

Consequently, the seats could only realistically have been used at selected points of re-entry, after the period of maximum atmospheric heating, and even then the astronauts' chances of survival were slim.

Young and Crippen again boarded Columbia on 12 April, lowered their visors and encountered their first problem: neither man could breathe properly. It turned out that a quick-disconnect fitting for the oxygen system, situated beneath the control panel, had been mispositioned. After this was resolved, the countdown proceeded smoothly.

RIDE OF A LIFETIME

The history of jokes and pranks between astronauts and the ground crews responsible for strapping them into the spacecraft before launch has become the stuff of legend, since the days of pioneering Mercury missions in the early 1960s. "John Young made a big deal about the size of the American flag on his suit," said KSC spacesuit technician Jean Alexander. "It came in with kind of a small version and they got several sizes before he was satisfied and it was kind of a joke. So on launch morning, there was a motel that we stayed at Cocoa Beach and they had this

STS-1 liftoff.

huge flag on a pole [outside] a real-estate office next door. One of the suit tech[nician]s that was down there for launch talked the real-estate people into letting him take that flag down and he took it to the suit room for suit-up morning and had it actually cover[ing] one whole wall! When John walked in, he said 'John, is *that* big enough?'." The mood was sufficiently lightened for what was to follow.

After almost four years training together for the most complex engineering and test flying challenge of their careers, the smoothness of the countdown on only their second attempt surprised both Crippen and his veteran colleague. As the clock ticked inside the final minute, their excitement began to build: despite Young's vast experience and four previous missions to his credit, both men were rookies as far as flying the Shuttle was concerned. Neither man fully knew what to expect.

Six seconds before midday, with a low-pitched rumble that soon turned into a thundering crescendo, Columbia's three main engines ignited. Young and Crippen would later recall that the Shuttle rocked perceptibly backwards and forwards, accompanied by a sharp noise increase in the cabin. Then, precisely on the hour, in front of an estimated three-and-a-half thousand media spectators at KSC, and doubtless hundreds of thousands more glued to television sets around the world, came the ear-splitting crackle of the two SRBs.

"We have liftoff of America's first Space Shuttle, and the Shuttle has cleared the tower," exulted the launch commentator over the public-address system as Columbia broke the shackles of Earth and lumbered off the pad. Crippen would later comment that, although the low-pitched roar of the main engines certainly grabbed their attention, it was the punch-in-the-back ignition of the SRBs that convinced them that they were really heading somewhere.

For the first few seconds, as the Shuttle cleared the tower and roared into the clear Florida sky on top of the two dazzling orange columns of flame from its boosters, the cockpit instruments were blurred by the vibrations, but according to the crew were still just about readable. By the time Columbia rolled onto her back under GPC control about 10 seconds after liftoff, setting herself on the correct heading for a 40.3-degree-inclination orbit, the two men reported that the vibrations had lessened to a point that allowed them to read their instruments without problems for the remainder of the ascent.

"When you get the vehicle going uphill and you're still in the 'sensible' atmosphere," said Neil Hutchinson, "there are tremendous aerodynamic pressures on it and you have to get the angle at which it is going through the airstream exactly correct. [The vehicle] has a *very* narrow performance corridor. In order to get the proper inclination, when the Shuttle takes off, it 'rolls'. What it's doing is getting itself oriented so [it] goes into orbit on its back. It goes upside down, with the crew *upside down*. You've got to get that roll out of the way and get that whole thing set up long before you get the max[imum] dynamic pressure, [which is] when the amount of atmosphere combined with the direction the vehicle's going and the velocity is the worst."

"As the Shuttle's main engines come up, you really feel the vibrations starting in the orbiter," said Jerry Ross, who has flown the reusable spacecraft a record-tying seven times since 1985, including one mission on board Columbia, "but when the

[SRBs] ignite, I describe it as somebody taking a baseball bat and swinging it pretty smartly and hitting the back of your seat, because it's a real 'bam'. The vibration and noise is pretty impressive! The acceleration level is not that high at that point, but there is that tremendous jolt and you're off!"

At the post-flight briefings, Young would tell engineers that Columbia's ascent was considerably more rapid that he had experienced during his two Saturn V launches to the Moon. Analysis also later showed that STS-1 had caused significant damage to Pad 39A which could have been catastrophic: the shockwaves produced by the Shuttle's engines and the SRBs had buckled a strut linking Columbia to the ET's liquid oxygen tank. Had the strut failed, it was determined, the result could have been the loss of the vehicle and crew and steps were taken to strengthen the struts in readiness for later missions.

"As [it accelerated] in the first 30 seconds or so, the wind noise on the outside of the vehicle became very intense," recalled Ross, "like it was *screaming*! It was *screeching* on the outside!"

A minute into the flight, as Columbia approached an altitude of 15 km, she passed through a period of maximum aerodynamic turbulence which required the GPCs to throttle the main engines back to just under two-thirds of their rated thrust. The passage through this period was described by the astronauts as marked by an increase in the noise and vibration of the engines, although their performance was within expectations. The sound from the SRBs remained sporadic and decreased to virtually nothing as the time approached, 2 minutes and 12 seconds into the flight, for their separation.

Shortly before the boosters burned out, the Capcom, rookie astronaut Dan Brandenstein, told the crew they were now "negative seats", meaning that Columbia was too high to use the ejection seats; questionable though their usefulness would have been. Fortunately, the vehicle was performing admirably. The SRBs actually turned out to generate more 'lift' than predicted and they separated at an altitude 2.9 km higher than anticipated. When the separation rockets fired and the SRBs fell away, Young and Crippen reported a bright, orange-yellow 'flash' which appeared to stream up in front of the Shuttle's nose and back above the front windows.

"As the [SRBs] tail off, like at 1 minute-45 or so [after launch]," said Ross, "it almost felt like you had *stopped accelerating*, like you'd stopped going up. At that point, [you are] already Mach 3-plus and well above most of the 'sensible' atmosphere, some 20 miles high or so. And at [SRB] jettison, then you're at four times the speed of sound and 25 miles high!"

The SRB separation was also accompanied by a harsh grating sound which Young likened to the noise made by the Saturn V's final stage. Both SRBs parachuted into the Atlantic Ocean, splashing down five minutes later about 250 km downrange of KSC. With the cumbersome boosters gone, the crew found it much easier to flip switches in the cockpit. At this stage, the so-called 'T-fail-pitchover' manoeuvre was executed, placing the horizon in their view for the first time, and the two men spotted penny- to fist-sized white particles flooding past the windows.

"What a view! What a view!" radioed a jubilant Crippen four-and-a-half minutes into the ascent.

"Glad you're enjoying it," replied Brandenstein. It was equally as exciting an experience for *him*, sitting at his console in Mission Control, as it was for his two colleagues rocketing into orbit. In fact, although Brandenstein would later fly four Shuttle missions of his own – including one on board Columbia in January 1990 – he has described STS-1 as the most exciting episode of his astronaut career. His main concern was the clearance between the vehicle and Pad 39A and he had listened carefully for the first few seconds, breathing a sigh of relief when the Shuttle cleared the tower safely.

Columbia flew on for six more minutes after SRB separation, reaching Mach 19 – close to 23,340 km/h – at which point her engines were throttled back to maintain a 3g environment in the cockpit. Throughout the ascent, not surprisingly as it was his fifth launch, Young's heart rate rose no higher than 90 beats per minute. That of first-timer Crippen, on the other hand, peaked at nearly 130. After the mission, Young would quip that "I was so old my heart wouldn't go any faster," but according to Ascent Flight Director Neil Hutchinson, "John was kinda asleep at liftoff!"

"As [you get] to about the seven-and-a-half-minute point, [that] is when you get to the 3gs of acceleration, *that's* a significant acceleration," said Jerry Ross. "It feels like there's somebody heavy sitting on your chest and makes it pretty hard to breathe. You have to grunt to talk, and you're just waiting for this 3gs to go away. [This period] is when the orbiter's three main engines start reducing their power output so that you don't exceed the structural limit of 3gs. And so for that last minute, the Shuttle's main engines are coming back. You're getting lighter and lighter. You're accelerating at 100 feet per second [*per second*], which is basically like going from zero to 70 miles per hour *every second*. So it's pretty good. And then at the time that the [orbiter's] computers sense the proper conditions, the main engines basically go from around 70% power, on a 3g acceleration, [then] shut off and you're in zero-g. And for me, I had the sensation of tumbling head over heels: a weird sensation."

At 12:08 pm, some 8 minutes and 32 seconds after leaving Pad 39A to the cheers of thousands of spectators, the main engines of America's first Space Shuttle were shut down and more than 2,000 kg of residual propellant was dumped into space through their nozzles. During the procedure to stow the engines for orbital operations, Columbia's nose unexpectedly pitched 'upwards' about five degrees. Nineteen seconds after the engines went out, the ET was jettisoned to follow a ballistic, suborbital re-entry and burned up over a sparsely inhabited stretch of the Indian Ocean.

The astronauts pulsed Columbia's Reaction Control System (RCS) thrusters to push themselves away from the now-useless ET; they later called it a very obvious "seat of the pants" manoeuvre. They also reported no noise associated with the separation of the tank and that, in fact, the only indication they had was that the red main engine lights on the control panel suddenly went out. Young and Crippen were in space; although still tightly strapped into their seats, the first trace of orbital flight came when bits of debris – washers, filings, screws and wire – began floating around the cabin.

CHECKING OUT COLUMBIA

Their next task was the first of two firings of the big OMS engines to establish Columbia in her correct orbit. This began 10.5 minutes after launch and lasted 90 seconds; it was described by Flight Director Jay Greene as "normal". The crew agreed: "We're looking good." A second OMS burn at apogee, about 35 minutes later, circularised the orbit. Although the burns were satisfactory, the instruments providing quantity readings for the OMS pods turned out to be erroneous. They showed sporadic propellant quantities throughout the flight, often staying constant for some seconds, then changing at faster-than-expected rates.

None of this had been seen in ground tests. Nevertheless, all evaluations of the system during STS-1 – using both thrusters in unison and singly – were performed without incident. To test an emergency procedure, Young and Crippen even successfully fired the right-hand OMS engine using propellant from the *left-hand tank*.

After establishing themselves in a stable, circular orbit around Earth, Young and Crippen turned their attention to opening Columbia's 18.2-m-long clamshell doors and expose the cavernous payload bay to space for the first time. The bay was empty for this first mission, but on subsequent flights it was expected that it would be crammed with commercial satellites, scientific instruments, laboratories – the most important of which was the European-built Spacelab – and major astronomical observatories. For STS-1, it carried sensors to record Columbia's performance and the stresses and strains endured at key points in the mission.

It was essential that the doors were opened within the first hours after reaching orbit, so that radiators attached to their inside panels could dump excess heat into space. If, for whatever reason, the crew had been unable to open the doors, flight rules dictated that they return to Earth at the end of their fifth orbit. Although extra systems were carried to dissipate heat, they could be used for a day at most. If, at the other extreme, difficulties were encountered closing the doors at the end of the mission, Crippen was spacewalk-trained to secure them manually.

Early plans called for STS-1 to fly with the doors closed throughout the mission – relying on Columbia's flash evaporators rather than the radiators – but it was soon realised that opening them was essential to dissipate her heat load.

Had a spacewalk been necessary, the entire cabin pressure would have been reduced from the normal 14.7 psi to 9.0 psi and after nine hours of 'pre-breathing', Crippen would have suited up and entered the payload bay. By lowering the pressure in this way, the crew's 'day' would have been shortened by two hours, and Young pre-breathed in case he needed to go outside to assist Crippen.

Fortunately, Crippen opened the doors perfectly at the end of Columbia's second orbit. He gingerly unlatched the starboard door first – "Here comes the right door and, boy, that is really beautiful out there" – then closed it again to verify the satisfactory performance of its seal. "All the latches work just fine," he told Mission Control, "and the door looks like she's doing her thing." Both doors would be opened and closed on several occasions during the next two days to evaluate not only the seals, but also their latches and actuators.

View of Columbia's payload bay, revealing missing tiles from her OMS pods. Note the two boxes, which form part of the DFI package.

Throughout this procedure, the astronauts worked at the rear of the cockpit, facing a pair of small square windows overlooking the payload bay; they reported that they could work there quite comfortably without needing any kind of foot restraints. Looking into the pristine white, insulation-enshrouded payload bay, there was not a great deal to see. There were no payloads to deploy and not much in the way of experiments to perform. In fact, the only real 'payloads' were the small, unassuming – yet vitally important – boxes of measuring devices, detectors and sensors.

It was known as the Development Flight Instrumentation (DFI), and although it was mainly destined to be used during the first four test flights, several of its components remained on board Columbia during later missions. Weighing 9,290 kg, it provided the first 'real' measurements of the Shuttle's performance and the stresses she endured during launch, ascent, in space and during re-entry and landing. Previous data had only been available through computer simulations and the DFI data was expected to provide the first hard details. It stored this information on three magnetic tape recorders, which were analysed after landing.

Other devices mounted on the DFI pallet were microphones to acquire acoustic data and an array of six different materials – including Teflon and gold – to assess their level of degradation in the harsh low-Earth-orbit environment.

Two more experiments were also carried on STS-1; one of which included actual hardware, and the other which took advantage of Columbia's re-entry flight path. The Aerodynamic Coefficient Identification Package (ACIP) complemented the DFI by collecting data during all flight phases, but particularly during the hypersonic, supersonic and transonic periods of re-entry, in order to help to validate wind-tunnel predictions. As well as helping to advance engineers' understanding of the thermal and structural dynamics of the Shuttle during its glide back to Earth, ACIP measured the positions of each flight surface and gathered about four hours of data.

The second experiment was a wholly passive one, known as the Infrared Imagery of Shuttle (IRIS). It featured nothing in the way of onboard equipment and involved NASA's Kuiper Airborne Observatory taking high-resolution infrared pictures of Columbia's belly and sides as she re-entered the atmosphere.

Young and Crippen had little direct involvement in any of these experiments, which operated autonomously throughout the mission. In any case, they had their hands full with the many engineering tests planned. It was during the first day of the flight, while Crippen was busy evaluating the performance of the payload bay doors, that they noticed several missing thermal protection tiles from one of the OMS pods.

"Okay," Young told Mission Control, "what camera are y'all looking at now, do you know?"

"We're looking out the forward camera," replied the Capcom.

"Okay. We want to tell y'all here we do have a few tiles missing off the starboard pod. Basically, it's got what appears to be three tiles and some smaller pieces; and off the port pod – looks like – I can see one full square and looks like a few little triangular shapes that are missing and we are trying to put that on the TV right now." Young's observations highlighted, for the first time, a problem that would become almost commonplace on Shuttle missions: tiles coming off certain areas of the vehicle during the violent climb into orbit.

He also commented, after a visual inspection, that no tiles seemed to be missing from Columbia's wings, vertical stabiliser or nose. However, it was impossible to determine if tiles had been lost from her belly. Back on Earth, managers watched the transmissions, but decided that none of the missing tiles was in a 'critical' area that might pose a hazard to the Shuttle's re-entry. The most that could happen, they said, was that after landing a patch of aluminium skin underneath some tiles may need replacing. Yet, even at this early stage, a potentially lethal problem with their integrity had reared its head.

A SPACIOUS, SPARTAN HOME-FROM-HOME

Three-and-a-half hours after launch, the astronauts doffed their bulky pressure suits and stowed them in the middeck. With the exception of a donning-and-doffing exercise on 13 April, they would not need them again until just a few hours before re-entry. For the remainder of the mission, Columbia circled the Earth with her topside and open payload bay facing 'down'. In general, the men found that the pristine new spacecraft was performing with very few problems – "The vehicle is performing like a champ, real beautiful," Young told Mission Control – and that life on board was comfortable and positively roomy.

Of course, compared to the cramped capsules Young had flown previously, Columbia *was* indeed voluminous, but conditions were basic and comparatively spartan compared to later Shuttle missions. The men slept in their seats on the flight deck, although plans were already well advanced to carry sleeping bags, bunks and phonebox-sized 'sleep stations' in the future. These would be particularly useful on Spacelab research missions, when crews would be split into two 12-hour shifts.

The 'middeck' was situated directly beneath the flight deck and, in space, astronauts accessed it by floating through a small 66 × 71 cm opening; there were actually *two* openings, but normally only one was used. Essentially, the middeck provided a living area for the crew, including storage lockers for experiments or equipment, sleep stations, a galley, toilet and the huge airlock module providing entry to Columbia's payload bay. Before launch and after landing, the astronauts entered and departed the Shuttle through a circular hatch in the middeck's port-side wall.

Above the middeck, the ten-windowed flight deck – once described by STS-107 astronaut Kalpana Chawla as "our favourite place" – was the location for controlling the Shuttle during ascent, re-entry and conducting the bulk of orbital operations. Its forward portion contained fixed seats for the Commander and Pilot, although on later missions two collapsible seats for Mission Specialists could be mounted directly behind them, and a bewildering array of displays, dials and switches. Six windows wrapped, airliner-like, around the front of the flight deck, with two more in the 'roof' and another two looking back over the payload bay.

Young and Crippen experienced a colder-than-normal first night in space, thanks to a temperature controller problem. Conditions improved on the second night. They prepared their meals using an onboard food warmer, although a larger and more elaborate galley was planned for later missions. Minor problems were experienced with a suction hose on the toilet, which stubbornly refused to work properly. When the time came to return to Earth, Young and Crippen – rather ignominiously for seasoned space explorers – were obliged to stuff paper towels into the hose to prevent it from overflowing and use the urine-collection devices in their pressure suits.

RETURN TO CALIFORNIA!

After a whirlwind two days, the time came to put the Shuttle to its ultimate test: knifing through the atmosphere, subjecting some of the tiles to thermal extremes up to 3,000 Celsius and performing an unpowered, 'deadstick' landing at Edwards Air Force Base in California. Although the last few minutes, from passing subsonic in the low atmosphere to the runway, had been rehearsed using Enterprise, the 45 minutes from the de-orbit burn of the OMS engines, through the searing heat of re-entry and the complex aerodynamic turns needed to 'bleed-off' Columbia's speed and align her for touchdown, were largely unknown.

To play things safe, NASA opted to use the wide expanse of dry lakebed at Edwards, deep in the Mojave Desert, for the first four test flights. This would provide Young and Crippen with a more forgiving runway and greater margins for error, although it was hoped that when the Shuttle became fully operational and its aerodynamic performance was better understood, precision landings on a narrower concrete runway at KSC would become the norm. Four hours before landing, at around 2:00 pm on 14 April, the crew closed and latched the payload bay doors for the last time.

Twenty minutes before the onset of the de-orbit burn that would drop Columbia out of space and into the upper reaches of the atmosphere, the astronauts oriented their spacecraft tail-first, and switched on two of the three Auxiliary Power Units (APUs). These essentially controlled the Shuttle's flight surfaces and hydraulics during re-entry. Fifty-three hours and 28 minutes into the mission, at 5:28 pm, as Columbia flew over the Indian Ocean, the OMS engines ignited in the vacuum, slowing her and beginning her perilous, high-speed glide home.

The burn, which lasted two-and-a-half minutes, was reported in typical matter-of-fact fashion by Young, who told Capcom Joe Allen, "Burn went nominal."

"Nice and easy does it, John. We are all riding with you." Allen's words echoed not only the prayers of the men and women in Mission Control, but also countless observers across the world.

Minutes later, Columbia was turned so that its nose was pitched 'up' at a 39-degree angle, and Young and Crippen removed the safing pins from their ejection seats and overhead hatches and switched on the third APU. As the spacecraft entered the denser portion of the atmosphere, the tracking station in Guam noted increasing amounts of static and the crew reported seeing the yellowish-orange bursts of Columbia's pulsing thrusters reflected in their cockpit windows. Travelling at close to 25,750 km/h, the spacecraft hurtled onwards, the colour of ionised atmospheric gases outside turning from light pink to pinkish-red, then reddish-orange, reminiscent of the inside of a blast furnace.

During this time, the public-affairs commentator at Mission Control reeled off a steady stream of updates: "We will be out of communication with Columbia for approximately 21 minutes. No tracking stations before the west coast and there is a period of about 16 minutes of aerodynamic re-entry heating that communications are impossible [due to the build-up of a plasma 'sheath' around the vehicle]." It was also during this time that the Kuiper Airborne Observatory snapped its infrared picture of the meteoric Columbia hurtling back through the atmosphere.

The aircraft had taken off from Hickam Air Force Base in Hawaii and established itself at an altitude of about 13.7 km, directly underneath Columbia's flight path, about an hour before the start of entry interface. It then recorded a single infrared image of the belly and side of the speeding spacecraft.

As they descended towards Edwards, the astronauts picked up UHF radio calls between Mission Control and one of the T-38 chase planes that would accompany the Shuttle down to the runway. "Hello, Houston," called Young, "Columbia's here. We're doing Mach 10.3 at 188 [thousand feet]." For the majority of this time, except for the 'roll reversals' – a series of S-shaped turns used to reduce speed – the computers flew the spacecraft. Shortly after Columbia had crossed the coastline near Big Sur – to which Crippen radioed excitedly, "What a way to come to California!" – Young took manual control of his ship.

Still travelling at well over four times the speed of sound, the Shuttle passed, as planned, over Bakersfield, Lake Isabella and Mojave Airport, enabling the astronauts to verify by glancing out of the cockpit windows that their ground track was "right on the money." Young then performed a sweeping, 225-degree turn to align Columbia with the lakebed Runway 23 at Edwards. Shortly thereafter, as their altitude dropped to around 12.2 km, he took the stick and later commented that the ship's controllability was crisp and precise.

Watching the arrival of America's first Space Shuttle from orbit were tens of thousands of people, including Larry Eichel of the *Philadelphia Inquirer*. His testimony perfectly illustrated the anxiety and nervous excitement of everybody awaiting the historic event: "The Shuttle appeared far above the north-east horizon, a white dot against a cloudless blue sky. That dot was dropping so fast that to an eye accustomed to watching the more gradual descent of commercial jets, it seemed inevitable that the Shuttle would crash to the desert floor."

As Columbia approached Edwards, the speedbrake – which flared out from the rear edge of the vertical stabiliser – was gradually retracted and was fully closed by the time the Shuttle was 600 m above the runway. Dropping at a precipitous rate, seven times steeper than a commercial airliner, it was understandable that Eichel should think it inevitable that the Shuttle was going to crash. It was at this stage, however, that Young pulled back on the stick, lifting the nose and transforming his ship at once from an apparently out-of-control falling brick into a graceful flying machine.

"The Shuttle has a very, very steep glide slope," said Neil Hutchinson. "It's about eight or nine degrees. That doesn't sound very steep, but if you were in an airliner doing that, you'd think you were headed for sure death!"

Weather conditions at the landing site were almost perfect, with Capcom Joe Allen telling the astronauts that "winds on the surface are calm".

"That's my kind of wind," replied Young.

At 6:20:35 pm, Crippen deployed the landing gear and all six wheels were down and locked within the required 10-second time limit. Columbia touched down 22 seconds later at 342 km/h, rolling for almost 2.7 km before coming to a smooth halt. The speedbrake was opened and full-down elevons were applied, giving the astronauts an impression of considerable deceleration. "As it touched down", reported Eichel, "at a speed 80 to 90 miles an hour faster than a commercial airliner does, the rear wheels nestled into the hard-packed sand, kicking a rooster-tail high into the air ..."

The countdown to landing was echoed by both the public-affairs spokesman at Edwards and by the crew of one of the T-38 chase planes, who were the first to welcome Young and Crippen home with a resounding, "Beautiful! Beautiful!"

Rookie astronaut John Creighton was on board a US Army helicopter at Edwards that day and he would later describe the remarkable efforts of some spectators to get a close-up view of Columbia's landing. "All kinds of people had camped out there for several days. There was a fence there and there'd been a patrol to keep people back there. As soon as the Shuttle rolled to a stop, these people charged forward, [this] fence went down and they got motorcycles and cars that went out racing. This was about five miles away from where the Shuttle actually landed and the only way you could see [it] was with binoculars, but boy, they wanted to get an up-front view! The security folks didn't know what to do, so they told the helicopters to try to get this crowd under control. So these helicopters would swoop down in front of these on-charging group of cars. The helicopter pilots loved it. They were having a great time trying to head off all of these people!"

Post-landing analysis showed that Columbia's right-hand inboard brakes suffered higher-than-anticipated pressure, which caused a slight pull to the right just before the wheels stopped. Young compensated for this by balancing the total braking to either side of the Shuttle, maintaining a near-perfect course straight down the runway centreline, stopping at the intersection of Runways 23 and 15. One notable surprise was the sheer amount of lakebed debris – pebbles and grains of sand – kicked up by the wheels.

"Do I have to take it to the hangar, Joe?" Young joked.

"We're going to dust it off first," deadpanned Allen.

Immediately after wheelstop, the astronauts unstrapped their harnesses and began safing the RCS and OMS switches before the arrival of the ground crew. When the latter arrived minutes later, they first hooked up sensitive 'sniffer' devices to verify the absence of toxic or explosive gases and attached coolant and purging lines to Columbia's aft compartment to air-condition her systems and payload bay and dissipate residual, potentially toxic fumes. Until this procedure had been completed, the ground crew operated in Self-Contained Atmospheric Protection Ensemble (SCAPE) suits. They then proceeded to roll an airport-type stairway over to Columbia's hatch.

John Young, who had remained totally cool throughout the re-entry, approach and landing, finally let his excitement get the better of him by asking the ground crew to hurry up so that he could leave Columbia. When finally allowed outside about an hour after touchdown, he bounded down the steps, checked out the tiles and landing gear and jabbed the air triumphantly with both fists. He also *kicked* the tyres, which gave Henry Pohl a scare.

"I was really worried about *that*, because those tyres have got 375 psi pressure in them," said Pohl, "and I knew the brakes [and tyres] got hot, and I was afraid the tyres were going to explode. It would have been a shame to do all that flying and a terrific landing and then have a tyre blow up because you went over and kicked it!" But Young, obviously, was over-excited. "I've often claimed John calmed down by [the time he got outside the orbiter]," Crippen would say later. "You should have seen him when he was *inside the cockpit!*"

A RACE AGAINST TIME

"Okay, the arm is out and it works beautifully," STS-2 Pilot Dick Truly told Mission Control excitedly on the afternoon of 13 November 1981. On only his first flight into space, the 44-year-old US Navy Captain had been given the enviable task of putting the Shuttle's Canadian-built robot arm – called the Remote Manipulator System (RMS) – through its paces. His enthusiasm, however, was tempered with disappointment, for yesterday evening, only hours after reaching orbit, Truly and Commander Joe Engle had lost more than half of their mission.

Their task now was now to cram five days' worth of scientific research and complex engineering tests into just over 54 hours. It had all seemed so much brighter yesterday and a perfect birthday present for Truly when he and Engle roared into orbit from Pad 39A, becoming the first team of astronauts to fly a 'second-hand' spacecraft. Despite the fantastic achievement of sending Columbia into space the previous April, bringing her home like an airliner and turning her around to fly again, NASA's promise of a Shuttle launch every two weeks was still a long way off.

Although, admittedly, the first four missions were considered to be test flights and not under tremendous schedule pressure, the space agency had hoped to have Columbia ready for STS-2 in considerably less than seven months. The delays had come partly from problems experienced during her first flight and partly from the fact that NASA had underestimated the sheer amount of attention that the Shuttle would require between missions. It was *not* an airliner and it was doubtful that it could ever be operated like one.

Still, the preparation time for STS-2 was much lower than for STS-1. When Columbia arrived in Florida in March 1979, she spent almost two years undergoing flight preparations; the work to ready her for her second mission, on the other hand, took a quarter of that time. She returned to KSC by 747 on 29 April 1981, two weeks after landing at Edwards Air Force Base, in need of a great deal of attention. The most pressing issue was repairing the damaged tiles: 350 were replaced, 818 removed and repaired and a further 2,000 serviced while still attached to Columbia.

Launch was originally targeted for 9 October 1981. A six-month turnaround was less than ideal, even at this early stage, but a significant amount of work was needed to prepare Columbia for her second flight. STS-2 would be the first Shuttle mission to carry a fully fledged scientific research payload – developed by NASA's Office of Space and Terrestrial Applications (OSTA) – and to make room for it in the payload bay, technicians needed to move the DFI pallet further aft. OSTA-1, as it was called, housed seven 'pathfinding' scientific instruments.

The history of OSTA-1 can be traced back to 1976 – the same year the author of this book was born – when a five-day Earth-observation mission was first sketched out. Shortly afterwards, NASA selected six experiments from a total of 32 proposals to fly on the Shuttle's second orbital mission; the seventh, a test of an investigation slated for the Spacelab-1 mission, was added later. Initial analyses showed that Columbia's payload bay could be pointed Earthward for up to 88 hours of the five-day mission, which would demonstrate its ability to hold major scientific instruments steady and acquire data.

OSTA-1 is installed into Columbia's payload bay.

The OSTA-1 experiments were mounted on an engineering version of the U-shaped Spacelab pallet, located at the midpoint of the payload bay. The DFI sat further back, near the aft bulkhead. The Spacelab pallet measured 3 m long by 3.9 m wide and was physically identical to those scheduled to be used on later Spacelab missions, but for STS-2 it was not fully equipped. Operational missions would also feature a cylindrical, temperature-controlled 'igloo' to provide cooling, power and data-management facilities. Pallets, it was intended, would be used to carry instruments requiring large, unobstructed fields-of-view, such as telescopes or radars.

Five of the OSTA-1 experiments were attached to the pallet, of which the largest and most visible was the Shuttle Imaging Radar (SIR), designed to assess the Shuttle's performance as a scientific research platform and further geologists' understanding of the radar signatures of various terrestrial features for mineral and petroleum exploration. Measuring 9.4 m long by 2.2 m wide and weighing 180 kg, SIR was a side-looking 'synthetic-aperture' radar that looked for all the world like an enormous rectangular dinner table filling almost half of Columbia's payload bay.

Assembled from spares left over from NASA's 1978 Seasat mission, it was mounted on its own truss structure which, in turn, was affixed to the Spacelab pallet, providing a viewing angle 47 degrees from nadir, so the Shuttle oriented itself to aim SIR at the ground. It covered the radio frequency of 2.175 GHz (L-band) and a wavelength of 23.5 cm and could construct two-dimensional radar images of the surface. It worked by transmitting microwave signals and receiving reflected 'echoes', recording data onto computer tapes for post-mission analysis. During STS-2, it would prove hugely successful and acquire some intriguing results.

Somewhat less visible on the Spacelab pallet, yet still capable of generating a tremendous amount of valuable scientific data, were the Shuttle Multispectral Infrared Radiometer (SMIRR), the Feature Identification and Location Experiment (FILE), the Ocean Colour Experiment (OCE) and the Measurement of Air Pollution by Satellite (MAPS).

Integration of these five experiments onto the Spacelab pallet was completed in the Operations and Checkout Building at KSC in the early part of 1981 and on 1 July the full OSTA-1 payload was loaded on board Columbia. A series of tests verified its compatibility with the Shuttle, and the ability of the crew to switch it on and off, and command the instruments, from the flight deck. On 10 August, she was rolled over to the VAB for attachment to her ET and SRBs and from thence to Pad 39A for final pre-launch preparations.

BIRTHDAY THRILLS, SPILLS AND DIRTY WINDOWS

Columbia's scheduled 9 October launch date was postponed by nearly a month following a spillage of nitrogen tetroxide as technicians were loading the highly toxic oxidiser into the forward RCS unit. Five-and-a-half litres splashed onto the Shuttle's nose, requiring 379 fragile tiles to be painstakingly removed, cleaned and reapplied. A revised date was set for 4 November and seemed to be going well until two days before, when, during the loading of the oxygen tanks of the fuel cells that were to power most of the orbiter's electrical systems, technicians noticed that one tank was losing pressure. A change in the oxygen-loading procedure seemed to rectify the situation.

Apart from some minor concerns about poor weather, the 4 November attempt seemed to be proceeding normally until nine minutes before launch. At that point, the countdown was halted following an indication of lower-than-allowable pressures in the oxygen tanks of the fuel cells. The problem was resolved within a couple of minutes and the countdown resumed, but quickly encountered further problems. Five minutes before launch, the three APUs were switched on and quickly started showing higher-than-normal oil pressures.

The clock continued ticking to T−31 seconds – just before the point at which, normally, the Shuttle's onboard computers would take over command of the remainder of the countdown from the Ground Launch Sequencer (GLS). At this stage, the oxygen tank pressures registered as 'low' and the clock was stopped. Unable to override this command, the APUs were shut down and the clock was recycled to the $T-9$ minute point, then $T-20$ minutes, to enable NASA and Rockwell technicians to assess the situation. During their deliberations, however, the weather closed in, further jeopardising Columbia's chances of launching that day.

Eitherway, it was decided that the APU oil pressures were too high – 100 psi instead of the maximum allowable 60 psi – and the Mission Management Team made the decision to scrub the launch attempt. Subsequent analysis revealed that the APUs' oil filters had become clogged by pentaerythritol, a crystal formed when hydrazine penetrated their gearboxes, which had caused the rise in temperature. Both gearboxes were flushed, their filters replaced and the launch was rescheduled for 12 November. As the astronauts had also reported that their visibility was marginal, Columbia's cockpit windows were cleaned in time for the next attempt.

Eight days later, the men returned to the pad for a second try. There had already been minor problems during the loading of the oxygen tanks, which meant that Tank 3 had to be loaded and pressurised separately from the other two tanks. This resolved the issue. Another glitch cropped up late on 11 November when one of four Multiplexer-Demultiplexers (MDMs) – which provide instrumentation measurements, commands and data to the Shuttle's cockpit displays – failed. A spare MDM was fitted, but also turned out to be faulty, requiring another one to be flown from California in the early hours of the 12th.

The new MDM, interestingly, came from the second space-rated orbiter, Challenger, which at the time was undergoing a final checkout at Rockwell's Palmdale plant in readiness for transportation to KSC. It was the first of many occasions in which parts were 'cannibalised' from one spacecraft to enable another to fly. Challenger's first orbital voyage was tentatively booked for some time early in 1983; unlike Columbia she would not need to perform a series of flight tests and would fly an operational mission to deploy an important NASA communications satellite.

A HALVED MISSION

"I would still call it a successful flight," said Truly early on 12 November, "even if we had to come home early." Presumably he was referring to the achievement of getting the Shuttle safely into space – becoming the first manned spacecraft to fly twice – but he did not know how prophetic his words would be. STS-2 was supposed to last more than twice as long as the first Shuttle mission, and to complete it Columbia needed additional modifications, the most significant of which was an extra set of cryogenic hydrogen and oxygen tanks for the fuel cells under her payload bay floor.

The mission got off to a good start with a picture-perfect liftoff at 3:09:59 pm, reaching space eight-and-a-half minutes later. This launch was much less damaging to Pad 39A; repairs to the ET struts proved effective and the shattering effect of the SRBs, which had caused excessive pressures during the STS-1 liftoff, were alleviated by improvements to the water-suppression system. Indeed, sensors attached to Columbia's base during STS-2 revealed that the pressures were less than one-tenth – 0.2 psi – of those experienced during her previous launch.

Moreover, thanks to these modifications, the DFI recorders operated without interruption during the ascent and no tiles were lost. Post-mission analysis of STS-1 had revealed that the missing tiles identified by Young and Crippen while in orbit were most probably shaken loose by the SRB over-pressure problem. The STS-2 ascent, on the other hand, suffered only from the need to shut down one of Columbia's APUs slightly earlier than planned due to higher-than-expected oil temperatures that exceeded the maximum allowable 143 Celsius.

After dumping residual propellant through the main engine nozzles – a process terminated 16 seconds earlier than expected due to the APU shutdown – Engle and Truly conducted two OMS burns to circularise Columbia's orbit at just 193 × 201 km, inclined 38 degrees to the equator. The low orbit was necessary to allow the OSTA-1 experiments to gather their required resolution of data. The astronauts also performed the now-customary opening-and-closing tests of the payload bay doors, exposing the Shuttle's first scientific cargo to space for the first time.

Although only the first two OMS burns were necessary to establish Columbia in her correct orbit, two extra firings were performed to raise the altitude to 220 km. The third burn was split into two halves in order to satisfy one of STS-2's flight test objectives: the ability to turn off an OMS engine and restart it a few minutes later in the vacuum of space with no ill effects. The fourth firing then demonstrated the ability of the OMS system to feed the right-hand engine with the left-hand pod and vice versa. All of these tests proved successful.

What did not turn out to operate satisfactorily, only two-and-a-half hours into the mission, was one of Columbia's electricity-generating fuel cells. This had damaging implications for the accomplishment of her planned mission, during which Engle and Truly were to conduct extensive tests of the new RMS arm in both manual and automatic modes and try out the new Shuttle Extravehicular Activity (EVA) spacesuit in the middeck. All that began to change dramatically at 5:37 pm when ground controllers noticed a high pH indication on the Number 1 fuel cell. Its overall performance, at least at this stage, remained normal.

The situation deteriorated significantly, however, and at 7:55 pm a sharp voltage drop was recorded on the troublesome cell, indicative of the possible failure of one or more stacks within it. If that was the case, it meant that the cell's capability to generate electricity and, as a byproduct, drinking water for the crew, might be compromised. With the likelihood of contamination to Engle and Truly's water supply, the Number 1 cell was switched off at 8:15 pm and in response to concerns that the water was being electrolysed – thus forming a potentially explosive mixture within the cell – it was depressurised.

Under proscriptive mission rules, laid down well before STS-1, all three fuel cells were required to be operational for a flight to continue. It was with disappointment that, early on 13 November, Capcom Sally Ride – later to become America's first woman in space – told the astronauts that they were going to come home the following day. "That's not so good," was all a quiet Truly could reply.

The shortened mission required them to front-load as many of their critical flight objectives as possible. Fortunately, this had already been timetabled into each of the four test flights in anticipation of just such an eventuality. The OSTA-1 payload had already been activated by Truly within hours of achieving orbit, and was collecting spectacular data. In fact, by the time they returned to Earth late on 14 November the astronauts had completed almost 90% of their pre-flight objectives. The fuel cell problem did have other impacts, not least of which was the slow dispensation rate of their drinking water.

FLEXING CANADARM'S MUSCLES

Just before 2:00 pm on 13 November, less than 23 hours into the mission, a major objective of STS-2 got spectacularly underway when Truly unlatched the $100-million RMS from its cradle along the port-side sill of the payload bay, rolled it out of its storage position and moved it to its outboard work station. It was Canada's contribution to the Shuttle – a contribution that dated to 1974 when Spar Space

Dick Truly (left) and Joe Engle in the Shuttle simulator's aft flight deck practise operating the RMS.

Robotics Corporation was contracted by the country's National Research Council to build a mechanical arm for deploying and retrieving satellites from orbit and ultimately assembling a space station.

The challenges involved in building an arm capable of such complexity and dexterity were enormous: it needed to operate automatically *and* under manual control and meet strict weight and safety requirements. Moreover, nothing quite like it had been built or used in space before, which made Spar's task yet more difficult. Although a horizontal floor rig was built to test the joints in 1977, the first real demonstration would not come until it was actually uncradled in orbit. The first space-rated RMS was delivered to KSC in April 1981 and, after a series of checks, installed on board Columbia on 20 June.

The arm was 15.2 m long, to enable it to reach the far end of the payload bay, and consisted – like a human arm – of shoulder, elbow and wrist joints, linked by two graphite-epoxy booms. Other components were made from titanium and stainless steel. To protect it from thermal extremes in space, the arm was covered in white insulation and fitted with heaters to maintain its temperature within required limits. Without a payload attached, the RMS could move at up to 60 cm/min, but this was reduced to a tenth of that time when fully loaded.

Ingeniously, the means by which the arm could 'pick up' and 'put down' objects was achieved by the so-called 'end-effector' – essentially a 'hand' that employed a kind of wire snare to capture a prong-like grapple fixture attached to deployable or retrievable payloads. Already, one of NASA's most important observatories – the Hubble Space Telescope, scheduled for launch in the mid-1980s – had an in-built grapple fixture that would enable it not only to be deployed, but also retrieved and repaired in space by future Shuttle crews.

During operational flights, astronauts would use two television cameras on the arm's wrist and elbow to guide the end-effector over a target's grapple fixture, before commanding three wire snares to close around it at just the right instant. When this was done, it would impart a force of 500 kg on the grapple fixture to allow the RMS to move the target into or out of the payload bay. No targets would be moved on STS-2, but on the next flight a desk-sized Induced Environmental Contamination Monitor (IECM) would be used to flex its robotic muscles.

"Its movements are much more flexible than they appeared during the training simulations," Truly told Sally Ride as the arm arched out of the payload bay. Although the RMS was controlled by the Shuttle's GPCs, its movements were commanded by Truly using a joystick in the aft flight deck. Under his control, he moved the shoulder and elbow joints up and down and left to right and pitched, yawed and rolled the wrist. As Truly issued each command, the GPCs examined them and determined the joints that needed to be moved, their direction and their speed and angle.

Meanwhile, the computers looked at each joint at 80-millisecond intervals and, in the event of a failure, could automatically apply a series of brakes and notify the astronauts. As Truly worked, a continuous flow of data on joint rates and speeds appeared on monitors in the flight deck. An hour after the RMS had been unfurled, Ride asked the men to establish the television feed, to which Engle replied,

"Transmitting. You people seeing anything down there yet?" After a moment, Ride confirmed that, indeed, the first pictures of the arm in an inverted V-position had appeared on Mission Control's screens.

The images drew spontaneous applause from the Canadian delegation. For the next four hours, Truly and Engle took turns operating the RMS and testing it in all five control modes, using both the primary and backup software. Four days of tests were, through necessity, now crammed into a brief 24 hours. As Truly 'flew' the arm, Engle fired a series of bursts from the RCS thrusters to assess its performance under stress. The arm's television cameras were also evaluated and, on one occasion, showed a grinning Truly peering through Columbia's aft flight deck windows with a sign that read, 'Hi Mom'.

Only a couple of minor problems were encountered. The first was a failure in the arm's primary control mode, which the astronauts managed to bypass using the backup electronics (the cause would later be traced to a broken wire). Also, the elbow camera – one of six cameras affixed to the RMS – suffered a short circuit and failed towards the end of the mission. Otherwise, the first demonstration of the arm in orbit was a spectacular success and it performed within expectations under a wide range of temperature variations.

A SCIENTIFIC BOUNTY

Although not intended to be manoeuvred by the RMS on STS-2, the IECM was carried in the payload bay attached to the DFI pallet. Its task was to help scientists to better understand the effect of the Shuttle on its local environment, as part of continuing efforts to assess its atomic 'cleanliness' before committing sensitive telescopes and detectors to future missions. IECM was carried on the last three test flights and the first two Spacelab missions, and was either fixed in the payload bay or moved around using the RMS. Its STS-2 data revealed contamination levels to be "within expected limits".

Promisingly for future flights, it confirmed that exhaust byproducts from the main engines had not leaked into the payload bay – demonstrating the integrity of the door seals – and revealed that the vast majority (more than 90%) of particulate contaminants tended to 'boil off' into space within the first day-and-a-half of the mission. In general, the payload bay environment was characterised as remarkably clean and debris-free. It did, however, show that after RCS thruster firings, small, short-lived 'clouds' of particles formed and hung over the bay.

In the meantime, the experiments in the payload bay were performing extremely well. SIR was in the process of gathering eight hours' worth of radar data, acquiring images with resolutions of just 40 metres of North America, southern Asia and Europe, Australia and the Pacific islands, North Africa and the northern part of South America. It enabled geologists to determine surface 'roughness' and picked out faults, drainage patterns and evidence of stratification, as well as making a truly unexpected discovery: ancient watercourses beneath the arid sands of Egypt, just as they might have appeared thousands of years ago.

Such data offered a promising preview of what might be achieved on operational research missions. The other instruments were also proving their worth. As Engle and Truly took photographs through Columbia's flight deck windows, SMIRR determined the best spectral resolution needed to identify and map rock or mineral deposits, while FILE evaluated new techniques to automatically classify surface features such as water, vegetation, bare land and snow, clouds or ice to better prioritise the timing of future Earth-resources missions. It complemented SIR and SMIRR by providing a means for them to be activated only when conditions were 'right' for data-acquisition.

Another experiment, MAPS, provided the first accurate indication of how severe the levels of atmospheric carbon monoxide really are. It surveyed the lower atmosphere – from the surface to an altitude of about 18 km – and began a series of missions that identified this uncomfortable trend. MAPS was flown again in 1984 and on two Space Radar Laboratory missions in 1994, revealing high pollution levels, particularly in the tropics, caused by seasonal biomass burning. Lastly, OCE evaluated a technique to map the colour patterns of plankton and chlorophyll as part of efforts to better identify schools of fish.

The last two OSTA-1 experiments were housed not on the pallet, but in Columbia's cabin. They were the Heflex Bioengineering Test (HBT) and the Night/Day Optical Survey of Lightning (NOSL). The first investigated the effect of microgravity and soil composition on the growth of the dwarf sunflower (*Helianthus annuus*) to study the relationship between its height and moisture content. It was a precursor of an experiment slated for the first fully fledged Spacelab mission, planned for late 1983, which would analyse the sunflowers in greater detail. The STS-2 test, however, achieved only partial success because of the shortened mission.

NOSL was also affected by the halving of the flight. It required Engle and Truly to take daytime and nighttime photographs of lightning flashes over land and water, in the hope that it might lead to the development of new systems to give early warnings of particularly severe storms. The astronauts removed the hardware from a middeck locker soon after reaching orbit, assembled it and successfully acquired nighttime images and motion-picture sequences of six large daytime thunderstorm systems. In recognition of the 'lost' data from STS-2, more tests of both HBT and NOSL were planned for STS-3.

After a jam-packed two days in space, on 14 November Engle and Truly began preparing their ship for the return to Earth. Columbia's re-entry profile as she hurtled through the atmosphere differed significantly to pre-flight plans, due to the shortened mission. The RCS thrusters at the rear of the spacecraft were commanded to fire over 1,000 times, consuming 815 kg of propellant – far more than planned – because the Shuttle's fuel-consumption rate at the end of two days in space differed from pre-flight estimates, which were based on the demands expected of a five-day mission.

Shortly before hitting the uppermost traces of the atmosphere, a large quantity of propellant was dumped out of the forward RCS unit to provide precise control of Columbia's centre-of-gravity during the descent. A series of flight tests were conducted during re-entry, the most important of which was a so-called 'push-over/pull-up' (POPU) exercise performed by Truly. As the Shuttle plummeted Earthwards at 22,500 km/h, he pushed her nose down from a 40-degree angle-of-attack to 35 degrees, lifted it to 45 degrees then returned it to 40 degrees. This provided extra data on the vehicle's aerodynamic performance during re-entry.

Engle took manual control of Columbia at an altitude of 94 m, pointing the Shuttle into a 37 km/h crosswind and Truly deployed the landing gear 18 seconds before touchdown. The first used spacecraft landed safely on Edwards' Runway 23 at 9:23:12 pm, wrapping up a journey that lasted just seven minutes short of STS-1. Although nosewheel steering was not yet available, differential braking was applied to maintain a straight course down the runway. Engle would later say that a fluctuating indicator in the cockpit made it difficult for him to maintain a constant deceleration rate.

AN AMBITIOUS MISSION

A Space Shuttle with a difference headed into overcast Florida skies at precisely 4:00 pm on 22 March 1982. For the first time, Columbia flew into orbit attached to a rust-coloured ET, the result of deleting a coat of white Fire Retardant Latex (FRL) primer that saved 270 kg in weight and $15,000. Although this was only the third Shuttle flight, and STS-1 and STS-2 had already demonstrated the reusability of the system and its viability as a scientific research platform, there was still much to prove and its powerful Congressional enemies continued their calls for its cancellation.

Fortunately, with the exception of a few minor technical obstacles – eagerly jumped upon by the press but, in reality, insignificant in terms of their effect on the mission – Columbia's third trip into space was a spectacular success. Not only did it almost quadruple the two-day endurance limit of the previous missions, it also conducted the first tests of the RMS arm's mettle by hauling 'real' payloads and carried another engineering prototype of the Spacelab pallet; this time outfitted for a series of experiments sponsored by NASA's Office of Space Science.

In fact, by the time the payload, dubbed 'OSS-1', was launched, the office itself had assumed a new name: the Office of Space Science and Applications. Originally, OSS-1 was meant to be the first in a series of missions that would ferry sophisticated astronomical telescopes and space plasma detectors into orbit. Before these could be flown, however, scientists needed to better understand the impact of outgassing, waste-water dumps and thruster firings on surfaces within the payload bay. Such waste products were known to deposit thin 'films' of debris that could cause the optics of very sensitive instruments to degrade.

One of the most important missions under consideration at the time would carry three ultraviolet telescopes for detailed surveys of the Universe. The knowledge gained as a result of the STS-3 experiments helped to get this mission off the drawing boards and – after a long wait – into orbit just before Christmas 1990. Known as ASTRO-1, it turned out to be one of the most complex, yet brilliant, astronomical missions ever attempted. Results from other OSS-1 experiments enabled NASA to build sophisticated solar and atmospheric-physics instruments which rode Shuttle missions late into the 1990s.

Eight of the nine OSS-1 experiments were mounted on the Spacelab pallet in Columbia's payload bay and were devoted to examining the near-Earth environment and measuring levels of contamination produced by the Shuttle herself. One of the most intriguing instruments was the University of Iowa's 158-kg Plasma Diagnostics Package (PDP), a small cylindrical canister of electromagnetic and particle sensors to 'sniff out' the environment surrounding Columbia. Its data on the 'cleanliness' of the payload bay would prove invaluable in allowing NASA to commit highly sensitive scientific instruments to future missions.

To support this important experiment, Commander Jack Lousma and Pilot Gordon Fullerton were extensively trained to use the RMS to lift the PDP from the OSS-1 pallet and manoeuvre it to various positions in the payload bay. According to their pre-mission press kit, they were to use the arm to hoist not only the PDP, but also the desk-sized IECM package, previously flown on STS-2. It was in eager anticipation for one of the most demanding missions ever attempted that the third pair of Shuttlenauts headed into orbit on that murky March day in 1982.

Although only on his first flight, Fullerton was in charge of the RMS for these tests. His training took place in several venues, including the robotic arm's home in Canada and in Houston. "I went a couple of times up to Toronto to see how [the RMS] worked," he said. "Then we had a full-size mockup at Houston with a 1g-capable arm driven by hydraulics. We had an electronic version of the arm, looking at screens in the windows and the simulator. There were a lot of tools to get the hang of working the arm. That was pretty cool."

Lousma, a 46-year-old US Marine Corps Colonel, had previously spent two months in orbit on board the Skylab space station in the summer of 1973. Ironically, had the proposed Shuttle repair mission to the ageing station gone ahead some time before 1979, it is possible that he would have flown it with Fred Haise. Fullerton, too, had amassed a wealth of experience both on the ground and above it, having piloted two of the five approach and landing tests on board Enterprise with Haise in 1977.

"A real barn-burner," was how Lousma described Columbia's launch; Fullerton was inclined to agree, although it did not go quite as intended. Firstly, it set off an hour late, following the failure of a heater on a ground-based nitrogen gas line. Then an APU overheated four-and-a-half minutes into the ascent, triggering a caution-and-warning alarm in the cockpit and forcing the astronauts to shut it down early. This left one of Columbia's main engines running at only 82% thrust for the last few seconds of the climb into space; its overall performance, however, was unaffected.

SPACE SICKNESS

The minor scare proved more than worth it; both Fullerton and his veteran colleague suffered bouts of space sickness during their first few days in orbit. Lousma's previous flight experience, it seemed, by no means guaranteed him immunity from the nauseous motion sickness which had been reported by countless astronauts and cosmonauts over the years. Both men took Dexedrine and Scopolamine medication and consumed the required amount of calories each day. The sickness did not affect their ability to work, however, and post-landing checkups would reveal both men to be in excellent health.

"Of course, everybody has their acclimation problems," Fullerton would later recall. "That's pretty consistent through the population. It takes about 24 hours to get to feel normal, at varying levels of discomfort. Most can hang in there and do their stuff, even though they don't feel good." Clearly, all astronauts, after entering a new and utterly alien environment, need time to gain their 'space legs' and this had posed particular problems with Fullerton and Lousma's predecessors, the STS-2 crew of Joe Engle and Dick Truly, neither of whom had flown into orbit before.

"They had not had time, in a two-and-a-half-day flight [to acclimatise, when] they were cut short," Fullerton continued. "By the time they got on-orbit and traced down the problem and the decision was made to come back early, they were getting ready to come back. So they had no time other than to kind of respond, do things that the ground was coming up [with] and they had some dizziness and orientation problems on entry that Jack and I worried about a lot.

"One thing that we had was a 'g-suit', like they wear in the F-18, except that for entry you could pump up the suit and just keep it that way, and so that helped you keep your blood flow up near your head. The other thing about the motion sickness is we're not sure there's a direct correlation to flying airplanes. I know if you go up and do a lot of aerobatics day after day, you get to be much more tolerant of it. So Jack and I flew literally hundreds of aileron rolls [in our T-38 jet trainers].

"If I did roll after roll, I could make myself sick, and I got to the point where it took hundreds of them to make me sick. For the first day or so [in space], I didn't ever throw up or anything; I never got disoriented, but I felt kinda fifty-fifty. You're pretty happy to just float around and relax rather than keeping on charging. And into the second day, this is really fun and great and you feel 100%. Whether the aileron rolls helped or not, I'm not sure, but it was relatively easy."

SUBSTITUTION: IECM 'OFF', PDP 'ON'

Luckily, it was on the second day of the mission, when Fullerton again felt "great", that he uncradled the RMS for the first of what would turn out to be a marathon 48 hours of tests. Soon after 3:00 pm on 23 March, he flexed the arm's robotic muscles, before returning it to its berth on the left-hand sill of the payload bay four hours later. One problem that cropped up during these tests was the failure of the wrist-mounted television camera which, crucially, would enable Fullerton to view the grapple fixture on the IECM and pick it up satisfactorily.

The IECM tests were a vitally important part of the mission because of the unit's relatively large size and weight of 385 kg, which would help to demonstrate for the first time the handling and manoeuvring capabilities of the RMS, before the Shuttle could be dedicated to carrying out more ambitious satellite deployments and retrievals. Already, a risky recovery and repair of NASA's Solar Max spacecraft, which would involve a pair of lengthy spacewalks and intensive RMS usage, had been provisionally pencilled into the Shuttle's calendar for 1984.

With the wrist camera effectively out of action, it was feared that faults in other cameras on the arm or in the payload bay itself would not provide the astronauts with acceptable views of the IECM on their monitors on the aft flight deck. This, in turn, might then have prevented them from safely reberthing it onto the DFI pallet after the tests, perhaps forcing them to discard the valuable payload. Nervous NASA managers decided, therefore, to defer the IECM deployment until the fourth and final Shuttle test flight, STS-4, in July 1982.

Instead, the IECM was substituted on this occasion for the smaller PDP, although the latter was less than half the mass. This obviously meant that the handling characteristics of the RMS with the PDP in its grasp would differ quite significantly from carrying the bigger IECM, but, in the event, the tests proved satisfactory. Original plans called for Fullerton to unberth the PDP for a series of eight-hour tests on both 24 and 26 March, allowing it to examine the electromagnetic and particle environment within a range of about 14 metres from Columbia.

In the event, the astronauts completed three PDP deployments and a good day-and-a-half of additional information was gathered while the canister remained attached to the OSS-1 pallet. Its data provided, for the first time, detailed insights into the strange ionospheric plasma 'wake' generated as the Shuttle passed, boat-like, through the electromagnetic environment of low-Earth orbit. This wake might, it was theorised, complicate the measurements of very sensitive scientific instruments and the STS-3 results proved beneficial for the planning and development of state-of-the-art space plasma detectors due to be carried on the Spacelab-2 mission.

The PDP was also used in conjunction with several other experiments affixed to the OSS-1 pallet. One of these was the Vehicle Charging and Potential (VCAP), provided by Utah State University to examine Columbia's electrical characteristics and her effect on surrounding ionospheric plasma. The experiment comprised a fast-pulse 'gun', which fired 100-volt bursts of electrons for durations ranging from 500 nanoseconds to several minutes. It investigated the extent to which electrical charges accumulated on the Shuttle's insulated surfaces and how 'return currents' could be established through a limited area of surface-conducting materials to neutralise active electron emissions.

PATHFINDER FOR SPACELAB

It was hoped that data from VCAP would provide practical experience of using electron accelerators on later missions, particularly Spacelab-1. Plans were also afoot in conjunction with the Italian Space Agency to build a revolutionary 'tethered satellite', which would be trawled through the upper-atmospheric plasma on the end of a 20-km-long conducting cable. The first tethered satellite mission took place in 1992, several years later than planned, and it flew again on board Columbia in February 1996. Such tethers, researchers argued, could provide a steady supply of electrical power for future spacecraft.

Several other OSS-1 experiments were also intended as forerunners of more advanced versions planned for later Spacelab missions. Two instruments – the US Naval Laboratory's Solar Ultraviolet and Spectral Irradiance Monitor (SUSIM) and Columbia University's Solar Flare X-ray Photometer (SFXP) – were devoted to observations of radiation emitted from the Sun, to better understand the processes responsible for them and their impact on Earth. To support them, the flight plan called for Lousma and Fullerton to orient Columbia to aim her payload bay directly at the Sun for several protracted periods of time.

In fact, positioning the Shuttle in a series of different attitudes also satisfied another in a long list of tasks that needed to be completed before the vehicle could be

Gordon Fullerton manoeuvres the PDP around Columbia's tail fin during STS-3.

declared operational. During their eight days in space, the astronauts oriented her in four 'inertial' attitudes to place different parts under maximum solar heating. Columbia spent 30 hours with her tail facing the Sun, 80 hours with her nose aimed at the Sun and 36 hours with her open payload bay facing the Sun. The men also performed several 'barbecue rolls' to passively thermal-condition the whole spacecraft.

During the course of these tests, Lousma and Fullerton exposed the payload bay to its coldest-yet environment as Columbia's tail was pointed at the Sun. The temperatures in the bay were so low that 'outgassed' condensation formed on the aft flight deck windowpanes! When this had been done, the radiators were stowed and latched and the port-side door was closed. In general, the doors performed as advertised under intensely cold conditions, with the exception of a problem when a 'latched' indication was not received for one of the aft bulkhead latches. A spell of passive thermal conditionings quickly resolved this.

The week aloft enabled the two men to indulge in taking photographs of Earth. This mission had, according to oceanographer Bob Stevenson, given them an opportunity to photograph a virtually cloud-free China and one of their shots almost got them into diplomatic hot water after landing. "Jack and Gordon were invited to China to speak to a huge audience [in] some auditorium," Stevenson said later, "and they showed this picture of [a] lake. It was such a beautiful picture that they had it enlarged and matted and framed and they signed it off to the Premier of China [as a gift]. When they got to this picture, [there was] silence. When [the talk] was over, there was [subdued] clapping and they didn't know what to think about this. So they turned to the [US] ambassador and said 'We want to give the picture to the Premier', and he grabs the picture, looks at it [and says] 'I think let's hold this for a while'. When they were leaving the stage, they said 'What's the problem?' [The ambassador replied] 'Well, see that built-up area [on the photo]? That's a secret nuclear facility in China that they didn't know anybody even knew about!'"

Lousma and Fullerton, it seemed, had inadvertently photographed the top-secret site while taking their Earth-observation photographs and, as Stevenson said later, "Jack wasn't sure [he] was going to come home [alive]!" After the flight, Stevenson and colleague Paul Scully-Power would arrange with a Chinese friend to put some important-looking comments and signatures in Mandarin on a blown-up copy of the photograph and presented it to Lousma and Fullerton. The faked inscription read: "If you damn Yankees ever come over China again …"

Meanwhile, as the astronauts continued to put Columbia through her paces, each of the OSS-1 experiments gathered its own treasure trove of scientific and engineering data. In addition to the instruments already mentioned, the pallet carried the Space Shuttle Induced Atmosphere (SSIA), Thermal Canister Experiment (TCE), Contamination Monitor Package (CMP) and – a boon for Britain's space ambitions – the University of Kent's Microabrasion Foil Experiment (MFE). The latter marked the first experiment built by researchers outside the United States to fly on board the Shuttle.

In effect, it was a square section of about 50 layers of tin foil. During the mission, it 'operated' in an entirely passive mode, measuring the numbers, chemical composition and density of tiny micrometeoroids in low-Earth orbit. Following Columbia's landing, the foil was removed from its place on top of the cube-shaped TCE and laboratory analysis enabled scientists to determine not only the depths to which the micrometeoroids had penetrated it, but, consequently, also their impact velocities. Heavier particles punched right through the foil and often left debris, while lighter icy ones left craters.

The TCE, to which the foil experiment was attached, was built by NASA's Goddard Space Flight Center in Greenbelt, Maryland, and evaluated a novel method of protecting scientific instruments from extremes of heat and cold – from 200 Celsius to minus 100 Celsius – in Earth orbit. It used a series of heat pipes which maintained several 'dummy' instruments at specific temperatures under various thermal loads and radiated waste heat into space. The canister actually performed *better* in orbit than it had done in ground tests and would later be used in the electronics module on the ASTRO-1 payload.

It also provided useful data for an ambitious experiment slated for Spacelab-2, which sought to better comprehend the physical properties of a peculiar substance known as 'superfluid helium' – the coldest-known liquid – and demonstrate its viability as a cryogenic coolant for future high-energy astronomical instruments. The Spacelab-2 experiment would build on data gathered during STS-3 by evaluating the behaviour of this strange liquid and testing a prototype containment vessel for it.

Within NASA, OSS-1 was known as the agency's Pathfinder mission. In many ways, several of its experiments would later find applications on 'operational' Shuttle missions and would fly late into the 1990s and beyond. Its last two pallet-mounted experiments (SSIA and CMP) assessed the impact of clouds and plumes of waste particles ejected from the spacecraft on scientific instruments. The first measured the brightness of particles emitted from the Shuttle, while CMP consisted of two mirrors – coated with magnesium fluoride over aluminium, commonly used in ultraviolet detectors – whose sensitivity was very carefully determined before and after the mission.

Scientific activity was also pursued inside Columbia's cabin, with several important experiments housed in middeck lockers. These were tended by Lousma and Fullerton throughout the mission. One of these experiments utilised a new, filing-cabinet-sized facility known as the Plant Growth Unit (PGU), which was so large that a middeck locker had to be *removed* in order to make room for it. The unit contained all the equipment necessary – growth lamps to provide 14 hours of artificial 'sunlight' each day, timers, temperature sensors, batteries, fans and a data-storage system – to grow almost a hundred plants in the weightlessness of space.

One of the key objectives of the PGU experiments on STS-3 was to test whether 'lignification' was a response to gravity or a genetically determined process with little environmental influence. Lignin is a structured polymer, which allows plants to maintain a vertical posture, despite the effects of gravity, and is thus highly important for the plant's ability to grow properly. The experiments tended by Lousma and Fullerton tried to find out if lignin was reduced in the microgravity environment and if this caused plants to lose strength and 'droop'.

Earlier experiments on board Skylab and the Russian Salyut space stations throughout the 1970s had revealed that the strange conditions in Earth orbit did indeed cause root and shoot growth to become disorientated, as well as increasing their mortality rates. However, little was known about the physical changes within them. Understanding how plants behave and grow in the absence of gravity was – and, with President George W. Bush's new vision for trips to the Moon and Mars, still is – essential for long-duration missions, in which astronauts will need to grow their own foodstuffs.

Chinese mung bean, oat and slash pine seedlings were chosen for STS-3 because all three could grow in closed chambers and under relatively low lighting conditions. Additionally, pine is a 'gymnosperm', which means that it is capable of synthesising large amounts of lignin, and it was believed that its growth was directly affected by gravity. Unlike the mung bean and oat seedlings, which were germinated only hours before Columbia's launch, the pine samples were germinated several days earlier.

The seedlings were used in three experiments. One looked at whether lignification was influenced by gravity or determined genetically within the plant. Several of the mung beans did indeed experience orientation problems, although the oats appeared to suffer no ill effects either on Earth or in space. The flight seedlings were all much shorter in stature than the ground control samples, but overall their levels of lignin reduction were only a few percent more than those grown on Earth. As such, although the results did point towards a reduction of lignin in space-grown plants, the difference was deemed statistically insignificant.

The second experiment used the mung beans and oats for chromosomal studies, revealing much fragmentation and breakage and confirming that their root cells had been affected by exposure to microgravity. A third experiment investigated how the organisation of the plants' gravity-sensing tissues, including the root cap, was affected by spaceflight. Within hours of Columbia's landing, the seedlings were removed from the PGU, immersed in fixative, thin-sectioned and stained for light and electron microscopy.

A MONSTER STORM

Unfortunately, the landing, like the launch, did not go entirely according to plan. STS-3 was originally supposed to come home, as the previous two missions had done, to Runway 23 at Edwards, on 29 March. However, unseasonal mid-March rain showers had left Edwards under several centimetres of water and four days before Columbia's launch the backup landing strip at White Sands in New Mexico was chosen instead. Ironically, despite having 90% near-perfect weather throughout the year, on the *very day* Lousma and Fullerton were to come home, it suffered its worst wind and sand storm for 25 years.

Although Edwards and KSC were the official end-of-mission landing sites, White Sands had been named as a 'contingency' strip in 1979. As well as offering an enormous runway, which provided the margins of safety needed by the Shuttle, it is virtually unobstructed and can be seen from space. It lies in a mountain-ringed, salt-

and-gypsum area called Alkali Flats and was first used by the Northrop Aviation Corporation in the 1940s to test their target-drone projects. The site quickly acquired the nickname 'Northrop Strip', which later became 'Northrup' due to a typo in a press release. The new name stuck.

By 1952, the site had become part of White Sands Missile Range and soon acquired a pair of 10.6-km-long runways, crossing each other in an 'X' shape. During the first two Shuttle missions, it would have been used if an emergency had forced them to return to Earth during their first orbit. Even today, although STS-3 is the only one to have ever come home to White Sands, it remains on NASA's list of contingency sites and astronauts continue to hone their flying skills there.

One person supporting STS-3's landing at White Sands was rookie astronaut Charlie Bolden, who would later fly Columbia in January 1986. "This dust storm was unlike anything I'd ever seen," he said later. "It's gypsum and it's very fine, like talcum powder. Everything was covered with plastic; the windows were sealed [but it] didn't make any difference. That was a hint that this was not a good place to land the Shuttle."

Blissfully unaware of the poor weather in New Mexico, on 29 March the astronauts proceeded smartly through their preparations to fire Columbia's OMS engines for the descent through the atmosphere. Then, with less than half an hour to go before the burn, Mission Control advised them that conditions were unsatisfactory and they would try again on the 30th. The reason given was based on higher-than-allowable surface gusts, but in fact high-altitude winds were also unacceptable. After seven hectic, work-packed days in space, Lousma and Fullerton were overjoyed at the chance to spend more time aloft.

"There was this realisation: 'hey, this is free time', and it was terrific," Fullerton said after the mission. "We got out of our suits, and then we got something to eat and watched the world and I wouldn't have had it any other way, if it had been my choice. In fact, we flew right over White Sands with [Columbia's] nose pointing straight down, and I could see this monster storm going on there. It looked like it was headed for Texas, the dust in the valley there. It was clearly a good decision. It looked really bad down there."

Already, astronaut John Young, flying weather reconnaissance runs over White Sands, had recommended that conditions were nowhere near acceptable for Columbia to land. "We had drifts blow sand into the [public-affairs] area and some areas up against buildings; it was about 18 inches deep!" said White Sands facility manager Grady McCright. "The runway got eroded [by the] wind. So we had people driving a road grader that night [to] grade it, compact it and get it ready for landing the next morning. The wind didn't quit blowing until dark that night."

Fortunately, by 30 March the sandstorm had subsided and the STS-3 crew repeated their preparations for re-entry without problems. One of the APUs was switched on just before the burn and the other two came online shortly after Columbia made contact with the uppermost limit of the atmosphere. Sixteen minutes into the fiery plunge back to Earth, the Kuiper Airborne Observatory successfully acquired its third set of IRIS data, photographing the Shuttle's glowing belly and sides as she hurtled home at 15 times the speed of sound.

Columbia descends towards landing at the end of STS-3.

Fullerton was clearly overwhelmed by his first flying experience where outside temperatures closely matched a blast furnace. "The entry was pretty cool," he said later, "because it was an early morning landing, meaning that the main part of the entry is at night, so we could see this glow from the ionisation really bright out there."

As Columbia continued to fall, passing over Edwards and heading for the mountains of New Mexico, Lousma prepared for the ultimate flying challenge of his career: landing a $2-billion spacecraft first time, with no opportunity to recover from a missed approach. When the Shuttle reached 3,000 metres, he tested the 'autoland' system – which NASA expected to use operationally on future missions – but took manual control 42 metres above the runway. After so many years preparing for this moment, there was no way the seasoned test pilot would allow a computer to land his ship.

That, however, was not the only concern. "The crews [are] very concerned that they have everything that they can at their control to make sure it goes well," said Arnie Aldrich, "and what they were worried about is not that the autoland system wouldn't fly the vehicle right, what they're worried about is if there was some glitch in the autoland system right at a critical [point] of approach, and they had to take control back over: the transient of getting off the autoland and back into manual control might be something they couldn't deal with."

Charlie Bolden, who had followed the development of the Shuttle's software during his first two years as an astronaut, was unhappy about using autoland so close to touchdown, especially on a test flight. "We developed the procedures that we would use for autoland, how they would manually take over at the very last second, to go ahead and land the vehicle," he remembered later. "We recommended [that] this was not a good thing to do. You're asking a person who's been in space to take over in this dynamic mode of flight and land the vehicle safely.

"Their physical gains, their mental gains, their balance; everything's *not there*. Not a smart thing to do. Everything we had seen at [NASA's] Ames [Research Center at Moffett Field in California] in the simulator, when we were in complete control of our faculties, told us you didn't want to do that. But the decision was made that 'we really need to demonstrate this, so we're going to do it and we're only going to go to 500 feet anyway."

It was decided before the flight to use airspeed, rather than altitude, as a cue to deploy the landing gear. The wheels began to lower 30 metres above the runway, but took longer than expected to deploy; they only locked into place a mere *two seconds* before Columbia touched down. To observers, it was nail-biting to see a Shuttle streaking in to land at over 320 km/h with her gear still in the process of coming down. Fortunately, the landing was successful, although NASA would revert to using altitude, rather than airspeed, as a cue on future flights.

The decision to use airspeed, rather than altitude, as the landing cue resulted in Columbia touching down 1.2 km past the runway threshold and required Lousma to apply differential braking to keep the Shuttle within 24 metres of the centreline. Although the vertical impact velocity of both the main and nose gears was within the limits set by mission rules, it was still harsher than expected and caused a 46-cm scrape in one of her tyres, a cracked beryllium rotor in one of her brakes and extensive contamination by the white gypsum dust.

So fine was the dust that it quite literally saturated the entire spacecraft and caused extensive damage that was not entirely resolved in time for her next flight – nor for the remainder of Columbia's career, according to Bolden. "That was Columbia! I flew it several flights later, on my first flight, and when we got on orbit there was *still* gypsum coming out of everything! They thought they had cleaned it, but I think probably some of the debris from Columbia that we gathered probably had gypsum in it. It was just unreal what it had done."

ROOM FOR IMPROVEMENT

Columbia's exact landing time was 4:04:46 pm on Runway 17, setting a new Shuttle duration record of just over eight days. As the gypsum-coated Shuttle sped down the runway, with her forward landing gear in the process of coming down, the nose unexpectedly pitched back up into the air, again providing a moment of shocked surprise among the assembled spectators at White Sands. The effect, as Fullerton would later remark, was "a kind of wheelie". The astronauts, it seemed, were trying to prevent what they thought was a premature nosewheel touchdown.

"It pointed out another flaw, or room for improvement, in the flight software," Fullerton would say later. "The gains between the stick and the elevons, that were good for flying up in the air, were *not* good when the wheels were on the ground. He [Lousma] kind of planted it down but then came back on the stick, and the nose came up. A lot of people thought this is a terrible thing [but] we improved the software and so people don't do that anymore; but we discovered a susceptibility."

Despite the concerns expressed by NASA management at the time, STS-3 was still a *test flight* – and such problems are commonplace on test flights – as well as only the third mission of the world's most advanced and complex spacecraft. The achievement was that the astronauts identified the problem before the Shuttle went operational and additional simulator runs by the STS-4 crew would use the 60-m altitude mark, rather than 500-km/h airspeed, as their cue to deploy the landing gear. The key points of STS-3's landing were that it was safe and it was successful.

Charlie Bolden watched the landing attentively. "Everything seemed to be going well until just seconds before touchdown, when all of a sudden we saw the vehicle kinda pitch up and then kinda hard nose touchdown. We found out that just as Jack Lousma had trained to do, you need to move [the control stick] an appreciable amount [to disengage the autopilot]. We didn't realise that. The way he had trained was just to do a manual download with a stick. When he did that, he disengaged the roll axis on the Shuttle, but he didn't disengage the pitch axis. So the computer was still flying the pitch, although he was flying the roll. Gordon Fullerton just happened to look at the eyebrow lights and he noticed that he was still in auto in pitch. He told Jack and so Jack just kinda really pulled back on the stick, and it caused the vehicle to pitch up. Then he kinda caught it and put it back down and he saved the vehicle."

As servicing vehicles encircled Columbia, the spacecraft sat on the runway, in Fullerton's words, "surrounded by white gypsum". So severe was the damage that the flow rate from the purge units attached to the Shuttle's forward fuselage was increased and the aft compartment's vent doors were closed to prevent further contamination. On hand at White Sands to greet the astronauts, in addition to their families, were New Mexico Governor Bruce King and former Apollo astronaut Senator Harrison 'Jack' Schmitt, as well as the missile range's commanding officer, Major-General Alan Nord.

Within hours of landing, Columbia's time-critical experiments, including the plants from the PGU and samples from the Electrophoresis Equipment Verification Test (EEVT), were removed from the middeck and returned to their research teams. The EEVT unit was a forerunner of later Shuttle experiments, which would employ 'electrophoresis' – whereby electric currents separate biological materials in fluids without damaging the cells themselves – to allow scientists to conduct studies of cell biology, immunology or for medical research. Electrophoresis on Earth is difficult because heat produced by the electric current introduces buoyancy and remixing of the cells and fluids, thus defeating the objective.

During STS-3, the unit held red blood cells and live kidney cells which the astronauts inserted into a series of glass columns for the separation process to take place. After an hour, the samples were removed and stored in a cryogenic freezer for the journey home. Unfortunately, at some point over the weekend of 3–4 April, as they underwent preparations to be flown back to NASA's Johnson Space Center in Texas for analysis, the freezer suffered a failure and thawed the samples, thus ruining them. It was an intense disappointment for the EEVT scientists.

Columbia, meanwhile, was towed to a huge crane known as the Stiffleg Derrick, which – with the assistance of a conventional crane – hoisted her on top of the 747 carrier aircraft for the return flight to Florida. At 2:00 pm on 6 April, the Shuttle left White Sands and, following a refuelling stop at Barksdale Air Force Base in Louisiana, touched down at KSC just six hours later. Had she returned to Edwards as originally intended on 29 March, her return to Florida was not anticipated before 9 April.

"IT WILL TAKE A HUNDRED FLIGHTS"

Columbia's early return to KSC after her eventful STS-3 landing was something of a double-edged sword for Shuttle technicians. On the one hand, it allowed NASA management – who were, by now, becoming increasingly confident of the reusable spacecraft's capabilities – to bring the target launch date of her next mission *forward* from some time early in July to the last week of June 1982. However, on the other hand, the Shuttle was not in particularly good shape following her White Sands touchdown and needed extensive repairs before she could fly again.

As Charlie Bolden commented, the vehicle was literally saturated with gypsum dust and, despite efforts to remove it, the powdery stuff would remain in small quantities, hidden in nooks and crannies, for the rest of Columbia's career. It was remarkable, therefore, that the OPF processing flow needed to meet a late June target for STS-4 was accomplished in just 42 days! That represented a significant reduction from the 610 days needed to prepare STS-1, the 104 days to ready STS-2 and the 68 days to process STS-3. It seemed that NASA was making headway with getting Shuttles ready to fly.

Launches every two weeks, however, were still a long way off and, privately, many doubted that they would ever be achievable. The sheer technical challenges facing NASA in preparing each payload for flight and tending to the refurbishment of the spacecraft – and, particularly, the thermal tiles – were almost overwhelming. Even at its peak flight rate in 1985, the year before NASA lost its first Shuttle, only nine missions were accomplished in a single calendar year. By 2002, the year before Columbia herself was lost in the STS-107 tragedy, that figure had fallen to just five flights.

With the benefit of hindsight, it was – and still is – naïve to suppose that a vehicle as complex as the Shuttle could ever come close to becoming the spacegoing equivalent of a commercial airliner. Yet that was the intention and it was a dream that persisted until the beginning of 1986: in fact, NASA's final Shuttle manifest, published just before the loss of Challenger, quoted plans for 14 flights in 1986 and almost twice that number in 1987. Even without the loss of Challenger, insiders doubt that such mammoth flight rates were realistically possible.

The plan to remove the ejection seats from Columbia was also controversial among many astronauts. Bryan O'Connor, who would later fly Columbia in June 1991, remembers a conversation he once had with fellow astronaut Ken Mattingly. "I told him I just didn't feel comfortable with how we could possibly get to a

The STS-3 crew: Gordon Fullerton (left) and Jack Lousma.

confidence level after such a short test programme, to be able to do that. He said, 'Well, don't worry about all the rhetoric [from NASA Headquarters]. In reality, you and I both know that it will take a hundred flights before this thing will be operational.'"

In the spring of 1982, however, it was a different story. On the outside, the Shuttle seemed to be prospering and it was with great anticipation that NASA set to work preparing Columbia for her fourth and final test flight. Thirty-six tiles and fragments of 14 others were found to have fallen from her nose and the aft body-flap underneath her main engines; fortunately, none of these areas was subjected to exceptionally high temperatures during re-entry. The tiles had been closely inspected after each flight and a process of 'densification' had been ongoing since before STS-1. This process involved the application of a silica solution to the tiles and was intended to improve their adhesion to a Nomex felt pad bonded to Columbia's aluminium skin. Since the airframe expanded when heated, the tiles –

which could not be allowed to open any gaps – were affixed to a 'dynamic' base. Most of the tiles in areas subjected to particularly high-temperature heating during re-entry – such as the belly – had been densified long before STS-1 and the remainder were gradually completed between each flight and finished during a scheduled year-long maintenance period after Columbia returned from her STS-5 mission.

Meanwhile, the payloads assigned to the STS-4 flight were being brought up to speed and by May 1982 most of them had been installed on board the Shuttle. The IECM unit, which Jack Lousma and Gordon Fullerton had been unable to deploy on STS-3, was returned to its manufacturer, inspected and replaced in the payload bay. An important new commercial facility called the Continuous Flow Electrophoresis System (CFES) – an 'operational' variant of the EEVT and destined to fly several Shuttle missions – was loaded into the middeck and the first 'real' Getaway Special (GAS) canister was mounted on Columbia's payload bay wall.

Getaway Specials were to become frequent passengers on the Shuttle. These 180-kg dustbin-sized canisters could be flown in numbers of up to 13 at a time and were part of a drive to encourage universities, government agencies, foreign nationals and even private individuals to develop their own scientific experiments for carriage into orbit. A 'practice' canister had been flown on STS-3 to demonstrate its viability as an operational facility: it was fitted with temperature, acceleration, acoustic-noise and pressure sensors to monitor the stresses that an experiment would be subjected to during launch, ascent, orbital flight and re-entry.

The test proved that the canisters could support even very sensitive experiments, including those housing living creatures such as insects, and on STS-4 a Thiokol manager named Gilbert Moore 'bought' one for several thousand dollars and donated it to Utah State University for student-built investigations. Nine experiments in total were placed in the canister, ranging from genetic studies of fruit flies and the growth of algae, duckweed and brine shrimp to a number of fluid physics and materials science investigations.

A QUIET MISSION

One passenger on board STS-4, which was not publicised as highly, was the first classified Department of Defense payload. The United States military had long harboured an active interest in both the development and usage of the Shuttle for their own purposes; in fact, a separate launch-and-landing facility had been built for the reusable spacecraft at Vandenberg Air Force Base in California. Already, the US Air Force was in the process of buying nine dedicated Shuttle missions to launch its classified spy satellites and conduct other military experiments for the bargain-basement price of just $268 million.

This remarkable deal – less than $30 million per flight – had been struck with NASA partly in recognition of the support offered to the Shuttle during its development by the Department of Defense, but also in anticipation of the latter's plans to fly regular Shuttle missions out of Vandenberg from 1986 onwards. One of

NASA's Shuttle fleet – most likely Discovery, due for completion in mid-1983 – would be exclusively detailed to Vandenberg to take either military payloads or polar-orbiting satellites into space. It would be the first time a manned spacecraft had launched from the West Coast.

History has shown that, in the wake of Challenger, it never happened. The loss of a Shuttle and its crew, together with the inevitable public scrutiny and interest in manned launches, eventually drew the Department of Defense back to using expendable rockets. By employing these, unlike the Shuttle, it could preserve the secrecy it needed for its highly sensitive spy satellites. Nevertheless, it did remain committed to NASA in terms of its nine agreed missions, and even added a tenth, which finally flew in December 1992.. However, no Shuttles ever flew from Vandenberg or into polar orbit.

In readiness for what was expected to be a flurry of military missions, the first-ever classified payload was carried on STS-4 and known rather cryptically as 'DoD 82-1', meaning it was the first – and only – Department of Defense experiment to be flown during financial year 1982. Some details of this payload have slipped out over the years and the centrepiece seems to have been a sensitive detector known as the Cryogenic Infrared Radiance Instrument for Shuttle (CIRRIS), which was also destined to fly on another military mission in April 1991.

It would appear that its objective was to test infrared sensors for an advanced surveillance satellite known as Teal Ruby which, at the time of the Challenger accident, was scheduled to be on board the first Shuttle mission out of Vandenberg sometime in July 1986. Interestingly, and perhaps pointing to Teal Ruby's significance, that mission would have been under the command of none other than STS-1 veteran Bob Crippen. In the wake of Challenger, and the near-three-year period of grounding that followed, Teal Ruby was eventually shifted onto the STS-39 mission and finally cancelled.

When STS-39 lifted off in April 1991, it carried not Teal Ruby, but an updated version of CIRRIS. Apparently, by the time it would have been ready to launch, the Teal Ruby technology – considered 'advanced' when it was built in the late 1970s – would be virtually obsolete. Had it flown, Teal Ruby would have been capable of detecting and tracking missiles passively from space.

"[Teal Ruby] was a prototype staring mosaic infrared sensor that was trying to be able to detect low-flying, air-breathing vehicles – things like cruise missiles – and a way to try to detect those approaching US territories," said Jerry Ross, who would have accompanied Crippen had the original Vandenberg mission not been cancelled.

"IT JUST GOES"

It was under an unusual shroud of quiet that Commander Ken Mattingly and Pilot Hank Hartsfield – the last two-man Shuttle crew – rode the bus to Pad 39A on 27 June 1982 for STS-4. Mattingly had previously flown on Apollo 16 with John Young, while Hartsfield was making his first space mission. Originally, when NASA intended to fly six test flights before declaring the Shuttle operational, Mattingly and

Hartsfield were assigned to the fifth (called the 'E') mission, and when the agency reduced the number to four they expected to be assigned an operational flight.

According to the initial plans, astronauts Vance Brand and Bob Overmyer – the 'D' crew – were in line for STS-4, but their roles were switched with those of Mattingly and Hartsfield and they were ultimately assigned as Commander and Pilot of Columbia's first operational mission, with the job of deploying two commercial satellites and overseeing the first Shuttle spacewalk. "The idea of trying to get on an early test flight", said Mattingly, "was what every pilot wants to do. Of course, none of us thought that it was going to take so many years before that first flight took place."

Concerns had been raised over their chances of launching safely, because on the previous night (26 June), a severe hailstorm damaged several thermal tiles and left water behind the covers of two RCS thrusters. Despite the chance of this freezing during ascent, it was decided that it would not present a problem.

Columbia lifted off precisely on time at 3:00 pm, right at the opening of a four-and-a-half-hour 'window', and was quickly arcing into the clear Florida sky, on course for a week-long mission. For Mattingly, who had endured a bone-jarring liftoff on board a Saturn V a decade earlier, the Shuttle, in comparison, was the smoothest ride in the world: "It [the Saturn] feels just like it sounds. You get this staccato cracking and all that from the engines. Inside, it's the same thing. It's shaking and banging and pushing hard and there's no doubt that something really gigantic is going on. ... The Shuttle [was] not noisy; it doesn't shake. It just goes."

Columbia's ascent, overall, was nominal, but several hydraulic sensors registered dramatic temperature drops in the nose landing gear wheel well – in one case from 25 Celsius to minus 15 Celsius – which took several hours to return to normal. Nothing like it had been seen on the three previous launches and it was attributed to rainwater from the storm having penetrated Columbia's wheel wells. Moreover, DFI measurements recorded moisture behind some thermal tiles.

This reinforced a decision by NASA managers, made before the launch, to orient the Shuttle in a belly-to-Sun attitude and so evaporate the water. This attitude was maintained for 12 hours, after which the astronauts began preparing for the 'normal' attitudes planned for their mission. With her belly facing the Sun and her payload bay and overhead windows aimed Earthwards, this presented some magnificent first views of orbital flight for both men. "All of a sudden," Mattingly said later, "it was like you pulled the shades back on a bay window and the Earth appeared!"

The ascent, however, had gone somewhat awry for Columbia's twin SRBs, whose parachutes had failed during their fall back to Earth. Both boosters sank after splashdown in the Atlantic Ocean and, although an underwater remote camera later photographed the wreckage, it was deemed too expensive to recover them. The cause of the failure was later traced to a new feature intended to separate the parachutes from the SRBs at the instant of splashdown; it was supposed to prevent the boosters being dragged through the water by their deflated canopies, and instead had prevented their deployment.

The system was active on the first three Shuttle missions, but was partially disabled for STS-4. During the launch of Mattingly and Hartsfield, frangible nuts

The desk-sized IECM is manoeuvred by the RMS on STS-4.

holding one of two risers for each parachute were replaced by two regular solid nuts that would not separate the riser. Preparations for Columbia's next mission, STS-5, would include the replacement of all frangible nuts with regular solid ones. As a result, when the Shuttle next lifted off in November 1982, both risers on each parachute remained attached to the boosters until they could be removed by recovery personnel. The fix was successful.

Meanwhile, despite the half-day 'hot soak' of Columbia's tiles in orbit, DFI data indicated that water still remained in several nooks and crannies. It was feared that, during cold periods, it could freeze and possibly crack the most sensitive tiles on areas of Columbia that endured maximum atmospheric heating during re-entry. As a result, another, 23-hour 'solar inertial' run was added to the astronauts' timeline for 29 June and resolved the problem. After the mission, several DFI-measured tiles were removed and checked for traces of water; none was found, thereby validating the solar inertial 'conditioning' attitude to resolve future incidents.

THERMAL TESTS

Manoeuvring the Shuttle into various attitudes to better understand its thermal characteristics before declaring it operational was a key objective of STS-4. "We had been assigned to do a bunch of thermal tests," Mattingly said later, "where you put the orbiter in an attitude and get one side hot, and then one side cold and then spin it around. They were collecting the data [because] after this flight we wouldn't have the instrumentation to do that. It was kinda something that had to be done, but was really not a glamorous kind of test that you can run."

Rookie astronaut Mike Coats, who was sitting in Mission Control monitoring these tests, called the Shuttle's thermal behaviour "the banana effect"; during orbital flight, the entire vehicle was bending like a banana, as its 'hot side' expanded and its 'cold side' contracted. During a practice opening and closure of the payload bay doors under extremely cold conditions, problems arose after one period in the belly-to-Sun attitude. A 'closed' indication on one of the doors was not achieved, prompting Coats to advise the astronauts to reverse their attitude and warm up Columbia's topside instead of her belly.

This procedure, as well as 10 hours of 'barbecue roll' and two hours in a tail-to-Sun attitude, rectified the problem and door opening and closure was demonstrated several times thereafter. A thorough understanding of factors affecting the satisfactory closure of the payload bay doors was essential before the Shuttle could be declared operational; if a crew was ever unable to close them properly, they could not return home. For the rest of her mission, Columbia spent 67 hours in a tail-to-Sun attitude and on 3 July undertook a 10-hour barbecue roll to thermally stabilise herself prior to re-entry.

In effect, the astronauts could place the Shuttle into the required attitude and leave her alone until it became necessary for another orientation change; in the meantime, they had a multitude of other tasks to accomplish. Chief among these were the final series of RMS tests and, to highlight the importance of the arm on this and future missions, a new console had been set up in Mission Control: that of the 'RMS, Mechanical Systems and Upper Stages Systems Officer', known by the callsign of 'RMU'.

TYING UP LOOSE ENDS

Although previous Shuttle crews had been assisted by an RMS officer in Mission Control, it was decided that the new RMU would also supervise the hydraulic systems, payload bay doors, the troublesome APUs and also the upper stages that would help to boost two communications satellites into geosynchronous orbits on STS-5. Mattingly and Hartsfield's primary work with the robotic arm was the deployment and manoeuvring of the large IECM unit, which had not been demonstrated on the previous mission because of a failed television wrist camera.

The IECM itself was somewhat different on STS-4 in that it had been fitted with an extra instrument to monitor bursts from Columbia's forward RCS thrusters. In total, Mattingly and Hartsfield completed two successful deployment and berthing runs with the device, lasting around nine hours in total, and commented that the handling characteristics of the RMS were "crisp and precise" with the IECM in its grasp. They also evaluated a new berthing device in the payload bay, known as a

Retention Engagement Mechanism (REM), and acquired good images of the payload in their monitors on the aft flight deck.

Images of the payload bay during the flight were minimised to keep the Department of Defense's secret package under wraps. It seems that, in addition to the CIRRIS infrared detector, there were six other experiments on board Columbia, all of which were attached to a cross-bay 'bridge' known as the Experiment Support Structure (ESS). These were the Horizon Ultraviolet Programme (HUP), the Autonomous Navigation and Attitude Reference System (ANARS), the Shuttle Effects on Plasma in Space (SEPS), the Sheath and Wake Charging (SWC), a set of passive cosmic-ray-collection panels and a pallet alignment modelling experiment.

Each of these payloads apparently functioned autonomously and without incident throughout the flight, except for CIRRIS itself, which refused to work when its lens cap would not open. Mission controllers in Houston discussed the possibility of knocking the cap off with the RMS or sending Mattingly out into the payload bay on the first-ever Shuttle spacewalk to open it manually, but it was ultimately decided not to complicate what was, after all, a test flight. Payloads were considered a bonus, rather than a necessity, and CIRRIS would have to wait for another chance to display its ability in space.

Mattingly, on the other hand, was eager to make a spacewalk, having already performed one during his Apollo 16 mission 10 years previously. Owing to the curtailed STS-2 mission, for which he and Hartsfield had served as the backup crew, the new-specification Shuttle spacesuit had not yet been used in orbit. Before a spacewalk could be tried, the procedures and timeline for donning and doffing the suit needed to be fine-tuned. That honour was completed admirably by Mattingly on STS-4, although the only problem, he said, was that "I didn't get to open the door".

VEIL OF SECRECY

The military experiments had already drawn sufficient criticism to drop a reconnaissance camera from Columbia's flight and the dismal failure of the high-profile CIRRIS only added to the embarrassment. In fact, the overly secret nature of the mission was often laughable, especially to the crew. "A funny thing happened on that flight," remembered Hartsfield. "Because it was highly classified, on this one experiment they had a classified checklist [and] because we didn't have a secure com[munications link], we had the checklist divided up in sections that just had letter-names like Bravo Charlie, Tab Charlie, Tab Bravo, that they could call out. When we talked to Sunnyvale [the US Air Force's Satellite Control Facility in California], they said 'Do Tab Charlie' or something. That way it was unclassified. We had a locker that we kept all the classified material and it was padlocked. So once we got on orbit, there was nobody going to steal it [so] we unlocked it and did what we had to do. When we'd finished the last part of that thing, and I remember I finally got it all stowed, I told Ken, 'I got all the classified stuff put away. It's all locked up.' He said 'Great.' It wasn't 30 minutes and [Mission Control] said the military folks needed to talk to us. So the military guy came on and he says he wanted me to do Tab November. Ken said 'What's Tab November?' I said, 'I ain't got the foggiest idea. I'm going to have to get the checklist out to see.' So I got the padlock off and got the drawer and dug down and got the checklist and went to Tab November and it says: 'Put everything away and secure it!' Ken and I really laughed about it!"

More visible in the middeck was the CFES unit, built by McDonnell Douglas, which would fly several times on the Shuttle in the early 1980s, accompanied on three occasions by an engineer representing the company. A full-scale version of the EEVT machine from STS-3, the CFES could process considerable quantities of biological materials in continuous streams. It was already known from primitive electrophoresis experiments dating back to the Apollo days that the absence of gravity could lead to the production of materials of greater purity than could be achieved on Earth.

It was hoped, for example, that scientists would be using CFES samples not only to develop advanced drugs in space, but to produce them at such a rate as to make them available on the world market. Some scientists speculated that by 1987 these could include beta cells to provide a single-injection cure for diabetes, high-purity Interferon, epidermal growth factor products for treating burns, hormonal products to stimulate bone growth and countless others. With the Shuttle still widely expected to achieve fortnightly launches, the world markets would soon be overflowing with drugs that could only be made in space.

At the time of STS-4, plans were afoot to fly CFES at least six times between 1982 and 1984 and, depending on their success rate, to expand operations using a pallet-mounted 'bridge' in the payload bay to support production from 1986 onwards. McDonnell Douglas engineer Charlie Walker flew three times with CFES – once in the summer of 1984 and twice in 1985 – and demonstrated its capabilities. The first pallet-mounted CFES, accompanied by another McDonnell Douglas engineer named Bob Wood, was scheduled for July 1986. When Challenger exploded in January of that year, however, all those plans were shelved.

DATE WITH THE PRESIDENT

Columbia's return to Earth on Independence Day, 4 July 1982, was interesting for Mattingly and different again from his memories of coming home after his lunar mission: "Apollo had aggressive forces on launch and entry, whereas the Shuttle has just really soft forces and entry is just a piece of cake." Their arrival at Edwards Air Force Base was being watched closely by President Ronald Reagan and the astronauts had already been briefed by NASA Administrator James Beggs and asked to think up some memorable words to mark the occasion.

"We knew they had hyped-up the STS-4 mission so that they wanted to make sure that we landed on the Fourth of July," Mattingly said later. "It was in no uncertain terms that we were going to land on the Fourth of July, no matter what day we took off. Even if it was the Fifth, we were going to land on the Fourth! That meant, if you didn't do any of your test mission, that's okay, as long as you land on

Hank Hartsfield (left) and Ken Mattingly (saluting) are greeted by President Ronald Reagan at Edwards Air Force Base on 4 July 1982.

the Fourth, because the President is going to be there. We thought that was kinda interesting."

Fortunately, Columbia's landing – the first on concrete Runway 22 – was precisely on time at 4:09:31 pm on Independence Day, completing a textbook seven-day mission. Unlike the experience of Lousma and Fullerton on STS-3, the Shuttle's landing gear was deployed at an altitude of 123 metres, a full 20 seconds before touchdown, allowing all six wheels to fully lock into place with plenty of time to spare. Columbia landed a couple of hundred metres past the runway threshold and Mattingly applied the brakes for 20 seconds to bring her to a smooth stop.

The astronauts had considered putting up a notice for the President, who they expected would want to come on board to look around the Shuttle's middeck, with words to the effect of 'Welcome to Columbia. Thirty minutes ago, this was in space.' Unfortunately, US Navy Captain Ken Mattingly's first meeting with his Commander-in-Chief – who used the occasion to declare the Shuttle fully operational – was done with a rather painful forehead ...

After the flight, Mattingly said, "I said [to Hartsfield, while on the flight deck] 'I am not going to have somebody come up here and pull me out of this chair. I'm going to give every ounce of strength I've got and get up on my own.' So I just got this mental set and I pushed and I hit my head on the overhead [control panel] so hard that blood was coming out! Oh, did I have a headache!" Nevertheless, both astronauts composed themselves, disembarked, descended the steps and Hartsfield and a sore-headed Mattingly smartly saluted the President.

2

Columbia delivers

AN UNPROTECTED CREW

When Vance Brand had his official astronaut portrait taken in 1979, wearing the new Shuttle pressure suit, he confidently expected to wear it on his first flight in a few years' time. Although it was questionable how useful Columbia's ejection seats would be as a means of emergency escape, the suit at least would keep him alive in case of a cabin depressurisation during the trip back to Earth. Little did Brand know that when he and his STS-5 crew lifted off on 11 November 1982, they would be wearing little more than overalls and a helmet ...

The reason was that, on Columbia's fifth mission into space, Commander Brand and rookie Pilot Bob Overmyer would be accompanied by two other astronauts: a physicist named Joe Allen – who had served as Capcom for the first Shuttle flight – and an electrical engineer named Bill Lenoir. It would be the first time that four astronauts had been launched together on the same spacecraft and would pose a difficult dilemma over how to eject from the Shuttle in the event of a major in-flight emergency. The problem was quite simple: not all of them would be able to get out alive.

There was only enough room at the front of the Shuttle's cramped flight deck to accommodate two ejection seats and their rails for the pilots. Allen and Lenoir, members of a new subset of astronauts called 'Mission Specialists' – men and women responsible for accomplishing in-flight tasks, such as satellite deployments and retrievals or scientific experiments – would have to make do with collapsible seats at the rear of the flight deck or on the middeck. Installing four ejection seats with rails would be difficult and, as Shuttle crews increased to a maximum size of seven, would become impossible.

"No-one wanted to fly with seven rockets in the cabin!" Arnie Aldrich has said, although others wanted to retain the ejection seats. "*That* would have restricted the size of the crew," admitted former assistant Shuttle director Warren North, "because [you] couldn't put seven ejection seats in there but we could leave the two pilot seats, [then add] four [abreast] behind the pilots that were of much lighter variety. The Yankee system, for instance, [was] a tractor rocket that pulls the pilot out in a prone position, where the seat [pan] collapses and the pilot is pulled out head first. That would have involved putting pyrotechnic escape hatches in four places – behind the flight crew and in the [overhead] payload deck – which would have involved redesigning the orbiter to some degree [including its] wiring. It could have been done, [but] would have involved a time delay, been a little [more] expensive [and added some] weight. We made mistakes along the way. We've got a vehicle today that has a moderate escape capability, but not nearly what some of the crew would like."

No ejection seats on STS-5 meant that Allen and Lenoir's chances of escape during an emergency would be limited to the 'long shot' of bailing out through Columbia's side hatch. Eitherway, even if ejection seats for the entire crew were practicable, they would not have provided greater advantages to the astronauts in terms of survivability. STS-1 Commander John Young once joked darkly that the seats' parachutes would open "about fifty feet *after* we hit the ground!" With this in mind, perhaps, Brand had elected to do away with the ejection seats and move to the blue NASA overall-like flight suits.

His launch on STS-5 would thus be totally different from his Apollo 18 mission in July 1975, when he and his crewmates had participated in a joint rendezvous-and-docking exercise with the Soviet Soyuz 19 spacecraft. Like Young, Lousma and Mattingly before him, Brand remembered first-hand the escape rockets that sat on top of the Apollo command modules, which were capable of whisking crews to safety in the event of a malfunction. The asymmetrical design of the Shuttle made such escape rockets impractical, and rendered the entire vehicle potentially far more hazardous than the Saturn rockets had ever been.

A civilian astronaut since April 1966, Brand was the only veteran among the STS-5 crew. In addition to his Apollo mission, he had trained for a daring rescue of three colleagues on board the Skylab space station and, owing to NASA budget cuts, had narrowly missed out on a flight to the Moon. He would, however, go some way in making up for this by flying three Shuttle missions and, in December 1990, would acquire distinction as the then-oldest man in space when he led a crew into orbit at the age of almost 60.

Seated alongside Brand was crew-cutted US Marine Corps Colonel Overmyer, who served as Pilot on the mission, before commanding his own Shuttle flight in April 1985. Occupying a collapsible seat behind the instrument panel, just behind and between the Commander and Pilot, sat Lenoir, to whom fell the task of serving as a flight engineer during the ascent portion of the mission. Meanwhile, the final member of the crew, Allen, sat alone 'downstairs' on the middeck for launch, trading places with Lenoir on the flight deck for re-entry. Both of their seats were folded away and stowed during the flight.

Although the bulky ejection seats in the forward part of the cockpit remained in place, and would not actually be removed for another couple of years, they were at least *deactivated* in time for STS-5. However, their very presence made Columbia's flight deck considerably more 'cramped' than the cockpits of later Shuttles. Indeed, STS-3 Pilot Gordon Fullerton later remarked that the flight deck of the second space-rated orbiter – Challenger, which he flew in July 1985 – seemed much more 'roomy' when he first climbed on board, because it contained more lightweight, collapsible seats.

SPACE TRUCK

Preparations for the STS-5 mission took place on three separate fronts. Firstly, there was the assembly of the boosters and External Tank, which took place in the VAB. Secondly, there was the processing – and a good deal of modification, too – of Columbia herself in the OPF. Finally, on this first operational flight of the Shuttle, the astronauts would be required to deploy two commercial communications satellites: Satellite Business Systems (SBS)-3 and ANIK-C3. Both arrived by aircraft at Cape Canaveral Air Force Station in the midsummer of 1982 and were quickly transferred to the Vertical Processing Facility for pre-mission checks.

Columbia, meanwhile, was having old parts removed and new parts installed. The IECM and the Canadian-built RMS mechanical arm – which would not be required on STS-5 – were removed, the pyrotechnics for the ejection seats were dismantled and two collapsible Mission Specialist seats were fitted. The final removal of the ejection seats was planned for an extended period of modification and refurbishment of Columbia, expected to begin after the Shuttle returned home from her STS-9 mission. Furthermore, the middeck floor was strengthened and parts of the DFI pallet were removed from the payload bay.

On 9 September 1982, Columbia rolled into the VAB and was attached to her tank and boosters. Following the now-customary verification of mechanical and electrical interfaces between each component, the butterfly-and-bullet stack was transferred to Pad 39A on the 21st. The two satellites and their attached solid-rocket motors – encased in a pair of lightweight sunshades – were moved to the pad on 12 October and loaded on board Columbia a week later. This procedure, which employed a device known as the Payload Ground Handling Mechanism, was the first time a piece of cargo had been loaded vertically into its launch vehicle.

It was a sign of things to come. The Shuttle, as advertised, was capable of carrying up to three communications satellites, together with their motors and sunshades, in her cavernous payload bay and NASA hoped that this could bring in millions of dollars of revenue each year. The launch of just two satellites by the STS-5 crew would net the space agency a first royalty payment of $18 million. Ultimately, it was hoped to out-compete for commercial contracts with the recently inaugurated European-built Ariane expendable rocket by offering 'free' seats on board the Shuttle to customers' representatives.

Looking 'down' on Columbia's payload bay, it seemed that two oversized versions of Pacman were sitting there, for when the protective sunshades opened to release the satellites inside, they looked just like a pair of jaws from the children's board game. Fortunately, unlike the real Pacman, these jaws were designed to release something, rather than gobble it up. Each cradle was composed of a series of aluminium tubes, covered with Mylar insulation, and measured 2.4 m long by 4.6 m wide across the width of the payload bay.

At the base of the cradle was a turntable to impart the required spin rate, which could vary between 45 and 100 rpm, depending on the payload, and a spring ejection system to release the satellite and its attached motor. During the Shuttle's ascent into orbit, a pair of restraint arms held the precious satellite steady inside the sunshade.

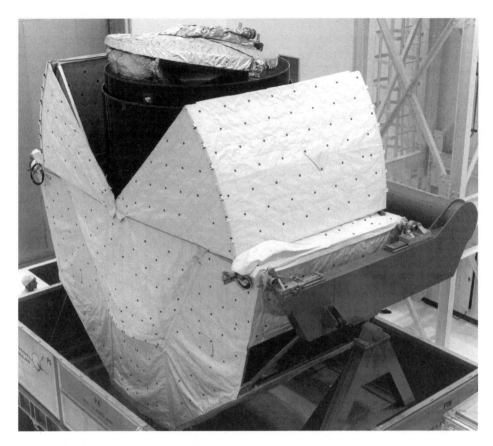

The ANIK-C3 communications satellite inside its 'Pacman' cradle before STS-5.

Soon after achieving orbit, the delicate shades were closed to protect the unpowered satellites from the space environment.

It was at this point that Lenoir and Allen would respectively supervise the deployment of 'their' individual satellite. The SBS-3 deployment, which would occur eight hours after launch, would be conducted by Lenoir and that of ANIK-C3 would take place under Allen's auspices almost exactly 24 hours later. As each astronaut performed 'his' deployment, the *other* Mission Specialist would photo-document the procedure, while Brand and Overmyer handled Columbia's systems and prepared to manoeuvre her safely out of the way before the satellites' motors were ignited to boost them into geosynchronous transfer orbits 36,000 km above the equator.

The 600-kg SBS-3 satellite was built by the Hughes company at its El Segundo plant in California and, when operational, provided all-digital voice, data, video and email facilities, a high-quality messaging service for business customers and a spare transponder for communications firms, broadcasters and cablecasters. Sitting just behind SBS-3 in Columbia's payload bay was ANIK-C3, which was more or less identical in size, shape and mass, but was owned by the Ottawa-based Telesat company. It offered, for the first time, rooftop-to-rooftop voice, data and video business communications, as well as Canadian television and other broadcast services.

Although its name implies that it was the third ANIK-C-series satellite to be placed into orbit, this was actually the *first* to be launched! Two others, ANIK-C1 and ANIK-C2, had been placed in storage until a suitable date could be established for their launch. By coincidence, ANIK-C3's completion occurred at the same time as Telesat's first contracted flight opportunity on the Shuttle, so it was decided to take it straight from the factory to the launch pad. The other two ANIK-Cs were launched on later Shuttle missions in 1983 and 1985.

Both SBS-3 and ANIK-C3 were cylindrical drums, measuring 2.7 m tall and 2.1 m wide when stowed on board Columbia, but increasing to more than twice that height in their final configuration. Both were coated with black 'skins' of 14,000 solar cells, which generated 1,100 watts of DC power to operate them over decade-long lifespans and also carried their own supplies of hydrazine fuel for station-keeping. Each had an onboard power system, including rechargeable batteries, to run its communications equipment. SBS-3 covered the entire contiguous United States and ANIK-C3 virtually all of Canada, including its remote northern regions.

SBS-3 was equipped with 10 transponder channels; ANIK-C3 had 16. Attached to the top of each satellite was a 1.7-m-wide shared-aperture grid antenna with two reflecting surfaces to provide both 'transmission' and 'reception' beams. In the case of ANIK-C3, which operated in the high-frequency radio bands at 14 and 12 GHz, a combination of high transmission power and high-frequency band usage meant that much smaller antennas just 1.2 m across could now be situated on rooftops or office blocks. This marked a significant reduction in size from the 3.6-m-wide reception dishes used previously, which had been viable only for hotels and office buildings.

Not only were commercial satellites new to the Shuttle, but so were the Payload Assist Module (PAM)-Ds, built by McDonnell Douglas, which were used by satellites on board both the Shuttle and the expendable Delta rocket. In essence, they and the payload bay-mounted cradle provided portable launch platforms, fitted with spin motors and deployment springs and could deliver satellites of up to 1,250 kg into geosynchronous transfer orbits.

THE 'ACE MOVING COMPANY'

It was with these two precious satellite cargoes tucked into her payload bay that Columbia and four of the most unprotected astronauts in history headed for orbit at 12:19 pm on 11 November 1982. Despite a minor problem during the countdown when one of the primary GPCs failed to synchronise properly with the other three computers, the Shuttle nevertheless lifted-off precisely on time at the beginning of a shorter-than-usual, 40-minute 'window'. The shortness of the time available was partly due to the need for daylight at Edwards and other contingency sites in case an emergency landing became necessary.

Predominantly, however, the window's short duration was because the deployment of the two satellites had to be timed in order to obtain the correct Sun-angle when they were ready to make the first use of their electricity-generating solar cells. Columbia's ascent into orbit was satisfactory, with no major problems, although the performance of the SRBs was slightly lower than expected. Nonetheless, corrections to their parachutes made in the wake of the STS-4 failure worked and both were spotted descending towards the Atlantic Ocean at a radar-clocked speed of 27 m/s. Both parachutes remained intact and attached to the boosters.

Immediately after Brand and Overmyer completed two OMS burns to establish Columbia in her correct orbit, the two Mission Specialists unstrapped and began to fold away and stow their collapsible seats. During the course of their first day in space, they activated the second 'real' GAS canister in the payload bay: an X-ray study of the solidification of liquid metals, sponsored by the West German Ministry of Research and Technology, which unfortunately failed completely. It would later become apparent that the GAS canister's battery had leaked its electrolyte and was unable to properly switch on the experiment.

Another problem noted by the crew, just two-and-a-half hours into the mission, was the failure of one of the Shuttle's three cathode-ray tube (CRT) monitors located on the forward flight deck control panel. In the early days, before the installation of advanced 'glass cockpits' in the late 1990s, there were four such monitors on board Columbia: three on the control panel in front of the Commander and Pilot, needed to display thousands of different functions and a multitude of data, and one on the aft flight deck for payload operations.

The aft-mounted CRT would not be needed for re-entry, so on 14 November Mission Control radioed up a procedure to switch the cables between the failed monitor and the healthy one at the back of the flight deck. This forced them to switch off the aft CRT, but ensured that Columbia was able to return home with three fully functional monitors on her forward control panel. Several other minor issues cropped up with a troublesome RCS thruster, which showed signs of a possible leak, and Brand and Overmyer kept a close eye on it for the rest of the mission.

These problems paled into insignificance, however, by the end of Day One, when Lenoir successfully accomplished the first commercial launch of a communications satellite from the Shuttle. Six hours before the historic deployment, updated computations of Columbia's orbit – including the spacecraft's attitude, velocity and orbital inclination – were radioed to Lenoir from the SBS control centre. Then, 40 minutes before deployment, Brand and Overmyer manoeuvred the Shuttle to the correct attitude with the open payload bay facing into the direction of travel.

The restraint arms pulled away from the satellite, Lenoir flipped a switch to open the Pacman jaws and imparted a 50-rpm spin rate on the payload. This steady rotation would help to stabilise SBS-3 during its deployment. Next, at 8:20:18 pm on 11 November, during Columbia's sixth orbit of Earth, he issued the command for explosive bolts to fire and release a Marman clamp that held SBS-3 and its PAM-D upper stage in place. Seemingly in slow motion, the spinning payload departed the bay at just 90 cm/s, as Allen enthusiastically fired off pictures.

Immediately after leaving the Shuttle, command of the satellite passed to the SBS control centre in Washington, DC. Fifteen minutes later, Brand and Overmyer backed Columbia away to a distance of 42 km, aiming their spacecraft's belly at the satellite to protect its delicate topside from the PAM-D's exhaust. At 9:05 pm, an onboard timer fired the upper stage's perigee motor for about 100 seconds to boost SBS-3 into the highly elliptical transfer orbit. Overall, the PAM-D's first Shuttle use was described as "satisfactory" and it separated from the satellite a few minutes after the completion of its burn.

Shortly thereafter, SBS-3's omni-directional antenna was successfully raised and, over the next two days, it used its own Thiokol-built solid-propellant apogee motor to insert itself into the required near-geosynchronous orbit and then manoeuvred onto its 'slot' at 94 degrees West longitude. During this time, it opened out to nearly twice its 'stowed' height and deployed its communications payload.

A little over a day into the mission, Brand and Overmyer recircularised the Shuttle's orbit with a burst of the powerful OMS engines. This set them up for the second satellite deployment – that of Telesat's ANIK-C3 – just a few hours later, under the supervision of Allen.

Launching the Canadian satellite followed much the same routine as that employed with SBS-3. During the afternoon of 12 November, Brand and Overmyer performed another two OMS burns in support of the ANIK-C3 deployment and would make yet another – their sixth in total – after the satellite had been set free to provide a safe separation distance before the PAM-D ignition. Following Lenoir's procedure of the previous day, Allen opened the sunshade, spun-up the satellite and ejected it from the payload bay at 8:24:11 pm, during Columbia's 22nd orbit as she flew over Hawaii.

A perfect firing of the PAM-D some 45 minutes later duly inserted ANIK-C3 into an elliptical transfer orbit that, like SBS-3, was a few hundred kilometres higher than its planned 'operational' orbital slot. The satellite then used its own motor to place itself into a near-geosynchronous orbit; by 16 November, it was on-station at 114.9 degrees West longitude, directly in line with Edmonton in Alberta. The astronauts celebrated by turning on music from the movie *2001: A Space Odyssey*, and unfurled a sign that would epitomise the growing maturity of the Shuttle as a 'space truck'.

The sign, which has been reproduced countless times in books about the reusable spacecraft and its achievements, is surrounded in a famous photograph by the four STS-5 astronauts. Held by Brand, it read: 'Satellite Deployment by The Ace Moving Company. Fast and Courteous Service. We Deliver.' Floating weightlessly around their skipper, head-to-head, were Overmyer, Allen and Lenoir. The idea for the sign, said Brand, actually came from Allen: "Joe coined the term 'Ace Moving Company', because we moved stuff to space. We put up a sign that [Joe had] made out ahead of time."

It was a triumphant moment for both the crew and NASA as a whole. After more than a decade of planning and development, maddening problems with thermal protection tiles and main engines, boosters and fuel tanks and billions of dollars poured into the programme, it seemed that the Shuttle was finally beginning to prove its commercial worth. It still, as many astronauts pointed out, had nowhere to shuttle *to* – certainly, no space station was officially on NASA's

SBS-3 departs Columbia's payload bay on 11 November 1982.

radar yet – but it seemed to be making access to space routine, which was one of its most important goals.

SPACEWALKING FROM THE SHUTTLE

This moment of triumph was expected to be extended still further by a three-and-a-half-hour spacewalk – or 'Extravehicular Activity' (EVA) – by Lenoir and Allen on 14 November, during which they would become the first American astronauts to leave their spacecraft since 1974 and give the world its first glimpse of the new Shuttle-specification spacesuit. Today, it has become increasingly familiar as missions have serviced the Hubble Space Telescope and begun building the International Space Station. It is a pity, therefore, that what should have been its first outing into the vacuum of space was frustrated.

Yet, originally, in the early development of the Shuttle, an EVA capability was considered unnecessary and was not provided. "The NASA perspective of a Shuttle was an airliner," said spacesuit engineer Jim McBarron, "and the people inside it wouldn't need suits. It was through prompting and questioning that Aaron Cohen, who was then the Shuttle project manager, finally accepted a contingency capability for closing the payload bay doors – which was an issue they were faced with – to put an EVA capability on the Shuttle."

The plan was for Lenoir, designated 'EV1' with red stripes on the legs of his suit for identification, and Allen ('EV2') to test the snow-white ensembles and practice techniques for a tricky repair of NASA's Solar Max satellite, planned for the spring of 1984. They had also trained to perform a manual opening and closure of Columbia's payload bay doors, in the event that a malfunction should make it impossible to complete this task automatically. Although the excursion was only intended to last a few hours, it would have occupied virtually the crew's entire day on 14 November.

Assisted by Overmyer, the men would have risen early and begun preparing their suits and the Shuttle's airlock at 7:00 am, before spending the next four hours 'prebreathing' pure oxygen to wash nitrogen from their bloodstreams and thus avoid attacks of the 'bends'. Their next step would have been to lower the airlock's pressure from the normal 14.7 psi to just 5 psi to check the integrity of the suits. Finally, a little under 73 hours into the mission, the two men would have opened the outermost hatch of the airlock and entered Columbia's payload bay.

The airlock was a cylindrical structure about the size of a Volkswagen Beetle, situated at the rear of the middeck. Its inclusion *within* the pressurised cabin was done to preserve the maximum amount of volume in the payload bay, and consisted of two hatches, one for the astronauts to enter from the middeck and another through which they would go 'outside'. The volume inside the airlock was cramped: astronaut Rich Clifford, who would make a spacewalk in March 1996, has described hanging there in his bulky spacesuit with his colleague, barely able to even move his arms.

Depressurisation and repressurisation of the airlock was, and still is, in the surviving Shuttles, controllable from the flight deck or from within the airlock itself. Normally, two spacesuits are stored inside the airlock, although there is room for up to four if needed. In fact, many Shuttle missions thoughout the 1990s and into the present century have involved four spacewalkers, working in two alternating pairs, and in May 1992 the airlock successfully demonstrated its ability to support three fully suited astronauts at the same time.

Later modification to most of the Shuttle fleet, in anticipation of planned dockings with the Russian Mir space station and the International Space Station in the mid to late 1990s, would lead to the airlocks being removed from the cabin and mounted instead in the forward part of the payload bay. It was decided, however, that one vehicle still needed a 'full' payload bay capacity for especially large cargoes, and Columbia retained her internal airlock. This proved beneficial when she carried the enormous Chandra X-ray Observatory into orbit in July 1999, which almost filled the entire bay.

Ironically, had she returned safely from STS-107 on 1 February 2003, Columbia's next mission in November of the same year would have been her first trip to the International Space Station. One of the most notable modifications planned in her preparation for *that* mission would have been the removal, after nearly a quarter of a century, of the internal airlock and its replacement in the forward part of the payload bay. Sadly, as everyone now knows, Columbia would never visit a space station ...

PRACTICE FOR THE SOLAR MAX REPAIR

The plan was that they should 'suit-up' and, assisted by Overmyer, enter the airlock on 14 November. The suits were somewhat different from the ensembles of previous Apollo and Gemini astronauts, yet was designed with the same objective in mind: venturing outside the pressurised confines of a spacecraft. During their spacewalk, Overmyer would have choreographed every move from the flight deck, while Brand photo-documented the whole event. Their first task would have been to tether themselves to slidewires running along the sills of the payload bay walls.

This safety procedure would have prevented them from floating away from the spacecraft in the event of an emergency. Then, with Lenoir tethered to the starboard sill and Allen attached to the opposite sill, both men would have moved themselves down the entire length of the payload bay until they reached the aft bulkhead. It was during this time that the men would have conducted the first 'real' evaluation of the new suits: their comfort, dexterity, ease of movement and the performance of the communications and cooling systems and the floodlights mounted in the payload bay.

They would also have looked for suitable locations from which future spacewalkers could best work on repairing the Solar Max satellite. Practising 'repair' techniques would dominate Lenoir and Allen's time: after these evaluations, they would have returned to the forward end of the payload bay to begin work on a 'dummy' set of equipment for the Solar Max repair. For more than an hour, it was intended that they should test and comment on a series of fixed and torsion-adjustable bolts, using a special wrench.

Lenoir would then have moved to the Solar Max equipment, which was basically a dummy main electronics box from the satellite's coronagraph. This was intended to be as complicated as it would be on the *real* Solar Max repair mission, as none of the components in the box had been designed for repair by astronauts. Assisted by Allen handing him parts, Lenoir would have worked a lengthy procedure to remove thermal blankets, take off mounting bolts and connectors, cut a grounding strap and reattach the connectors. All of this would have been done while encased in a bulky pressure suit.

No mean feat, it seemed. Yet this kind of work was essential, not only for the successful reactivation of Solar Max, but also for future servicing missions to the still-to-be-launched Hubble Space Telescope, planned for later in the 1980s. Although the Solar Max tests would have occupied a large fraction of their time, they would also have evaluated the manual system for closing Columbia's payload bay doors in the event of an emergency. This involved using a winching system, attached to the forward bulkhead, both with and without foot restraints.

The astronauts would also have practised moving a large bag of tools halfway down the length of the payload bay and, sitting on top of Allen's helmet, was a small black-and-white television camera with a postage-stamp-sized lens. Lighter, less complex and requiring less power than earlier suit cameras, its performance would have been evaluated and should have yielded some impressive pictures of the first spacewalk outside the Shuttle. Lenoir and Allen would also, if time had permitted, have practised using Velcro fasteners, tape and thermal gloves.

SPACEWALK CANCELLED

None of it happened; at least, not on STS-5. In fact, no astronaut would leave Columbia's airlock for a spacewalk until STS-87 in November 1997! The excursion was postponed by 24 hours until 15 November – the day before the Shuttle was due to return home – when both Lenoir and Overmyer suffered a particularly severe dose of space sickness. More trouble was afoot, however, when they tried again. When the men finally donned their suits and ran through the pre-spacewalk checks, a problem was noted with Allen's ventilation fan: it sounded, said the crew, "like a motorboat".

In effect, the fan was starting up, running unexpectedly slowly, surging, struggling and finally shutting itself down. Nor was Allen's suit the only one causing headaches. The primary oxygen regulator in Lenoir's suit – which would have been used during his 'pre-breathing' exercises and throughout the spacewalk itself – failed to produce enough pressure; regulating to 3.8 psi instead of the required 4.3 psi. Some of the astronauts' helmet-mounted floodlights also refused to work properly. After several fruitless attempts to troubleshoot the problems, the spacewalk was cancelled and deferred to STS-6.

This proved disappointing to Columbia's crew, particularly Lenoir. According to Brand: "I guess I was the bad guy. As much as I hated to, I recommended to the ground that we [cancel] the EVA, because we had a unit in each spacesuit fail in the same way. It was a little pump [and] one of its [functions] was to be a cooling pump. So it looked like we had a generic failure there. It was the first time out of the ship. We didn't want to get two guys or even one guy outside and then have [another failure]. We could have taken a chance and could have done it, but we didn't. I'm not sure Bill Lenoir was ever very happy about that, because he and Joe, of course, wanted to go out and have that first EVA."

Fortunately for Joe Allen, he would perform not one, but *two* spacewalks on a later satellite-rescue mission, but Lenoir would not fly again and would depart the astronaut corps for a NASA management job in 1984.

Immediately after Columbia's landing on 16 November, a task force, headed by Richard Colonna, was set up by NASA to investigate the problems. The fault in Lenoir's suit was traced to two missing 'locking' devices – each the size of a grain of rice – in the primary oxygen regulator. According to paperwork provided by the suit's manufacturer, Hamilton Standard of Hartford, Connecticut, the devices *were* fitted in August 1982, but actually had not been fitted at all and were not checked. Their absence allowed the pressure in Lenoir's suit to drop back half a pound from 4.3 to 3.8 psi.

The problem in Allen's suit, on the other hand, was a faulty magnetic sensor in the fan electronics. Colonna's report, released in December 1982, pointed out that "even with no improvements, if the regulator were fabricated properly, the PLSS [Portable Life Support System – the suit's backpack] would function properly". It also listed ways to test and inspect regulators and motors and recommended testing inside the Shuttle's airlock on the day before launch. Additional plans were set out, though not in time for STS-6, to provide sensors with better moisture-resistant coatings for future motors and new tests to pick up defects.

A HELLISH RIDE HOME

By the time the spacewalk was finally cancelled, Columbia's crew were already setting in motion their preparations to come home on 16 November. It was hoped, initially, to land in a crosswind at Edwards and so evaluate the Shuttle's handling characteristics under duress, but instead the spacecraft would touchdown in California under calm weather conditions. The re-entry was particularly dynamic from the astronauts' point of view and photographs taken by Allen from the rear Mission Specialist seat on the flight deck recorded the hellish pinkish-orange glow outside Columbia's windows.

The differences between re-entering the atmosphere in a Shuttle, rather than an Apollo command module, were profound to Brand. "[There were] very large windows and you weren't looking backwards at a doughnut of fire [as with Apollo]. You were able to see the fire all around you [and] you could look out the front. First the sky was black, because you were on the dark side of the Earth, but as this ion sheet began to heat up, you saw a rust colour outside, then the rust turned a little yellowish. [Eventually, around Mach 20], you could see white beams or shockwaves coming off the nose. If you had a mirror – and I did on one of my flights – you could look up through the top window and see a pattern and the fire going over the top of the vehicle. It was really awesome! At Mach 18, I had to take over [manual control] from the autopilot and do a [flight-test manoeuvre]. First, I pushed down from 40 degrees to 35 degrees angle-of-attack, then up to 45 and then back."

This change in the Shuttle's angle-of-attack during re-entry was a repeat of the POPU exercise practised during earlier flights; it was performed satisfactorily, but a few other flight-test manoeuvres had to be cancelled because there was an insufficient amount of RCS propellant remaining to support them.

"That's [the POPU pitch change] just a few degrees," continued Brand, "but when I did that Joe Allen was watching what was coming off the nose and he said a shockwave came from the nose and it came up and attached to the window right in

The 'Ace Moving Company' of STS-5. Clockwise from top-right: Bob Overmyer, Joe Allen, Vance Brand and Bill Lenoir.

front of us. That was a little worrisome, because he knew it was hot. But then about as soon as it got there, why, I was on my way to a higher angle-of-attack, so it 'walked back' to the nose. There were a lot of interesting things like that happened.

"Eventually, we were coming over the coast of the United States. You [are] pitch[ing] down to more of a [normal] airplane angle-of-attack. [Then] you can see out the front and see the countryside. I guess when we first landed at Edwards, it was dawn, but you could see the sun lighting the [desert] way up ahead. When you get down to about Mach 3 or 2 you're getting into thick atmosphere and it's rumbling outside: you can *hear* it rumble and you're decelerating such that it's pushing you into your [seat] straps. At Mach 1, you feel [a lurch, an] increase [and decrease] in deceleration, and a decrease as you go through. Eventually you're over the field. We went through a very thin cloud deck. I was on instruments flying and circled down and landed at Edwards."

The touchdown, which occurred at 2:33:26 pm, was immediately followed by a test of Columbia's maximum braking capabilities, achieving a peak deceleration of 3.9 m/s^2. However, the left-hand main inboard wheel 'locked-up' during the final 15 m of the rollout due to a brake failure.

"We completely ruined the brakes," remembered Brand. "I had to stomp on them as hard as I could during the rollout, which points out that we had a lot of flight test[ing] on the mission. Even though it was [billed as] the first commercial flight, I think we had 50 flight-test objectives. That braking test was just one of them. We ruined the brakes – completely ruined them – but it was a test to see how well they would hold together if you did that." The brake failure was evidenced by skidmarks on the runway, cracked stators and badly damaged brake rotors.

Despite this dramatic conclusion to her first operational mission, Columbia was in good shape, although pictures of her on the runway – reflected in a pool of rainwater from a recent shower – revealed her to be extremely dirty. She had completed five missions in 19 months, spent three-and-a-half weeks aloft and truly showed the world what she could do. Columbia deserved, and would be granted, a rest. Not until November 1983 would she again journey into space: and *that* mission would exceed all of her previous achievements with a record-sized crew, record-length flight and first-ever use of a European-built space laboratory.

SPACELAB: A MULTIPURPOSE RESEARCH FACILITY

Towards the end of 1969, not long after Neil Armstrong and Buzz Aldrin became the first humans to set foot on the Moon, NASA outlined a number of key directions for the United States space programme after Apollo. President Kennedy's challenge to land a man on the lunar surface before the end of the decade had been met, but it was not until this time that serious consideration – and dollars – were given to what would be done next. Of course, Earth-orbital space stations, Moon bases and trips to Mars before the end of the century were envisioned but considered unlikely.

Establishing some kind of permanent, or at least frequent, human presence in space was of major importance to NASA. During the first eight years of sending people aloft, from Al Shepard's pioneering suborbital 'hop' to Apollo 11's triumphant landing at the Sea of Tranquillity, fewer than two dozen astronauts had gazed down on the Earth. Some of them, admittedly, had flown as many as three times, but two dozen of a United States population of a quarter of a billion was insignificant. If NASA was to maintain public interest, it had to make access to space routine.

A space station called Skylab was already being built, but the most important long-term aim was the development of the Shuttle, for which the United States sought foreign cooperation. Already, Canada had agreed to develop the Shuttle's RMS mechanical arm and, in December 1972, just as the final Apollo crew prepared to leave the Moon, the European Space Research Organisation (ESRO) agreed at a ministerial conference in Brussels to develop a modular space laboratory for the Shuttle, known as 'Spacelab'.

ESRO, which in 1974 merged with the European Launcher Development Organisation (ELDO) to become today's European Space Agency (ESA), would design, develop and manufacture the Spacelab system for carriage in the Shuttle's payload bay, in return for flying its own astronauts into space on specific missions. In September 1973, the contracts with NASA were finally signed and the largest international cooperative venture in space so far was set in motion. Spacelab would be a multipurpose scientific research platform, capable of supporting some experiments in a pressurised, 'shirt-sleeve' environment and exposing others to the vacuum of space.

As has already been seen, the segment of Spacelab designed to expose experiments directly to the space environment was flight-tested during the STS-2 and STS-3 missions. Known as 'pallets', they were essentially U-shaped rigid metal frames measuring about 3 m long by 3.9 m wide and covered with aluminium panels onto which large telescopes, antennas or sensors requiring unobstructed large-fields-of-view could be attached. On STS-2, for example, a single engineering version of the pallet was used to carry a synthetic-aperture radar, as well as a number of other experiments.

Up to five pallets, bolted together to form a rigid 'train', can be flown on board a single Shuttle mission. The most pallets actually flown on one mission has been three – on board Spacelab-2 in July 1985 – and these versatile platforms continue to be used into the International Space Station era. On missions that employ only pallets, a 2.1-m-tall cylinder called an 'igloo' stands vertically at the forward end of the 'front' pallet. Igloos are temperature-controlled, pressurised containers which house subsystems and equipment needed to support the pallet-mounted instruments.

The second major component of Spacelab, originally called the 'sortie can' but later renamed the 'module', was a bus-sized aluminium cylinder in which experiments that needed hands-on attention from astronauts could be accommodated. It came in two sections: a 'core' module, which carried data-processing equipment, a workbench and a set of air-conditioned experiment racks lining its walls; and an 'experiment' module, which provided additional space for scientific activities in orbit. Although the core could be flown on its own, this configuration was never used, and all module flights employed both sections joined together.

With a pressurised volume of 75 m^3, this so-called 'long module' – of which two flight units were built – flew 16 times between November 1983 and May 1998 and was used to carry experiments ranging from life sciences, to technology and fluid physics, to materials processing. Of those flights, 11 were on board Columbia. The version using only the core section, known as the 'short module', was due to have flown an Earth-observation mission in 1986, but the Challenger disaster ended that. Although the mission was flown under the new name of 'ATLAS-1' in March 1992, it was as a pallet-only configuration.

When one considers the dimensions of the short module, it is clear why NASA opted to use the longer version: the latter was 7.1 m long, almost double that of the former and thus virtually doubled the amount of 'rack space' in which to store experiments. These 'racks' were essentially refrigerator-sized facilities that could be loaded with experiments and 'rolled' into the module's cylindrical shell. The long module also offered a long central aisle of floor space, onto which additional experiments could be affixed, and provided two ceiling openings for a viewing window or scientific airlock.

"The racks were pretty much standard," remembered NASA's former director of space life sciences, Gene Rice. "You either had a drawer in a rack or you had a whole rack, or you might have a double rack, depending on the magnitude or size of the experiment. We would help [experiment customers] through the process of designing their experiment, integrating it into a Spacelab rack, doing the testing that they needed to do [and] getting it to a [NASA] centre. They would have to show that they met the safety requirements to put it into the Spacelab and to fly it."

The racks contained air ducts to cool experiments and power-switching panels. On the first 'dedicated' Spacelab mission, flown on Columbia on STS-9 in November 1983, the module contained 12 racks, of which the two nearest the entrance were devoted to control subsystems. The ceiling of the core section provided a 0.3-m-wide opening for a high-optical-quality Scientific Window Adaptor Assembly (SWAA), through which Earth-observation cameras could be aimed, and that of the experiment section provided a Scientific Airlock (SAL), into which samples requiring exposure to space could be easily inserted and later retrieved.

As with other payload bay-mounted hardware, the exterior of the modules and pallets were covered with a layer of passive thermal insulation material to protect them from extremes of sunlight or frigid orbital darkness. The module was situated in the midpoint of the bay, to avoid jeopardising the Shuttle's centre-of-gravity constraints during landing, and was linked to the crew cabin's airlock hatch by a 5.8-m-long tunnel. Also, because the Spacelab's hatch was 1.5 m 'higher' than the airlock hatch, a 'joggle' section was included in the tunnel to compensate for this vertical offset.

Artist's concept of the Spacelab-1 module and pallet in the Shuttle's payload bay.

Of course, astronauts might still need to perform emergency space-walks, perhaps to close the payload bay doors if it became impossible to do this from the flight deck. With this in mind, a 'mini-airlock' was built into the tunnel and a pair of spacewalkers could essentially close off the tunnel without having to depressurise the crew cabin or the Spacelab. Although a spacewalk was never actually performed during a module flight, on one mission in June 1991 such action was briefly considered when a problem with some thermal insulation threatened to interfere with the closure of Columbia's payload bay doors.

SPACELAB-1: A 'FREE-FOR-ALL' MISSION

For Spacelab-1 – the first 'dedicated' flight – a long module and single pallet would be flown together in the payload bay. However, this mission did not require an igloo because all of the subsystems needed to run the pallet-mounted experiments were housed inside the pressurised module. Also, the payload was not dedicated to one scientific discipline, but was a 'free-for-all' covering virtually all possible areas of research for which the system had been designed. The module and pallet would play host to more than 70 life sciences, technology, astronomy, solar physics, Earth-observation, plasma physics and materials science investigations.

It was in anticipation of this ambitious make-or-break mission to demonstrate the new space laboratory's capabilities that Columbia returned to KSC a few weeks before Christmas 1982. Spacelab-1 would be the most ambitious flight in her career so far. Scheduled to run for nine days – easily surpassing the previous record set by STS-3 astronauts Jack Lousma and Gordon Fullerton – the mission, designated STS-9, would also carry a crew of six; the largest yet flown into orbit on a single spacecraft. Of this close-knit team of astronauts, two in particular stood out.

One was a 42-year-old West German solid-state physicist named Ulf Merbold. He had been chosen as one of the first two 'Payload Specialists' to fly on the Shuttle and was the first European representative to accompany Spacelab into orbit under the contract signed a decade earlier. Payload Specialists, unlike Mission Specialists, were 'career' scientists who were chosen by their respective government, agency or – in some cases – from within private industry or elsewhere to operate a specific experiment. "They know more about the science and what's going than we do," said Merbold's colleague on STS-9, Mission Specialist and astrophysicist Bob Parker. "We're generalists. I'm an interstellar matter supernova remnant man. We don't do too much of that stuff on the Shuttle."

Merbold would be joined on the flight by a second Payload Specialist from the United States, a bioengineer named Byron Lichtenberg. Together with Parker and Mission Specialist Owen Garriott – an electrical engineer who had previously flown on board Skylab in 1973 – this quartet of astronauts would work in two 12-hour shifts to run the myriad of experiments on Spacelab-1. Meanwhile, Commander John Young and Pilot Brewster Shaw would monitor and run Columbia's systems from the flight deck.

The assignment of Young, the first person to fly into space for a sixth time, to command the mission epitomised perfectly its importance to both NASA and ESA. During STS-9, he would be a member of the Red Team with Parker and Merbold, while rookie astronaut Shaw joined the Blue Team with Garriott and Lichtenberg. It was intended that this practice of running Spacelab missions around-the-clock, with a pilot managing the Shuttle, and Mission and Payload Specialists operating the experiments, would prove useful as NASA strove towards its ultimate goal for the 1990s: developing and operating a long-term space station.

'SPACELAB ONLY' MODIFICATIONS TO COLUMBIA

The longer-than-usual time needed to prepare Columbia reflects the mission's complexity. Originally scheduled for launch on the last day of September 1983, it eventually set off eight weeks later and would prove to be a magnificent success. In addition to the normal processing flow in the OPF, a series of so-called 'Spacelab Only' modifications were effected. These included adding extra seats for the large crew, fitting an onboard fax machine, improving the Shuttle's brakes and tyres, installing structural and electrical provisions for Spacelab-1, strengthening the payload bay floor and placing phonebox-sized sleep stations for the astronauts into the middeck.

Admittedly, not all of the improvements were in support of Spacelab-1, but the focus of the work was upon this particular mission. Columbia spent the first three months after STS-5 in OPF Bay 1, where her propellants were drained and her OMS and RCS units removed for repairs. Some of the early Spacelab Only modifications also got underway during this time, and on 27 February 1983 the vehicle was transferred to OPF Bay 2 to complete the improvements. To support the planned nine-day mission, three 'substack' fuel cells, using five (rather than three) cryogenic tank sets were installed.

Preparations for STS-9 took place in a range of locations, not just Florida. A new set of upgraded main engines were assigned to the mission and they completed their certification testing at the National Space Technology Laboratory's facility at Bay St Louis in Missouri in the early summer of 1983. They were capable of achieving a thrust of 104%, significantly higher than had been possible using the original specification models. After test firings and leak checks, the new engines arrived at KSC and had all been installed on Columbia by 20 July.

ASSEMBLY OF SPACELAB-1

Elsewhere at KSC, in the Operations and Checkout Building, the Spacelab-1 hardware had been arriving in dribs and drabs since October 1981. The two showpieces of the mission – the long module and pallet – arrived shortly before Christmas of that year and an official unveiling ceremony took place in February 1982, attended by Vice-President George Bush. By the end of summer, all the experiments had arrived and were fully installed; in November 1982 a Mission Sequence Test verified their compatibility with one another and with 'dummy' Spacelab systems.

"The experiments were brought in by their various scientific teams," remembered Spacelab-1 Mission Manager Harry Craft of NASA's Marshall Space Flight Center at Huntsville in Alabama. "We would let them check the experiment out initially in an off-line capability and then we'd bring them into a room and just make sure the instrument had met the transportation environment and still worked. [Then] they'd turn it over to us." Six months later, in May 1983, Spacelab-1 was hooked-up to the Cargo Integration Test Equipment (CITE) stand, which duplicated the Shuttle's systems and verified that the hardware would be compatible with Columbia.

On 16 August, the 15,265-kg integrated Spacelab-1 system was moved to the OPF and loaded into the Shuttle's payload bay. Two weeks later, the tunnel was connected between the middeck airlock and Spacelab hatches and fit and leak checks were completed. Early the following month, more tests were conducted to verify the compatibility of the 'real' payload with the 'real' Columbia and the Spacelab-1 experiments were briefly operated via remote control from the Payload Operations Control Center (POCC) at Johnson Space Center.

By this time, problems with an important communications satellite needed to support Spacelab-1 meant that the launch of STS-9 had slipped until late October and Columbia did not move to the VAB for stacking until 24 September. Then on 14 October, two weeks after her smooth rollout to Pad 39A, NASA and ESA jointly decided to postpone the mission again because of potentially hazardous problems uncovered with an SRB exhaust nozzle. The STS-9 stack was returned to the VAB, dismantled and the nozzle replaced; because it was the lowermost portion of the booster, the entire SRB had to be disassembled.

While this work was going on, Columbia returned to the OPF on 20 October. She was back in the VAB two weeks later for restacking and the Shuttle returned to the pad on 8 November. Nearly a year after STS-5, she had now been joined by two new vehicles, the first named Challenger – which had already flown three times in the spring and summer of 1983 – and the newly completed Discovery, due to make her maiden voyage some time in 1984. A fourth Shuttle (Atlantis) would be delivered to KSC in about a year's time to begin making the promise of routine spaceflights a reality.

A COMPLEX MISSION

By the time Columbia finally lifted off at precisely 4:00 pm on 28 November, kicking off the ninth Shuttle mission and her own sixth trip into space, her Mission and Payload Specialists had been training for the best part of five years and the entire crew had become a close-knit team. "When we were training the science crew," said Parker, "we had prime and backups for the American Payload Specialist and for the European Payload Specialist, and then Owen and myself. The six of us almost always travelled together and did the training together. At the very end, in the last year or so, when it became obvious this is going to be on 'your' shift and this is going to be on their shift, we separated a bit in that sense, but still you could do an experiment on both shifts. The Payload Specialists might have felt separated from the orbiter crew [Young and Shaw], because they weren't participating in these trips. I served as the flight engineer, so at the same time as we were training on the experiments, I was training with Brewster and John on ascents and entries. For a good last six months or so, I'd be training on Mondays with them, fly to Huntsville for experiment training, fly back and do ascents and entries on Wednesday, and back and forth. So I got to keep in touch with them a lot better than the others."

Flight engineers had become an increasingly familiar member of Shuttle crews since STS-5: they were chosen from among the Mission Specialists and called 'MS2'. They sat behind the centre console in between the Commander and Pilot and provided assistance during both ascent and re-entry, helping to monitor the Shuttle's instruments. In fact, one up-and-coming astronaut was looking to the new flight engineer's seat as the next challenge in his career. Carl Walz, who joined the astronaut corps in 1990 and served as Columbia's flight engineer in July 1994, described the role as "one of the most important jobs that a Mission Specialist can have on a Shuttle flight. During launch and landing, the MS2 [makes] sure that all the checklist steps for normal and emergency procedures are performed flawlessly by the Commander and Pilot [and] keeps a log, recording all the systems that have had problems during launch and landing. Afterwards, he or she then figures out how those problems affect future Shuttle procedures. The training for MS2 is very similar to the training for the Commander and Pilot. In orbit [he or she] is the 'traffic cop' who makes up the plans [to get the vehicle ready for operations] and makes sure [these are] executed properly." No mean feat, it seemed, especially on a particularly complex mission like Spacelab-1, which would involve not only a vast number of scientific experiments but also a large number of engineering demonstrations and tests.

Spacelab-1 was one of two so-called 'Verification Flight Tests' of the new system: it would evaluate the performance of the long-module-and-single-pallet configuration, while the Spacelab-2 mission would put the pallet-train-and-igloo version through its paces. Only when these two pathfinding missions had been successfully accomplished would the Spacelab system be declared fully operational. For STS-9, more than 200 sensors were installed throughout the Shuttle and Spacelab, providing data on their combined performance during launch, ascent, orbital flight, re-entry and landing.

This equipment verified that Spacelab-1's passive thermal control system kept temperatures within the module and on the pallet within required limits and prevented condensation or heat leaks. The thermal, acoustic and structural responses of the entire payload during the most

The Life Sciences Minilab for Spacelab-1 arrives in Florida for processing.

critical and dynamic portions of the mission were closely monitored, and the astronauts checked the satisfactory operation of the SAL and the tunnel, evaluated their ability to communicate and transmit data through NASA's first Tracking and Data Relay Satellite (TDRS) and confirmed that the Spacelab module was indeed habitable and comfortable to work in.

In general, the system performed admirably and the crew even demonstrated the optical-quality of a small window in the rear end-cone of the module by shooting television pictures through it of the experiments on the pallet. The two-team, 'Red and Blue' system also worked extremely well: the astronauts' sleep cycles had been deliberately adjusted a fortnight before launch to better support their separate 12-hour shifts. Although shifts would not be necessary on *all* Spacelab missions, they were an important demonstration of how to maximise the crew's valuable time in orbit.

After reaching orbit, the crew unstrapped and set about activating their payload. It was during this time that Shaw, who was making his first spaceflight, remembers a minor scare: "We couldn't get the [Spacelab] hatch open, so we couldn't get into the [module] at all! Looking at Owen and Bob's faces, I remember thinking, 'Boy, these guys are seeing their lives pass in front of their face, if we can't get in there and do the stuff they've been training to do'." Fortunately, the problem proved temporary and by the evening of 28 November Spacelab-1 was up and running.

Typically, during shift operations one of the pilots – Young or Shaw – would spend their time on Columbia's flight deck monitoring her systems and communicating with the science crew in Spacelab-1 by means of an intercom. Although undoubtedly awestruck by his first space experience, Shaw recalled the monotony of this job: "We had a few manoeuvres to do once in a while with the vehicle, and then the rest of the time you were monitoring systems. After a few days of that, boy, it got pretty boring. You spent a lot of time looking out the window and taking pictures and all that. But there was nobody to talk to, because the other guys were back in the Spacelab working away and you just [thought] 'Gosh, I wish I had something to do.' But the flight went fine and we did a great bunch of science." The manoeuvres that Shaw was referring to were a number of so-called 'thermal attitudes' to expose the Spacelab hardware to 24 hours of extreme cold and 12 hours of fierce solar heating to test its performance.

The only real problem of note that emerged from Spacelab-1 was a temperature glitch in the Remote Acquisition Unit (RAU)-21, which served all of NASA's instruments on the pallet. Subsequent analysis by ground-based engineers found that the temperature of the Shuttle's freon coolant was partially to blame and subsequent use of the unit at 22 Celsius apparently solved the problem. However, as part of these efforts to work around the RAU-21 problem, a 'patch' inserted into Spacelab-1's software caused the module's experiment computer to crash and temporarily affect data-gathering from the pallet-mounted instruments.

Otherwise, as Shaw commented, the mission gathered a great deal of scientific data. Its experiments were divided into five disciplines: (1) astronomy and solar physics, (2) space plasma physics, (3) atmospheric physics and Earth observations, (4) life sciences and (5) materials sciences and technology. These were jointly sponsored by NASA and ESA, although they featured cooperation from Canadian and Japanese scientists. NASA's part was run by its Marshall Space Flight Center, while ESA's portion was controlled by a team called SPICE: the Spacelab Payload Integration and Coordination in Europe, based at Cologne in West Germany.

In recognition of ESA's immense contribution, Garriott and Parker spent a great deal of their training with the two Payload Specialists working in continental Europe. "[It was the] first European mission," recalled Parker, "so we had to go to every European country that had a piece of it in ESA. We went to Denmark to look at this experiment, [then] two or three places in France, but they were centralised in Germany, so we spent most of our time [there]." The genesis of these experiments went back to 1976, when 400 proposals were jointly solicited by NASA and ESA.

After each of the 70-plus experiments had been carefully selected, its Principal Investigator formed a working group and, together with a NASA Mission Scientist, guided it through tricky waters from initial design and manufacture to ensuring that it met each of the stringent demands needed for integration into the Spacelab-1 payload. During the course of the mission, science operations were coordinated in real time from the POCC at JSC and it was possible – for the first time – for ground-based scientists to communicate directly with the crew, rather than indirectly via the Capcom in Houston.

Six experiments were dedicated to Astronomy and Solar Physics. These included a NASA-provided far-ultraviolet telescope on the pallet, which acquired spectra of high-temperature celestial sources as part of efforts to gain clearer insights into the life-cycles of stars and galaxies. Unfortunately, a fogged film ruined many of its images and a post-mission investigation recommended that on its next flight it should record photons electronically, rather than on film as time exposures, to better pinpoint the cause of the fogging. It did, however, achieve 95% of its objectives and took the first far-ultraviolet picture of the Cygnus Loop, a relatively nearby supernova remnant.

The module's roof-mounted SAL was used to support ESA's Wide Field Camera, which took a series of photographs of 10 astronomical objects on 3 December. Among its notable results were impressive ultraviolet images of a 'bridge' of hot gas linking the Large and Small Magellanic Clouds, suggestive of an interaction between them. Sadly, the month-long delay to Columbia's launch from late October until late November resulted in shorter-than-expected orbital 'nights', which reduced the camera's viewing opportunities by about 60%. The third astronomical instrument was a highly successful ESA-built X-ray spectrometer on the pallet.

The other three experiments – the Active Cavity Radiometer (ACR), Measurement of Solar Constant (SOLCON) and Measurement of Solar Spectrum (SOLSPEC) – tracked with extreme precision the total amount of solar energy received by Earth's atmosphere and its impact upon our planet's environment to further the study of the solar–terrestrial relationship. Modified versions of each of these instruments would fly throughout the 1990s on a series of Shuttle research missions in order to calibrate similar instruments on a variety of spacecraft and NASA's Upper Atmosphere Research Satellite (UARS) which were to investigate environmental changes over time on a global scale.

Five experiments came under the Space Plasma Physics banner, all of them mounted on the Spacelab pallet at the rear of Columbia's payload bay. Of these, the most noticeable was the Space Experiments with Particle Accelerators (SEPAC), designed by Japan's Tokyo-based Institute of Space and Astronautical Science (ISAS). SEPAC consisted of an 'electron gun' to investigate the dynamics of Earth's ionosphere. During the mission, it fired streams of gas and high-intensity electrons into the ionosphere to better understand the production of aurorae, the nature of Earth's magnetic and electric fields, and the effects of plasma on the Shuttle's structure. Despite the failure of its electron-beam assembly to run in a high-power mode, the device was highly successful and achieved almost all of its planned objectives. A similar electron-gun experiment, provided by France, was the Phenomena Induced by Charged Particle Beams (PICPAB), which contained an 'active' unit attached to the pallet and a 'passive' recorder inserted into the SAL. Like SEPAC, its electron gun was capable of producing and examining artificial aurorae. Some data was lost when one of its gas bottles failed, but it nevertheless achieved a 60% success rate.

Two other experiments worked in conjunction with SEPAC and PICPAB by measuring atmospheric constituents using emissions from their particle-beam firings, as well as examining real aurorae under ultraviolet and visible light. They also measured particulate contamination in the vicinity of Columbia herself and observed the effects of the Shuttle's emissions – from water dumps, outgassing and thruster firings – on the artificial aurorae generated by the electron guns. Several of these investigations required Young and Shaw to orient Columbia to acquire the necessary data. A cosmic-ray detector was also carried to measure their abundances in low-Earth orbit.

Six experiments were covered by the Atmospheric Physics and Earth Observation category: five provided by ESA and one – the Imaging Spectrometric Observatory (ISO) – from NASA. This latter device, consisting of *five* spectrometers housed in a single unit, examined the presence and relative abundances of oxygen, nitrogen and sodium in the 'middle atmosphere' (or 'mesosphere') between 80 and 100 km above the Earth's surface. It was part of a project to build the first comprehensive database of the vertical structure of the atmosphere. During the mission, ISO also aided studies of the emissions produced by Columbia, which were of concern to potential future customers.

The European experiments included a Metric Camera, attached to the roof-mounted SWAA optical-quality window in the forward section of the Spacelab-1 module. Loaded with black-and-white, colour and infrared films, it achieved an 80% success rate and took a wide range of high-resolution Earth-mapping photographs. Three experiments mounted on the pallet examined oxygen and hydrogen emissions from the atmosphere and tested equipment which, it was hoped, might lead to the development of advanced, all-weather remote-sensing systems.

The only Atmospheric Physics experiment that did not perform as advertised was the Grille Spectrometer, so-called because it employed a specially designed 'grille' as a window for one part of its optical system and as a mirror for the other. This was supposed to measure atmospheric constituents in the stratosphere, mesosphere and thermosphere, as part of efforts to understand their dynamic behaviour. Although it achieved some promising results, the month-long delay to STS-9 meant that observing conditions were unfavourable and it only completed 16% of its objectives. Nevertheless, it was reflown with greater success on a dedicated Earth-observation mission in March 1992 and later installed on the Mir space station to make long-term observations.

PILOTS: AVOID THE SPACELAB!

The bulk of the Spacelab-1 experiments, however, were housed inside the bus-sized, cylindrical module and comprised 16 life sciences investigations and three dozen studies of the behaviour and processing of materials and fluids in the microgravity environment. The predominance of the latter clearly reflected the fact that this mission was the first time extensive materials research could be conducted, in-depth, on board a manned spacecraft. Although some experiments had been performed on Skylab, they were comparatively primitive and offered little more than a taster of what could be achieved in space.

It had long been anticipated that, in the absence of gravity and the complications caused by buoyancy and other factors, it should be possible to produce lighter and stronger building materials, more reliable and less costly electronic components, new alloys, plastics, ceramics, composites and glasses for industrial machinery and household products and, perhaps, 'grow' crystals of important proteins, analysis of which could lead to the development of new drugs to combat cancer and other diseases. The first efforts to set this 'microgravity revolution' in motion had already begun on STS-3 and STS-4, with the preliminary tests of advanced electrophoresis equipment.

Thirty of the materials sciences experiments were accommodated in ESA's floor-to-ceiling Materials Science Double Rack, a refrigerator-sized facility that was activated on 29 November, less than a day after Columbia reached orbit. It housed four furnaces and materials-processing chambers: a Fluid Physics Module to examine the behaviour of liquids in space; a Gradient Heating Furnace and Mirror Heating Facility, which conducted crystal growth experiments in support of pharmaceutical and industrial research; and an Isothermal Heating Facility for studies of the solidification and casting of new metals, alloys, ceramics, glasses and composites.

All four members of Spacelab-1's science crew had undergone extensive training with this rack, but in some cases they were little more than lab technicians. "We put materials sealed in cartridges into furnaces, heated, melted, solidified the materials, pushed buttons and started computer programs," remembered Merbold. All of the experiments using the fluid physics and gradient heating units were successful, but the other two furnaces suffered partial power failures on 30 November and achieved limited results. The mirror furnace was, however, later restored through a maintenance procedure carried out by the crew.

The STS-9 crew inside Spacelab-1. Clockwise from top: Brewster Shaw, Byron Lichtenberg, Bob Parker, John Young, Ulf Merbold and Owen Garriott.

The life sciences experiments were conducted by the entire crew, including, much to Young's chagrin, the two pilots, who would be pounced-upon by the science team for blood draws whenever they dared enter the Spacelab module. Nine of the experiments were provided by ESA and seven by NASA and primarily focused on the effects of the microgravity environment and high-energy cosmic radiation on the human body. One particular experiment was remembered well by Shaw: "Helen's balls! Helen [Ross, from the University of Stirling in Scotland] had a bunch of little yellow balls that were different mass [and] different weight. What you were supposed to do – since there's no weight, only mass in zero-g – we had to differentiate between the mass of these balls. You would take a ball in your hand and shake it and feel the mass of it by the inertia and the momentum of the ball as you would start and stop the motion. Then you'd take another one and try to differentiate between them, and eventually try to rank-order the balls. They were numbered as to which was the most massive to the least massive. We did that several times."

A number of vestibular experiments, some provided by Canadian scientists, were also conducted to examine the behaviour of the vestibular system in the inner ear – which controls our balance and orientation – and identified a relationship between astronauts' sense of balance and eye movements. These experiments also provided invaluable new insights into the effect of head motions on the onset of space sickness. Other investigations studied the role of microgravity in the reduction in red blood cell mass and its effect on the astronauts' immune systems.

Overall, with a few minor problems, the mission proceeded smoothly; so smoothly, in fact, that on 3 December NASA and ESA agreed to extend the crew's time in orbit by 24 hours when it became clear that Columbia's cryogenic consumables could comfortably support 10 days aloft. Overall, Spacelab-1 power consumption rates averaged about 1.2 kilowatts less than had been conservatively predicted before launch. Even at the halfway mark, Mission Scientist Rick Chappell was describing the flight as "a very successful merger of manned spaceflight and space science".

The crew members, too, were performing well, although, at least in Shaw's eyes, it was clear who had been here before and who were novices. "John and Owen were the experienced guys and they were mentors of the rest of us. It was fun to watch Owen back in the module, because you could tell right from the beginning he'd been in space before. He knew exactly how to handle himself, how to keep himself still, how to move without banging all around the place. The rest of us were bouncing off the walls until we learned how to operate. [For] Owen, it was just like he was here yesterday and it really had been years and years [since his 1973 Skylab mission. But] your body remembers that stuff. The human body is remarkable in its ability to remember adapting to a previous thing."

Although the crew adapted well to microgravity, three of them suffered space sickness during the first two days of the mission; interestingly, data from one of the experiments revealed that they were most susceptible after several hours of sustained physical exertion. They likened their symptoms to prolonged motion sickness and fluids shifting in their bodies.

PROBLEMS BEFORE RE-ENTRY

After a spectacular week-and-a-half in space, late on 7 December the science crew began the process of shutting down the experiments, storing and securing the samples and closing down the Spacelab module and pallet. Meanwhile, Young and Shaw, assisted by Parker, worked their way through the now-routine procedures to prepare Columbia for her descent to Edwards Air Force Base next day. It was at 11:10 am on the 8th – only five hours before their scheduled landing, as they were configuring the Shuttle's computers for re-entry – that gremlins hit the mission: a GPC failed.

When any of the five computers encounters a problem, the prognosis is bad, because they are responsible for monitoring literally thousands of separate functions during the dynamic phases of the mission. Worse was to come, however: six minutes later, a second GPC stopped working! Young and Shaw successfully restarted the second computer, but all their efforts to bring the first one back on-line proved fruitless.

The mood inside Columbia's cockpit was undoubtedly tense. "About the time that we were reconfiguring the computers," remembered Shaw, "we had a couple of thruster firings, and the big jets in the front fired, and these big cannons – boom! boom! – shook the vehicle. You can really feel it if you're touching the structure. We had one of these firings and we got the big 'X-pole' fail on the CRT [display], meaning the computer had failed. So I got out the emergency procedure checklist and said 'Okay, first GPC fail. Here's what we do.' And in just [a few] minutes we had another one fail the same way: a firing of the jets and the computer failed."

Post-mission analysis revealed that one of the failures was due to a sliver of solder that had become dislodged during the RCS thruster firings. Young later told reporters at the post-landing press conference: "My knees started shaking. When the next computer failed, I turned to jelly. Our eyes opened a lot wider than they were before!" Re-entry was postponed until the computer problem was resolved. Eventually the 'restored' GPC and two others were used to support re-entry.

Then, at 11:42 am, one of Columbia's Inertial Measurement Units (IMUs) – used for navigational purposes – failed. Shaw remembered waking Young up with the bad news: "It was the end of John's shift and [he] went down to take a nap. As I recall, Bob came up and was up there with me part of the time. During that timeframe, all of a sudden there starts this kind of [banging] noise. The next thing, one of our IMUs fails and we didn't know why. So John comes back upstairs and says, 'I really appreciate you guys making all that banging noise when I'm trying to sleep.' I said, 'Jeez, John, I've got some bad news. We lost an [IMU]. And John's eyes got this big again, because we've had two GPC failures and now an IMU failure." Eventually, after nearly eight hours of troubleshooting – during which the crew was kept prepared for re-entry – at 10:52 pm the de-orbit burn was performed and Columbia began her descent back to Earth. Overall, the Shuttle's performance was nominal, until about four-and-a-half minutes before touchdown, when the temperature of one of the APUs rose sharply.

"Then we had another lesson: 'Never let them change the software in the flight-control system without having adequate opportunity to train with it'," said Shaw. "There were 'gains' in the flight-control system, and [these] changed depending upon what phase of flight you're in. When you're flying a final [approach], there are certain gains that make the vehicle respond a certain way to the inputs the pilot makes on the stick.

"Then, when the main gear touch down, the gains change and the gains are set up so you can de-rotate the vehicle and get the nose on the ground in an appropriate way. We had done all our training in a simulator with a certain set of gains, and then they changed the flight software and these gains so that when it came time for John to land the vehicle in real flight, the gains were different than he'd done all his training on. Certainly, when John started to de-rotate the vehicle, it responded differently than he had trained on.

"So John's flying the vehicle [and] I'm giving him all the altitude and airspeed calls and everything and you feel this nice main gear gently settling onto the lakebed. From downstairs, the rest of the guys [clapped] when the main gear touched the ground very gently. Then John gets this thing de-rotated and we're down to about 150 knots or so when the nose hits the ground and it goes 'smash'. So [the cheers from the rest of the crew] change from 'Yay' to 'Jesus Christ. What was that?' "

Immediately after touching down at 11:47:24 pm, the GPC previously restored by Young and Shaw failed again. Six-and-a-half minutes later, one of the APUs shut itself down, as did a second unit shortly afterwards. The astronauts did not know it at the time, but one APUs was on *fire* as Columbia sped down the runway.

"We got called the next day," recalled Shaw, "because we had an APU fire. The reason the first one shutdown was it was on fire and the fuel wasn't getting to the catalyst bed, and so it 'undersped' and automatically shut itself down. Then, in response to that, I configured some of the systems and APUs. So the next one didn't shut down until we actually shut it down. But two of them were burning and we had a generic failure of a little tube of metal where the fuel went through and was injected into the catalyst bed and it cracked. When we shut [the APUs] down and shut the ammonia off to them, the fires went out. We had some damage back there, but the fires stopped. But we didn't know anything about that till the next day!"

NEW MISSION, NEW NUMBER

By the beginning of January 1986, Steve Hawley had developed something of a reputation for himself within NASA's astronaut corps. The boyish-looking, 34-year-old astronomer had already made one previous flight in August 1984 – the maiden voyage of Space Shuttle Discovery – and that had sprinkled the first seeds of his reputation. His forthcoming second mission, this time on board Columbia for a commercial satellite-deployment flight saddled with the peculiar designation of 'STS-61C', would cement it still further and finally set it in stone.

"My approach", Hawley would say years later, "has always been [that] I'd go out to the launch pad every time expecting not to launch. If you think about all the things that *have to work*, including the weather at several different locations around the world, in order to make a launch happen, you would probably conclude, based on the numbers, that it's not even worth trying. So I always figured that we're going to turn around and come back. I'm always surprised when we launch. My mindset was always that we'll go out there and try and see what happens."

Hawley's first flight should have taken place in June 1984, but was halted a few seconds before liftoff when Discovery's three main engines roared to life and then abruptly and dramatically shut themselves down. Two months later, after extensive repairs, he and his five crewmates for the mission returned to Pad 39A to try again and were thwarted by bad weather, but they finally got off at their third attempt on

Columbia returns to KSC on top of the 747 carrier aircraft.

30 August. The mission itself was successful and shortly before the end of that year Hawley was named as a Mission Specialist for another flight called 'STS-51I'.

His colleagues on STS-51I would be Commander Robert 'Hoot' Gibson, Pilot Charlie Bolden and Mission Specialists George 'Pinky' Nelson and the first Hispanic astronaut, Franklin Chang-Diaz. They began training for a week-long mission scheduled for August 1985 to deploy two communications satellites (Leasat-4 for the US Navy and the American Satellite Company's ASC-1) and run a series of automatic experiments on the Materials Science Laboratory (MSL)-2 in the Shuttle's payload bay. By the time the five astronauts and the two extra crewmates they had picked up along the way finally lifted off in January 1986, their mission had changed quite significantly.

The first point to make was the new numbering system for Shuttle missions. After STS-9, for reasons best known within the higher echelons of NASA management, the agency redesignated its flights with a cryptic and somewhat clumsy combination of numbers and letters: in the case of '51I', for example, the '5' denoted the financial year under whose budget the mission would take place (1985), the '1' pointed to the launch site of KSC in Florida ('2' was reserved for flights staged from Vandenberg Air Force Base in California) and the 'I' meant the ninth scheduled Shuttle mission for that year.

Of course, as 1984 wore into 1985 and each mission took to the skies, all they really demonstrated was how best to confuse the public! Some were delayed or cancelled, higher-priority ones were pushed forward in the launch pecking-order and frustrating problems with upper stages caused some payloads to be dropped. The result was a roster that read as follows: STS-41B, 41C, 41D, 41G, 51A, 51C, 51D,

51B, 51G, 51F, 51I, 51J, 61A, 61B and 61C! Nor would STS-51I – although it *did* fly – carry Hawley and his crew. They, like the payloads, would be bumped to a later flight: STS-61C.

"Back in those days," remembered Hawley, "it wasn't very unusual to change flights several times. We were assigned to two or three different flights before it stabilised out. Today, you get assigned to a flight and that's the flight you'll fly. If the payload's delayed, then you delay with the payload, and a lot of that's driven by the specific training requirements. Back in those days, a lot of the flights were similar: they were launching satellites or running some experiments that could be quickly learned and it wasn't as important to stick with your payload."

A 'NEW' COLUMBIA

If Space Shuttle Columbia had been a sentient, thinking person, she would undoubtedly have been just as confused by the new numbering system as everyone else. However, following her record-breaking, 10-day STS-9 mission at the end of 1983, she was removed from active flight status for more than a year-and-a-half for major repairs and refurbishment. Unlike the Spacelab Only modifications, which were undertaken at KSC, these repairs would be done at the Shuttle's prime contractor, Rockwell International, at Palmdale in California, and would involve a complete overhaul and detailed structural inspection of the entire spacecraft.

The Shuttle flights planned for 1984 and most of 1985 would be performed by two other spacecraft – Challenger and Discovery – which, although outwardly identical to Columbia, were much lighter and lacked ejection seats. A fourth Shuttle, named Atlantis, would then join the fleet in the autumn of 1985, followed by the return of Columbia to active flight status to begin a full roster of ambitious missions in 1986. On 26 January 1984, meanwhile, Columbia was mounted on the Boeing 747 carrier aircraft for the cross-continental trip from Florida to California for her modification period.

After a stopover at Kelly Air Force Base in Texas, she arrived at Edwards on 30 January and was then transported overland to Palmdale. Subsequent modification periods would involve flying the Shuttles directly into Palmdale, but at this time Rockwell's plant was not equipped with a Mate/Demate Device to remove the spacecraft from the top of the Boeing. Removal from the carrier aircraft therefore had to take place at Edwards, which *did* have such a device, and Columbia was towed – again to the amazement of open-mouthed motorists – down the highway to what would be her home for the next 18 months.

Many have commented that the product of these modification periods, which have increased in frequency as the Shuttle grows older – and now take place every three or four years for each vehicle – is not so much an *overhauled* spacecraft, but a *brand-new* spacecraft! Already, by the time she flew to Palmdale, Columbia had changed a great deal since her maiden mission in April 1981. Parts of her ejection seats had been dismantled, the DFI pallet was more-or-less gone, sleeping bunks had been added, a new Ku-band communications antenna fitted in the payload bay, improved brakes and tyres installed and the payload bay floor strengthened.

Much of the data from Columbia during the four test flights also led engineers to the conclusion that they had *over-designed* the Shuttle. "It was *too* strong, *too* beefy," remembered Arnie Aldrich, "and what we could actually do was take about 1,800 lb out of the orbiter by redesign. That was very desirable because that would be directly related to payload. Both Columbia and Challenger were built to this heavier design, but Discovery, Atlantis and Endeavour weren't yet created. They could take advantage of this knowledge of areas where we could take some of the weight out."

Out of each modification period would emerge a spacecraft which outwardly looked unchanged, but inside was brimming with new, state-of-the-art equipment. In addition to the structural inspections, Columbia also received a Heads-Up Display (HUD) for her pilots, lost her ejection seats altogether in exchange for lighter versions and was fitted with a myriad of sensors to monitor her performance during ascent, orbital flight and throughout re-entry. One device in particular was the Shuttle Infrared Leeside Temperature Sensing (SILTS) instrument, which took the form of a 50-cm cylindrical pod attached to the top of Columbia's tail fin.

It was intended that SILTS would acquire high-resolution infrared images of the upper surfaces of the Shuttle's port-side wing and fuselage during the high-temperature portion of re-entry. This was expected to highlight those parts of the spacecraft that experienced maximum amounts of heating during Columbia's fiery plunge back to Earth. A hemispherical dome at the forward-facing end of the cylinder contained two windows – one forward-facing, the other at an oblique angle – for the infrared camera. Throughout re-entry, a constant supply of room-temperature nitrogen gas would flow across the windows to protect them from the onslaught of atmospheric heating.

If the windows were not protected in this way, the infrared camera would only have been able to see the window itself, rather than what lay beyond. To accommodate SILTS, the pod and the top three metres of Columbia's tail fin were covered with black High Temperature Reusable Surface Insulation (HRSI) tiles, which would make her easily identifiable from a distance. It was intended that SILTS would be activated by Columbia's computers at an altitude of around 122 km – the point of 'entry interface' – when two plugs would be jettisoned from the windows.

For the next 20 minutes, the camera would monitor the port-side wing and fuselage, alternating between them every 11 seconds until the spacecraft had descended to 24.3 km and was through the worst of atmospheric heating. It would then be turned off. With the benefit of hindsight, it is a pity that SILTS did not remain operational through the remainder of Columbia's career, for it might have shed some light on the causes of the disaster on 1 February 2003 ...

Other new devices included a brand-new nosecap for the Shuttle Entry Air Data System (SEADS), which consisted of pressure sensors to assess the spacecraft's aerodynamic performance at various stages in the high atmosphere. The new nosecap contained 14 tiny holes, through which the pressure of the outside airflow could be determined; this provided data about Columbia's attitude in relation to the airflow, but also allowed predictions of atmospheric density at different altitudes. To ensure that the 'holes' in the nosecap would not cause leaks and destroy the vehicle, the inside bulkhead 'behind' SEADS was covered with protective HRSI tiles.

The third device was the Shuttle Upper Atmosphere Mass Spectrometer (SUMS), which took air samples through a small hole on the underside of Columbia's nose, just between the nosecap and nosewheel well. These samples were then used to pinpoint the quantity and nature of gas species at different altitudes and the density of the atmosphere. With all of these devices fitted – and a thorough inspection that had involved detailed examinations of corrosion and wear-and-tear after six spaceflights – Columbia was finally flown back to KSC on 11 July 1985.

Her time at Palmdale was, according to astronaut Dave Leestma, much longer and more frustrating than anticipated. He had been named as a crew member for Columbia's next flight after STS-9 – along with Commander Bob Crippen, Pilot Jon McBride and Mission Specialists Sally Ride and Kathy Sullivan – but the delays in getting the spacecraft flight-ready meant that their mission was ultimately shifted onto Challenger. "It wasn't going well," Leestma said of Columbia's modification period, "and it was real hard to come back and be real upbeat when you know that your orbiter's not going to make your flight date."

A POLITICAL FLIGHT

After temporary storage in the VAB and OPF, on 26 September 1985 Columbia finally began processing for her next flight, which by now had been redesignated STS-61C, and was scheduled for launch shortly before Christmas. Although Hoot Gibson's crew had retained the MSL-2 payload, the remainder of their planned five days in space would be relatively 'roomy', with comparatively few experiments to perform and just one commercial satellite – the Radio Corporation of America's Satcom Ku-1 – to deploy into geosynchronous transfer orbit.

Hawley would say later that one of Columbia's passengers, Congressman Bill Nelson of Florida, may have had something to do with it. "Frankly, our payload wasn't very robust [and] were it not for [Congressman Nelson's] presence on the flight, we might have been cancelled. We had one satellite and some other experiments. It was almost kind of a clearing-house sale. We had a lot of GAS cans and Hitchhiker payloads and a bunch of stuff that hadn't been able to fly previously, and here came a flight that we only had one satellite and nothing else on board. So they were able to put some of this other stuff – which was important in that they had commitments – but in the grand scheme of things, after we got into delays you could conceive of somebody saying, 'Well, you know, I'll bet we can put that satellite somewhere else and just not fly this flight.' We wondered about that and always thought that might have happened if we hadn't had a congressman, but this was *his* flight and so we had some guarantee that it would happen."

By the end of 1985, all four Shuttles were fully operational and more than 20 missions had been successfully flown by Columbia, Challenger, Discovery and Atlantis. Thirty satellites had been placed into orbit, including three top-secret ones for the US Department of Defense, and representatives of Canada, France, Saudi

The STS-61C crew. Bottom row (left to right) are Steve Hawley, Franklin Chang-Diaz, Pinky Nelson (background) and Hoot Gibson. Top row (left to right) are Bob Cenker, Bill Nelson and Charlie Bolden.

Arabia, West Germany, the Netherlands and Mexico had accompanied their payloads into orbit. Two commercial satellites whose upper stages failed had been salvaged by the Shuttle and two others – Solar Max and Leasat-3 – had been successfully repaired in orbit by spacewalking astronauts.

On the scientific side, the reusable spacecraft had amply demonstrated its capabilities as an Earth-orbiting research platform. Since the pioneering flight of STS-9, the Spacelab module had flown twice – including a dedicated mission sponsored by West Germany – and the pallet-train-and-igloo combination had been satisfactorily tested. The outlook for the Shuttle project seemed bright: nine missions were accomplished in 1985 – transporting 54 astronauts into space, four of whom actually flew *twice* that year – and 14 were on the books for 1986, including the long-awaited deployment of NASA's scientific showpiece: the $1.5-billion Hubble Space Telescope.

NASA's decade-old promise of achieving routine access to space and sending Shuttle crews aloft once every two weeks seemed to be drawing closer. All four operational vehicles would be required to accommodate the intensive schedule for 1986: of those 14 missions, Columbia was slated to fly at least four times. Her return-to-flight on STS-61C, after more than a year of modification and refurbishment, finally received a firm launch date: 18 December 1985. It was supposed to be

NASA's tenth mission of that year, capping a triumphant 12 months of Shuttle achievements and heralding an even-brighter 1986.

'MISSION IMPOSSIBLE'

It did not happen. Launch on the 18th, scheduled for 12:00 midday at the start of a 49-minute 'window' for that day, was routinely postponed by 24 hours when technicians required additional time to finish closing out Columbia's aft compartment. The second effort to get the flagship spacecraft off the ground ended dramatically just 14 seconds before launch, when the flight controllers received an indication that the hydraulic power unit on the right-hand SRB had exceeded maximum-allowable turbine-speed limits. After finally deciding that the signal was erroneous, the window was missed and the seven astronauts disembarked from Columbia.

"We were happy as clams," recalled Bolden, "thinking 'We're going to go now'. All of a sudden, everything stopped and the countdown clock went back to $T-9$ [minutes] and kinda ticked there. We had no idea what had happened. As they started looking at the data, they had an indication that we had a problem with the right-hand [SRB]. As it turned out, when we finally got out of the vehicle and they went in, they determined that there wasn't really a problem. It was a computer problem, not a physical problem and it probably would have functioned perfectly normally."

Launch was rescheduled for 6 January 1986, but was under some pressure to get underway because the next flight – STS-51L, further demonstrating the chaos that NASA's Shuttle-numbering system could cause – was due to fly from KSC's newly refurbished Pad 39B on 22 January. This latter mission would be a public-relations boon for the agency, as Challenger's crew included the first private citizen to fly into space: teacher Christa McAuliffe from Concord, New Hampshire. NASA also wanted Columbia back from STS-61C as soon as possible, because the flagship was already booked for an important mission to observe Halley's Comet on 6 March.

An 18 or 19 December launch would have allowed just enough time to refurbish Columbia and load three astronomy telescopes – collectively called 'ASTRO-1' – to observe the comet's 75-yearly journey through the inner Solar System. Delaying the STS-61C mission into January was a headache that NASA could not afford; but there was worse to come. For the crew, however, the downtime over Christmas and the New Year was a chance to relax after more than a year of intensive training in the simulators and uncertainty over when they would finally fly.

"We stayed in quarantine a lot of the time," remembered Hawley. "When you're in launch mode down in Florida, the pace is not very hectic. You're not in training typically like you would be if you're in Houston and going to the simulators every day. You're reviewing procedures and checklists and having a nice time, because you have the opportunity to sort of sit back without the pressure of having to be in a sim[ulator]. I've always enjoyed the time in quarantine, although, because of the launch time, we were getting up at two in the morning every day!"

Columbia's launch attempt on 6 January proved to be one of the most hazardous yet in the Shuttle's five-year history: the countdown was halted at $T-31$ seconds, following the accidental draining of more than 1,800 kg of liquid oxygen from the ET. The tank's liquid oxygen fill-and-drain valve, it seemed, did not close when commanded to do so. Launch controllers reset the clock to $T-20$ minutes and efforts were made to re-initiate the liquid oxygen loading, but by now it was realised that the window for that day would close before the vehicle was ready. Another 24-hour delay was enforced.

Next day's attempt was scrubbed due to bad weather at two Transoceanic Abort Landing (TAL) sites in Spain and Senegal; which had long runways to accommodate a Shuttle in the event of an emergency during ascent. Yet another try on 9 January was called off when a liquid oxygen sensor on Pad 39A broke off and lodged itself in the prevalve of one of the three main engines. "That would have been a bad day!" Bolden would recall grimly years later. "[It] would have been catastrophic, because the engine would have exploded had we launched."

Heavy rain put paid to their next opportunity on 10 January. All seven astronauts, ironically on this occasion, were relieved *not* to launch. "We went down to $T-31$ seconds," remembered Bolden, "and they went into a hold for weather, and it was the worst thunderstorm I'd ever been in. We were really not happy about being there, because you could *hear* the lightning! You could hear stuff crackling in [your] headset. You're sitting out there on the top of two million pounds of liquid hydrogen and liquid oxygen and two [SRBs]. None of us were enamoured with being out there."

THE CURSE OF STEVE HAWLEY

When the crew left their quarters early on 12 January, Steve Hawley had gained the unenviable record of having ridden the bus to the launch pad on 10 occasions for only two 'real' liftoffs. To this day, he thinks a conversation and agreement he had with Commander Hoot Gibson may have helped to finally get Columbia into space: "I decided that if [Columbia] didn't know it was me, then maybe we'd launch, and so I taped my name-tag with grey tape and had the glasses-nose-moustache disguise and wore that." It worked and the STS-61C crew lifted off safely at 11:55 am.

Even the first few seconds of Columbia's ascent did not go well, as recounted dramatically by Bill Nelson in his book about the flight, *Mission*. Shortly after leaving the pad, Bolden – whose job it was to monitor the main engines – noticed an indication of a possible helium leak. "[We] had an alarm go off within seconds after lifting off," he said later. "I looked down at what I could see, with everything shaking and vibrating, and we had an indication [of] a helium leak in the right-hand engine. Had it been true, it was going to be a bad day.

"We called the ground and said, 'We've got a helium leak. We're going to work the procedure.' The ground didn't see anything. It was [a glitch] in one of the computers. [We] tried to isolate the first system; no luck. Still looked like we had a leak. Tried to isolate the second system; no luck. Then I looked down and it looked like we were making helium, so I told Hoot, 'I think we've got a sensor problem.' So we called the ground [and] reconfigured the system back to normal. This [was] all inside a minute after we lifted off!"

Other than the malfunctioning sensor which provided this momentary scare, Columbia's ascent was normal – although, in Bolden's words, his first launch into space "went by *really* fast!" – and was quickly established in a 340-km orbit. Bolden has downplayed his role in isolating the sensor problem: "To this day [Bill Nelson] thinks that Charlie Bolden saved the crew and Columbia, and I didn't. We *had* a problem, but it was an instrumentation problem." Whatever the problem, the mission was at last underway and the crew prepared to deploy their commercial payload.

For Bolden, one of relatively few black members of NASA's astronaut corps at that time, there was one particular place on Earth that he *really* wanted to see immediately after reaching orbit: Africa. Years later, he would describe it as "awe-inspiring and, in fact, it brought tears to my eyes." However, Bolden's first glimpse of his ancestral homeland from space would mirror the reactions of many astronauts upon seeing the Earth from the Shuttle's altitude: there were *no lines* to demarcate the different countries and he found it difficult to orient himself and realise what he was looking at.

The payload bay was sparsely occupied for STS-61C. Reminiscent of STS-5 – the first commercial Shuttle flight – Columbia carried a Pacman-type cradle, in which sat the Satcom Ku-1 communications satellite, mounted on an uprated PAM-D2 upper stage. Just in front of the cradle was a cross-bay bridge called a Mission-Peculiar Equipment Support Structure (MPESS), onto which were affixed three experiments forming the second Materials Science Laboratory (MSL-2). Behind the Satcom cradle, near the end of the payload bay, was yet another bridge, fitted with no fewer than *twelve* privately sponsored GAS canisters.

Although the Pacman cradle closely resembled those flown on STS-5, its content on STS-61C was somewhat different. The satellite, for starters, was a different shape. Unlike SBS-3 and ANIK-C3, which were both cylindrical 'drums', Satcom was cube-shaped and was part of a network of three satellites that would provide commercial communications services in the Ku-band of the electromagnetic spectrum. Owned by RCA – which had paid NASA $14.2 million to launch the satellite – the Satcoms were equipped with 16 transponders to cover the 48 members of the 'continental' United States or the eastern and western 'halves' of the nation.

Confusingly, the *second* Satcom, known as 'Ku-2', had already been placed into orbit by Space Shuttle Atlantis in November 1985. After Columbia's delivery of Ku-1, a third member of the series was scheduled to be sent aloft in 1987. Satcom Ku-1 was a three-axis-stabilised satellite, carrying its own electricity-generating solar cells in a pair of deployable solar panels, attitude-control thrusters, thermal-control system and command-and-telemetry equipment. Like the other Satcoms, its 45-watt transponders were considerably more powerful than the 12–30 watts used in C-band transponders. This was expected to allow users to employ dishes as small as a metre across.

The Satcom Ku-1 deployment. Note the MSL-2 payload in the foreground.

Since Ku-band frequencies were not shared with terrestrial microwave systems, dishes served by the Satcoms could be located within major metropolitan areas characterised by heavy terrestrial microwave traffic. The deployment of Satcom Ku-1 was supervised by Franklin Chang-Diaz and Pinky Nelson towards the end of the first day of the mission. "I had done [a deployment] on a previous flight," remembered Hawley, "[so] we kind of divided up the responsibilities so that we could maximise the resultant training and experience of everybody on the crew."

After several hours of checkout in the payload bay, Satcom and Columbia finally parted company at 9:26:29 pm, some nine-and-a-half hours into the STS-61C mission. It was a textbook deployment, similar to those performed by Bill Lenoir and Joe Allen on STS-5. The main difference was that Satcom was more than three times heavier than either SBS-3 of ANIK-C3: weighing some 1,920 kg. For this reason, it was bolted to a newer version of the PAM upper stage, which was capable of placing larger satellites into geosynchronous transfer orbit.

The PAM-D2 could accommodate payloads up to 3 metres in diameter, as opposed to just 2 metres for the PAM-D. Thanks to its upgraded Thiokol-built Star 63D engine, it could send satellites weighing up to 1,920 kg into geosynchronous transfer orbits, as opposed to the 1,270 kg achievable with the PAM-D. The larger size of the satellite and the upper stage meant that the Pacman cradle used on STS-

61C was also bigger. Fifteen minutes after Satcom left Columbia's payload bay, Gibson and Bolden fired the OMS engines for 12 seconds to provide a safe separation distance before the PAM-D2 fired.

A TRIUMPHANT START TO A TRAGIC YEAR

The satellite was finally inserted into an operational slot at 85 degrees West longitude, where it remained until April 1997. It was then moved to 86.5 degrees West and finally 'retired' in July of that same year, to be replaced by GE Americom's GE-2 communications satellite. By the time Satcom retired, so too had many of Columbia's STS-61C crew. Hoot Gibson would leave the astronaut corps in November 1996, after completing five Shuttle missions, including command of the first docking with the Russian Mir space station. He also served a few years as chief of NASA's Astronaut Office.

Bolden left NASA in August 1994 after four space trips, including the Hubble Space Telescope deployment flight and command of the first Shuttle mission to carry a Russian cosmonaut. Hawley would accompany Bolden on the Hubble flight, before entering NASA management, then returning to active status to fly two Shuttle missions in 1997 and 1999. Congressman Nelson and the second Payload Specialist, an RCA engineer named Bob Cenker, returned to their pre-flight jobs and would not fly again. Chang-Diaz would fly six more missions after STS-61C and now jointly holds the record for having been launched into space seven times.

The last member of the crew, Pinky Nelson, would fly once more: but it would be particularly poignant, as it was the first Shuttle mission after the Challenger disaster. On 28 January 1986, a few days after Columbia returned to Earth, Challenger lifted off with a crew of seven – including teacher Christa McAuliffe – for a week-long mission to deploy NASA's second TDRS communications satellite. A minute after launch, a leak of hot gas through rubberised O-ring seals in one of the SRBs acted like a blowtorch, and severed one of the struts that held it to the ET, the SRB broke free, pivoted around its upper strut, and breached the ET, igniting the volatile liquid oxygen and liquid hydrogen inside in an enormous fireball.

Fifteen kilometres above Earth, the boosters were sent spiralling away like wild Roman candles until they were blown up by remote control. The Shuttle was torn apart in the most public disaster ever witnessed at that time. The cockpit, still containing the remains of the astronauts, was recovered from the Atlantic Ocean a few months later and as late as the winter of 1996, bits of the wrecked spacecraft continued to wash up periodically on Floridian shores. The greatest impact, however, was to the United States space programme.

In the summer of 1986, a presidential commission presented its final report. It pointed to serious problems not only pertaining to the safety of the O-ring seals inside the two boosters, but also highlighted management failures which ignored or did nothing to prevent flights by a system that was known to be 'unsafe'. Commission members heard that, on the night before Challenger's fatal flight, engineers from SRB manufacturer Morton–Thiokol had begged managers to postpone the launch because they knew the O-rings could not withstand extremely cold weather conditions. At 36 Celsius, Challenger's launch was the coldest yet attempted. Their pleas to keep the Shuttle on the ground were ignored.

Schedule pressure was also blamed. Nine missions were flown in 1985, but at a cost. One top-secret Department of Defense flight in January resulted in O-ring damage so severe that it should have grounded the fleet. Another mission in July got off the ground only after a dangerous on-the-pad shutdown of the three main engines; even when the Shuttle finally headed into space, an engine failed during ascent and almost forced the crew to make an emergency landing in Spain. Moreover, parts were being cannibalised from other Shuttles to maintain the façade that the vehicles were flying 'routinely'.

Many analysts have remarked privately since Challenger that, even if the disaster had not happened, the project would have ground to a halt or the flight rate severely curtailed before the end of 1986. According to a Shuttle manifest published at the end of 1985, an incredible two dozen missions were pencilled-in for 1987! It was inconceivable that launch rates of two missions each month could be achieved, when even one mission a month was stretching NASA to its limits. Yet the mindset of many was to deny the problems and go on as if the Shuttle was an operational spaceliner.

The immediate consequences of the Challenger disaster in terms of the Shuttle's future direction were that all commercial launches were forbidden by the White House and transferred instead onto expendable rockets. New laws dictated that future missions should only be flown if they explicitly *required the presence of a human crew* to assure success, required the Shuttle's unique capabilities or were in response to "compelling circumstances". Such circumstances could, at the discretion of NASA's Administrator, involve the launch of national-security satellites or payloads that supported the United States' foreign policy interests.

STS-61C: "THE END OF INNOCENCE"

Essentially, therefore, the Satcom launched by Columbia on STS-61C was her last fully commercial payload. It was also, as Hoot Gibson remembered years later, the "end of innocence" for the Shuttle. After deploying Satcom, he and his crewmates settled down to what should have been five days of experiments and observations of Halley's Comet, which was making its closest approach to Earth since 1910. The Comet Halley Active Monitoring Programme (CHAMP) was a camera which involved astronomer Pinky Nelson burying himself under a black shroud to eliminate cabin-light interference, before shooting a number of high-resolution pictures of the comet.

The camera failed due to a battery problem and did not return even one good image of the famous comet. Another attempt was scheduled during Challenger's mission and then, during Columbia's ASTRO-1 flight in March, the most detailed observations would be conducted, utilising a battery of ultraviolet telescopes in the payload bay. The loss of Challenger and stalling of the Shuttle until September 1988 meant that NASA had lost its chance to see Halley's Comet from space. Although

Nelson and his colleagues tried to fix CHAMP, their efforts were in vain and by 14 January they were told to press on.

Franklin Chang-Diaz, on the other hand, was having more luck with the experiments under his watchful eye. The 35-year-old physicist was making his first flight and had been given responsibility for running three experiments on MSL-2, a bridge-like structure that straddled Columbia's payload bay. It was a reflight of a similar facility that had flown on Challenger during the STS-7 mission in June 1983 and carried an Electromagnetic Levitator (EML), an Advanced Directional Solidification Furnace (ADSF) and a Three-Axis Acoustic Levitator (3AAL). All three were dedicated to materials processing in space.

EML examined material flow during the solidification of melted materials. Six samples were suspended in the electromagnetic field of a cusp coil and melted by induction heating from its electromagnetic field. ADSF used four furnaces to melt and solidify several materials. Lastly, 3AAL carried 12 liquids suspended in sound pressure waves, which were rotated and oscillated to study bubble behaviour in microgravity. All three were up and running by the end of 12 January and 3AAL activities were completed the next day. The other two experiments encountered problems, however; they did not power-up properly and were terminated earlier than intended.

The second bridge in the payload bay was hidden behind Satcom's Pacman cradle, but was the first time that as many as 12 GAS canisters had been flown on a single Shuttle mission. An extra canister was also attached to the payload bay wall. The decision to fly the so-called 'GAS bridge' came about following the cancellation of one of two satellites originally assigned to the mission. The US Navy's Leasat-5 military communications satellite was supposed to fly on Hoot Gibson's mission along with Satcom Ku-1 and MSL-2, but it was withdrawn for inspections following the failure of its predecessor after reaching its orbital slot, and shifted onto Columbia's STS-61L mission, scheduled for November 1986. NASA then opted to

Columbia performs a nocturnal landing on 18 January 1986.

fly the GAS bridge instead to help to clear the backlog of small experiments awaiting launch. The Challenger disaster kept Leasat-5 grounded for longer than planned, but it finally reached space in 1990. The GAS canisters were activated by 14 January and included an ultraviolet instrument, a US Air Force payload, numerous student experiments looking at materials processing, fluid physics, crystal growth and brine-shrimp behaviour and an amateur radio package.

Other experiments, in what Bolden, like Hawley, described as "a year-end clearance sale", included the first of NASA's Hitchhikers. Unlike GAS payloads, Hitchhikers were controlled from the Goddard Space Flight Center in Greenbelt, Maryland, and allowed experimenters to actively interact with their investigations. The mounting hardware for the Hitchhikers also allowed them to draw on some of the Shuttle's power supply. At the time of STS-61C, two versions of the Hitchhiker were under development: one, called 'Hitchhiker-G', was attached to the payload bay wall and another, called 'Hitchhiker-M', consisted of a cross-bay bridge.

For its first flight, the Hitchhiker-G carried three experiments: a US Air Force package to record particle distribution in Columbia's payload bay, as part of preparations to fly sensitive infrared detectors on future Department of Defense missions, together with an investigation to examine the Shuttle's effects on a set of coated mirrors and a test of a capillary pumped-loop experiment for heat transport purposes. All three Hitchhiker experiments were successfully activated within the first four hours of the mission and performed satisfactorily.

Other activities included Chang-Diaz conducting a live televised broadcast in Spanish, in which he gave Latin American audiences a guided tour of Columbia. Meanwhile, Bill Nelson operated the Hand-held Protein Crystal Growth (HPCG) experiment, which 'grew' large crystals of sufficient purity to permit scientists to analyse their structure through X-ray diffraction (crystals grown on Earth tend to be irregular and difficult to study). He also participated in several medical tests which observed bodily fluid shifts, electrolyte changes and pharmacokinetics. Each of these tests helped to build a clearer picture of the effects of microgravity exposure on the human body.

TOUGH TO LAUNCH, TOUGHER TO LAND

If the factors that conspired to delay Columbia's ascent into space were maddening, those that kept her in orbit and hampered her return to Earth proved equally so. Originally supposed to touch down at KSC's 4.6-km-long Shuttle Landing Facility (SLF) on 17 January, the return was brought *forward* a day to allow additional time to prepare the spacecraft for her next mission. It was imperative that Columbia be ready to meet a 6 March target launch date for STS-61E, in order to acquire the best-quality images and spectra of Halley's Comet.

In readiness for the 16 January landing, Gibson and Bolden checked out their ship's RCS thrusters, but were waved-off due to bad weather at KSC. A further 24-hour postponement was enforced on the 17th, for the same reason. The two extra days were spent reactivating MSL-2 and an infrared camera that had spent most of the mission photographing storms, aurorae and volcanic activity. "It seemed to be equally hard to land [as launch]," recalled Hawley. "We were supposed to be the first flight to go back to KSC after [STS-51D in April 1985] and the weather just didn't cooperate. So they kept waving us off and making us wait another day to try to get back into KSC. I remember by the third day we had run out of almost everything, including [camera] film, and part of our training had been to look for spiral eddies near the equator, and Charlie [Bolden] was looking out the window and claimed to see one, and I told him he'd better draw a picture of it, because we [didn't] have any film!"

During this time in orbital limbo, Gibson wrote a song to the tune of 'Who Knows Where or When?' that he and Bolden sang to Mission Control over the communications loop. The two pilots sang it in two-part harmony:

> *It seemed that we have talked like this before*
> *The de-orbit burn that we copied then*
> *But we can't remember where or when*
>
> *The clothes we're wearing are the clothes we've worn*
> *The food that we're eating's getting hard to find*
> *Since we can't remember where or when*
>
> *Some things that happened for the first time*
> *Seem to be happening again!*
>
> *And so it seems we will de-orbit burn*
> *Return to Earth and land somewhere*
> *But who knows where or when?*

Mission Control's humorous response was to design a 'Wanted' poster for all seven astronauts, instructing would-be alien captors to return them to Earth immediately if found.

The crew's first chance to land in Florida on 18 January was also delayed by one orbit and, when NASA finally ran out of time, Columbia was diverted to Edwards in California. "Everything worked except God!" joked Bolden. "Finally, on our fifth attempted landing, in the middle of the night on [the 18th], we landed at Edwards, which was interesting because with a daytime scheduled landing, you'd have thought we wouldn't be ready for that. Hoot, in his infinite wisdom, had decided that half our landing training was going to be nighttime, because you needed to be prepared for anything."

Columbia touched down on Edwards' Runway 22 at 1:58:51 pm, wrapping-up a mission of just over six days. Subsequent inspection would reveal severe thermal damage to the right-hand main gear inboard brake, and it was decided that major improvements to withstand higher energy wear would be incorporated before STS-61E. All this made the planned launch target date for ASTRO-1 just seven weeks later increasingly untenable. "The landing", remembered Bolden, "was otherwise uneventful, other than the fact that it really upset Congressman Nelson, because he really had these visions of landing in Florida and taking a Florida orange! The [recovery] crew that picked us up was unmerciful, because they came out with a peck-basket of *California* oranges and grapefruits! Even having come back from space, [Nelson] was not in a good mood. That was a joke he really did *not* appreciate."

Ten days later, as he watched Challenger explode in the skies above KSC, Nelson would be giving thanks that he returned to Earth at all. Not only that, but he would find out during the course of the inquiry that, on the STS-61C ascent, Columbia's SRBs also suffered severe O-ring damage, similar to the problem that doomed Challenger.

3

Recovery after Challenger

A BUSY YEAR CUT SHORT

Had Challenger not exploded, Columbia was set to fly a further four missions during the course of 1986. It would have been the most she had ever flown in a single calendar year and, judging from the impossibly short seven weeks expected to get her ready for STS-61E, probably could not have been realistically achieved. On the 6 March flight, she and another seven-man crew would have been launched to spend eight days in orbit with the ASTRO-1 observatory of three ultraviolet telescopes affixed to a pair of Spacelab pallets in the payload bay.

This would have been Columbia's first use of the pallet-train-and-igloo combination, as well as putting a telescope-aiming device called the Instrument Pointing System (IPS) through its paces. The IPS had been tested, with mixed results, during Spacelab-2 and will be discussed in more depth later, but was essentially a platform onto which the ASTRO-1 telescopes were mounted. The STS-61E crew – Commander Jon McBride, Pilot Dick Richards, Mission Specialists Jeff Hoffman, Dave Leestma and Bob Parker and Payload Specialists Sam Durrance and Ron Parise – would have worked around-the-clock in two teams to operate the telescopes.

As well as an extensive programme of astronomical observations, the crew's time would have been consumed by studies of Halley's Comet using the CHAMP hardware and the ultraviolet telescopes. Following her return from STS-61E, her next mission was STS-61H, a seven-day flight scheduled to launch in the last week of June. As well as carrying her heaviest load of communications satellites so far – Indonesia's Palapa-B3, the Western Union's Westar-6S and the British Ministry of Defence's Skynet-4B – she would also have carried Indonesian and British Payload Specialists.

Like STS-5, the three drum-shaped satellites would have been housed inside Pacman-like cradles and deployed atop PAM-D upper stages. Since 1985, as part of efforts to compete with the European Ariane rocket for commercial launch contracts, NASA had begun offering customers an extra incentive: a seat for a Payload Specialist representative. On STS-61C, RCA's Satcom Ku-1 had been accompanied by RCA employee Bob Cenker, and Saudi Arabian and Mexican satellites sent aloft in June and November 1985 had been watched intently by representative Payload Specialists from those countries.

On STS-61H, Briton Nigel Wood, a Royal Air Force squadron leader, and Indonesian microbiologist Pratiwi Sudarmono would have observed the deployment of 'their' satellites. The remainder of the crew consisted of Commander Mike Coats, Pilot John Blaha and Mission Specialists Anna Fisher, Jim Buchli and Bob Springer; Fisher and Pratiwi becoming Columbia's first female passengers.

A top-secret Department of Defense mission (STS-61N) in September and another commercial flight (STS-61L) in November would have completed Columbia's roster for 1986. On the first of those missions, Commander Brewster Shaw would have led a five-man crew – Pilot Mike McCulley, Mission Specialists Jim Adamson, Dave Leestma and Mark Brown and US Air Force Payload Specialist Frank Casserino – into space to deploy a classified military satellite. On the second, which would have deployed the Leasat-5 and GStar-3 satellites and operated MSL-3, only one of the seven crew members was ever named: Payload Specialist John Konrad of the Hughes Aircraft Company.

One point to make from these assignments is that, in spite of Challenger and the almost three-year stand-down of the Shuttle fleet, the crews remained more-or-less intact. Most of the original STS-61E crew – Hoffman, Parker, Durrance and Parise – were members of the post-Challenger ASTRO-1 mission when it finally reached orbit in December 1990. So, too, might McBride have been, had he not resigned from the astronaut corps in May 1989. With the exception of Fisher, who took leave-of-absence to bring up her children, the core STS-61H crew of Coats, Blaha, Buchli and Springer remained together for the STS-29 mission in March 1989.

Similarly, the STS-61N crew – minus Mike McCulley, who was replaced by Dick Richards, and Frank Casserino, who returned to the US Air Force – would fly Columbia's first mission after the Challenger disaster. That, however, could not have been further from their minds in 1986 as each became immersed in the plans to return the Shuttle fleet to operations. A huge number of modifications, implemented on the recommendations of the presidential commission, had to be made to Columbia before she could fly again and her managers quickly found their ship in 'third place' in the pecking order behind Discovery and Atlantis.

It was at around this time that, again for reasons best known to NASA management, the agency reverted to its original numbering system for Shuttle missions. The first flight after Challenger would be the 26th and, logically, was labelled STS-26. Atlantis would fly STS-27 and Columbia was scheduled for STS-28 to deploy a classified Department of Defense payload. Early in 1988, she also received her crew for the mission: Shaw, Richards, Adamson, Leestma and Brown. Of the quintet, Shaw had flown twice before – including the Spacelab-1 mission – and Leestma once. The others would be making their first trips into orbit.

"At the time," remembered Leestma, "Discovery was going to fly first, then Atlantis and then Columbia. [I was] assigned to Columbia and they were having a hard time. This was its big 'down' period to make it look like the other orbiters that had been built, and since Columbia had been built earlier, there were a lot of differences. So it was still in the process of being modified and put back together, so it didn't make 28. It slipped too, when they flew Discovery and Atlantis one more time each before we finally flew in the summer of 1989."

With her forward RCS unit, OMS pods and a large number of thermal protection tiles conspicuously absent, Columbia rolls over to the OPF in early 1989 to begin processing for STS-28.

Columbia entered OPF Bay 2 on 24 June 1986 and from thence was transferred to the VAB's High Bay 2 for storage in mid-March of the following year. From September 1987 until July 1988 she underwent low-key attention in the Orbiter Maintenance and Refurbishment Facility (OMRF) – later to become the third OPF bay – and at the end of January 1989 she finally began STS-28 processing in earnest. Her External Tank was mated to its boosters on 23 May and, after the addition of Columbia to the stack, the complete STS-28 vehicle was rolled out to Pad 39B on 15 July.

More than 250 modifications were incorporated into Columbia during the post-Challenger period: among the most important were the addition of a reinforced carbon–carbon 'chin' just behind and underneath her nose, installation of a new telescopic pole to enable astronauts to escape from the vehicle during the later phase of re-entry, and improvements to her wiring, power-distribution system and thermal protection tiles and blankets. She also received upgraded GPCs, new fuel cells and APU controllers. Although these modifications added 1,130 kg of weight to Columbia, some of this was saved by replacing more than 2,300 protective tiles with newer-specification thermal blankets.

"There was an opportunity to look at the whole system and make it as good as it could be," said Arnie Aldrich, who became director of the Shuttle programme in 1986, "and I think the Shuttle benefited tremendously from that. What we found was that [with] all the pressure to get to [the] first flight, there'd been a lot of decisions made about, 'We'll live with this for now, but we'll fix it later', and so there were a lot of things that troubled people, that we wanted fixed, but there was never time. And once we started flying, the flights came so quick one after another, you couldn't stop and fix anything. So we got all of this on the table and we made a lot of changes and made the Shuttle a lot better. In fact, some of the changes to the main engine[s had] literally taken over a *decade* to make! They were just completed some 10 years later! So it was a time to take a lot of stock."

Although the technical cause of the disaster was corrected, and many other potential problems were also addressed, the problem of a two-and-a-half-year downtime worried many within NASA. "I was concerned about the long delay between flights," said Henry Pohl, "and primarily the technicians that put [the Shuttle] together. If you're doing something every day and it's kinda routine, you remember what to do, but now if I don't do it for nine months or a year and then try to start up, you don't exactly remember how you went about doing those things."

THE 'REAL ASTRONAUT' LOOK RETURNS

The astronauts themselves also received an upgrade of sorts. Since STS-5, and up to the Challenger disaster, they had only worn light-blue overalls and a clamshell-like 'crash helmet' with a limited supply of breathing air. Afterwards, however, NASA decided to revert to sending its crews into space with the best kind of personal protection possible. When the STS-26 crew lifted off in September 1988, they wore bulky, bright-orange partial-pressure ensembles known as the Launch and Entry Suit (LES). From STS-28 until STS-73, all of Columbia's fliers wore an LES.

Veteran astronaut Jerry Ross, who flew both before and after Challenger, has described the differences in wearing the different launch-and-entry suits. "The cloth flight suits and the 'motorcycle' helmet gave you a much more dynamic sense of what was going on. With the launch-and-entry suits that we have now, you're in a pressurised suit. There's a *lot* more bulk around you. It kind of cushions things and deadens the sounds and the vibrations."

The intention was to provide astronauts with hyperbaric protection during ascent, as well as cold-water-immersion protection in the event of an emergency egress over water. Designed and built by the David Clark Company – which also produced the brown-coloured suits used on the first four Shuttle missions – the LES bestowed a look of 'real astronauts' on crews once more. It was, however, a partial-pressure garment and further modifications would later be made, culminating in 1993 with the commissioning of a lighter, less bulky Advanced Crew Escape Suit (ACES). On STS-73 in October 1995, and all later missions, each Columbia flier wore this improved suit.

"Those partial-pressure [LES] suits utilise a system where you only pressurise certain critical portions of the body," said former suit technician Troy Stewart, "and the way that you do this is inflate a bladder which pushes against a restraint layer on the suit and also exerts pressure at the same time on certain areas of the body, such as the calves and thighs and of course the head. You maintain pressure in these bladders, which gives you the counter-pressure to be able to get the oxygen back into the lungs where you need it. The full-pressure [ACES] suit, of course, puts them right back into a full-encapsulated environment [like that provided by the suits on the first four test flights] that protects the entire body."

The STS-28 crew in a flight deck simulator, wearing new-specification partial-pressure suits. Left to right are Dick Richards, Jim Adamson, Dave Leestma, Mark Brown and Brewster Shaw.

Whether or not the provision of the LES or ACES would have helped to save the lives of Challenger's astronauts is, said Stewart, "purely academic at this point", but undoubtedly would have given them the benefit of better body protection and a capable life-support and parachute mechanism.

SECRET MISSION

With pressure on getting Discovery and Atlantis into space before the end of 1988, Columbia's launch was delayed until July and eventually the second week of August in the following year. However, as STS-28 drew closer, the shroud of secrecy covering it showed no sign of being drawn back. Not until years later would details of exactly *what* Columbia's crew did while in space begin to trickle out. Leestma, whose previous Shuttle flight had been a scientific one, described preparations for his top-secret mission as unusual and very cloak-and-dagger in nature.

"Sometimes you had to disguise where you were going! You'd file a flight plan in a T-38 [training aircraft] from one place and go somewhere else, to try to not leave a trail for where you were going or what you were doing, who was the sponsor of this payload or what its capabilities were or what it was going to do. You had to be careful all the time of what you were saying." Already, three Department of Defense missions had been flown, the payloads for which were rumoured to be advanced military communications, signals-intelligence and radar-imaging satellites.

STS-28 would transport the Department of Defense's fourth major Shuttle-launched payload into orbit. In the wake of the Challenger disaster, the US military and intelligence community began to rely to a lesser extent on the reusable spacecraft and began moving towards expendable rockets. Only payloads that were too large, heavy or awkward to be easily reconfigured for a rocket launch remained on the Shuttle. The presence of a human crew and Atlantis' Canadian-built RMS mechanical arm had already, in December 1988, aided the repair of one multi-billion-dollar spy satellite when it experienced difficulties shortly after deployment.

"It was a big plus for the DoD," said Leestma, "because they had some failures before [which] they might have been able to fix if people had been there or they had been able to spend some attention [on] it. They also were able to get pictures of everything that's going on and see that it was configured just right before it was let go. [But the DoD] did not like dealing with NASA. It was a kinda constrained arrangement, but it worked very well and the DoD was happy with the product that they got in the end."

It was already standard practice in the build-up to such missions that the countdown was conducted virtually in complete secrecy, until the public-affairs commentary began at T−9 minutes before launch. Only after that point could the public listen in to intercom exchanges between the astronauts and launch controllers. A software problem caused the countdown to be held for longer than planned and a combination of haze and fog over the SLF runway – used in the event of an emergency return to KSC – meant that the mission set off 40 minutes late at 12:37 pm on 8 August 1989.

As Columbia left Pad 39B – her first launch from this pad – and headed for a 290 × 307 km orbit, NASA's Deputy Administrator J.R. Thompson commented, "We're off to a good start on this mission." The high-inclination, 57-degree orbit needed for STS-28's primary payload was achieved by a 140-degree roll manoeuvre 10 seconds after liftoff. Brewster Shaw could be heard making the now-familiar "Roll program" call as the vehicle rolled over onto its back. Eight-and-a-half minutes later, the flagship of NASA's Shuttle fleet was back in space for the first time in more than three-and-a-half years.

A MYSTERIOUS, 'FLASHING' SATELLITE

Considering that STS-28 was such a historic mission, the official announcement from NASA spokesman Brian Welch, a couple of hours after launch, was a flat, businesslike "The crew of Columbia has been given a 'go' for orbital operations." The primary payload was deployed at 8:06 pm, about seven-and-a-half hours into the mission; at the time, John Pike – a space policy analyst for the Federation of American Scientists – speculated that it was a massive 14,500-kg 'KH-12' satellite, one of the latest generation of 'Key Hole' photographic-reconnaissance platforms that can trace their lineage back to the 1960s.

Pike commented at the time that the KH-12 was the Pentagon's most expensive payload ever orbited, with an estimated price tag at close to a billion dollars. Other observers, including *Aviation Week & Space Technology*, suggested that it was a somewhat-lighter Strategic Reconnaissance Satellite, known as 'SRS'. Still more civilian sources speculated that the satellite, whatever it might have been, was capable of manoeuvring itself to an orbital altitude of about 480 km, from which vantage point it could take photographs with a resolution as fine as a metre.

More recently, it has come to light that STS-28's payload was most likely a member of the second-generation Satellite Data System (SDS)-B family of US Air Force military communications satellites. Doubts over whether it was a KH-12 were raised within weeks of its launch, when ground-based observers noted that it 'flashed' – as sunlight reflected from its solar panels – at regular intervals, a phenomenon not usually consistent with a spying platform.

Certainly, it was not – unlike the huge Lacrosse radar-imaging satellite placed into orbit by Hoot Gibson's STS-27 crew in December 1988 – deployed with assistance from the RMS, which apparently was not carried on STS-28. The first photographs of what an SDS-B *looked* like were not actually made public until the spring of 1998, almost a full decade later, when the National Reconnaissance Office released pictures and videotapes of two military satellites. One was identified as an SDS-B, built by Hughes, and looked physically similar to the drum-shaped Intelsat-VI series of communications satellites.

The US Air Force began to develop the first-generation SDS in 1973, granting the contracts to Hughes. The satellites provided the US intelligence community with a network of Earth-orbiting relays capable of transmitting real-time data and images from spy satellites that were out-of-range of ground stations. Another of their responsibilities was to support voice-and-data communications for covert military operations. The second-generation SDS-B series – the first of which flew on STS-28 – operated in high-apogee and low-perigee Earth orbits, as close as 400 km and as far as 38,000 km, at steep inclinations, achieving apogee over the northern hemisphere.

This enabled them to cover two-thirds of the globe, relay KH-11 satellite data of the entire Soviet land mass and cover the entire north polar region in support of US Air Force B-52 aircraft communications. Such wide coverage was not available to geostationary satellites 36,000 km above the equator. The SDS-B featured two 4.5-m-wide dish antennas and a third, smaller, 2-m-diameter dish, which provided a Ku-band downlink. Overall, the satellite was 4 m long and 2.9 m wide with a launch mass that has been estimated at close to 3,000 kg.

In total, three of these cylindrical, solar-cell-covered SDS-B satellites were deployed by the Shuttle: on STS-28, STS-38 in November 1990 and STS-53 in December 1992. Although it is unclear *how* they were deployed, they were possibly released like the Leasats: in effect, they 'lay down' in the payload bay with their 'top' facing the crew cabin and were 'rolled out'. Alternatively, during its pre-Challenger 'commercial' days, the Shuttle was booked to carry Hughes HS-393 satellites, including members of the Intelsat-VI family, which occupied 'cradles' in the payload bay and would have been 'tilted' upwards before deployment.

In whatever manner the satellite was deployed, it is certain that Columbia performed a separation manoeuvre at 8:58 pm on 8 August. A second payload, weighing just 125 kg, was also deployed during the STS-28 mission and has been rumoured to have been some kind of 'ferret' satellite for radio and radar signals-intelligence-gathering purposes. The remainder of the mission went well and the crew tended a number of military experiments in Columbia's middeck and two others in GAS canisters. A few minor problems were experienced with RCS thruster failures, but none was sufficient to impact the mission's objectives.

COLUMBIA BACK IN BUSINESS

At 1:37:08 pm on 13 August, a little over five days after leaving Pad 39B, Shaw guided Columbia smoothly onto the dry lakebed Runway 17 at Edwards Air Force Base in California. "We hadn't tried to land [at KSC]," recalled Leestma, "because the first [post-Challenger] flight of each of the vehicles after the big down period was always the lakebed [at Edwards]." In fact, it had been more than four years since the Shuttle had last returned to land at its home base in Florida; not until the end of Atlantis' STS-38 mission in November 1990 would KSC landings resume.

"Super team and great machine!" radioed Capcom Frank Culbertson as Columbia rolled to a graceful stop. Shaw, however, later referred to this landing as demonstrating that he "wasn't such a hot pilot. When I landed [STS-]61B, it was on the [concrete] runway at Edwards that's got defined boundaries and it's easy to judge sink rate and your height. On STS-28 we landed on the lakebed, which has stripes painted on [it]. It's like oil that they put down there so it outlines the runway, but it's not a well-defined thing and you don't have the same kind of depth perception.

"So when we came down and I flared the orbiter, I [didn't] know how high we were. Looking at the photographs, we weren't very high, but I basically levelled the vehicle off and then it floated. So instead of landing at 195 knots, the way we were supposed to, we landed at 155. This was Columbia again and so here we are on the main gear, decelerating fast and I've got to get the nose on the ground. The same thing that happened to John [Young] on STS-9 happened to me and the nose [went] 'bam' on the ground. I felt terrible because I let the thing float for 40 knots' worth of deceleration. [We] got a lot of great data about low-speed flying qualities on the orbiter, but it wasn't supposed to work out that way."

Almost an hour after touchdown, the five astronauts emerged from their spacecraft. "Getting Columbia back [in space] was a key milestone," said JSC Director Aaron Cohen. "This crew did an outstanding job in doing that." Columbia herself looked very dirty after her eighth trip into space, but according to the US Air Force's Shuttle support manager, Ed Jenner, she had suffered minimal damage.

AN URGENT MISSION

When Space Shuttle Challenger reached orbit at the beginning of April 1984, the main objective of her week-long STS-41C mission had been the retrieval, repair and redeployment of NASA's Solar Max satellite. This was accomplished in a pair of spectacular spacewalks and deft handling of the Canadian-built RMS mechanical arm.

However, in order for the Solar Max retrieval to go ahead – and to make room for it in Challenger's payload bay – another satellite had first to be deployed. That satellite was a 12-sided, bus-sized structure called the Long Duration Exposure Facility (LDEF), which, as its name implies, was intended to accommodate experiments that required exposure to the hostile environment of low-Earth orbit for a protracted length of time. No one could possibly have known, at the time of the LDEF's launch, exactly *how long* the huge satellite would remain in space before being retrieved by another Shuttle and returned to Earth. NASA originally intended to retrieve the LDEF during Brewster Shaw's STS-51D mission in February 1985, but that was delayed. By the time Challenger exploded, the retrieval had been scheduled for STS-61I in the autumn of 1986.

Almost three years *after* the loss of Challenger, the LDEF was in a precarious state as it entered its sixth year of operations. Trajectory planners predicted that by March 1990, at the latest, it would be unable to maintain itself in orbit and would tumble back to Earth, burning up in the atmosphere. As many of the experiments on board the LDEF were expected to provide invaluable data as NASA developed new materials to build its Earth-orbital space station – at the time known as 'Freedom' – it was imperative that a Shuttle bring it home.

That was the job of Dan Brandenstein's STS-32 crew on Columbia's second post-Challenger mission. As chief of the Astronaut Office – a position he had taken over from John Young in mid-1987 – he had already flown two Shuttle missions, including the first nighttime liftoff and landing on STS-8. Privately, some astronauts have speculated that Brandenstein's position enabled him to pick for himself the two best missions of the early 1990s: the LDEF retrieval and the ambitious first flight of the new Space Shuttle Endeavour, which had been built to replace Challenger.

Joining Brandenstein for STS-32 were Pilot Jim Wetherbee and Mission Specialists Bonnie Dunbar, Marsha Ivins and David Low; of these, Dunbar was on her second spaceflight and the other three were first-timers. Interestingly, Dunbar was the only member of the *original* LDEF-retrieval crew from STS-61I to fly on STS-32. It was Dunbar's job to operate the RMS: plucking the satellite out of space and planting it safely into Columbia's payload bay. Brandenstein and Wetherbee would conduct the delicate orbital ballet to reach the LDEF and Ivins would photo-document the entire procedure.

A LEASED SATELLITE

David Low, meanwhile, had the task of deploying Columbia's own primary payload: a US Navy communications satellite called Leasat-5. Just as the STS-41C crew had done, the STS-32 crew would launch one satellite and pick up another. Leasat had originally been scheduled for launch on Hoot Gibson's STS-61C mission but by the time of Challenger had slipped until STS-61L in November 1986. Leasat (or 'leased satellite') was the fourth generation of Hughes-built synchronous communications satellites known as 'Syncoms'. As such, it had not one, but *two* names: in some quarters it was called Leasat-5 and, in others, Syncom 4-F5.

In shape, it took the form of a 2,400-kg 'drum' to provide worldwide, high-priority communications between ships, aircraft, submarines and land-based stations for the US military, as well as the Presidential Command Network. It measured 4 m tall and 4.6 m wide when 'stowed' in the Shuttle's payload bay, but after full deployment in geosynchronous orbit – when its Ultra-High Frequency (UHF) and omni-directional antennas had been swung out – its height increased to 6 m. It was a spin-stabilised satellite, covered with solar cells capable of generating 1,500 watts of electrical power.

Two helical antennas on top of the satellite provided receive-and-transmit capabilities in the 240–400-MHz UHF band, and it also carried two X-band 'horns' and the omni-directional antenna for tracking and control. The antennas were separately attached to a collapsible boom to allow them to be folded during ascent and sprung open in space; this helped to keep Leasat's costs down. When operational, the satellites provided 13 UHF communications channels, including a 500-kHz wide-band channel. They provided their own attitude-control functions, using onboard Sun and Earth sensors and thrusters, and carried nickel–cadmium batteries for backup power.

The Leasat project began in 1978, when the US Navy awarded contracts to Hughes Communications Services – today part of Boeing Satellite Systems – to build five Shuttle-deployable communications satellites, of which one would serve as a 'spare', for the Department of Defense. It was intended that Leasat would augment the US military's Fleet Satellite Communications (FLTSATCOM) network and the Hughes contract also provided for the construction of a control centre at El Segundo in California and a series of fixed and movable ground stations. The US Navy acted as the project's executive agent, working on behalf of the Department of Defense.

The 'lease' stipulated that the US military would pay to use communications channels on board the Leasats, at a cost of $84 million per year, per satellite. The first two Leasats were launched on Shuttle missions STS-41D and STS-51A in August and November 1984, followed by a third on STS-51D in April 1985. Leasat-3 suffered an electrical failure that stranded it in low orbit and was later retrieved, 'hot-wired' and successfully redeployed by the STS-51I crew in August 1985. Ironically, the crew of that mission also deployed a fourth Leasat whose UHF system failed completely.

The situation at the time of STS-32's launch, therefore, was that four Leasats had been launched but only three were fully operational; and because the fourth had failed after entering geosynchronous orbit, it was too high for a repair by Shuttle astronauts. The fifth Leasat, which would be deployed by Brandenstein's crew, would thus be essential in completing the 'minimum' of four satellites needed by the Department of Defense. Its frisbee-like deployment from Columbia's payload bay was scheduled for the second day of the mission but, as with STS-61C, the Shuttle required several 'tries' before getting off the ground.

PUSHING THE ENVELOPE

Launch was originally planned for 18 December 1989, which – judging from its planned 10-day duration – would have made the STS-32 astronauts the first-ever Shuttle crew to spend Christmas in orbit. This evidently played so much on their minds that they privately arranged for an impromptu crew portrait to be taken, in which they dressed in Santa suits, hats and dark glasses. Fortunately, they also wore their NASA name-tags to make them identifiable. It was not to be the only prank that Brandenstein's team would play ...

Since the Shuttle's return to flight after the Challenger disaster, most missions had lasted around five days. STS-32 would push the envelope by getting close to – or even exceeding – the 10-day record set by John Young's STS-9 crew at the end of 1983; according to the pre-mission press kit, Brandenstein's mission was planned for 9 days and 21 hours. Although the deployment of Leasat and retrieval of the LDEF did not require the unusually long mission, an increasingly confident NASA wanted to demonstrate the Shuttle's capabilities as it was planning to modify Columbia for missions lasting up to a month.

The STS-32 crew with a model of the LDEF. Left to right are David Low, Marsha Ivins, Bonnie Dunbar, Jim Wetherbee and Dan Brandenstein.

The processing of STS-32 involved modifications to support the longer-than-normal mission: a fifth set of cryogenic oxygen and hydrogen tanks was installed under Columbia's payload bay floor. This enabled the fuel cells to produce electricity and, as a byproduct, drinking water for the crew. By the end of November 1989, the Shuttle stack had been rolled out to Pad 39A – becoming the first to use this pad since STS-61C – in anticipation of a launch a few days before Christmas. Unfortunately, problems getting Pad 39A ready resulted in a delay until 8 January 1990, so the Santa joke fell flat.

The next obstacle was weather. "Our main concern", said US Air Force meteorologist Ed Priselac on 6 January, just after the countdown started, "is that low-level cloudiness will not clear out of here very quickly." This threat also included showers obscuring high-altitude cloudiness associated with a slow-moving cold front, which reduced the prospects of acceptable weather on the 8th to just 40%. The chances of successfully setting off that day were reduced still further by a comparatively short, 54-minute launch window, which had been precisely timed to allow Columbia to rendezvous with the LDEF early on the mission's fourth day.

NASA engineers also expressed concerns that pad hardware used to load propellants into the External Tank might leak, although these concerns proved unfounded. Otherwise, the attempt on 8 January proceeded relatively smoothly: the crew were on board Columbia by mid-morning, although the hold at T−9 minutes was extended due to unsatisfactory weather at the SLF. In an effort to keep launch options open, the countdown continued to T−5 minutes, before being held again. For a time, it looked like the weather might cooperate, but then a faulty electronics component signalled a potential problem with the water-fed sound-suppression system.

This system is activated when the main engines and SRBs are ignited and floods the launch pad surface with water from a series of giant 'rainbirds' to reduce the reflected soundwaves at the moment of liftoff. A team of engineers was sent out to the pad to check the electronics component and were satisfied that it was aligned correctly. However, ultimately, the weather closed in and led NASA managers to scrub the attempt. "Even if they had not had that mechanical problem," said US Air Force meteorologists' spokesman Ron Rand, "it was still 'no-go' for weather."

ONE DOWN, ONE TO GO

Brandenstein's crew had more luck next day and Columbia thundered into space right on the opening of the hour-long launch window at 12:35 pm. A perfect ascent established them on an orbital racetrack to catch up with the LDEF and retrieve it on 12 January. In the meantime, the astronauts spent their first day in space concentrating on two major objectives: checking out the RMS – which Dunbar called "a beautiful piece of hardware" – and preparing to deploy Leasat. Affixed to a support structure called the Airborne Support Equipment (ASE) in Columbia's payload bay, the deployment was under Low's supervision.

At 1:18:39 pm on 10 January – a little under 25 hours after launch and on Columbia's 17th circuit of Earth – the satellite was released as the Shuttle flew over

Africa. It had been attached to the ASE by means of five attachment points, which were released when locking pins were retracted. An explosive device then released a spring which pushed Leasat over the payload bay sill like a frisbee at a velocity of about 4.6 cm/s, spinning at a couple of revolutions per minute for stability.

"Houston, we had a good deploy," Low radioed from his station on Columbia's aft flight deck.

"David, we copy and congratulations," responded Capcom Tammy Jernigan.

A few minutes after the satellite had left Columbia's vicinity, Brandenstein and Wetherbee performed a separation manoeuvre to create a safe distance before Leasat's first engine burn. The satellite's Hughes manufacturers were pleased: "It was as good as you can get. Everything looks great," said spokesman Tom Bracken later that day. A series of manoeuvres by the satellite's own propulsion system were required to achieve its geosynchronous orbital slot. The first, at 1:53:48 pm, involved Leasat firing its solid-fuel motor to boost itself firstly into an elliptical transfer orbit with an apogee of 13,200 km, which was later circularised and raised to geosynchronous altitude.

During this time, Dunbar again uncradled the RMS and used one of its cameras to photograph the first Leasat burn. Several additional manoeuvres were made by the satellite to achieve geosynchronous orbit, but it had settled into its correct 'slot' over the Pacific Ocean at 177 degrees West longitude by 13 January. After a month-long period of checks, it was declared operational and joined the other three Leasats. Later in 1990 and early 1991, it was employed to support military communications during Operations Desert Shield and Desert Storm in Iraq.

The satellite finally retired from US government service in February 1998, bringing to an end more than a decade of Leasat operations. However, its usefulness was not yet over. Under a multi-million-dollar contract with the Australian Defence Force, in May of that year it was moved into a new orbital slot at 156 degrees East longitude for use by the Royal Australian Navy. It was a reprieve for the satellite: according to Hughes' spokesman Ronald Swanson, "[it] was literally within days of being propelled into [a] useless ['graveyard'] orbit, since its service to the Department of Defense had been completed".

After the successful deployment of Leasat, Columbia's crew turned their attention to the retrieval of the LDEF. At the time of their launch, they were trailing the satellite by about 2,730 km, closing in at some 60 km per orbit. Three flawless manoeuvres were performed by Brandenstein and Wetherbee on 9 and 10 January to help to close the gap between the crew and the LDEF, and an excited Flight Director Bill Reeves exulted, "Everything is 'go' for the rendezvous. The crew and the ground feel very excited and are looking forward to it."

On the morning of 12 January, the astronauts were awakened by music from Mission Control: 'Bring It Home', to the tune of 'Let It Snow'. Due to the sensitivity of the LDEF's experiments, their approach was as unobtrusive as possible to minimise contamination. From a distance of 40 km down to 1.6 km, Columbia's radar and star-tracker locked onto their target and guided Brandenstein closer. He then took manual control, passing 'below' and 'ahead' of the satellite, then pitching the Shuttle's nose 'upwards' to achieve a position directly 'above' the LDEF.

Leasat-5 drifts away from Columbia on 10 January 1990.

"You don't want to close too fast with something *that* big and massive," Wetherbee would say later. "With the risks involved, you want to keep the closure very controlled and Dan was able to do that."

WHAT YOU SAW WAS WHAT YOU GOT

This profile resulted in Columbia flying 'upside down' in an Earth-facing orientation, with the satellite hanging 60 m 'above' her payload bay. By this point, the astronauts – who had watched computer-generated views of the LDEF in simulators for more than a year – saw the real McCoy sitting there waiting to be captured. It was a strange device, measuring 9.1 m long by 4.2 m wide and weighing just over 9,520 kg. At its most basic, it was a bus-sized structure made from aluminium rings and longerons and loaded with trays for 57 materials experiments.

Shortly after the formation of NASA in 1958, scientists began to give serious consideration to developing a satellite that could carry material samples and assess how the harsh environment of low-Earth orbit caused them to degrade over time. By 1970, these ideas had acquired a name: the Meteoroid and Exposure Module (MEM), which it was proposed would be carried aloft by the Shuttle – then scheduled to make its first flight in 1978 – and retrieved a few months later. As the name implies, its primary focus was the impact of micrometeorites on satellites and how best to protect them.

By the mid-1970s, the MEM had been renamed the LDEF and contracts for its development were granted to NASA's Langley Research Center in Hampton, Virginia. The basic 12-sided structure was complete by 1978 and, after tests, was kept at Langley until a Shuttle flight became available. By this point, its objectives had expanded from micrometeoroid research to studies of changes in material properties over time, performance tests of new spacecraft systems, evaluating power sources and conducting crystal growth and space physics investigations.

The satellite was designed to be reusable and adaptable for differing lengths, if desired, although ultimately it would fly only once. Its length was divided equally between six bays for the experiment trays, with a central 'ring' at the midpoint connected by longerons to the endframes. Aluminium 'intercostals' linked each longeron to adjacent rows of longerons on each side and removable bolts joined the longerons to the endframes and intercostals. This meant that the LDEF could be made 'shorter' or 'longer' if a mission required it. Experiment trays were then clipped into the rectangular openings between the longerons and intercostals.

Two RMS grapple fixtures were provided on the satellite: one to allow it to be picked up by the robotic arm for deployment and retrieval and a second to send signals to initiate the experiments. The LDEF had no attitude-control system of its own and what you saw was what you got: a passive experiment container with no manoeuvring capabilities. It was designed to remain in orbit by being placed into a 'gravity-gradient' attitude, with one end facing Earthwards, which made an onboard propulsion system unnecessary. This also freed it from acceleration forces or contamination caused by thruster firings.

The orientation of the satellite also meant that the two 'ends' would be subjected to a unique thermal environment, although *all* parts of the LDEF were subjected to daily temperature changes as the Sun 'rose' and 'set' every 90 minutes and solar angles changed annually. Heat management was accomplished by coating the interior surfaces with high-emissivity black paint, which kept thermal gradients across the structure to a minimum and maximised heat-transfer across the LDEF's body. The experiments were also spread evenly to equalise thermal properties across the satellite.

Bonnie Dunbar gently lowers the LDEF into Columbia's payload bay after retrieval.

Eighty-six trays – 72 around the circumference, six on the Earth-facing end and eight on the space-facing end – accommodated 57 investigations. The 1.3 m × 86 cm trays came in several depths and housed experiments weighing up to 90 kg. These covered four disciplines: (1) Materials and Structures, (2) Power and Propulsion, (3) Science and (4) Electronics and Optics. They captured interstellar gas atoms to understand the Milky Way's formation, observed cosmic rays and micrometeoroids, studied shrimp eggs and tomato seeds and investigated the impact of atomic oxygen on different materials, including solar cells.

It was therefore with added urgency that Columbia returned to flight operations in August 1989, because she had been designated as the Shuttle that would retrieve the wayward satellite. By the time Brandenstein's crew finally set off in January of the following year, the LDEF was within weeks of being lost forever, having reached an altitude of less than 290 km. During its almost six years in space, it had covered 1.3 billion km – roughly the distance between Earth and Saturn – and completed more than 32,000 orbits.

LDEF COMES HOME

As they gained their first glimpse of the satellite, Columbia's crew noticed it had suffered some damage during its six years in orbit: a small solar panel had apparently dislodged itself and was flying in formation and a number of holes – apparently caused by micrometeoroid impacts – were evident. After retrieval, the Interim Operational Contamination Monitor (IOCM), an instrument attached to the starboard payload bay wall, close to the forward bulkhead, would reveal that the LDEF also emitted a large amount of particulate debris.

Dunbar focused the RMS wrist camera on the LDEF's starboard side and prepared to grapple it. Brandenstein then performed a yaw manoeuvre to align the wrist camera with the LDEF's grapple fixture. By this point, the satellite was 'above' Columbia's cabin as the pilots maintained formation with their quarry in an inverted orientation. When she saw the grapple fixture in the monitor for the wrist camera, Dunbar went for the kill. She rotated the camera 180 degrees to the correct retrieval position and, at 3:16:05 pm, as Columbia flew over the Atlantic Ocean near Brazil on her 50th orbit, grasped the satellite.

"Houston, Columbia," radioed Brandenstein, "we have LDEF," as wire snares in the Canadian-built arm's end effector closed around the satellite's grapple fixture.

"You've made many scientists very happy their LDEF experiments are finally coming home," replied Capcom Jernigan over the sound of applause in Mission Control. It was the first of many accolades for Columbia's crew that day. Lead Flight Director Al Pennington called it "the culmination of a lot of work by a lot of people", while NASA Administrator Dick Truly – who had first put the RMS through its paces as an astronaut himself more than eight years earlier – expressed his admiration as he "watched America's space programme at its best".

According to the LDEF's chief scientist, William Kinard, "The investigators are scattered around the world and we heard comments from them as far away as Australia and Europe. They are extremely excited." Immediately after the retrieval, Columbia's computers commanded the RMS to align the LDEF with the payload bay walls' berthing guides and Dunbar took control of the arm to lower the satellite gently into position at 8:49 pm. "It looks like LDEF is going to join us for the ride home," said Bill Reeves from Mission Control after it had been successfully anchored in the payload bay.

Marsha Ivins, meanwhile, had spent the past four-and-a-half hours painstakingly photographing every surface of the satellite for the engineers' benefit. To help her, Dunbar rotated the LDEF slowly on the end of the RMS. By the time the day's work was completed and the RMS stowed, the crew had been awake for almost 17 hours, but according to Brandenstein, "all the faces up here are smiling and happy". Clearly, the triumph had lifted their spirits; to such an extent, in fact, that next morning they transmitted a cartoon picture to Mission Control showing the LDEF literally imprisoned by overgrown tomato seeds!

"We saw something strange," grinned Brandenstein, "so we got it on the video recorder and thought we would show it to you." The cartoon was a light-hearted reference to more than 12 million tomato seeds flown by students on board the LDEF, which had seemingly overgrown after their longer-than-planned stay in orbit. With both primary objectives of their mission – deployment of Leasat and retrieval of the LDEF – now satisfactorily accomplished, the crew settled down to what would turn out to be just over a week of scientific and medical experiments.

A MINI-SPACELAB MISSION

"Now we've got to get down to work," said David Low on 13 January. "This is the second-longest Shuttle mission we've had so far, so we can do some good science experiments up here and get some very good medical data." His words would prove ironic, if not a little prophetic, for Columbia would break her own endurance record set during STS-9 over six years earlier. One avenue of study for the astronauts was materials processing in microgravity and Dunbar spent a great deal of her time tending to the Fluids Experiment Apparatus (FEA) in one of Columbia's middeck lockers.

This device was capable of heating, cooling, mixing, stirring or imposing centrifugal forces on gaseous, liquid or solid material samples, and had been carefully designed to meet industrial requirements. Dunbar supervised the processing of seven samples of indium – chosen because it was a well-characterised material with a relatively low melting point – to assess disturbances caused by Shuttle thruster firings or the movement of crew members. It was anticipated that results from the experiments could lead to more advanced, industrial-standard versions for the planned space station.

The FEA was switched on a few hours after Columbia reached orbit on 9 January and ran successfully for almost a week, until a sensor indication showed that it had exceeded its touch-temperature limit. The unit deactivated itself as programmed, but the astronauts reported that its front surface was not hot. Nevertheless, after six days of operations, it achieved more than three-quarters of its objectives.

Ivins and Low, meanwhile, tended to the Protein Crystal Growth (PCG) experiment – by now a frequent flier on board the Shuttle – which was part of an ongoing effort to 'grow' crystals in space of key proteins which could some day be used in the production of new drugs to combat AIDS, cancer, high blood pressure, organ-transplant rejection, rheumatoid arthritis and other diseases. However, as biochemist and crystal-growth researcher Larry DeLucas – who would fly on Columbia in June 1992 – has said, such experiments are part of a long process and their benefits would probably not be seen for a decade or more.

Like snowflakes, protein crystals are structured in a regular pattern; by closely examining this pattern under powerful scanning microscopes, scientists have been able to better study their molecular architecture and design drugs to either block or enhance their normal functions in the human body. The main problem with such crystals produced on Earth is that the gravitational factors, such as sedimentation and convection, can lead to imperfections in the crystalline structure; such imperfections and their causes are largely side-stepped in a microgravity environment. Overall, Ivins and Low conducted more than 120 protein experiments during the course of the STS-32 mission.

Another of Ivins' responsibilities was the American Flight Echocardiograph (AFE), an off-the-shelf ultrasound device specially modified for the Shuttle. It could non-invasively generate three-dimensional, cross-sectional images of the heart or soft tissues and display them on a CRT screen. Previously flown in April 1985, it was hoped that the echocardiograph might lead to the development of new counter-measures for cardiovascular changes experienced during spaceflight. It was also used in conjunction with another experiment called Lower-Body Negative Pressure (LBNP), a collapsible set of 'trousers' that drew fluids into the legs to counteract the adverse effects of returning to Earth's gravity.

Other experiments included studies of whether the circadian rhythms – daily repeating 'biological clocks' – of pink bread mould persisted in the absence of terrestrial gravity. Samples of the mould were kept in darkness inside a middeck locker and were examined by the astronauts 10 hours after launch, midway through the mission and shortly before re-entry. The results generally indicated that the mould's circadian rhythms did indeed persist in space, although its biological clock may have been affected and 'reset' by daily repeating changes in cabin temperature and carbon dioxide levels on Columbia's middeck.

Another interesting experiment of note has been described by Wetherbee as "kind of a sextant in reverse", known as the Latitude–Longitude Locator, or 'L-cubed'. This had its genesis in October 1984, during Challenger's STS-41G mission, when Australian-born oceanographer Paul Scully-Power observed many interesting features from orbit, but was unable to plot their exact geographical locations. To address this problem, NASA, in cooperation with the US Department of Defense, began to develop an instrument that could determine a location's latitudinal and longitudinal coordinates from orbit.

During STS-32, this 'L-cubed' instrument was used by crew members to take repeat photographs of a geographical feature every 15 seconds; the data was then fed into an onboard computer, which calculated two possible sets of latitudinal and longitudinal coordinates. The crew, by knowing whether the target was 'north' or 'south' of their flight path, could then determine which set was correct. The instrument, which utilised a modified Hasselblad large-format camera with a wide-angle lens, proved extremely successful.

Despite the impressive nature of the flight, it was not entirely smooth sailing. On 11 January, before the LDEF retrieval, some 7.5 litres of water oozed from a leaking dehumidifier on the middeck into a compartment where the air-purification equipment was kept. The astronauts switched the leaky unit off and used a backup, to which Brandenstein joked, "We get the plumber-of-the-year award, but not the housekeeper-of-the-year award." They then vacuumed up the waste water and dumped it overboard.

A more serious problem arose as the crew slept during the night of 14 January, when Brandenstein was awakened by Mission Control following indications of a problem with one of the IMUs that kept track of Columbia's acceleration. The device was reset and worked normally, but Al Pennington pointed out that permanent problems with the unit could lead to an early end to the mission. Another key factor in a possible early landing was the weather outlook at Edwards, which was predicted to feature overcast skies and a chance of snow flurries on 17 and 19 January.

A RECORD-SETTING FLIGHT

"In the event [the IMU] fails, what we would do is execute a minimum duration flight," said Pennington. Fortunately, an early landing was not necessary and revised weather forecasts predicted dry conditions, scattered clouds and light northeasterly winds – coupled, however, with frigid temperatures – for the 19th. With the LDEF in her payload bay, Columbia would weigh a mammoth 103,400 kg at touchdown, meaning that the dry lakebed at Edwards might be 'too soft', with fears that the nose gear could dig into the runway surface and make the vehicle difficult to steer properly.

NASA managers decided, therefore, to land on the 4.8-km-long concrete runway, although even that presented significant challenges for Brandenstein. The presence of the LDEF shifted Columbia's centre-of-gravity further forward, meaning that without the veteran Commander's deft handling of his ship the nose could 'slap down' too hard onto the runway surface. In a press conference conducted while he was in space, Brandenstein told journalists that he needed to maintain sufficient speed after main gear touchdown in order to gently lower the nose.

Despite hopes that the weather would cooperate for a 9:59 am landing on 19 January, a light 'dusting' of snow at Edwards the previous day and the presence of water on the runways – coupled with a possibility of fog – was expected to force NASA to wave-off the attempt. According to space agency spokesman Kyle Herring on the evening of the 18th, "the weather is right on the edge of the flight rules right now". Nevertheless, Brandenstein's crew stepped smartly through their pre-landing checks of Columbia's flight control systems; so smartly, in fact, that they finished the work three hours early.

As feared, fog at Edwards conspired to delay STS-32's landing until the 20th. "We're looking and watching the weather," said Bill Reeves on the evening of the 19th. "Edwards is improving for tomorrow." In fact, four landing opportunities were available for Columbia on 20 January, followed by three more on the 21st and the Shuttle crew had enough consumables to remain aloft until the 22nd if necessary. They *could* have landed at KSC or White Sands, but NASA elected to hold out for Edwards, whose wide runway provided a more forgiving environment with the heavy LDEF on board.

The astronauts followed a light-duty day for the remainder of 19 January, quietly breaking the STS-9 duration record at 8:23 pm. Brandenstein also became the most experienced Shuttle flier, having notched up 573 hours in space on three missions, surpassing the previous record held by Bob Crippen after his fourth flight in October

1984. "Every hour is more than enjoyable," he told Mission Control when informed of his feat. On 17 January the Commander had celebrated his 47th birthday aloft with an inflatable plastic cake smuggled on board by his crewmates.

He also received a chorus of pre-recorded birthday greetings from the rest of the Astronaut Office and a message from basketball star Larry Bird, who congratulated him on the "slam dunk with LDEF". When questioned about his age by Mission Control, Brandenstein told them, "I was hoping travelling at Mach 25 [orbital speed of 28,000 km/h], you wouldn't age!"

Columbia's return to Earth on 20 January was successful, but a switch failure in one of her GPCs during de-orbit preparations led Mission Control to wave-off the first landing opportunity. This particular GPC [the Number 5 unit] contained the all-important backup set of flight software that would be used if her four other computers failed or became corrupted during re-entry. To play it safe, the backup software was loaded into the Number 2 GPC and the failed unit was shut down for the remainder of the mission. Fortunately, GPCs 1, 3 and 4 ran the show perfectly during re-entry.

It was a nail-biting time; Brandenstein and Wetherbee were just 18 minutes away from firing Columbia's OMS engines to begin the irreversible de-orbit burn, for an anticipated 8:00 am touchdown, when the GPC 5 problem arose. The engines finally ignited for almost five minutes – the longest so far in the Shuttle programme – at 8:30 am, slowing Columbia by almost 152 m/s and dropping her into the atmosphere. No radio 'blackout' was experienced during re-entry because constant communications were maintained through NASA's TDRS communications satellite network.

Soaring through the darkness of a pre-dawn Edwards, Columbia touched down on concrete Runway 22 at 9:35:35 am. "Welcome home. Outstanding job!" radioed Capcom Mike Baker from Mission Control, as Brandenstein brought the vehicle to a halt. "You showed the Shuttle at its best: deploying and retrieving satellites." All six wheels stopped by 9:36:39 am, giving a new duration record of 10 days, 21 hours, 1 minute and 39 seconds that would not be broken for another two-and-a-half years. "Records are there to be broken," said Dunbar, "[but we were] just glad to get another day [in space]."

For Dunbar, records would characterise the rest of her astronaut career. On her next flight – also on board Columbia – in June 1992, she would break her record from STS-32 by spending two weeks aloft. *That* would be the first in a series of long-haul missions using a new system called the Extended Duration Orbiter (EDO), and would be possible following a protracted series of modifications to Columbia that were scheduled to begin in the summer of 1991. Before those modifications could begin, however, the veteran ship had two more missions to fly, including some unfinished business.

WELL-WEATHERED

When Columbia arrived in the OPF on 26 January 1990, a few days after her return from the record-breaking STS-32 mission, one of the first tasks was removal of the

LDEF satellite. It had remained inside her payload bay during the long ferry flight back to KSC from Edwards Air Force Base, closely monitored by a set of environmental sensors. Although excited LDEF scientists were able to photograph their precious payload through Columbia's aft flight deck windows while at Edwards, it was not until it was back at KSC that they could get their hands on it properly.

After removal from the Shuttle, it was placed in a specially built LDEF Assembly and Transportation System (LATS) which trundled it firstly to the Operations and Checkout Building and later, on 1 February, to the Spacecraft Assembly and Encapsulation Facility (SAEF). Once it was safely inside the latter, the laborious job of removing the 86 experiment trays could get underway in earnest. For the next two months, engineers and scientists measured radiation levels, conducted an infrared video survey, observed contaminants – including micrometeoroids – and took over 10,000 photographs of the LDEF's exterior.

The satellite was intact, but certainly well-weathered after almost six years in space. Clear evidence existed of 'pitting', in which micrometeoroids had punched into her outermost surfaces, and some erosion to a Kevlar-foil thermal-protection cover on her space-facing end. "I think the conclusion that we all came away with," said William Kinard, "is that you have to be cautious in designing a spacecraft." Organic materials such as Mylar, Kapton, paint binders and bare composites showed severe erosion caused by exposure to atomic oxygen.

'Coated' composite materials generally survived and maintained their mechanical properties, but due to the extended mission a few of the thin polymeric films and blanketing materials were virtually destroyed; these had deposited debris onto adjacent surfaces of the satellite. A low-density particulate cloud was also spotted by Columbia's crew, trailing in the LDEF's wake. Many of the satellite's surfaces had a light-brown discoloration, although the leading face tended to be cleaner than the others. The data would aid designers when planning the structure of the space station's habitable modules and electricity-generating solar-cell arrays.

UNFINISHED BUSINESS

Columbia, however, remained in the OPF, where work gradually picked up on removing the STS-32 flight hardware and preparing her for another 10-day jaunt in early May 1990, during which she would tend to some business that had been left unfinished four years before. This was the completion of the ASTRO-1 mission, which was originally her next flight after STS-61C but had been indefinitely postponed following the Challenger disaster. Now renumbered STS-35, even after the resumption of Shuttle flights, the mission would prove a bear to get off the ground and already her crew had changed several times.

Although the 'science crew' – Mission Specialists Jeff Hoffman and Bob Parker and Payload Specialists Sam Durrance and Ron Parise; the team who would actually operate the ASTRO-1 telescopes during two 12-hour shifts – remained intact from the original STS-61E crew, the other three astronauts were relatively new. When the 'new' STS-35 crew was named in November 1988, Commander Jon McBride remained in charge, but chose to resign from the astronaut corps just six months later. His replacement was three-flight veteran Vance Brand, who flew Apollo 18 in July 1975, Columbia's STS-5 mission and another Shuttle flight in early 1984.

The Pilot for STS-35 was Guy Gardner, who made his first flight on a top-secret Department of Defense mission in December 1988, deploying an advanced radar-imaging satellite known as Lacrosse. The third Mission Specialist – and incumbent of the flight engineer's seat – was Mike Lounge, who would be making his third trip into space. Interestingly, Lounge had also flown on STS-26, the first post-Challenger Shuttle mission, in September 1988. The original STS-61E Pilot, Dick Richards, and Mission Specialist, Dave Leestma, were already, by this time, well-immersed in their training for Columbia's return-to-flight STS-28 mission.

Hoffman and Parker would both be making their second trips into space – the former flew on Discovery in April 1985 and performed the Shuttle's first unplanned spacewalk, and the latter had been on board Spacelab-1 – while both Payload Specialists were first-timers. Parker would later recall his activities on the morning of the Challenger disaster: "We were preparing to fly in 40 days to observe Halley's Comet. That morning [28 January 1986], we were training [for] launches and entries [and] stopped to watch the Challenger launch on TV. Obviously, we didn't fly 40 days later!"

Original plans called for three flights of the ASTRO observatory, each with two Payload Specialists. Durrance and Parise would fly the first mission; then Parise would join another Payload Specialist, Ken Nordsieck, for ASTRO-2 and Durrance would fly with Nordsieck on ASTRO-3. All three missions were expected to be completed by July 1987. Hoffman and Parker, it seems, would have flown *all three* missions! It seems remarkable today, when astronauts typically wait three or four years between flights that NASA was planning to fly them into space in such rapid succession.

Parker has since expressed disbelief at the sheer number of missions planned up to the time of the Challenger disaster. "It's amazing when you look back at that [schedule pressure] and the rate at which we thought we had to keep pumping this stuff out. You'd have thought the world was going to end [if we didn't meet our launch targets]. My favourite expression is: Guess what? The Sun kept on rising and setting! The Sun didn't even notice [if we missed our launch targets]."

One important factor that *was* missed, however, was Halley's Comet, which only visits the inner Solar System every three-quarters of a century, and by the time ASTRO-1 finally set off in 1990 had continued its cyclical journey to more distant parts of the Sun's empire. Fortunately, a flotilla of other unmanned spacecraft – Europe's Giotto, the NASA-led International Cometary Explorer (ICE), Japan's Suisei and Sakigake and the Soviet Union's two Vega missions – had close encounters with the comet.

ASTRO-1: FOUR POWERFUL EYES ON THE UNIVERSE

The ASTRO observatory can trace its origins back to 1978, when NASA issued an Announcement of Opportunity for advanced astronomical instruments for carriage on future Shuttle missions. Three were ultimately chosen – the Hopkins Ultraviolet Telescope (HUT), provided by the Johns Hopkins University of Baltimore, Maryland; the Wisconsin Ultraviolet Photopolarimeter Experiment (WUPPE), built in the Space Astronomy Laboratory at the University of Wisconsin at Madison; and the Ultraviolet Imaging Telescope (UIT), sponsored by NASA's Goddard Space Flight Center. The project was to be managed by the Office of Space Science.

By 1982, however, control had passed to NASA's Marshall Space Flight Center and the missions were renamed 'ASTRO'. Two years later, the first flight of the series was tentatively scheduled for the spring of 1986 – exactly the same time that Halley's Comet would visit the inner Solar System – and a special Wide Field Camera was added to permit detailed observations of the celestial wanderer. By the end of January 1986, ASTRO-1 had completed its final checkout and was ready for installation into Columbia's payload bay, when Challenger was lost.

For the next 32 months, the Shuttle and ASTRO-1 were grounded. The telescopes were removed from the Spacelab pallets and stored. Although periodic health checks were conducted, NASA decided to recertify them before clearing them for flight. This included replacing more than 300 bolts in mid-1987. The Wide Field Camera, no longer needed as Halley's Comet was by now long gone, was deleted. In its place, NASA added the Broad-Band X-Ray Telescope (BBXRT), which had also been chosen in 1978 and had hitherto been assigned to another mission called the Shuttle High-Energy Astrophysics Laboratory (SHEAL).

NASA hoped that the addition of BBXRT would shed some new light on a major supernova – known as 1987A – which had first been spotted by astronomers in February 1987. Stars 10 to 100 times more massive than our Sun burn a succession of chemical elements rapidly until their cores collapse and they explode as 'supernovae'; these are among the most powerful events in the Universe. The 1987A event spewed debris into space from a distance of 170,000 light years and one of BBXRT's aims was to examine the different elements present.

Originally, BBXRT had not been expected to fly until 1992, but as it was finished ahead of schedule it was quickly added to the ASTRO-1 payload. A second SHEAL instrument, the Diffuse X-ray Spectrometer (DXS), was eventually shifted onto another Shuttle mission scheduled for the winter of 1992. As work on readying the BBXRT for flight shifted into high gear, preparations were also underway for getting the other three instruments out of storage and back on the Shuttle.

Of these, HUT was kept at KSC throughout the post-Challenger downtime, although its spectrograph was removed and returned to the Johns Hopkins University in October 1988. Checks had confirmed that, although it was protected from air and moisture by a continuous supply of gaseous nitrogen, its ultraviolet sensitivity had degraded and the spectrograph was replaced. When it reached orbit, the 3.6-m-long and 1.2-m-wide HUT, which weighed over 770 kg, was intended to explore objects such as quasars, active galactic nuclei and 'normal' galaxies at far- and extreme-ultraviolet wavelengths.

This region of the electromagnetic spectrum was inaccessible from Earth and even to the instruments on the forthcoming Hubble Space Telescope. To achieve far- and extreme-ultraviolet sensitivity HUT's mirrors were coated with iridium. When the ultraviolet spectrograph returned to KSC in the spring of 1989, it failed its first acceptance test and was again changed; then an ageing television camera had to be removed and replaced.

Meanwhile, the other two ASTRO-1 instruments also underwent recalibration and testing. WUPPE was not shipped back to the University of Wisconsin, but instead a portable vertical calibration facility was built and delivered to KSC. The telescope passed its checks with flying colours in April 1989. UIT also remained in Florida, where the power supply for its onboard image intensifier was replaced in the summer of 1989. By the beginning of autumn, all three instruments had been declared 'flight-ready' and just before Christmas were installed onto the two Spacelab pallets in the Operations and Checkout Building.

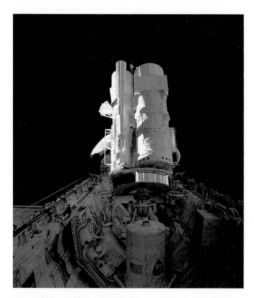

The ASTRO-1 observatory at work during STS-35. Note the cylindrical igloo at the front of the first Spacelab pallet.

When in space, WUPPE would examine the ultraviolet polarisation of hot stars, galactic nuclei and quasars. Any star – with the obvious exception of our Sun – is so distant that it only appears as a far-off point of light in a telescope eyepiece. If its light is polarised, however, it is possible to derive more information about its geometry and physical composition. UIT took wide-field-of-view images of star clusters, planetary nebulae, supernova remnants and galactic clusters. Although Hubble was expected to have higher magnification, UIT could cover much larger areas of the sky at once.

Major targets for ASTRO-1 included red giants which, at the end of their lives, shrink to become dense, hot embers no bigger than Earth, known as 'white dwarfs'. Since this is believed to be the final destination for many stars, they are an important area of study; they also emit most of their radiation at ultraviolet wavelengths, thus placing them squarely within ASTRO-1's capabilities. The observatory was also directed to examine suspected 'black holes' – stars whose density is so high that they collapse under their own gravitational influence to such an extent that nothing, not even light, can escape.

It was hoped that ASTRO-1's sensitive ultraviolet instruments, coupled with BBXRT's capabilities, would be able to 'see' hot, swirling material being dragged into a black hole's clutches. Other studies would focus on existing stars: 'binary' systems, in which two stars reside close to one another and sometimes exchange material, and stellar clusters, in which anything up to a million stars reside. In 'visible' light, it is difficult to distinguish the light from each celestial source, but under ASTRO-1's ultraviolet and X-ray gaze they were expected to blaze as individual stars.

More broadly, the observatory would examine the 'interstellar medium' – the enormous expanse of gas and dust *between* stars – which actually provides building material for future objects. Although the interstellar medium chiefly comprises hydrogen and has a typical density no higher than one atom per *thimbleful* of space, it was expected that ASTRO-1 would measure its physical properties more closely and explore 'pockets' of it which are much hotter than normal. A wide range of observations of galaxies and objects known as 'quasars', which are supermassive black holes in the nuclei of 'active' galaxies, were also among the mission's primary targets.

AN AMBITIOUS VENTURE

As already mentioned, ASTRO-1 marked the first time Columbia had carried the pallet-train-with-igloo combination and was also her first 'operational' Spacelab mission, following the inaugural test flight of the system on STS-9. It would also be the first operational mission of the Instrument Pointing System (IPS), which was attached to the Spacelab pallets and provided a base for the three telescopes. The IPS allowed the telescopes to be pointed with an accuracy of just two arc-seconds and could move them backwards and forwards, side-to-side and 'roll' them in a 22-degree circle about its 'straight-up' position.

It also featured a clamp to hold the telescopes horizontally in Columbia's payload bay during ascent and re-entry and its movements were commanded from the aft flight deck's control panel. For safety reasons, there was provision for the emergency jettisoning of the telescopes if they were unable to retract properly before the closure of the payload bay doors. The IPS had been used to support a battery of solar-physics instruments during Challenger's Spacelab-2 mission in July 1985, but ASTRO-1 would be the first time it had been employed for deep-sky objects.

As well as the precision-pointing afforded by the IPS, an additional image motion-compensation system had been provided by NASA's Marshall Space Flight Center to better stabilise UIT and WUPPE. This was capable of sensing crew- or thruster-induced movements of the Shuttle and sent data to the telescopes, which automatically readjusted themselves to achieve a stability finer than a single arc-second. This would prove particularly helpful for UIT, which would record its images on sensitive astronomical film.

The telescopes were attached to two Spacelab pallets, which had been joined together to form a short 'train'. Electrical power, cooling and command and data-acquisition services were provided by the igloo, mounted on the front pallet. Behind the pallets was a separate support structure, known as a Two-Axis Pointing System (TAPS), which held the BBXRT. Like the IPS, it was capable of manoeuvring its telescope backwards and forwards and side-to-side; but unlike the ASTRO-1 instruments, it was controlled remotely from the ground. The astronauts' involvement would be little more than turning it on and off and monitoring its health.

Built by NASA's Goddard Space Flight Center – from where the telescope was operated – the BBXRT observed high-energy celestial targets, including active galaxies, quasars and supernova remnants; particularly that of Supernova 1987A. Prior to each observation 'run', it was intended that stored commands would be loaded into the telescope's computer and when the crew manoeuvred Columbia to face a celestial target, TAPS automatically aligned the BBXRT with that target.

ASTRO-1 would become the first scientific mission to be managed from the Marshall Space Flight Center's new Spacelab Mission Operations Control Facility, replacing JSC's Payload Operations Control Center (POCC) used on earlier flights. The new facility was capable of sending commands directly to the Shuttle, monitoring the ultraviolet telescopes, directing their observations, receiving and analysing data, adjusting schedules to take advantage of unexpected discoveries and working with the crew to resolve hardware problems. Each of these capabilities would prove instrumental in the mission's success when it eventually took place.

The complete ASTRO-1 payload – including the three ultraviolet telescopes, pallets, IPS and igloo – weighed a whopping 7,830 kg, while the BBXRT and TAPS were a little over half as much at 3,920 kg. This would prove to be one of the heaviest research facilities ever carried into orbit by the Shuttle. Further tests of the combined system were conducted throughout the early spring of 1990 and on 20 March it was loaded into Columbia's payload bay. Four years after ASTRO-1 should have been observing Halley's Comet, it seemed that the mission would finally get to fly.

HYDROGEN LEAKS

By late April, the Shuttle was on Pad 39A and on-target for a mid-May launch. However, NASA managers deferred on setting a firm date when it became clear that a freon coolant loop proportioning valve needed replacement. After this had been completed, 30 May was chosen as the new launch date. Thus far, three Shuttle missions had flown satisfactorily during 1990: Columbia had retrieved the LDEF in her spectacular January mission, Atlantis had launched a top-secret Department of Defense reconnaissance satellite in February and Discovery had delivered NASA's scientific showpiece, the Hubble Space Telescope, in April.

It was from this point that things began to go wrong and set off a chain of events that would lead to more than six months of delays before STS-35 finally got off the ground in early December 1990. The problems arose late on 29 May, as Columbia's ET was being loaded with 1.9 million litres of cryogenic propellants to supply her main engines during the climb to space. Sensitive detectors picked up an ominous hydrogen leak coming from the vicinity of the massive 43-cm disconnect valve in Columbia's belly.

When one views the Shuttle from 'underneath', the disconnect is easily visible: a pair of openings in her belly roughly at the junction between the aft compartment and the wings. It is through these openings that propellant flows from the ET and is channelled into the three main engines. Following the separation of the tank eight-and-a-half minutes after launch, a pair of disconnect 'doors' are closed. The leak indication on 29 May came from this area and, as a precaution, the next day's launch was scrubbed and the ET drained of propellants.

A 'tanking test' on 6 June confirmed the presence of a hydrogen leak and pointed to the disconnect as its source, although its findings were inconclusive and repairs could not be conducted at the launch pad. Protective covers were therefore installed over ASTRO-1 and the BBXRT and the STS-35 stack was rolled back to the VAB on 12 June where engineers could conduct a more detailed examination. Columbia returned to the OPF three days later and all parts of the disconnect that could possibly have caused the leak were replaced.

By now, the launch had been put off until at least early August, as replacement hardware had to be 'borrowed' from the Challenger-replacement orbiter, named Endeavour, which at the time was undergoing final assembly and checkout in Palmdale prior to its delivery to KSC in mid-1991. Columbia returned to Pad 39A on 9 August – passing Atlantis, which was returning to the VAB after her *own* spate of hydrogen leaks while being readied for her STS-38 mission – and on the 29th the launch countdown began again.

NASA hoped to get STS-35 off the ground on 1 September, not only because the space agency wanted the high-priority mission flown, but also to avoid interfering with the next Shuttle mission. Columbia's sister ship Discovery was, on 5 October, scheduled to take the Ulysses probe into orbit and set it on a mission to explore the Sun's polar regions. However, in order to achieve this ambitious feat, Ulysses had only a narrow, two-and-a-half-week launch window. If Columbia could not be launched in early September, her flight would have to be postponed until *after* Discovery's mission.

The 1 September attempt for STS-35 was scrubbed when telemetry from the BBXRT – used to provide critical temperature and pressure data on its argon-filled dewars – was lost. After correcting this problem, a second try on 5 September ran smoothly until the External Tank was being 'fast-filled' with propellants, when an unacceptably high level of hydrogen (6,500 parts per million as opposed to the maximum-allowable 600 ppm) was detected in the aft compartment. This highlighted to NASA that there was not one, but *two*, hydrogen leaks: one in the disconnect and another in a recirculation pump deep within Columbia's aft compartment.

As part of a series of isolation tests to pinpoint the new leak, the recirculation pump was replaced and damaged seals on the engine prevalves were inspected and repaired. Yet another launch attempt on 18 September was called off when leak sensors again picked up concentrations of hydrogen in excess of 6,700 ppm. After this latest scrub, a special 'leak team' was established to definitively identify the cause of the problems. Under its auspices, many more leak checks were performed, the prevalve seals were again replaced and Columbia's entire liquid hydrogen system was retorqued to ensure that everything was tight.

Finally, on 8 October STS-35 was transferred from Pad 39A to Pad 39B – from which Discovery had lifted off two days earlier carrying Ulysses – to make room for Atlantis, which NASA opted to fly next. Pad 39A was needed for Atlantis' top-secret payload, because its stringent levels of cleanliness could not be met by Pad 39B. Tropical Storm Klaus then forced a temporary rollback of Columbia to the VAB on 9 October. After returning to the pad, a tanking test was performed on the 30th, carefully scrutinised by video cameras and sensors, which confirmed that the hydrogen leaks had been eliminated.

After months of exasperation, NASA triumphantly announced that Columbia was once again 'flight-ready', but still at least four weeks away from launching because she had first to wait for Atlantis, whose own mission had now been delayed following problems with her payload. Originally set to fly on 30 November, STS-35 was finally rescheduled for 2 December after astronomers argued that a late November launch would adversely impact several of ASTRO-1's important scientific objectives. In fact, after so many delays, the astronomers were so cautious in their planning that they did not anticipate a launch before 1 January 1991!

Fortunately, Columbia would complete her mission long before the year's end. It would also provide a useful (and rare) opportunity to study a mysterious celestial object known as a 'blazar', which had been spotted in late November. "This will be high priority now," said Johns Hopkins astronomer Art Davidsen. "This is something you always hope will happen, but cannot plan for." Blazars are 'quasars' – distant, extremely luminous objects now known to be the cores of active galaxies – which have suddenly flared several dozen times brighter than normal, often remaining so for some months.

"We're right on the timeline," said NASA Test Director Mike Leinbach as the STS-35 countdown got underway on 29 November, "and not tracking any problems at all." In the wake of so many delays, his words might have heralded further bad luck, but not so on this occasion. Weather for 2 December was expected to be 70% favourable, although forecasters kept a close eye on a tropical storm just south of Cuba. The seven astronauts arrived at KSC in their fleet of four T-38 jet trainers on the evening of the 29th in a jubilant mood.

"We're back! We're ready!" Commander Vance Brand excitedly told the gathered journalists. By this time, he had been an astronaut for a quarter of a century and with this trip into space – his fourth overall – he would become the oldest person to venture into space, only five months shy of his sixtieth birthday. After so many delays, STS-35 had been pushed down the line and was now the 13th Shuttle mission after the Challenger disaster. Payload Specialist Sam Durrance, however, was convinced that it was not a bad omen, telling the press, "I think we're going to go this time."

Jeff Hoffman, meanwhile, was sure *he* knew why Columbia would definitely fly this time. His original ASTRO-1 was destined to observe Halley's Comet and, as STS-35 encountered delay after delay throughout 1990, the teams of astronomers timetabling the mission's scientific targets were forced to reshuffle their plans

The STS-35 crew eat breakfast on launch morning. Left to right are Bob Parker, Ron Parise, Guy Gardner, Vance Brand, Sam Durrance, Jeff Hoffman and Mike Lounge.

because the range of visible celestial objects changed month-by-month. A number of lesser-known comets were among the targets during that period, but by December there were none. "We all know comets are harbingers of bad news and bad luck," said Hoffman. "This time, we have no comet. So we are going to go."

And go they did. At 6:49:01 am on 2 December, lighting up the nighttime Florida sky for hundreds of miles, Columbia roared aloft on her tenth mission. The launch suffered a short, 21-minute delay due to concerns about clouds beneath the 2,430-m 'ceiling' needed to monitor the first two minutes of ascent. With Discovery's Ulysses mission on 6 October and Atlantis' top-secret Department of Defense flight on 15 November, NASA set its second-best record for launching three Shuttles within just 57 days. This slightly missed a 54-day record set by three missions launched in October and November 1985.

"When we were not prepared to launch we stood down and solved our problems," said Shuttle director Bob Crippen, a former astronaut. From Columbia's cockpit, meanwhile, Brand was experiencing the most dramatic of all four of his launches: his first nighttime liftoff: "You had the feeling you were lighting up all that part of Florida," he recalled later. "It was like night flying in an airplane. You had to really have your lights adjusted and pay attention to your gauges. You couldn't really tell much about what was going on outside. You couldn't see the horizon very well."

The nocturnal launch oriented Columbia to ensure that her passage through the so-called 'South Atlantic Anomaly', where the Van Allen radiation belt dips down towards the ionosphere, occurred mainly during orbital 'daytime'. High-energy particles were already known to affect instrument operations and increased the 'background' levels in sensitive detectors. Since this 'natural' background, which consists of scattered light and ultraviolet atmospheric airglow emissions, was also higher on the daytime portion of each orbit, it preserved the nighttime passes for allowing ASTRO-1 to observe faint celestial targets.

ASTRO-1 COMES ALIVE

Activation of the observatory began almost immediately after the pilots had established Columbia in her 300-km orbit; the Red Team – consisting of astronomers Parker and Parise and led by Gardner – took charge of activating the telescopes and their support equipment. Meanwhile, the Blue Team of astronomers Hoffman and Durrance and shift leader Lounge bedded down for a shorter-than-usual sleep period. They would awaken for their first 12-hour work shift at 6:00 pm on 2 December. Although not specifically assigned to either shift, as Commander of the mission, Brand's schedule was more flexible and he coordinated both teams.

ASTRO-1 marked the first use of the new Spacelab Mission Operations Control Facility at NASA's Marshall Space Flight Center; this was heralded at 10:56 am on 2 December when Parker opened his space-to-ground communications lines with the words, "Huntsville, this is ASTRO". He was quickly answered by Michelle Snyder, the crew interface coordinator at Marshall, who radioed, "Bob, we just want to let you guys know that everyone here [at] Huntsville is really excited and we're looking forward to a great ten-day mission and a lot of terrific astronomy. And we've got a lot of smiles."

"Michelle, this is Ron," Parise then called. "We know there's a lot of people down there that did a lot of work on this mission, and we're hoping to make it a real success for everybody. So let's get this show on the road."

By this time, the Red Team had turned on the BBXRT and at 12:36 pm, Parker received a go-ahead to unlatch and raise the IPS and ASTRO-1 telescopes from their horizontal position in the payload bay. This was completed within just seven minutes and by the time STS-35 was 16 hours old and the Blue Team had taken charge of the flight deck, the ASTRO-1 telescopes had finished their initial checks and were ready to begin calibratory observations. "That is a real star, folks," Durrance radioed as he performed the first stellar observation of the star Beta Doradus using the HUT.

This particular star, in the constellation Dorado (the Swordfish), was chosen because of its suitability when aligning and focusing the telescopes. The sighting was part of the so-called 'joint focus and alignment' process, a lengthy procedure in which the three ultraviolet telescopes were focused on a common target as a prelude to upcoming observations. Unfortunately, a computer failure in the WUPPE prevented it from participating in this alignment. Ground-based engineers diagnosed the problem as having been caused by an unactivated heater and, when this was switched on early on 3 December, the telescope's checkout could get underway.

Overall, however, ASTRO-1 operations were running smoothly. Typically, the 'duty' Mission and Payload Specialists, stationed on Columbia's aft flight deck, used two Spacelab keyboards and two Data Display Units (DDUs) to command the IPS and the telescopes. Closed-circuit television monitors then provided the astronauts with images of the starfields being observed by the HUT and the WUPPE and allowed them to check their data. Meanwhile, the shift leader – Gardner for the Reds and Lounge for the Blues – was responsible for firing the RCS thrusters to orient Columbia properly. Observations took between 10 minutes and a full hour to complete.

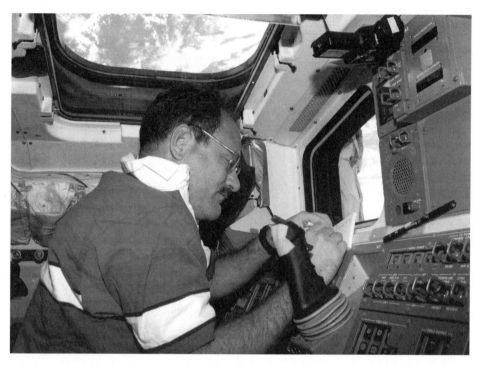

Mike Lounge records data during his shift in charge of Columbia's flight deck.

More problems, of potentially greater severity, were afoot, when the IPS experienced difficulties 'locking-on' to guide stars. An alternative plan was worked out to help the astronauts to manually point the telescopes and track targets on the HUT's television camera using a 'hand paddle'. This allowed them to aim the telescopes with an accuracy of within three arc-seconds. "The mood is one of concern," said Flight Director Bob Castle. "We'd certainly like the system to work perfectly, but there is no panic. People are working to solve the problem and we have confidence we will solve [it] in a fairly short time."

Worse, though, was to come. Late on the afternoon of 2 December, while working on Columbia's flight deck, Brand picked up the scent of warm electrical insulation; it turned out that one of the two DDUs had overheated. It was bad news because these units enabled the crew to point the IPS and also the ASTRO-1 telescopes. By 3 December, Castle was telling journalists that the IPS problems and the resultant work-around had thrown ASTRO-1's observation timetable at least six to eight hours behind schedule.

By midday on the 4th, when observations resumed, Flight Director Al Pennington described the outlook as brighter, although two dozen celestial targets scheduled for that day had been lost. "What we have to do," said Mission Scientist Ted Gull, "is make sure we reallocate what is left to the higher-priority objects that have been lost." By the day's end, the pace had picked up – with ASTRO-1 scrutinising the bright galaxy NGC 4151, thought to contain a massive black hole at its centre, and the BBXRT acquiring spectra of the Crab Nebula – but the crew was still well behind schedule.

With its computer woes finally rectified, WUPPE came on-line around 10:30 pm on 4 December and was directed to observe a 'variable binary system' of two stars called HR-1099 and HD-22468; these consist of a pair of close-by stars, one of which 'stirs up' the other and causes massive blobs of material to spiral away from it. Next it was used to study a rapidly rotating star known as 21 Velpecula. The telescopes, however, were still being handled with kid gloves. "We're still learning to use the observatory," said Gull. "It's taking longer to make the observations that we'd like."

As the telescopes found their feet, efforts were underway to obtain full capabilities from the IPS's Optical Sensor Package (OSP), whose star trackers provided one means of locking onto celestial objects. Thanks to support from Houston and Huntsville, successive refinements were made to its pointing geometry. These proved successful on 4 December, when Durrance accomplished the first operational identification of a desired celestial target: a white dwarf star. "Intensive efforts continue in trying to get good optical holds with the [IPS] to obtain the desired science targets that we have selected," said ASTRO-1's Assistant Mission Manager Stu Clifton.

On the whole, the mission seemed to have recovered from its IPS problems and the DDU failure and at a press briefing on the afternoon of 4 December, Mission Manager Jack Jones told journalists that "we have a good healthy payload. All the instruments are up and working and all the pointing systems appear to be pointing nominally at this time. Eliminating any unforeseen events, I think we're off and ready to go. There may be some minor refinements, but I believe we're in the mode to start getting science."

Gull agreed. "We have an observatory that's really coming alive," he said early on 5 December. "I can smile now!" After exhaustive troubleshooting by members of the ASTRO-1 ground team, a stream of successively more refined calibrations of the star-tracker optics had been sent up to Columbia. This now enabled the observatory to perform, at last, *automatic* acquisitions of celestial targets. Already, late on 4 December, Hoffman and Durrance had demonstrated the observatory's new capabilities by acquiring one target, followed immediately by another, with no need for recalibration of the instruments between each sighting.

The pointing stability continued to be lower than expected, because control of the Spacelab pallets, the IPS and the telescopes were being conducted from the only-remaining DDU on the aft flight deck. It was hoped that both situations should improve, leading to an increase in the quality of ultraviolet data, but such hopes seemed to be dashed at 12:15 pm on 6 December when the second DDU overheated and failed. Initial efforts to restart it were unsuccessful and Mission Control asked the crew to remove its lower panel, look at the internal components and check the air-intake filters for lint.

"No joy," Hoffman radioed dejectedly, after vacuuming a small amount of lint from the vents which cooled the DDU. At the time of the failure, ASTRO-1 had completed 70 of its scheduled 250 celestial observations; it was a success thus far, but a devastating blow for the remainder of the mission. "I'm not sure whether to smile from ear-to-ear or cry," admitted the HUT's Assistant Project Scientist William Blair.

Parker would recall years later that the prognosis for Columbia's crew was grim. "Suddenly, we couldn't point [the telescopes]. The ground came up with a scheme where they could control the telescopes. It was not part of the plans, but unlike the Apollo Telescope Mount [on Skylab] they could push a little 'go' button on the ground as well as the crew can push it up there, so they basically controlled the instruments. What they *couldn't* do in near-real-time was guide the telescopes. I was used to sitting there for hours, guiding a star on a cross-hair. We really had insisted all along that 'we need to be sure we have that capability, because if all else fails, we need to be able to do that'. We observed maybe a third of what we had intended to. Everybody put a good face on it, but it was a far cry from what it was supposed to be." In fact, for several hours on 6 December, after the second DDU failure, Durrance was left with nothing to do. "We're just sitting here enjoying the view," he told Mission Control.

Yet, still, the remarkable ground team managed to bounce back from the computer failures and by 7 December were able to command all but the final movements of the telescopes. It was then left to the astronauts to fine-tune them for each observation run. "We've had a lot of setbacks, but success is at hand," said Art Davidsen. Ted Gull, after watching his colleagues practising commanding the telescopes during simulations, agreed and thought the successful acquisition of celestial targets should be possible with JSC controlling the IPS and Marshall the instruments.

He added, however, that "it is going to be a close teamwork effort. Instead of one Mission Specialist and one Payload Specialist on the flight deck, there are going to be a lot of people on the loop, each having to do something in sequence to get the task accomplished." The revised procedures took the form of operating the ASTRO-1 telescopes sequentially – first the UIT, with the largest field-of-view, then the HUT and lastly the WUPPE – and even affected the BBXRT which, although mounted on a separate pointing system, was still compromised because it had to cease operations whenever Columbia entered a 'safe' attitude.

"Looks great!" radioed backup Payload Specialist Ken Nordsieck just after 11:00 pm on 6 December, as he guided Durrance to acquire the long-awaited Supernova 1987A with the UIT. As the astronaut on board Columbia controlled the telescope manually with a joystick, Nordsieck provided him with second-by-second pointing instructions. A total of six minutes of high-quality ultraviolet spectra were collected. More was to come and by the following evening – after ASTRO-1 had successfully observed a radio-quiet, extremely luminous quasar called Q1821 + 643 – Stu Clifton would be describing the day's observations as the proudest achievement of the mission so far.

"Following the loss of the [DDUs], ASTRO-1 has recovered substantially in less than 22 hours," Clifton told journalists. "All experiments are recording excellent data." By this time, the astronauts had established and were efficiently executing a smooth working pattern with their counterparts on the ground to maximise their observation time. Typical clipped exchanges between Nordsieck and Durrance on the space-to-ground communications loop ran tersely: "Sam, you're within an arc-minute. Okay, give me a 'mark' when you're happy. The data's looking real good, Sam; we're seeing lots of photons down here … "

Columbia's pilots – Brand, Gardner and Lounge – were fascinated by the research being conducted by the four professional astronomers on the crew. "We would look over their shoulder," remembered Brand. "A funny thing about the telescopes was that it wasn't like a telescope with an eyepiece. What you looked at was a TV-type display, so we could watch them use a hand controller that you might liken to a Nintendo controller, to move the telescopes manually."

The result of these pointings was a total of 231 observations of some 130 different celestial targets, with ASTRO-1 operating for 143 hours and accomplishing some 70% of the mission's original objectives. Despite its own pointing problems, the BBXRT also returned an enormous amount of valuable data: one of its most important targets was Markarian 335, a bright, compact object 325 million light years from Earth and considered a possible contender for having a black hole at its core. The telescope allowed astronomers to 'see' for the first time X-ray emissions from material being sucked into the hole.

Although the BBXRT was isolated from ASTRO-1, its TAPS support structure encountered difficulties and the original observation timetable had to be dropped in favour of a new, less efficient, day-by-day version. No stable pointings of the telescope were possible for the first 60 hours of the mission because of an improperly compensated gyro drift rate; this was corrected by the evening of 4 December and the BBXRT satisfactorily observed the Crab Nebula. However, it was then found that, although TAPS could *point* the telescope quite stably, its ability to acquire targets was slow, and this required several tweaks by ground controllers.

SUCCESS FROM FAILURE

Nevertheless, in light of its multitude of problems, the mission proved a fantastic success, achieving far more than expected after so many failures. "We just had a wonderful mission," said Art Davidsen at the last ASTRO-1 ultraviolet science gathering on 10 December. "We've spent 13 years getting ready for this opportunity. There were many times when we feared it would never come to pass, and other times when we thought it wouldn't work, but it actually worked spectacularly well."

To illustrate how successful the mission had been, Gull asked Davidsen what the final results were in a 'match' between 'The Huntsville ASTROs' and 'The Universal Secrets'. Davidsen replied, with a grin, that "the ASTROs won by a mile!"

Although the focus of STS-35 had been the observatory in Columbia's payload bay, the crew participated in several other activities. One of the most noteworthy was the 'Space Classroom', which gave Hoffman the dubious honour of becoming the first person to wear a tie in orbit. "All the male teachers wore ties," he told Mission Control. "I know nobody has ever worn a tie in space, so I thought I'd give it a try and see what it looks like." Then, with the tie secured by a piece of Velcro, he introduced the first school lesson ever transmitted from space.

It was part of a project entitled 'Assignment: The Stars', intended to spark students' enthusiasm for science, maths and engineering. On 7 December, the four astronomers gathered on Columbia's flight deck and, speaking over a two-way televised link with school pupils in Huntsville, Alabama, and Greenbelt, Maryland, told them all about the electromagnetic spectrum and its central role in ASTRO-1's mission. At one stage, Durrance played two taped versions of the same musical theme: the first, he explained, was unrecognisable because the high and low notes had been removed. The second, however, proved to be the theme from 'Star Wars'.

"You need to hear *all* the notes", Durrance told the pupils, "to appreciate the sound." Then, waxing lyrical somewhat, he told them that, in a similar way, "the heavens are playing a symphony of light". Hoffman turned the children's attention to ASTRO-1 itself and revealed how, operating high above Earth's hazy atmosphere, its sensitive telescopes could reveal more about the Universe than ever before. Later, the children were given the opportunity to pose questions to Parise and Parker.

AN EARLIER-THAN-PLANNED LANDING

It was already becoming clear by this stage that the STS-35 mission would probably not run to its planned length of 9 days and 21 hours; this time, however, the reason could not be traced to mechanical problems, but to weather conditions at Edwards Air Force Base, which were predicted to be unfavourable for a landing anytime between 12 and 14 December. NASA managers therefore decided to bring Columbia home a day earlier than originally intended, on 11 December. The early return forced astronomers to hastily reprioritise their remaining celestial targets.

ASTRO-1 was finally deactivated 12 hours prior to landing and the BBXRT was shut down six hours later. Meteorologists continued to track a West Coast cold front, which had been expected to produce rain and low clouds over Edwards, and the final decision was made to bring Columbia home early. "You've all had a fantastic mission," Capcom Story Musgrave told the crew from his seat in Mission Control, "but all good things have got to come to an end and you're coming home tonight." At 4:48:31 am on 11 December, the OMS engines fired to drop Columbia out of orbit.

A little more than an hour later, at 5:54:08 am, the flagship of NASA's Shuttle fleet completed her tenth mission in spectacular fashion by swooping out of the darkness and alighting on Runway 22 at Edwards. "We're home," announced Brand, "and glad to be back." Although he had brought two previous Shuttle missions in to land before this one, STS-35 was the veteran astronaut's first opportunity to return in darkness. "When you're landing a Shuttle in the daytime," he said later, "you land it like an airplane and see all the visual cues as you're coming in. If you're landing at Edwards in the daytime, out of the corner of your eye you can see the sagebrush and it's giving you an impression of how high you are above the ground, but at night you're really relying on light patterns that you see." Despite the problems with ASTRO-1 and the BBXRT – as well as other niggling glitches on board Columbia herself: such as a text-and-graphics machine that jammed time and time again, a blocked waste-water line in the toilet and a faulty RCS thruster – the mission was, nonetheless, a spectacular demonstration of teamwork triumphing over adversity.

STS-35 was the first in a stream of Spacelab missions which, until the vehicle's fiery loss during re-entry on 1 February 2003, would be increasingly devoted to scientific research. Although a handful of satellite-deployment flights would be reserved for her, the majority of these missions would be assigned to her sister ships Discovery, Atlantis and Endeavour. NASA's new direction of only using the fleet for missions which *absolutely required* its unique capabilities was now increasingly clear: no more commercial satellites would fly and the Shuttle would be rededicated to science and eventually the construction of a space station.

Shortly before Christmas 1990, at a post-mission press conference, the STS-35 crew told journalists they were convinced that the ASTRO-1 and BBXRT data would be highly prized after it had been fully examined. "The story of this mission is not finished," said Hoffman. "I think it's up to the scientists who have the data to give the answer. *They* are the people who write the final chapter."

Lounge, an astrogeophysicist by profession but assigned as one of the shift leaders on the crew, agreed: "What made this mission successful and possible was having human experts who could deal with the technical problems. If this observatory had flown unmanned, we would not have had a mission."

It was a fitting comment, although clearly it applied not only to the seven astronauts on Columbia, but also to thousands of others on the ground who had spent much of their professional lives preparing the observatory and working around each successive problem. The overall success of ASTRO-1 was acknowledged and rewarded five months later, in May 1991, when NASA announced its intention to refly the HUT, UIT and WUPPE on a second mission some time in the winter of 1994.

RETURN OF SPACELAB

By the early summer of 1991, Columbia was well on the way to flying another scientific mission, STS-40 – although this time, it would have a very different flavour: dedicated entirely to medical research. NASA already had plans to send the flagship of the Shuttle fleet back to Palmdale later in 1991 to begin a lengthy series of modifications that would enable her to fly longer missions of up to 16 days. That could only take place, however, after the satisfactory completion of the STS-40 mission which, like ASTRO-1, had been waiting many years for its chance to fly.

Unlike the pallet-train-and-igloo combination flown on STS-35, Columbia would this time be transporting the bus-sized Spacelab module into orbit for the first Spacelab Life Sciences mission, known as 'SLS-1'. It would be the first time either of the two ESA-built pressurised modules had been used since the Challenger disaster, only its fourth outing overall and the second time it had been carried on board Columbia. Originally, in pre-Challenger days, the mission had been known as Spacelab-4 and, like ASTRO-1, its four-person science crew had remained attached to it for many years.

The STS-40 crew in Columbia's middeck. Clockwise from the mannequin at bottom are Millie Hughes-Fulford, Jim Bagian, Rhea Seddon, Sid Gutierrez, Tammy Jernigan, Drew Gaffney and Bryan O'Connor.

Although, as Steve Hawley has said, pre-Challenger crews changed quite frequently, one general point that has remained consistent throughout the Shuttle programme is that Spacelab science crews do remain intact and attached to 'their' payload. Therefore, it was always clear to Mission Specialists Jim Bagian and Rhea Seddon and Payload Specialists Drew Gaffney and Bob Phillips that they could confidently train for the mission, reasonably safe in the knowledge that they would one day fly it. Unfortunately, for one of them at least, that would not be the case.

Phillips, the oldest of the four, was removed from the crew in mid-1989 after being medically disqualified. His replacement was biochemist Millie Hughes-Fulford who, like Gaffney, would be making her first trip into space. Bagian and Seddon, on the other hand, each had completed one previous mission: the former flew on STS-29, which launched an important communications satellite for NASA in March 1989, while the latter had been on STS-51D in April 1985, which featured an unplanned spacewalk to attempt to overcome a problem with the newly released Leasat-3 satellite. Interestingly, Seddon was also married to fellow astronaut Hoot Gibson.

Following a similar pattern to that blazed by STS-35, the crew also had a three-member team who would be responsible for monitoring Columbia's systems and keeping the spacecraft in generally good shape while their counterparts handled the bulk of the medical experiments in the Spacelab module. The Commander for STS-40 was Bryan O'Connor, who had flown once before, in November 1985; his Pilot, Sid Gutierrez, and the third Mission Specialist – and 'flight engineer' – Tammy Jernigan, on the other hand, were both first-timers.

Gutierrez was a late addition to the crew; the original Pilot was John Blaha, but he moved onto STS-33 to replace the Pilot of *that* mission, who had tragically died in an air crash. However, Blaha's ties with Spacelab Life Sciences were not to be severed and he later commanded the second SLS flight late in 1993. Jernigan became, at just a few weeks past her thirty-second birthday when STS-40 lifted off, one of the youngest women to fly in space. The presence of three women on the crew – Seddon, Hughes-Fulford and Jernigan – set yet another record.

When O'Connor, Gutierrez and Jernigan began their training in earnest in the summer of 1989, they confidently expected to launch the following June, but a multitude of delays caused by hydrogen leaks which affected two of NASA's three Shuttles pushed the mission into the autumn of 1990 and eventually well into the next summer. After Columbia returned from the long-delayed STS-35 mission, she was ensconced in the OPF where the ASTRO-1 and the BBXRT payloads were removed and electrical and mechanical provisions made to install the Spacelab module and its support equipment.

By the end of March 1991, the module was in Columbia's payload bay and on 3 April the tunnel was installed to connect it to the middeck airlock hatch. Meanwhile, problems arose with the preparation of STS-40's SRBs: after stacking, a set of gauges affixed to the boosters' aft skirts (their lowermost segments) started giving anomalous 'load' readings. Engineers suspected that their hold-down posts were misaligned and a decision was made to destack, and then *restack*, them. By mid-April, the boosters were complete, the External Tank was attached and Columbia arrived in the VAB before the end of that month.

After a smooth rollout to Pad 39B on 2 May, the complete seven-member crew arrived in Florida for their Terminal Countdown Demonstration Test (TCDT); while they were performing it, the previous Shuttle crew – that of STS-39 – returned home to KSC after an eight-day Department of Defense mission. Following a now-customary emergency escape exercise, the STS-40 astronauts returned to Houston to complete the final stages of their training. It would not be long before they were back for what they hoped would be the real thing.

On 19 May, O'Connor led his crew into Florida in a fleet of four T-38s and alighted on the SLF runway to greet the waiting journalists. "We're all ready to go. Light 'em," said Jernigan excitedly. Behind her in the distance, standing majestically on the pad, was Columbia, just three days away from her scheduled 22 May launch. "We hope you all have your fingers crossed for clear skies and smooth sailing," added Seddon. Weather forecasters were anticipating a 90% chance of favourable weather and Test Director Mike Leinbach was reporting a smooth countdown so far.

He spoke too soon. By 20 May, NASA managers opted to postpone the launch when a leaking liquid hydrogen transducer in the Shuttle's aft compartment – which had been removed and replaced during the exhaustive leak tests in the summer of 1990 – failed an analysis by its vendor and revealed weld defects. This increased fears that, if a weld failed during Columbia's ascent, one or more of the nine liquid hydrogen and oxygen transducers projecting into the fuel and oxidiser lines could crack, break off and be ingested by the main engines' turbopumps.

"I'm discouraged that it happened so late prior to our launch," said O'Connor. "But, on the other hand, they are talking about a potentially very serious problem, and the last thing we want to do is launch with a very serious problem on board. The temperature probes just upstream of the turbopumps for those engines have to be in good shape." Indeed, the possibility of disaster had the defects gone undetected does not bear thinking about for, almost certainly, the engines would have failed or even exploded if a transducer had broken away and lodged inside one of them.

Moreover, one of Columbia's five GPCs – the Number 4 unit, part of the 'redundant' set – failed, as did an MDM, which controlled her hydraulic ordnance and the functions of her aft-mounted RCS and OMS thrusters. A new GPC and MDM were fitted and one liquid hydrogen and three liquid oxygen transducers were replaced. Launch was provisionally booked for 1:00 pm on Saturday 1 June but this attempt was also postponed at the $T-20$ minute mark after several unsuccessful efforts to calibrate one of Columbia's IMUs.

Over the weekend, the unit was replaced and tested and the countdown clock again began ticking down, this time towards a projected launch at 12:00 midday on 5 June. With no mechanical obstacles in their way, the crew made their final preparations to go. One of these was the insertion of a catheter in Drew Gaffney's arm to monitor cardiovascular changes and fluid shifts in his body during Columbia's ascent and first few hours in orbit. The clock was held for almost an hour and a half when the weather did not cooperate, but STS-40 set off at 1:24:51 pm.

A SINGLE-SHIFT MISSION

After a problem-free ride into orbit, O'Connor and Gutierrez fired the Shuttle's OMS engines for two minutes to establish their spacecraft in a 260 × 240 km orbit, inclined 39 degrees to the equator. "Thanks for a great ride," O'Connor told Mission Control. "We appreciate it." Normally, the first few hours of any mission are a flurry of activity, and this one was no exception; but on most Spacelab missions, which operate a dual-shift system, half of the crew generally scurries to the middeck to begin their first sleep period very soon after reaching orbit.

That did not happen on STS-40, which operated just a single-shift system, although the astronauts averaged 14-hour workdays during their time aloft. Their 'circadian rhythms' – sleep/wake cycles – were strictly followed in this way to provide for a more uniform set of biomedical experiment results from the whole crew. This mission was the first time that a Spacelab fully dedicated to medical research had been sent aloft, together with a team of medical professionals; as the launch commentator had remarked as Columbia speared for the heavens, it was "the first dedicated medical research flight".

SLS-1 consisted of 18 major experiments to investigate the fundamental problems affecting the biology of humans, animals and fish in microgravity. Ten of the experiments required human subjects, seven used rats and one used almost 2,500 jellyfish. During the mission, Columbia's crew explored how the heart, blood vessels, lungs, kidneys and hormone-secreting glands responded to a microgravity environment, studied the causes of space sickness in greater depth and examined minute changes in muscles and bones while in orbit.

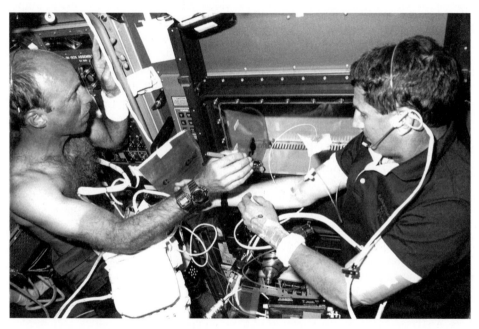

Jim Bagian (left) removes a catheter from Drew Gaffney's arm shortly after Columbia reached orbit.

Since this mission would include the most detailed medical studies since Skylab, scientists were particularly interested in the astronauts' physiological responses during their first few hours in microgravity. In support of these objectives, the science crew – Bagian, Seddon, Gaffney and Hughes-Fulford – underwent an extensive battery of demanding medical tests both before launch and after landing. They went through a full 24 hours of 'head-down' bedrest to evaluate whether or not this ground-based analogue to spaceflight produced similar cardiovascular responses and a series of vestibular tests were also conducted to assess their sensitivity to linear acceleration.

The mission also had a pronounced international flavour, with researchers from France, Russia, Germany and Canada participating in a biospecimen-sharing programme. This kind of cooperation would feature with increasing prominence in subsequent Spacelab missions, helping to lay the groundwork for today's International Space Station. Like ASTRO-1, the SLS project got underway in 1978 when NASA issued an Announcement of Opportunity for medical and biological experiments to fly on future Spacelab missions. Testing of SLS equipment began on Spacelab-3 in April 1985 and highlighted several operational flaws, notably contamination, leaks and odour problems from animal-housing facilities in the Spacelab module.

Originally assigned to fly on STS-61D some time in early 1986, by the time of the Challenger disaster SLS-1 had slipped until at least the spring of 1987. The almost-three-year post-Challenger downtime provided SLS-1 investigators with ample opportunity to modify their experiments and facilities, in particular the Research Animal Holding Facility (RAHF), which would accommodate the rats. It was, therefore, five years overdue and 13 years after the initial plans were laid, when Jim Bagian finally entered the Spacelab module on the afternoon of 5 June 1991, turned on the lights and began activating the experiments for nine days of intensive research.

A MODERN-DAY TORTURE CHAMBER

For the science crew, at least, to say that SLS-1 was their 'home' for the next week would not be an over-exaggeration, for as well as *working* in the Spacelab module, they also *slept* there, finding it more comfortable than the cramped middeck. "They all thought it was a great place to sleep," O'Connor told Mission Control. Moreover, all seven astronauts typically put their all into their work, starting their science activities earlier than timetabled and working through mealtimes. The result was a dramatic increase in the overall scientific yield from the mission.

One of the most important experiments focused on the heart, lungs and blood vessels and was slightly delayed on 5 June when a crucial piece of equipment – the Gas Analyser Mass Spectrometer (GAMS) – experienced difficulties and needed additional time to stabilise itself. Detailed studies were conducted into the cause of light-headedness reported by many astronauts upon standing after landing; one experiment tested the theory that it may arise because the normal reflex system regulating blood pressure behaves differently after adapting to microgravity. Crew members wore a specially designed neck cuff, akin to a 'whiplash collar', that detected blood-pressure levels in their necks.

Other experiments evaluated how rapidly astronauts adapted to microgravity through prolonged expiration and 'rebreathing' – inhaling previously exhaled gases – while at rest or pedalling a stationary bicycle. This yielded data on the amount of blood being pumped out by the heart, oxygen usage and the amount of carbon dioxide produced, as well as heart contractions, lung function and blood pressure. Other tests using different gaseous mixtures examined the influence of gravity on lung function. Measurements of blood pressure in the 'great veins' close to the heart were conducted using the catheter inserted in Gaffney's arm during Columbia's ascent to orbit.

The catheter was inserted in his arm on 4 June, a full day before launch, and was removed about four hours into the mission. Its data indicated the degree of body fluid redistribution and the speed at which this redistribution occurred. The catheter revealed that Gaffney's blood pressure – not surprisingly – rose while on the launch pad, increased during ascent and then dropped closer to 'normal' within a minute of reaching orbit. This appeared to refute earlier theories that a rise in blood pressure results from fluid shifts to the upper body due to weightlessness.

A large number of blood-related experiments were also performed to investigate mechanisms responsible for decreasing numbers of circulating red blood cells – known as 'erythrocytes' – in space and a subsequent reduction in the oxygen-carrying capabilities of the blood. Samples were taken from the crew members before, during and after the mission and their volume to plasma was measured to check the rate of production and destruction of blood under 'normal' and microgravity conditions. Although the SLS-1 results did not provide conclusive answers, they indicated that a drop in red blood cell production *was* a contributory factor.

Televised downlinks from the Spacelab module were, by 7 June, showing what appeared to be a modern-day torture chamber, thanks to all the blood draws taking place. On one occasion, Gaffney calmly offered his arm to Hughes-Fulford for a needle and, within an hour, submitted to another test by Bagian. After one draw, Seddon told Mission Control that it was taking a little longer than expected to collect the blood samples. "It's still running a little ragged because everything takes longer and veins are not cooperating," she said. "But I think we're getting it all done."

RODENT STUDIES

Still more investigations focused on changes in the muscles and bones of 74 male rats (*Rattus norvegicus*), 28 of which actually flew on Columbia, while another 45 served as ground-based 'control' specimens. Originally, 29 rats were supposed to fly, but a clogged water line in one of the cages failed early on 5 June, just a few hours before launch, and so it was flown 'empty'. "I assume he's having a happy life now," Test Director Mike Leinbach said of the sole rat dropped from the mission.

The rats weighed around 275 g and were about nine weeks old at the time of launch; the 20 (soon to become 19) specimens were loaded into the RAHF a few days before STS-40 lifted off. This facility had been greatly improved since its first flight on Spacelab-3 in April 1985, when escaped monkey faeces left the Spacelab module, entered Challenger's flight deck and floated under Commander Bob Overmyer's nose ...

The new RAHF flown on SLS-1 required an entire Spacelab rack and contained cages that provided the rats with food, water and waste-management functions, as well as controlling temperature and humidity, lighting and ventilation. The rats' movements were monitored and recorded by infrared light sources and sensors mounted inside each cage. During the mission, the facility was extensively tested by the science crew, who demonstrated its ability to capture escaped food crumbs, rodent hair and simulated faeces (black-eyed peas) and avoid contaminating the cabin. General results were favourable and showed that it maintained high levels of containment.

The facility's containment was so good, in fact, that it was possible for the astronauts to move several rats from the RAHF to a workbench called the General Purpose Workstation (GPWS), using a special transfer unit. This marked the first time that rats had been allowed to float freely in space and also provided scientists with an opportunity to assess their behaviour and performance outside their cages. Scientists would use the GPWS on SLS-2 to conduct rat dissections in space. The only problems with the facility were failures of the activity sensors and a fault in the drinking-water pressure transducer.

Nine other rats were flown in two Animal Enclosure Modules (AEMs) in one of Columbia's middeck lockers. Like the RAHF cages, these also provided food, water, waste-management, air and lighting for the rodents. Although the AEMs did not allow the astronauts to actually handle the rats, they were able to view them through a clear plastic window on the front of each unit. These nine rats rode in the middeck because they were loaded on board the Shuttle so late in the countdown – just 15 hours before launch – which was too late for technicians to easily access the Spacelab module.

A focal point of the rat research on SLS-1 were studies of the effect of microgravity on their muscles and bones, particularly by measuring changes in circulating levels of calcium-metabolising hormones and the uptake and release of calcium in their bodies. These changes were a cause for concern, because they appeared to be similar to those observed in human patients with osteoporosis – a condition in which bone mass decreases and bones become porous and brittle – and understanding the mechanisms responsible for them was deemed vital when planning future long-duration space station missions.

Bone growth in the rats' legs, spines and jaws was closely monitored and the loss of calcium and phosphorus was measured, revealing decreased skeletal growth and reduced leg-bone breaking strengths and spinal masses. Other experiments investigated decreases in the strength and endurance of their muscles. In general, the rats returned from space much more lethargic than when they left Earth, with reduced muscle tone, and were found to use their tails much less frequently as a balancing aid. Their red and white blood cell quantities decreased during the mission, although overall they were in much better shape than expected.

The crew participated in the calcium experiments, too, by carefully measuring their daily intakes of food, fluid and medication and weighing themselves frequently. The science team also took blood samples to better determine the role of calcium-regulating hormones on the observed changes in their calcium balance.

FIRST JELLYFISH IN SPACE

Another subset of SLS-1 investigations focused on the brain, nervous system, eyes and inner ear and included experiments from a joint US/Canadian project to investigate the impact of space sickness on the performance of the crew. One intriguing piece of hardware involved an astronaut placing his or her head inside a rotating dome, which induced a sense of self-rotation in the direction *opposite* that of the dome's own rotation. The subject then used a joystick to indicate his or her perception of self-motion.

Awareness of position in space is important, particularly during re-entry and landing when astronauts need to reach levers and switches. The general results of the SLS-1 motion experiments pointed to a loss of sense of orientation and limb position in the absence of visual cues. On several occasions, astronauts were also blindfolded and asked to describe the position of their limbs in reference to their torsos and point towards familiar structures within the Spacelab. Other experiments investigated changes in the inner ear, which has long been known to be highly sensitive to gravity and responsible for causing disorientation in space.

The nervous-system experiments, meanwhile, involved the first-ever flight of jellyfish on board the Shuttle: 2,478 moon jellyfish (*Aurelia aurita*), to be precise, which were housed in two containers in one of Columbia's middeck lockers. These were chosen because they are one of the simplest organisms known to possess a nervous system; they employ structures known as 'rhopalia' to maintain their correct orientation in water, akin to mammalian otoliths. The main aim of the jellyfish experiment was to determine their reproductive abilities and the impact of microgravity on their gravity-sensing organs and swimming behaviour.

The jellyfish were videotaped throughout the mission and finally 'fixed' on 12 June to preserve them for their return to Earth. Overall, the jellyfish polyps, which developed into sexually reproductive ephyrae in space, proved 'normal' in most respects, although they did exhibit hormonal changes and abnormalities in their swimming behaviour after landing. Differences were also noted in the gravity-sensing organs of terrestrial-born jellyfish, compared to their space-born counterparts.

While this work was underway in the Spacelab, 12 GAS canisters conducted their own research on a bridge at the rear of Columbia's payload bay. This structure was identical to that flown on STS-61C in January 1986 and its experiments included a European-built accelerometer so sensitive to movement that it could only be turned on while the astronauts were asleep. Others grew gallium arsenide crystals – some up to 9 cm long – that might prove useful in high-speed semiconductors for solar cells, produced harder-wearing, 'hollow' ball bearings, experimentally soldered metals and studied atomic oxygen effects on flower and vegetable seeds.

A CONTINGENCY SPACEWALK?

However, the problem-free appearance of the flight so far was deceptive. Scientifically, STS-40 was already proving itself to be a great success – "The

Close-up of the damaged payload bay seal.

mission has exceeded our expectations," exulted Arnauld Nicogossian, the head of NASA's space life sciences division – but a potentially dangerous problem with some of Columbia's thermal insulation cropped up soon after the crew reached orbit. Within minutes of opening the payload bay doors on the afternoon of 5 June, cameras revealed that several thermal 'blankets' attached to the aft bulkhead had become unfastened. Moreover, part of the payload bay door seal strip was displaced.

The danger, of course, was that the seal and the insulation could hamper the successful closure of the Shuttle's payload bay doors prior to re-entry. Lead Flight Director Randy Stone told the press there was no cause for alarm. "We don't believe this to be any issue with respect to safety or mission duration," he said. "The latches on these doors are very strong and we believe that, even if the seal *was* in the way, we could collapse the seal and close the doors safely with no problem."

Nonetheless, it was decided to ship a section of seal to JSC where astronaut Kathy Sullivan simulated a spacewalk to evaluate the procedures and tools needed to remove it manually. Encased in a bulky spacesuit, she was immersed in a huge water tank called the Weightless Environment Training Facility (WETF) and verified managers' beliefs that, despite the loose seal, the doors could close properly without requiring a spacewalk. If Bagian and Jernigan *did* need to venture outside their spacecraft, it was considered straightforward to either cut off the broken seal or push it back into its retainer.

By 8 June, Mission Control told the crew they did not believe a spacewalk would be necessary, and were surprised by a rare note of disagreement. O'Connor was worried that the seal might snarl on a mechanical 'fork' that assists in the closure of the payload bay doors. However, after several discussions during the day and the transmission of a printed explanation of the procedures to the Shuttle, he seemed to acquiesce, telling them, "If you've been through this and you still think there's a pretty good chance of the door latching, then you've answered our big questions."

In the meantime, the mission was proceeding so smoothly that, on the evening of 9 June, a relaxed Bagian – an amateur magician – decided to play a trick on Capcom Marsha Ivins and Flight Director Al Pennington. With the assistance of apprentice Gutierrez, Bagian asked Pennington to pick a card from a new deck. He also had a deck of cards in space and said before the flight that he had selected a card and turned it opposite to the other cards. Bagian predicted that the card he had selected before launch would match the one Pennington chose on the ground.

The flight director picked the four of spades, as did Bagian. "Truly incredible," was all Ivins could say as she then succumbed to the same trick.

After a highly productive mission, O'Connor, Gutierrez and Jernigan began preparing Columbia's systems for re-entry on 13 June. Then came the moment of truth early the following morning: the closure of the payload bay doors. To play things safe, for the last half-hour before the closure, the Shuttle's nose and the troublesome seal were oriented towards the Sun to 'thermally condition' the seal. Meanwhile, Bagian entered the Spacelab module for the last time, partly to store the last few blood and urine sample bags and also to videotape the payload bay door closure.

The port-side door was satisfactorily closed and latched at 11:20 am with no ill effects for the seal, followed by the starboard door at 12:08 pm. "We see both doors closed and latched," Jernigan, stationed on the flight deck, told Mission Control. "Roger Tammy. We see both doors closed and latched," replied Capcom Steve Oswald. "Nice job." Then, at 2:37 pm, Columbia's OMS engines ignited for two-and-a-half minutes to begin the descent back to Earth.

An hour later, at 3:39:11 pm, O'Connor and Gutierrez guided their spacecraft crisply onto Runway 22 at Edwards, wrapping up a mission of just over nine days – eclipsing STS-35 to become the third-longest Shuttle flight to date. "Congratulations on a super flight," Oswald radioed the crew. More than 7,500 people had gathered at the California landing site to watch Columbia's return, but they would not – unlike previous missions – see the seven astronauts disembark. On this occasion a specially designed airport-style 'people mover', known as the Crew Transport Vehicle (CTV), had been commissioned to whisk them away for medical checks.

"It's like a big trailer house on high that can be jacked-up," spacesuit technician Jean Alexander said of the new vehicle. "They [the astronauts] come inside and take the suits off and kind of get their land legs back and the doctors check them out. If there's any medical experiments that have to be done after landing, there's gurneys and stuff in there. They do blood draws [and] whatever [else] they require for that particular mission."

The rats, too, were quickly taken away, but to a somewhat different fate; 10 of the RAHF occupants and five of the AEM occupants, together with their ground-based 'control' counterparts, were killed and dissected about a week later for detailed analysis of their inner ear mechanisms. Their first flight had proved a remarkable success and would blaze the trail expected to be set on the SLS-2 mission some time in 1993 which, as well as lasting for almost two weeks, would controversially involve the first dissections of rats in space.

4

Extended capabilities

A PERSONAL CONNECTION

When Ken Bowersox boarded the International Space Station as its sixth incumbent skipper on 25 November 2002, he had – in a way – already flown two 'miniature space station' missions in his 15-year astronaut career. Nicknamed 'Sox' throughout the astronaut corps, he flew four highly successful Shuttle missions during the early-to-mid 1990s, two as Pilot and two as Commander, before joining the space station project as backup skipper for its first long-duration crew. Now, teamed with fellow astronaut Don Pettit and Russian cosmonaut Nikolai Budarin, he was about to spend what he thought would be a routine four months in space.

It turned out to be almost six months, in fact, and far from routine, for halfway through Bowersox's expedition, on 1 February 2003, Space Shuttle Columbia was lost as she plummeted through the atmosphere, bound for a Florida touchdown. The news hit the station crew hard. Not only would the Shuttle fleet be grounded for more than two years as NASA sought to repair the technical – and human – faults behind the tragedy, but during this time station crews would rely on Russian Soyuz capsules to deliver them to and bring them home from the outpost. Bowersox, however, was hit particularly badly.

Only days earlier, he, Pettit and Budarin had spoken to Columbia's astronauts over two-way radio; both crews were enjoying their time aloft and having successful missions. Now, the station crew were grieving for their fallen friends. Bowersox, too, had a close personal connection with Columbia herself, for he had flown her twice during his career: once in June 1992, on his first foray into space, and again in October 1995, this time as her Commander. Both flights lasted around two weeks, giving Bowersox his first taste of what life might be like with a long-duration space station crew.

THE BENEFIT OF LONGER SHUTTLE MISSIONS

Extended Shuttle missions, possibly lasting up to a month, had been in the back of NASA's mind since long before the Challenger disaster. They were considered a useful way of evaluating group dynamics in a closed environment for extended

The new EDO pallet being installed into Columbia's payload bay.

periods of time, as well as conducting research for longer than the week or so that had previously been possible. In fact, not since Skylab in the early 1970s had NASA had the facility to have astronauts tend experiments in space for more than a few days.

By the beginning of 1990, the plans had reached fruition: Columbia, the longest-serving Shuttle, would be decommissioned for six months in the summer of the following year to be extensively modified for long-duration trips. Initially, these were planned for up to 16 days, although the option of longer, month-long stays was kept open for the future. The modifications began almost immediately after Columbia returned to Earth from her highly successful SLS-1 mission in mid-1991. After three refuelling stops, the 747 carrier aircraft dropped her back at KSC on 21 June. She would not, however, remain in Florida for long.

After several weeks of 'deservicing' activities – draining her APUs, repairing her damaged thermal insulation material, taking out her forward RCS unit and removing her tyres – she was finally transferred from OPF Bay 2 to the new OPF Bay 3 (formerly the Orbiter Maintenance and Refurbishment Facility) for fit checks in preparation for her ferry flight to Palmdale. On 10 August 1991, secured to the top of the 747, she left KSC and, after two refuelling stops and two days on the ground due to bad weather, arrived at Rockwell International's Shuttle plant on the afternoon of the 13th.

For almost six months, she underwent over 150 modifications; the most notable of which were the addition of a new Regenerative Carbon Dioxide Removal System (RCRS) and the 1,630-kg Rockwell-built Extended Duration Orbiter (EDO) pallet.

Measuring 4.6 m wide, the wafer-like pallet was to be situated in an upright configuration at the rear of Columbia's payload bay on extended missions and be equipped with additional oxygen and hydrogen tanks to supplement her electricity-generating fuel cells. Eight tanks were attached to the pallet – four for oxygen, two for hydrogen and two for helium – although more could be fitted for longer missions.

When fully loaded, the pallet could store 1,420 kg of liquid oxygen and 167 kg of liquid hydrogen, which pushed its weight to almost 3,180 kg. Coupled with the cryogenic tanks already on board each Shuttle – which can ordinarily support missions of around 10 days – this provided Columbia and, for a time, Endeavour, with the option of spending 16 days aloft. Some consideration was given to adding a *second* EDO pallet to Endeavour, to permit even longer missions, but such plans were abandoned to save weight as the International Space Station effort gathered pace.

Although Endeavour *did* fly an EDO mission lasting almost 17 days in the spring of 1995, the system has since been removed and she is no longer capable of such flights. Similarly, Atlantis was equipped with some of the electrical provisions needed to make her EDO-capable, but ultimately NASA decided not to proceed with the final changes. The result was that Columbia became the only vehicle to routinely employ the EDO pallet and, to date, she holds – and, most likely, will continue to hold – the record for having flown the longest Shuttle mission.

A SCIENTIFIC AND PSYCHOLOGICAL CHALLENGE

The RCRS provided a means for removing the crew's exhaled carbon dioxide from the Shuttle's cabin in a more efficient way than had been done on earlier missions with replaceable lithium hydroxide canisters. "It [the RCRS] uses amine, coated on very small beads, almost like a powder," said NASA's environmental control and life support systems branch chief Frank Samonski, "but it pours like sand into a multilayered 'bed', like a heat exchanger, and when you pass gas containing carbon dioxide and moisture over it, the moisture activates this coating and the carbon dioxide molecules stick to the coating [by adsorption]. It's not a chemical bond and can be broken by exposure to vacuum. One bed is 'online', collecting CO_2, and the other is 'desorbing' to space vacuum. Through a series of valves, you switch those beds and dump the CO_2 overboard in a cyclic manner. But you've got to have the vacuum there, so it [doesn't] work on the ground and you have to be willing to throw away the water that you collect. So it's not good for long-term missions like the space station, but it's just right for the Shuttle. It saves all the weight of lithium hydroxide."

As well as the pallet, two extra nitrogen tanks were added during the six-month modification period to support the crew cabin and several more storage lockers were installed in Columbia's middeck. Other changes included a drag chute – which would be deployed from a pod just underneath the tail fin after touchdown to help to slow the Shuttle and reduce stress on her brakes – and beefed-up synthetic-rubber tyres to replace the earlier, natural-rubber versions. She returned to KSC on 9 February 1992 and moved into the OPF to begin preparations for her first EDO mission, a record-breaking 13-day trip, scheduled for June.

"I think we've all been aware that living together and working together for 13 days is certainly a challenge," said astronaut Dick Richards, who would command the first EDO mission, known as STS-50. "We'll be sharing the same sleeping quarters. We have only one restroom on board. We all have to cooperate to make life as good as possible." His words highlighted a key aim of the EDO project: to study astronauts' physical and psychological performance in close quarters on long missions. Such information, as already mentioned, would be vital for planning space station expeditions or journeys to Mars.

Richards' crew had already been training for almost a year and a half by the time Columbia was finally rolled to Pad 39A on 3 June 1992, for a launch anticipated later that month. The Commander would be making his third spaceflight overall, having flown Columbia as Pilot on STS-28 and then commanded another mission in the autumn of 1990 which deployed the Ulysses solar probe. Ken Bowersox, interestingly, had actually been assigned to STS-50 in *two* different capacities: first as a Mission Specialist and then, in August 1991, as Pilot.

Originally trained as a Shuttle Pilot, Bowersox knew from previous flights that some pilots flew first as a Mission Specialist, occupying the flight engineer's seat to gain experience. They were then promoted to Pilot for their second mission and commanded their third. Generally this was done on missions which required a 'third pilot' – such as dual-shift Spacelab flights – although exceptions to the rule were frequent. Some pilots, like Steve Nagel and Charlie Precourt, flew first as a Mission Specialist and then as Pilot, while others, like Charlie Bolden and Guy Gardner flew twice in the Pilot's seat.

When the STS-50 crew was named shortly before Christmas 1990, Bowersox was assigned as the flight engineer, occupying the centre seat behind Richards and a veteran Pilot named John Casper. Circumstances changed, it seems, when NASA 'ran out' of Commanders in August 1991. Within days of each other, astronauts Mike Coats and Bryan O'Connor – the latter of whom had led Columbia's STS-40 crew – left NASA to enter industry. The agency's response was to give Casper command of another mission in early 1993, promote Bowersox to fill his shoes and name a third astronaut, Ellen Baker, as STS-50's 'new' flight engineer.

The remainder of the crew, fortunately, stayed together. They were Mission Specialists Bonnie Dunbar and Carl Meade and Payload Specialists Larry DeLucas and Gene Trinh. As the 'science crew', they would supervise 31 crystal growth, combustion science, fluid dynamics and biotechnology investigations in the bus-sized Spacelab module. The versatility of the latter as a platform for many different activities had by this point been well demonstrated: after a 'pathfinding' mission on board STS-9, it was used operationally on Spacelab-3 for materials research, on Spacelab-D1 for West German technological research and on SLS-1 and the International Microgravity Laboratory for biomedical research.

For STS-50, the Spacelab was designated as the first United States Microgravity Laboratory (USML-1) and more than one scientist would liken it to the work expected to be performed on the space station, at that time still known as 'Freedom'. Around-the-clock investigations – with the Red Team of Dunbar and DeLucas, led by Bowersox, and the Blue Team of Meade and Trinh, led by Baker – would utilise the microgravity environment of orbital flight to produce new materials and protein crystals in support of new drugs with a purity far greater than could be achieved on the ground.

Moreover, the USML-1 experiments were expected to help the United States to maintain its 'lead' on worldwide microgravity research and applications. Terrestrial spinoffs of this research were expected to lead to the production of newer, faster semiconductors to help to build the next generation of high-speed computers and construct more efficient chemical catalysts for converting petroleum into gasoline. Several of USML-1's experiments would be conducted within a laboratory-style 'glovebox' – an enclosed compartment to minimise contamination of the Spacelab environment – while others were mounted in refrigerator-sized racks lining the module's walls.

By this time, the growing importance of the Shuttle for scientific missions such as USML-1 was becoming clear, to such an extent that, for each research flight, NASA named one of the Mission Specialists as a 'Payload Commander'. He or she – Dunbar was given the job on STS-50 – would be responsible for the overall planning and execution of the research performed aloft. To gain some headway and enable them to smooth out operational issues relating to the experiments, it had become common practice to assign Payload Commanders a few months in advance of the rest of the crew.

Dunbar joined STS-50 in October 1990, followed by the other astronauts in December. "We think we're ready," she told journalists as the crew arrived at KSC on 22 June 1992, in preparation for launch three days later. "We're really looking forward to an aggressive 13-day flight." On this mission, she would actually break her *own* record, jointly set as a member of the STS-32 crew more than two years earlier, when she spent nearly 11 days aloft. When she returned from STS-50, she would set another record by having the most number of hours in space for a female astronaut.

Bowersox's eyes were firmly focused on the future space station, although even before his first flight he was looking forward to longer missions. "I think it will be a big boost," he said, "but I've got to admit two weeks doesn't seem like a really long time. I kind of look at this as a test of doing things for future flights on board a space station, when someone is up there for a *really* long time."

After rollout to Pad 39A, Columbia underwent routine checks, as well as several new ones in which technicians rehearsed loading cryogenics into her storage tanks and EDO pallet to establish 'realistic' timelines for how long these procedures would take. The spacecraft destined to carry Richards' crew into space looked, as she always did after upgrades, identical outwardly, but internally she had changed significantly even in the last year. "Columbia may be the oldest," Shuttle Test Director Eric Redding said, "but with all the modifications, it's the best orbiter we have. I don't really expect an increase in frequency of problems."

The only anticipated problems over the weekend of 20–21 June seemed to be weather-related: concerns about lightning, hail and heavy rain materialised, although Air Force meteorologists predicted a 70% chance of favourable conditions in time for launch on 25 June. However, with iron-clad, overcast skies, the US Air Force's Mike Adams could not be sure. "We might be able to get the Sun to burn off those clouds and give us a break," he told journalists. "To be honest, it's too close a call. I'd have to put my two dollars on Mother Nature, I'm afraid."

Then, on 23 June, the attention of the engineers was drawn to a sensor problem that threatened a delay of up to a week. The sensor was located near the combustion chamber on Columbia's Number 2 main engine and monitored the temperature of the supercold liquid oxygen as it was pumped into the External Tank. More importantly, it provided an early warning of possible leaks. Engineers were concerned that it was generating erratic readings; if *it* was at fault, it could easily be replaced in time for 25 June, but if the problem lay with the *main engine*, a week-long delay would become necessary.

Fortunately, the sensor proved to be at fault and was replaced and tested on 24 June. "That problem is behind us," said Test Director Al Sofge, whose team was supervising part of Columbia's countdown. Meanwhile, engineers replaced two of three Tactical Air Navigation (TACAN) units on board the Shuttle, which would be needed during the approach and landing phase of her mission; they, too, had displayed inaccurate readings during pre-launch tests. Other minor difficulties with wiring in Columbia's aft compartment and a problem with a freon-circulating coolant pump were also rectified.

Nevertheless, all was ready by the morning of 25 June. "We'll see you in a couple of weeks," Richards told launch controllers from Columbia's cockpit a few minutes before liftoff. For a time, it seemed that the Florida weather might, then might *not*, cooperate as bands of cloud and rain marched across Pad 39A, alternately raising hopes, dashing them, then raising them once more. Ultimately, though, the weather only forced a short five-minute delay and Columbia blasted off at 4:12:23 pm, quickly disappearing behind a bank of clouds as she headed out over the Atlantic Ocean.

However – in an event that would portend her loss on STS-107 a decade later – a 60 × 25 cm piece of foam fell off the left bipod ramp of the External Tank during ascent, producing a 20 × 10 cm divot in a thermal protection tile. Otherwise, the climb to orbit was "nominal" and by 5:00 pm the crew established themselves in a 257-km orbit. Almost immediately, the Red Team began activating the Spacelab module for 13 days of research, while their Blue counterparts started an abbreviated, seven-hour sleep period.

"We're going to find out whether those kids are really friends or not," said former astronaut Brewster Shaw, who had flown with Richards on STS-28 and was now the chairman of NASA's Mission Management Team. "They will be so busy that I'm pretty sure they won't have time to get mad at each other." He was right not to worry. Echoing the feelings of the remainder of the crew, Bowersox would later describe his colleagues as "a joy to be around", which helped them immeasurably to work well as a team.

The Red Team, plus Richards, who – although not attached to a single shift –

The STS-50 crew gather in the Spacelab module for their in-flight portrait. On the front row (left to right) are Carl Meade, Ellen Baker and Gene Trinh and on the back row are Larry DeLucas, Dick Richards and Bonnie Dunbar, while Ken Bowersox reclines on a United States flag at top.

tended to align his work schedule with that of the Reds, was stepping smartly through the Spacelab activation procedures until late on the evening of 25 June. Columbia's payload bay doors were open and their radiators deployed by 5:37 pm and, 10 minutes ahead of schedule, at 7:42 pm Dunbar opened the hatch to the module, floated inside and turned on the lights. She had been here before, having made her first flight on West Germany's Spacelab-D1 mission in October 1985, which also used the pressurised facility.

CRYSTAL GROWTH

Activation of USML-1 was complete by 9:00 pm and, while Dunbar busied herself with preparing it for research, Bowersox and DeLucas unstowed, initiated and photographed two samples in the Protein Crystal Growth (PCG) unit, housed in a locker on Columbia's middeck. It was hoped that data from these experiments, which had already flown on several occasions, would yield new knowledge about the proteins' molecular arrangement and assist in the production of more nutritious foods, highly resistant crops and better medicines with fewer adverse side effects. Moreover, the extended nature of USML-1 meant that slower-growing crystals could be produced.

After their return to Earth, the three-dimensional structures of such crystals would be carefully mapped. Crystals of proteins grown in terrestrial laboratories *are* large enough to study, but usually they include numerous flaws, caused by gravity-induced convection, buoyancy and sedimentation. Space-grown crystals, on the other hand, are of greater purity and have more highly ordered structures which significantly improves their X-ray analysis. For USML-1, protein crystal growth experiments were conducted both within an incubator in the middeck and inside the glovebox on the Spacelab module.

The importance of this research to the pharmaceutical industry was illustrated by the inclusion of DeLucas, a biochemist from the University of Alabama at Birmingham, on the crew. He was already recognised as one of the United States' foremost experts in protein crystal growth and USML-1 would be the first time such experiments had been conducted with a human expert in attendance to enhance the production process using a special controllable furnace. Previously, the experiments had largely been automated and the only astronaut involvement had been to mix the chemical solutions.

"Someday," DeLucas said before the flight, "I feel confident we will find a drug that might help to prevent the complications of diabetes, maybe prevent hypotension, maybe find a cure for some types of cancer." It was considered possible that such research may eventually lead to treatments for emphysema and HIV-AIDS. Yet, despite the promising outlook for the experiments, DeLucas noted it would be a decade or more before a breakthrough could be made. "We need a space station," he said. "My investigators don't need one crystal so we can make a breakthrough. We need a constant supply of high-quality crystals."

In addition to the protein crystal growth experiments, a Crystal Growth Furnace (CGF) in the Spacelab module was used to solidify a wide range of materials – mainly semiconductors – which form the basis of electronics devices, including computers, timepieces and audio, video and communications equipment. The furnace could process materials at temperatures of up to 1,260 Celsius and allowed researchers to investigate the factors affecting crystal growth and explore the best methods for producing better crystals. Post-flight analysis of the CGF crystals from USML-1 revealed distinguishable differences from terrestrial-grown samples and showed them to be of much higher purity.

During the mission, the CGF grew crystals using two different methods: 'directional solidification', whereby the solidification front proceeded in a specific direction along the sample, and 'vapour crystal growth', in which part of the sample was heated to make it sublime and the vapour was then allowed to flow towards, and condense upon, a substrate base in a cooler part of the sample ampoule. The USML-1 experiments grew crystals of cadmium telluride, mercury–zinc telluride, gallium arsenide and mercury–cadmium telluride, all of which also found applications in infrared detectors for medical equipment, night-vision goggles and telescope sensors.

The furnace itself, meanwhile, could handle up to six samples simultaneously, each of which was housed in its own quartz ampoule in a rotary changer; these samples were processed under computer control, although the experiments' principal investigators on the ground could uplink changes or adjustments where necessary. A flexible 'glovebox' provided access to the interior of the furnace if any of the ampoules failed, but handling of the samples was only done if absolutely necessary to prevent the risk of contamination to the Spacelab module's environment. It was, however, demonstrated successfully during the flight.

Dunbar activated the CGF during her first shift, late on 25 June, and started the first processing run of a mercury–cadmium telluride crystal by loading six samples into the rotary changer and setting it up for an 18-hour solidification run. "This is the culmination of all those years of study," said Principal Investigator Heribert Weidemeier, whose team had spent a decade developing it. "Nobody can grow as good a crystal on Earth as it is possible to grow in space."

Unfortunately, a circuit breaker tripped early on 29 June and, although Meade followed a procedure to successfully revive it, a mercury–zinc telluride crystal had to be stopped halfway through its growth cycle. Instead, Meade inserted a sample of cadmium–zinc telluride for a 92-hour processing run. The furnace was later used to grow the longest-ever gallium arsenide crystal – measuring almost 16.5 cm – on 4 July. This latter compound contained a 'dopant' – a trace impurity deliberately added to a semiconductor to allow scientists to precisely engineer its properties.

"It is amazing to me that we can replace 10 atoms in a million with 10 atoms of a *different substance* and drastically change a crystal's electronic properties," said Principal Investigator David Matthiesen. "But that tiny impurity makes electronic switching much faster, requiring less power and therefore allows more circuits to be packed into a given area." Gravity-driven convection on Earth makes it very difficult to control placement of these dopants and if too many are concentrated in one spot the crystal could have inconsistent material properties.

Another mercury–cadmium telluride sample was also removed from the CGF on 5 July after six hours of growth. Unlike earlier samples, it was only three-thousandths of an inch thick. Weidemeier told journalists that the wafer-thin sample would allow scientists to study the crystal with as little processing as possible. Several samples were loaded into the furnace using a flexible glovebox unit and, in total, the CGF ran for 286 hours during the course of the mission, processing seven semiconductor crystals – three more than planned – and reaching a maximum temperature of 870 Celsius.

Meanwhile, crystals of zeolites were being grown in the middeck. These complex arrangements of silica and alumina, which occur both naturally and synthetically, have an open crystalline structure that is selectively porous, enabling them to serve as molecular sieves, and they have found terrestrial applications as 'sponges' or 'filters' to selectively absorb elements or compounds. They are used as highly efficient catalysts, absorbents and ion-exchange materials. Scientists sought to grow them as part of efforts to refine gasoline, oil and other petroleum products more cheaply. The zeolites were grown both in middeck lockers and inside the Glovebox in the Spacelab module.

The aim of the USML-1 zeolite experiments was to study the effects on the crystals' morphology of subtle changes in the chemical solutions and the overall reduction of defects. On several occasions, demonstrating the usefulness of having a trained observer present, Dunbar spoke directly to Zeolite Crystal Growth (ZCG) Principal Investigator Al Sacco – who would himself fly on Columbia as a Payload Specialist on USML-2 in October 1995 – to discuss the best mixing procedures for a particular experiment run. "A person up there optimising the mixing of the crystal solution," said USML-1 Mission Scientist Don Frazier, "is a milestone in research."

RCRS PROBLEMS

The EDO system, meanwhile, was performing admirably on its first outing and began supplying cryogenic reactants about 24 hours after launch. Four additional tanks were mounted underneath Columbia's payload bay floor, as well as the four attached to the EDO pallet, and on 26 June mission controllers detected a slight leak in one of them. The quantity of oxygen escaping from the tank was too small to present any problem to the success of the mission, but it was decided to use the leaking tank first to minimise wastage. This procedure was completed on 27 June.

Problems also occurred with the RCRS, which had been switched on shortly after Columbia reached orbit and operated successfully for the first 25 hours of the mission. Then, beginning at 8:26 pm on 26 June, it suffered a series of six failures, forcing the crew to shut it down 90 minutes later. Fresh lithium hydroxide canisters – which absorbed carbon dioxide and were replaced when they became saturated – were installed in the cabin and the Spacelab module. It was found that faulty sensors showing the positions of internal valves caused the unit to prematurely switch itself off. A maintenance procedure to revive the RCRS was devised and rehearsed on the ground, before being transmitted to Columbia's teleprinter for Richards and Bowersox to perform on the afternoon of 30 June.

The 32-step procedure took the two men almost five hours to complete. First, they pulled out four middeck lockers to gain access to the Shuttle's lower deck, under which the RCRS was situated. Next, Bowersox unscrewed its top cover, unplugged one electronic wire connector and spliced two sets of wires together to bypass the faulty sensors. Finally, the system was reactivated at 5:00 pm and performed normally for the rest of the mission. "As it recovers, it's belching and wheezing," Richards told Mission Control as the repair drew to a successful conclusion.

A STEADY MISSION

In general, the mission was proceeding extremely well, both in terms of the research in the USML-1 module and the performance of Columbia herself. A few minor problems did surface, however: the troublesome text-and-graphics system – similar to a fax machine – suffered a paper jam and printed messages, including the RCRS repair instructions, had to be sent up to the Shuttle's teleprinter instead. The crew

Dick Richards and Bonnie Dunbar eat a meal on Columbia's middeck.

also tried out a new space-to-ground modem, which provided an email link between themselves and Mission Control.

While the science crew was busy in the Spacelab module, Richards and Bowersox kept their ship in a 'gravity gradient' attitude – whereby Columbia's tail was pointed Earthwards and her nose 12 degrees 'off' her direction of travel – which provided a steady platform for the many sensitive microgravity experiments on board. This attitude had been used on previous missions and used the natural flow of trace gases in low-Earth orbit to steady the Shuttle and obviate the need to make thruster firings, which could ruin the experimental results.

They did, however, slightly adjust Columbia's attitude on a couple of occasions to expose different parts of her airframe to the Sun to keep them warm and within minimum-temperature requirements. This included, on 2 July, pointing her belly at the Sun for several hours to help to maintain the air pressures and temperatures in her landing gear tyres at acceptable levels. Overall, the crew seemed to be enjoying their longer-than-normal stay in space and each astronaut was given two half-days off-duty to avoid being overworked.

"We're trying to pace ourselves and our timeline," Dunbar had told journalists before Columbia left Earth. "We will come home, hopefully, as refreshed as we left." Now, midway through their mission, Richards was more than happy with USML-1's progress. DeLucas, too, described himself as "relaxed" after his first period of off-duty time. Then, at 1:14 pm on 6 July, the entire crew quietly exceeded Columbia's previous endurance record of just under 11 days, set at the end of the STS-32 mission two-and-a-half years before. Mission Control even played them the Zodiacs' 1960s song 'Stay' in honour of their achievement.

FLUID PHYSICS

Remarkably, in spite of their relaxed state, the crew continued to enthusiastically pursue their hectic schedule of around-the-clock research in the Spacelab module. In addition to the crystal growth experiments already mentioned, a wide range of fluid physics investigations were carried out. These examined the behaviour of fluids under different influences, including the application of heat, in the hope that they could one day be used to produce high-technology glasses, ceramics, semiconductors, metals and alloys from ingredients mixed as fluids and then cooled into solids.

It was already well known that fluid motions on Earth – 'convection', caused mainly by buoyancy – often introduced defects that restricted certain materials from meeting their full potential as lenses, computer chips, turbine blades and other products. In the microgravity environment, the influence of this buoyancy-driven convection is drastically reduced and other, more subtle forces begin to have greater importance in fluid motions. The region at which a fluid interacts with another material is called its 'interface' and is particularly subject to forces causing convection in the fluid.

On 28 June, DeLucas began the Interface Configuration Experiment (ICE) inside the Glovebox. This unit very closely resembled the kind of gloveboxes used in ground-based research laboratories: it contained two openings, through which a crew member could insert his or her hands, and a third port through which experimental samples could be placed inside. Rugged gloves and finer surgical gloves were used, depending on the degree of precision needed when handling each sample. The Glovebox helped to reduce contamination to the Spacelab environment by keeping potentially hazardous materials in an enclosed space.

Most of the sample ampoules destined for use in the Glovebox had magnetic bases, which enabled them to 'stick', and a large plastic window in the top of the unit allowed the astronaut to view the interior. Four cameras provided still and video coverage, which could be transmitted to Earth in real-time. "The Glovebox takes some of the rigidity out of space science," said Robert Naumann of the University of Alabama at Huntsville. "It gives crew members the flexibility to adjust the experiment on-the-spot the way they would in a lab on Earth."

After mounting video and still cameras over the Glovebox for the ICE investigation, DeLucas filled an experiment vessel with an immersion fluid composed of hydrogenated terphenyl and an aliphatic hydrocarbon. This allowed scientists to determine the behaviour of fluids in different-shaped containers in microgravity. On Earth, fluids behave in predictable ways, but it was considered important that their flows in microgravity had to be understood to better design fuel tanks and containers for biological materials or human waste.

Two other major facilities in the Spacelab module were used for fluid physics research: the Surface Tension Driven Convection Experiment (STDCE) and the Drop Physics Module (DPM). The former investigated how fluids reacted in microgravity when there were temperature differences along their interfaces. A lightweight silicone oil was used as the test fluid, because it was not susceptible to surface contamination, which could otherwise ruin such experiments. In total, almost 13 hours of video footage of the fluid were gathered from more than 38 test runs during the mission.

"Understanding this unusual phenomenon", said STDCE Principal Investigator Simon Ostrach of Case Western University, "is important to fundamental science. Fluid physics is the underlying science for everything that happens in microgravity research. We're building a knowledge base to help us to understand the unique laboratory environment of space."

Three STDCE runs, each lasting eight hours, were planned, but in fact it became possible to add a fourth when an opening appeared in the crew's timeline. Typically, the experiment got underway when a command to start its computer was uplinked from the ground; one of the science crew members then powered up the experiment, inserted the heater and filled the fluid chamber with oil. When it had been determined that the oil surface was 'stable', the computer started the experimental run. The science crew could then adjust the settings or pause the experiment where necessary.

For most of the mission, Trinh supervised the bulk of the STDCE runs. The only problem was an inability of the carbon dioxide laser to reach the required power level of 3 watts in 10 of its 12 tests. Furthermore, several tiny bubbles were introduced into the test chamber during the first fill, due to a very small air gap around the heater base. Although these bubbles turned up in the experimental data, they did not adversely affect its outcome. Rather, they actually proved *beneficial*, because their motion – recorded on video – perfectly demonstrated the 'jitter' caused by Columbia's thruster firings.

"Our diagnosis", said Ostrach, after watching 'live' video of the fluid on 30 June, "had an even higher resolution than we expected. I was amazed at how clearly we could see the fluid." Forty-eight-hour intervals were set between each STDCE run – with three conducted on 28 and 30 June and 2 July – to enable researchers to examine and assess their data from each one and choose the most interesting phenomenon to study next. They could also see that fluid motions in both STDCE and the Solid Surface Wetting Experiment (SSWE) pulsated each time Columbia fired her thrusters.

"They correlate!" cried backup Payload Specialist Joe Prahl in delight on 3 July, as he sat at his console at Huntsville watching a video of a 'bridge' of water between the two SSWE injectors. "Every time the thrusters fire, the fluid wiggles! That's amazing!"

Meanwhile, in the DPM facility, containerless studies of droplet behaviour were underway. Materials processed on Earth are often constrained by containers, which can introduce their own impurities and adversely influence their properties. The DPM experiments investigated how water, glycerin and silicone oil drops and bubbles responded to certain forces, including their behaviour when dropped into immiscible (non-mixable) fluids. Using speakers mounted within the chamber, droplets were rotated, oscillated, merged, split and suspended under acoustic pressure.

It was hoped that such experiments could lead to new ways of 'encapsulating' living cells within membranes to protect them from harmful antibodies, which could lead to revolutionary drugs to cure diseases. Encapsulation studies conducted with the DPM on this mission included injections of sodium alginate droplets into a calcium chloride drop. The movement of the fluid was closely monitored by motion-picture cameras. "We saw some things we never expected," said Yale University's Robert Apfel, a DPM principal investigator, after watching one run on 26 June. "We're still flexing the facility's muscles, learning how to manipulate drops in microgravity."

After initial concerns that the amplitude of the speakers was lower than expected, Trinh succeeded on 27 June in suspending drops of water and glycerol – a mixture 50 times more viscous than water alone – for observations of their shape and internal flow. "It's an early study and the rotation is slowing down," Principal Investigator Taylor Wang – who had flown as a Payload Specialist on Spacelab-3 in April 1985 – excitedly told Trinh. "I think we are on the right track, my friend." Later, Dunbar used the DPM's speakers to cause droplets to 'bounce' using soundwaves.

Still other experimental runs succeeded in contorting droplets into squares, rippled on one surface, which drew applause from ground-based scientists. "You made my day," a pleased Dunbar told them. In fact, once control of the levitation speakers had been mastered by the astronauts, Trinh was able to maintain single drops for up to three hours at a time. "We're learning a lot of new things," said

Gene Trinh tends to the Drop Physics Module (DPM).

Wang. "When the time comes to go home, we'll do the hard work – learning what our observations have to teach us. Nature will reveal its secrets to us if we are diligent enough."

COMBUSTION RESEARCH

Other investigations focused on flame behaviour in microgravity. As with many other physical processes on Earth, buoyancy affects the spreading patterns of flames as heated gas rises, drawing in gas from below to replace it. In space, however, such gravity-induced distortions are minimised and their shape and behaviour alter markedly. Studies of how fire behaves in space, and how to control it, have obvious applications in the design and testing of fire-fighting tools for use on board spacecraft and on Earth. On 26 June, DeLucas started the Smouldering Combustion in Microgravity experiment and began videotaping its progress for ground-based investigators.

Meade completed the experiment with its fourth scheduled run on 28 June, igniting polyurethane foam cylinders, sealed in clear Lexan, as sensors recorded temperatures and the spreading rate of the combustion process. "The experiment went very smoothly and we expect to get a lot of information as we analyse our data," said designer Carlos Fernandez-Pello of the University of California at Berkeley. "From what we've seen so far, the results seem to confirm a possible theory about the relative roles of cooling and airflow in smouldering combustion."

After performing another test, the Wire Insulation Flammability Experiment (WIFE) inside the Glovebox, which heated and burned electrical wires, Meade was described as "the coolest guy off Earth" by experiment designer Paul Greenberg of NASA's Lewis (now Glenn) Research Center in Cleveland, Ohio. Other investigations studied the progress of flames along samples of Plexiglas and a special filter paper in an artificial 'atmosphere' containing nitrogen and argon. On 2 July, Dunbar successfully ignited the ashless filter paper in the Glovebox and downlinked video footage to experimenters.

Still more tests were conducted using a set of 10 candles, each measuring 2.5 cm long – similar to the kind that adorn birthday cakes – to examine the behaviour and shape of their flames in microgravity. Composed of about 80% paraffin and 20% stearic acid, the candles were lit by science crew members by pushing electrically controlled igniters through an opening in each candlebox. Immediately after ignition, the candles flared into spherical balls with bright yellow cores; within eight to 10 seconds, however, presumably due to soot, they turned blue and assumed hemispherical shapes about 1.5 cm in diameter.

This data was consistent with short-duration studies on board parabolic aircraft, which were able to simulate microgravity for up to 30 seconds, and even briefer drop-tower tests in the United States and Japan. It was speculated that the change in shape from spherical to hemispherical indicated that the flame was providing heat to the wick. Generally, the flames extinguished themselves within about a minute. "We're overjoyed," said experiment designers Howard Ross and Donald Dietrich of Lewis Research Center. "The observations [the crew] called down were invaluable, since they were able to report things we could not see on the video downlink."

BIOTECHNOLOGY

More than one scientist would be praising not only the excellent efforts of the crew, who Ross and Dietrich said "went far beyond the call of duty", but also the fact that many of the experiments would not have been possible or nearly as successful without humans in attendance. The last major group of experiments on board USML-1 concerned biotechnology, the most important of which were the Generic Bioprocessing Apparatus (GBA) and Astroculture. The former, housed in the Spacelab module, supported more than a hundred separate studies on biological samples ranging from molecules to small organisms.

One very significant experiment studied how collagen – the protein in the fibres of connective tissue, bones and cartilage – forms and how to artificially synthesise it.

Medical researchers hoped that this could lead to the ability to construct strong materials for use as artificial skin, blood vessels and other body parts. The GBA was also used to build spherical structures that might be used to encapsulate pharmaceuticals, study the development of brine shrimp and wasps in space and examine the germination and development of plant seeds.

The Astroculture experiment, on the other hand, was stored in a middeck locker and demonstrated a prototype plant-nutrient system with exciting possibilities for future long-duration trips to the Moon and Mars. It was already known that, on Earth, water flows *downwards* through the soil and is absorbed in plant roots, but an artificial means of providing it would be necessary to grow food for longer missions. During his shifts on the flight deck, Bowersox darted back and forth to keep a close eye on Astroculture, setting up the apparatus to test its water and nutrient delivery, lighting and humidity control systems.

"We're testing the hardware to make sure it is functioning optimally," said Astroculture team member Robert Morrow of the University of Wisconsin. "Before we add actual plants, we want to be sure we have a well-controlled chamber so we can separate gravitational effects from other variables such as temperature and light levels." Two runs of the experiment, both supervised by Bowersox, successfully provided a steady flow of water to 'simulated' plant roots.

A FIT AND HEALTHY CREW

"A technological triumph and a scientific success" was the alliterative description made by Mission Manager Charles Sprinkle as Meade finally deactivated the last of the USML-1 experiments and closed the Spacelab module's hatch early on 8 July. The crew's scheduled landing later that day, after nearly 13 days aloft, was not so assured, however, due to changes of rain associated with Tropical Storm Darby, which lay just southwest of the California coastline. Undeterred, Richards, Bowersox and Baker ran through the routine checks of Columbia's systems, before Mission Control finally waved off their landing attempt until the 9th.

The seven astronauts were surprisingly fit and healthy after two weeks in space and this had much to do with the rigorous exercise regime they followed. The new EDO Medical Project included a special suit to redistribute fluids to their legs and lower torsos and help to counteract the punishing onset of gravity. Known as the Lower Body Negative Pressure (LBNP) suit, it was an inflatable cylinder, 1.2 m tall, which sealed around their waists and drew fluids down into their legs to offset the 'upward' fluid shift that occurs on entrance to the microgravity environment.

"The crewmembers' responses were as expected for this phase of the mission," said LBNP team member Sheila Boettcher after initial runs on 27 June. Later, Meade and Trinh also instrumented themselves with portable cardiovascular monitors to measure their heart activity and the blood pressure in their arms. "On my previous two flights, I didn't exercise at all," Richards had said before STS-50 lifted off. "This time, I'm taking exercise a *lot* more seriously. During entry, you have to have a strong cardiovascular response, particularly during the final phase."

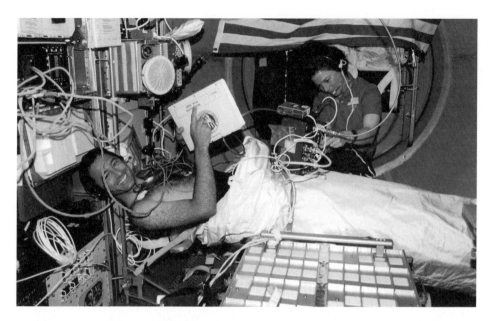

Assisted by Bonnie Dunbar, Larry DeLucas prepares for a Lower Body Negative Pressure (LBNP) experiment.

Bowersox agreed. "What we like here," he said in a pre-flight interview, "are perfect landings. We like to have things very predictable and very controlled, within very tight limits." In addition to their physical exercise, the astronauts drank Florinef, a mineral solution that helped them to retain fluids and a medication prescribed for people who regularly faint when they stand up too quickly – a condition known as orthostatic intolerance. All crewmembers drank a litre of heavily salted water shortly before re-entry to help them to hold bodily fluids in place until they could adapt to terrestrial gravity at a more leisurely pace.

After being waved-off on 8 July, Richards and Bowersox fired Columbia's OMS engines for 30 seconds to better align their orbital path for subsequent landing opportunities over the ensuing days. Flight Director Jeff Bantle had already pointed to his preference to bring the crew into Edwards and land on the long concrete runway, which was much better suited to handle the heavier-than-normal 102,500-kg Shuttle with the Spacelab module and EDO pallet in her payload bay. When the rain at Edwards did not clear by 9 July, however, Bantle opted to land at KSC.

It would be Columbia's first landing in Florida, within sight of the launch pad she had vacated two weeks earlier. Weather was ideal, with light winds and a slight chance of patchy ground fog. The payload bay doors were closed and Richards performed a three-and-a-half-minute OMS burn at 10:41 am to begin the hour-long glide home. The Shuttle could be seen streaking, meteor-like, over Houston at 11:30 am as she headed for Florida. Touchdown on Runway 33 was picture-perfect at 11:42:27 am, with Bowersox deploying the new drag chute as soon as all six wheels were firmly on the concrete.

STS-50 touches down on 9 July 1992, marking Columbia's first landing in Florida and her first use of a drag chute on the runway.

Together with the new tyres, this brought Columbia to a smooth halt within a minute. Lasting 13 days, 19 hours, 30 minutes and 4 seconds, STS-50 established itself firmly as the fourth-longest spaceflight in United States history. It was surpassed only by the three Skylab missions in 1973–74, which spent 28, 59 and 84 days aloft. "I think we got past Gemini 7, but with a *lot* more comfort!" Richards said, referring to Frank Borman and Jim Lovell's arduous mission in December 1965 in a capsule that they likened to the front of a Volkswagen Beetle car.

"You sort of forget what it's like to be back on Earth," Richards told journalists. Flanked by Bowersox, Dunbar and DeLucas, he excused Blue Team crewmates Baker, Meade and Trinh, who had not slept for more than 24 hours. "They just need to get home for a well-deserved rest because it's a very tiring process, not just staying up over 24 hours, but getting used to [gravity] again." Summing up STS-50, he was philosophical: "The human performance in space has always fooled us. The whole history is filled with anecdotal experiences that we were expecting something worse than actually occurred. I think we are just being cautious now as we go into this era of longer duration flights. The smart thing to do is do them incrementally, so we don't bite off more than we can chew and run into something we weren't expecting."

HALF-FULL OR HALF-EMPTY?

Much to the chagrin of her crew, thousands of satisfied scientists and Shuttle director Bob Crippen, Columbia's 13th mission would be forever saddled with the age-old question: 'Is the cup half-full or half-empty?' It has often been posed in situations which, at least in some quarters, are perceived as unworthy of the risks involved. To be fair, the cargo trucked into orbit by the STS-52 crew on 22 October 1992 was a mix of different payloads: an Italian satellite, three experiments on a cross-bay bridge and a myriad of other investigations; even a *quilt* affixed to Columbia's RMS mechanical arm.

The major payload – if it could be called that – for Commander Jim Wetherbee and his crew of five was the second Laser Geodynamics Satellite (LAGEOS-2), part of a collaborative venture between NASA and the Italian Space Agency (ASI). It was essentially a large aluminium ball, 60 cm in diameter, with a dense brass core which gave it a heavier-than-it-looked mass of 410 kg. It had been built in Italy, at a cost of $160 million, using blueprints from NASA's Goddard Space Flight Center, which had launched an identical satellite almost two decades earlier.

Like LAGEOS-1, it was a passive spacecraft dedicated to 'laser ranging'. In other words, laser beams transmitted from Earth would impinge on 426 nearly-equally-spaced, cube-corner retroreflectors, or prisms, embedded in LAGEOS-2's shell. By measuring the round-trip travel time of the beams, it was expected to very precisely determine the distances between ground stations with an accuracy of just a few centimetres. Data from LAGEOS-1 had already revealed that the Pacific tectonic plate, upon which lies the Hawaiian island chain, is slowly drifting, at a rate of a few centimetres per year, northwestwards in the direction of Japan. Additionally, it helped to confirm that the East Pacific Rise – an 'axis' of sea-floor spreading – is separating the Hawaiian islands from South America at approximately the same rate. Movement, or *lack* of movement, on this scale, is important as it enables geophysicists to understand what goes on beneath our feet, providing insights that may allow them to predict the occurrence of earthquakes or volcanic activity. "The satellite may be small, but the data returned is bigtime science," said LAGEOS Project Scientist Miriam Bartuck, adding that it also had applications in understanding our planet's 'wobble' and determining more accurately its size and shape.

The solid, outermost layer of Earth – measuring up to 100 km thick in places – is known as the 'lithosphere' and is composed of vast, rigid 'plates' which float on a semi-liquid region called the 'mantle'. These plates move with respect to one another, but generally at speeds no higher than a few centimetres per year. This motion, dubbed 'plate tectonics', is the means by which North and South America are presently moving away from Europe and Africa, and although these motions are incredibly slow, they can be catastrophic when collisions occur.

Occasionally, tectonic plates bump into one another, spread apart or scrape horizontally past each other; many of the world's great mountain chains, earthquake-prone areas and volcanic hotspots are the consequence of this activity. One tectonic model which, it was hoped, LAGEOS-2 would address was the theory that the coastal strip of southern California is migrating slowly northwards and will cataclysmically collide with Alaska in about 150 million years' time. This research has enabled geologists and geophysicists to better understand *how* Earth's continents arrived at where they are today, where they were in the past and where they might be in the distant future.

A SATELLITE FOR THE PAST, PRESENT AND FUTURE

Present theories argue that, around 200 million years ago, a giant 'super-continent' known as 'Pangaea' (Greek for 'all lands') existed in the middle of a vast ocean which covered 70% or more of our planet's surface. Within the next 20 or 30 million years, this gigantic landmass gradually began to break apart into several 'smaller' continents and by 150 million years ago cracks had begun to emerge in the most northerly of these new landmasses. Molten rock, coming from deep within the mantle, poured up through the cracks and pushed the plates further apart.

Carrying the fragments of the super-continent with them, these plates gradually began to drift apart, leaving newly created sea-floor between them. Moving no faster than a few centimetres per year, the continent we now know as North and South America is today 4,800 km away from the one we call Eurasia. Rifting is still going on. For example, the East African Rift, where the continent is in the process of splitting apart, will ultimately form an island from what is presently its 'east coast'.

Modern maps offer tantalising hints of Pangaea's existence, suggesting that parts of West Africa might slot neatly, jigsaw-like, into the coastline of North America and Brazil. However, since coastlines are affected by sea-level and hence are subject to change over extended periods of time, the lines of the continental 'shelves' have become a more reliable marker of where the landmasses were pulled apart.

In order to track these slow tectonic movements, LAGEOS-2's entire aluminium skin was literally covered with the 426 retroreflectors, which gave it the appearance of a gigantic golf ball. Each one had a flat face and a prism-shaped 'back'; the bulk of them were made from suprasil – a fused silica glass – and four from germanium. Its aluminium-and-brass design evolved from a series of trade-offs: on the one hand, it had to be heavy enough to minimise the effects of non-gravitational forces, yet light enough to be placed into a high, stable orbit around the Earth.

Preparations to launch the satellite had been going on for some years, but moved into high gear on 29 September 1992 when LAGEOS-2 was installed into Columbia's payload bay at Pad 39B. Attached to Italian Research Interim Stage (IRIS) and LAGEOS Apogee Stage (LAS) boosters, it proved an unusual sight as it was transferred from its transport canister to the Shuttle. The whole ensemble was housed in a Pacman-type cradle like that used on STS-5 and STS-61C. Almost seven years after Challenger, there was a sense that STS-52 harked back to the Shuttle's early days of launching commercial satellites.

TESTING NOBEL PRIZE SCIENCE

In fact, right up until the eve of Columbia's 22 October liftoff, NASA would continue to defend its decisions – many of which, to be fair, had been made long before the Challenger disaster – to launch such a small cargo. The questioning did not just relate to the presence of LAGEOS-2, either, for the Shuttle also carried the first United States Microgravity Payload (USMP-1). Not to be confused with the large Spacelab flown on Columbia's previous mission, this, too, was a small cargo which some critics argued should have been flown alongside a 'major' payload, or not at all.

Did the presence of what were essentially two *secondary* payloads require an entire mission, they asked, with all the risks involved, not to mention the enormous pricetag, which ran into several hundred million dollars? NASA's reply was that its commitment to ASI to launch LAGEOS-2 had been signed before Challenger, and the agency intended to honour it. Questions over the worthiness of USMP-1 to fly on Columbia as a primary payload were best answered by Al Diaz, NASA's deputy associate administrator for space science, who rhetorically asked journalists, "How do you determine how much Nobel Prize science is worth?"

Diaz was referring in particular to one of USMP-1's three investigations: the Lambda Point Experiment (LPE), which sought to test a complicated theory of the thermal conductivity of liquid helium which had won mathematician Kenneth Wilson the Nobel Prize in 1982. It predicted that, under the right pressure and conditions, liquid helium would conduct heat a thousand times more efficiently than copper, but it could not be realistically tested on Earth because of gravitational interference. Wilson's theory was considered to have wide-ranging applications from studies of hurricane dynamics to superconductivity.

When activated, the LPE examined liquid helium as it changed, through a transitional phase known as its 'lambda point', from a 'normal' fluid into a substance called a 'superfluid'. In this latter phase, the helium moves freely through small pores that block other liquids and conducts heat far more efficiently than copper. The LPE investigation cooled a liquid helium sample far below its minus 268 Celsius lambda point and then, during a series of two-hour, computer-controlled experiment runs, its temperature was slowly raised and measured with an accuracy of less than a billionth of a degree throughout the transitional phase.

"To visualise how precise that is," said LPE designer Reuben Ruiz of NASA's Jet Propulsion Laboratory (JPL) in Pasadena, California, "think of the distance between Los Angeles to New York City as one degree. On that scale, one billionth of a degree would be about the thickness of a human hair!" Joining the LPE investigation on the USMP-1 payload were two other experiments: the Materials for the Study of Interesting Phenomena of Solidification on Earth and in Orbit (a French-language acronym which spelt 'MEPHISTO') and the Space Acceleration Measurement System (SAMS).

The former was a cooperative US/French project to study the behaviour of metals and semiconductors as they solidified to help determine the effects of gravity on the point at which the liquid met the solid (its so-called 'solid/liquid interface'). It was hoped that such research would lead to the development of more resilient metallic alloys and composites for future aircraft engines and turbine blades. For its first trip into space, MEPHISTO was loaded with three identical, 15 cm-long tin–bismuth alloy samples, which were processed by one fixed and one moving furnace.

On Earth, buoyancy-induced convection and differences in hydrostatic pressure affect how materials solidify, as well as masking several of the key processes in the solidification process. During the USMP-1 mission, sensors attached to the furnace accurately measured the temperature and shape of the solid/liquid interface and determined the speed at which it moved through the tin–bismuth sample. The third experiment, SAMS, was really a sensitive accelerometer which recorded and tracked the effect and severity of Columbia's thruster firings on LPE and MEPHISTO, in order to assess the purity of the microgravity environment during the experiments.

All three were attached to a pair of Mission-Peculiar Equipment Support Structure (MPESS) carriers, which straddled the Shuttle's payload bay like a bridge and was derived from the Microgravity Science Laboratory flown on STS-7 and STS-61C. The 'front' MPESS provided electrical power, data, communications and thermal-control services to the payload, while the experiments were mounted on the 'rear' one. Managed by NASA's Marshall Space Flight Center, the USMP system was essentially autonomous and required little crew interaction, other than switching it on and off; moreover, it also provided exciting opportunities for scientists to control their experiments from the ground, via 'telescience'.

"This is an excellent use of the Shuttle to perform microgravity experiments that are primarily operated remotely from the ground," said USMP Project Manager David Jarrett, adding that the ability to exercise such control would provide scientists with useful experience as NASA prepared to build a permanent space station in the mid-to-late 1990s. In terms of the kind of research being conducted, USMP could be envisaged as an 'automated' version of USML-1 flown on Columbia's previous mission, since it too was designed to capitalise on the United States' lead on worldwide microgravity research.

FLIGHT RULES OVERRULED

So it was, with this 'half-full-or-half-empty?' question hanging over the mission, that Wetherbee led his crew to Pad 39B early on 22 October 1992. Launch was originally planned for a week earlier, but had slipped because of problems with Columbia's Number 3 main engine; concerns arose over possible cracks in its liquid hydrogen coolant manifold. It was decided that it would be less time-consuming to simply replace the entire engine, rather than conduct a painstaking X-ray analysis of the damaged area. The engine was pulled out on 29 September and a replacement had been installed and tested by 2 October.

The mission would mark Wetherbee's second trip into space; his previous flight, STS-32 in January 1990, set a record of almost 11 days. Yet he was still at the start of a glorious career that would see him fly six missions in total, including two flights to the Russian Mir space station and *another two* to the International Space Station. In doing so, he also became the first – and, so far, only – American astronaut to command *five* space missions.

Not until the end of 2002, only weeks before Columbia's fiery loss during re-entry, would he hang up his helmet for the last time. His Pilot on STS-52, Mike Baker, was also making his second spaceflight, having occupied the same position on a nine-day mission the previous summer, which had deployed an important communications satellite for NASA. Mission Specialists were Lacy Veach, Bill Shepherd – later to become the International Space Station's first skipper – and Tammy Jernigan. All three had flown before: Shepherd twice, Veach and Jernigan once each.

The only member of Wetherbee's crew who would be savouring his first experience of flying into space on this mission was a Canadian physicist named Steve MacLean. He had been selected as a Payload Specialist only about eight months before Columbia was due to liftoff, and the bulk of his time on the planned 10-day mission would involve operating a battery of experiments provided by his own country, including an important space-vision system which was being evaluated for use during the construction of the International Space Station.

Even though their launch had been delayed by a week, when the crew finally set off at 5:09:39 pm on 22 October, they still set a new post-Challenger record for the shortest interval between two flights by the same Shuttle. Columbia had returned to Earth from her previous mission, STS-50, only 105 days earlier. Nevertheless, launch on the 22nd was delayed by almost two hours due to unacceptable crosswinds at the SLF runway, which would have been used if problems during the early part of Columbia's ascent had forced Wetherbee and Baker to perform an emergency landing in Florida.

However, after discussions between members of the Mission Management Team, chaired by former astronaut Brewster Shaw, it was decided that although the 37-km/h SLF crosswinds exceeded flight rules (which stipulated speeds no higher than 27 km/h), they were 'safe' enough for Columbia to go. Ascent Flight Director Jeff Bantle's reservations were overruled and Shaw decided, based on simulations that showed that Wetherbee and Baker *could* brake the Shuttle safely to a halt if necessary in the faster-than-desired winds, to proceed with the launch attempt.

"We accepted Jeff's recommendation," Shaw said later at a post-launch press conference, "based on his interpretation of the guidelines, and made a management decision that went in a different direction." Fortunately, an abort landing was not necessary and Columbia followed a picture-perfect ascent trajectory, inserting herself into a 302 × 296 km orbit after a single OMS burn. The veteran flagship's 13th mission, which, by chance, would also be aloft over Hallowe'en, was underway as the crew doffed their pressure suits and got ready for 10 days of intensive activity.

LAGEOS-2 DEPLOYMENT

Although the mission was not quite as long as the two weeks Columbia spent aloft during STS-50, several EDO components remained on board: most notably the

LAGEOS-2 leaves Columbia's payload bay on 23 October 1992.

RCRS carbon dioxide-removal system, which had behaved erratically during its maiden outing. On STS-52, by complete contrast, it operated perfectly: the astronauts switched it on at 6:11 pm, an hour after reaching orbit, and it flawlessly scrubbed the cabin's atmosphere of exhaled carbon dioxide for the remainder of the mission. During the first day of their mission, the crew activated several scientific experiments stored in Columbia's middeck.

The first major objective, however, was deployment of LAGEOS-2, under Jernigan's supervision. "Houston, we see a good deploy," she told Mission Control as the satellite and Columbia parted company at 2:57:24 pm on 23 October. Forty-five minutes later, the IRIS booster – built by the Italian Alenia Aerospazio company – fired to deliver LAGEOS-2 into a 297 × 5,923 km orbit, inclined 41 degrees to the equator. Then, at around 5:30 pm, the satellite's own LAS motor fired to insert itself into a near-circular orbit of 5,617 × 5,950 km, inclined at 52.6 degrees.

Overall, the procedure was almost identical to that followed for the deployment of the SBS-3, ANIK-C3 and Satcom Ku-1 satellites in Columbia's early days: the Pacman-like sunshade was opened and a turntable imparted the required 60-rpm rotation rate on LAGEOS-2 before pyrotechnic bolts fired and spring actuators pushed the payload free. Meanwhile, Veach used cameras on the RMS to monitor the ignition of the IRIS and Wetherbee and Baker fired the Shuttle's OMS engines to lower her orbit to 287 km to maintain a safe separation distance and better support USMP-1 operations.

Entirely appropriately, the six astronauts were awakened on 24 October to the strains of Joe Turner's 'Shake, Rattle and Roll', in honour of the plate motions that LAGEOS-2 was expected to measure. The satellite, according to Miriam Bartuck, had already been successfully tracked by four ground-based laser stations by 2:00 pm that day. During the course of its mission, it has been tracked on a more-or-less-daily basis by 10 NASA-operated Satellite Laser Ranging (SLR) stations and 35 National Satellite Laser Ranging Committee coordinated sites around the world.

Unlike LAGEOS-1, which had been inserted into a 109.9-degree orbit, the inclination of its successor was chosen to enable it to provide better coverage of the seismically active Mediterranean basin, as well as California. It was also hoped to clear up several irregularities that had cropped up with LAGEOS-1's orbital position; these seemed to be linked to erratic spinning of that satellite. After reaching its correct orbital slot, LAGEOS-2 underwent a month-long checkout to precisely calculate its orbit by laser ranging, after which scientific operations got underway in earnest.

"[The satellite] will reduce the problems caused by cloudy weather and will enhance our modelling of Earth's gravity field," said Bartuck before the mission. "Its orbit has been designed to improve coverage of the Mediterranean, a poorly understood and geologically complex area that is naturally of great interest to our Italian partners."

In fact, LAGEOS-2's orbit is so stable that it will not be dragged into Earth's atmosphere for perhaps eight million years! It should remain 'operational' for about half a century and the only limiting factor is the degradation of its retroreflector prisms. In view of the long time it would spend aloft, NASA asked planetary scientist Carl Sagan – who chaired the team which produced the Voyager plaques – to make a stainless-steel disk for LAGEOS-2. This contained three maps of Earth: as it appeared 268 million years ago, how it looks 'today' and how it may be eight million years hence.

Interestingly, the 'future' map includes the eventual drift of California way out into the Pacific Ocean. "A being need only compare its view of Earth to the depictions of Earth's past, present and future to determine its place in time relative to the Earth's," said NASA spokesman James Hartsfield before STS-52 set off. Plans already exist for a third LAGEOS, which will involve collaboration between the

United States, Italy, France, Germany, the United Kingdom and Spain, although a launch date remains to be set and, almost certainly, it will ride an expendable rocket rather than the Shuttle.

'TELESCIENCE' PUT TO THE TEST

When asked by journalists during a space-to-ground press conference what he felt about the risks associated with launching such a small cargo, Wetherbee replied, "If we want to get this great science, we must take a certain amount of risk. I think this science is worth going after." With the LAGEOS-2 deployment behind them, the crew could now focus their entire attention on more than a week of medical and scientific experiments in Columbia's payload bay and her middeck. Although those attached to USMP-1 were the most visible, the astronauts actually had little involvement in their operation.

In fact, their job, two-and-a-half hours after reaching orbit, was simply to turn the payload 'on'. The two experiments and SAMS then operated autonomously, with regular adjustments sent up from ground controllers using 'telescience'. "They're getting better data than they expected," said David Jarrett on 25 October, in reference to the excellent performance of both LPE and MEPHISTO. "The key element of USMP-1", added Assistant Mission Scientist Martin Volz, "is that we've been getting data from our experiments while they're in orbit and the science teams have been able to make adjustments on that basis."

After being activated, LPE underwent a day of calibration of its high-resolution thermometers, which verified their ability to measure heat capacity with pulses of a billionth of a watt. "This is a very critical step," said Reuben Ruiz as he watched LPE begin operations, "since it demonstrates our ability to conduct the experiment." Early results showed that the unit had far higher data resolution than could be achieved on Earth. "The operation of the experiment was initially in the automatic mode, with very little commanding from the ground," said LPE Principal Investigator John Lipa of Stanford University on 26 October.

"Now that the 'transition region' [between the 'fluid' and 'superfluid' states] has been reached, we are operating almost entirely in the interactive mode. This allows us to respond to events in near-real time." He noted that, even before the halfway mark of the 10-day mission had been reached, more than 600 commands had been transmitted to the experiment. By 30 October, as USMP-1 operations drew to a close, Lipa was telling journalists that his team had acquired more than 300% of the high-resolution temperature data they needed to rigorously test Kenneth Wilson's Nobel Prize-winning theory.

Meanwhile, the MEPHISTO researchers also calibrated their experiment for 24 hours, heating it up to 343 Celsius and checking the tin–bismuth samples to ensure that they were flawless and which segments of them were best-suited for subsequent analysis. On 24 October, co-investigator Andre Rouzard said his team had observed some phenomena that were entirely different in space from those seen in comparable terrestrial experiments. "At first look, they seem to confirm theoretical models of what we thought would be happening in microgravity," he said. "We are very pleased with the evolution of the experiment."

As LPE and MEPHISTO continued their own research, SAMS was also gathering data on the impact Columbia's minute accelerations and thruster firings were having on the delicate experiments. Data-collecting 'heads' for the device were mounted close to both experiments. Early in the mission, Wetherbee and Baker performed a series of manoeuvres to calibrate the accelerometer and, during one OMS burn late on 23 October, the data from MEPHISTO's sensors clearly detected a 'blip' caused by vibrational disturbances. Accelerations were also recorded during LAGEOS-2's deployment.

"The data will help to define how extensively Shuttle manoeuvres affect crystal growth," said MEPHISTO's Assistant Mission Scientist Don Gillies on 27 October. "That information will be extremely valuable in planning future experiments." In addition to measuring Columbia's vibrations, the accelerometer picked up virtually every other movement the crew made too, as the SAMS team told Wetherbee on 30 October. "We've been keeping an eye on you during your exercise programmes, waving the RMS, launching the satellite and", making a light-hearted reference to the Commander's exploits as drummer in the all-astronaut rock band 'Max-Q', "your drum solos!".

PREPARING FOR SPACE STATION OPERATIONS

As USMP-1 operated autonomously in the payload bay, Columbia's astronauts were busy with a myriad of other experiments in the middeck. MacLean, for example, had his hands full throughout the mission with the second set of Canadian Experiments (CANEX-2) – the first set had been conducted by Payload Specialist Marc Garneau on STS-41G in October 1984 – to such an extent that Wetherbee asked Mission Control to give him some free time. "Steve's doing a great job," he told them on 27 October. "He wants to do it all. It's my call to offload him a little bit."

CANEX-2 consisted of 10 space science, technology, materials processing and medical experiments, provided by the Canadian Space Agency (CSA). Of these, one of the most exciting as the construction of the International Space Station loomed on the horizon was the Space Vision System (SVS), which helped to improve astronauts' perception of large structures under unfavourable viewing conditions. It was already known that, in space, there are frequent periods of extreme darkness and lighting and very few reliable points of spatial reference. This had led astronauts to comment that it was difficult to accurately judge distances and speeds.

During STS-52, MacLean evaluated a prototype vision system for space station proximity operations and assembly tasks, which provided him with data on the precise positions, orientations and motions of small 'target' objects. It employed a television camera on Columbia, which monitored a series of dots on what looked like a giant, 2-m-long 'domino', known as the Canadian Target Assembly (CTA). As the object moved, the computer in the SVS measured the ever-changing positions of its dots, providing MacLean with a real-time display of their locations and orientations.

The domino-like Canadian Target Assembly (CTA) is manoeuvred by the RMS.

"If our tests are successful, we could perhaps speed up the assembly of the space station," the Canadian astronaut had told journalists before Columbia lifted off. "What we have tried to do is mirror the assembly of the station. The controllability of the [RMS] arm is very good now, but the position knowledge of the end of the arm is only good within two inches. That makes it difficult to do many of the operations that are planned for the Space Station." Veach added that "this could be as revolutionary as 'instrument flying' was to aviation".

The comments from both men, in hindsight, were interesting; one was never to see the construction of the International Space Station, while the other will be on board one of its pivotal missions. Veach, sadly, would die of cancer within three years of returning from STS-52, while MacLean would join NASA's astronaut corps in April 1996 to train as a fully fledged Mission Specialist. At the time of writing, he is set to fly on STS-115, some time early in 2006, which will attach a massive girder and new set of electricity-generating solar arrays to the station.

During *that* mission, expected to last around 10 days, MacLean will operate 'Canadarm2' – the International Space Station's 'own' RMS – and use a descendant of the space vision system he first tested on STS-52 to very precisely pick up, manoeuvre and install the girder and arrays onto the outpost. All that, of course, was far in the future when Veach uncradled the RMS on 24 October 1992 and used it to grapple, move and replace the 82-kg CTA in Columbia's payload bay.

Similar tests were performed over the next few days and, on 29 October, Veach lifted the domino and rolled it from side-to-side, then up-and-down, while MacLean closely monitored its movements with the SVS. These evaluations demonstrated the clearance-measuring abilities of the system, which was expected to be particularly useful during Hubble Space Telescope repair missions or station assembly tasks. Things went smoothly until Veach tried to lower the CTA back onto its berthing plate in the payload bay, when he met resistance.

After inspecting the target's berthing latches, he successfully lowered it into place on the second attempt. NASA spokeswoman Barbara Schwartz said, however, that the incident would not prevent further SVS runs planned for the end of the mission. Indeed, on 30 October, Jernigan joined Veach and MacLean in a set of tests to judge the amount of 'flex' in the RMS, guiding the CTA through a series of very precise manoeuvres. Then, at 11:05 am the following day, Veach finally jettisoned the domino and Wetherbee flew Columbia in formation at a distance of 42.6 m.

MacLean, who watched the departing target with the SVS, described the unit's performance as "excellent" and accurately measured the increasing distance between the Shuttle and the domino. As it drifted away, destined finally to burn up in the atmosphere on 1 November, Wetherbee 'blasted' the CTA with Columbia's RCS thrusters.

Other CANEX-2 experiments supervised by MacLean included a high-temperature furnace called the Queens University Experiment in Liquid Metal Diffusion (QUELD), capable of reaching 900 Celsius, which examined the 'diffusion' of bismuth and tin into one another. Knowledge of the rates at which atoms of different substances move around and between each other – their 'diffusion rates' – had long been considered important for industrial purposes. Such rates are difficult to examine in ground-based experiments, because convection masks the actual degree of diffusion taking place, but in microgravity is was possible to obtain more precise data.

Another intriguing experiment, the Material Exposure in Low-Earth Orbit (MELEO), took the form of a 'quilt' of about 350 different samples attached to the RMS. These materials were then analysed after Columbia's landing to assess the adverse impact of atomic oxygen exposure. It was already known from earlier materials-exposure experiments flown on the Shuttle that plastics and composites could become degraded. Materials flown on STS-52, which were exposed to space on three extended occasions, were candidates for use on the space station's robotic arm and also Canada's Radarsat remote-sensing platform.

The crew also tended a variety of other experiments. A Canadian-provided Spectrophotometer measured the light-absorption characteristics of Earth's atmosphere, while MacLean took photographs of the Shuttle's tail and flight surfaces during thruster firings in an effort to better understand the mysterious atomic oxygen-induced 'glow' around spacecraft. Heat pipes were tested as part of efforts to develop lighter, more efficient thermal-control systems for future spin-stabilised satellites and Bill Shepherd monitored a Boeing study called the Crystals by Vapour Transport Experiment (CVTE), which produced large, high-quality, electro-optical cadmium–telluride crystals.

"This experiment is important to the semiconductor industry because the ability of semiconductors to process and store information is dependent on the quality of the crystals used," said Boeing's CVTE Project Manager Barbara Heizer before the flight. "Large, uniform crystals grown during spaceflight may lead to greater speed and capability for computers, sensors and other electronic devices."

Medical experiments were also conducted during the longer-than-normal, 10-day mission. One, the Physiological Systems Experiment (PSE), studied the effects of hormone therapy on changes in rats in the microgravity environment. "The goal of this experiment", said Roy Walker of designer Merck and Company of Pennsylvania, "is to see if an experiment compound we're developing will prevent or slow osteoporosis from developing in microgravity during spaceflight. If it does, the compound may be a useful treatment for many people on Earth who suffer bone loss from being bedridden for long periods of time due to accidents or paralysis."

Twelve adolescent male albino rats (*Rattus norvegicus*), accommodated in a pair of Animal Enclosure Modules (AEMs) in Columbia's middeck, were the test subjects for the PSE experiment. They were matched by body mass into pairs, and one member of each pairing was given subcutaneous injections of an anti-osteoporotic protein two days before launch, followed by a second injection of a bone marker called calcein. The other rats were left untreated. Other medical investigations included a Canadian study of back pain, which had been experienced by several astronauts.

During a Spacelab mission in January 1992, Canadian Payload Specialist Roberta Bondar conducted the first in-depth studies of this back pain and found it to be at its worst during the first few days in orbit. Her research suggested that it may be attributable to spinal increases of up to 7.5 cm, caused by a lengthening of the spinal column and of the normal spinal 'curves'.

This, investigators theorised, may itself have been due to increased water content and thus the height of the discs between the spinal vertibrae. The resulting tension on soft tissues such as muscles, nerves and ligaments might then lead to the occurrence of back pain. Throughout STS-52, MacLean carefully measured his height and used a special diagram to record precise locations and intensities of back pain.

COLUMBIA RETURNS 'HOME' TO KSC

After what Shuttle director Bob Crippen called a mission "chocked full of work", Wetherbee, Baker and flight engineer Shepherd ran crisply through the routine checkout of Columbia's systems on 31 October. The payload bay doors were closed at 10:32 am the following morning, and two-and-a-half hours later Wetherbee performed the two-minute OMS burn to drop his spacecraft into the upper reaches of Earth's atmosphere. There were, however, a few drama-tinged moments during the Shuttle's fall towards Florida, when a cockpit indicator stopped working.

The indicator was particularly critical, because it was used to display information for the pilots on the positions of Columbia's elevons and other steering surfaces. Wetherbee and Baker switched to small digital displays as a backup measure and temporarily recovered the use of the indicator, before it again failed. Post-flight analysis of the indicator would blame a faulty fuse. Fortunately, the crew's return to Earth was otherwise uneventful and the Shuttle touched down on Runway 33 at KSC at 2:05:52 pm, wrapping up a mission that had lasted 9 days, 20 hours, 56 minutes and 13 seconds.

Despite being Columbia's 13th trip into space, and having been in orbit over Hallowe'en, STS-52 had proved a spectacular success. In fact, the mission lasted almost exactly the same length of time predicted in the pre-flight press kit. It was the second landing in Florida for NASA's flagship Shuttle; for more than a year, KSC had officially been on 'equal footing' with Edwards as the primary end-of-mission landing site. In June 1991, NASA had announced its intention to routinely use KSC: among other reasons it saved the agency the million-dollar cost of ferrying Shuttles back from California.

"We will treat them [Edwards and KSC] as equal landing sites," spokesman James Hartsfield had said at the time, "with neither preferred over the other." The overwhelming use of Edwards since the post-Challenger resumption of missions in

The STS-52 crew on Columbia's aft flight deck. Front row (left to right) are Bill Shepherd, Steve MacLean and Mike Baker and back row are Jim Wetherbee, Tammy Jernigan and Lacy Veach.

September 1988 had been primarily due to its long lakebed runways and stable weather, which offered a more forgiving landing environment. Missions which experienced technical or mechanical difficulties while in orbit, as the crew of Atlantis did on STS-44 in November 1991, would continue to be directed to Edwards, however, to provide their pilots with greater safety margins for error.

The downside of landing at KSC, on the other hand, was the often-unpredictable Floridian weather and the fact that its narrow all-concrete runway had contributed to brake failures and substantial tyre damage on previous missions. Nevertheless, beginning with Atlantis' STS-43 landing in August 1991, KSC played host to no fewer than 6 out of 10 Shuttle touchdowns, including that of Wetherbee's crew. There were still problems, however, with the new drag chute, which was meant to assist braking on the runway. During Endeavour's STS-47 touchdown in September 1992, it had 'dragged' the spacecraft to one side during rollout.

In an effort to acquire extra data on the chute's performance, Baker deployed it a couple of seconds before Columbia's nose gear hit the runway. The crew detected a perceptible tug into the wind as the canopy billowed out, pulling the Shuttle 4.6 m to the 'left' of the 90-m-wide runway's centreline. At the post-flight press conference, Wetherbee told journalists there were still problems with the chute, but that overall "I would put my stamp of approval on [it]. As soon as we work out these minor difficulties, the system will be ready to go operationally."

His remarks would lead NASA to restrict the use of the chute to near-calm wind conditions. "Certainly, if we had crosswinds of only five knots or less, it would be okay to live with," said Wetherbee. "It didn't cause me any concern. If we were landing on a very narrow runway like an [emergency] runway over in Africa, and it pulled even more, then it would be cause for a little bit more concern."

"YOU WOULDN'T BELIEVE ME!"

Columbia's next flight, STS-55 in the spring of 1993, was an international one which featured the cooperation and involvement of many hundreds of scientists, engineers, technicians and managers from the United States, Europe and elsewhere. However, although it might be expected for this cooperation to extend to space centres, universities and scientific laboratories around the world, one would probably be forgiven for not adding to that list of prestigious educational and astronautical establishments ... *the carpark of a Safeway supermarket!*

Yet that is exactly where the progress and scientific accomplishments of the STS-55 mission was discussed one day in May 1993 between a shopper and one of his friends. Not too much unusual about that, bar one thing: the shopper's 'friend' happened to be astronaut Steve Nagel, the Commander of Space Shuttle Columbia, who had arranged before blasting off to give him a quick call on his hand-held amateur radio. Although chatting to his friends was by no means the main part of Nagel's mission, it was a free-time activity that the licensed amateur radio operator enjoyed while in space.

"I looked at the flight plan [before the mission]," he told an interviewer in

December 2002, "[and told my friend] 'If I get time, I'll call you on the radio'. I gave him the number for Mission Control, to call a contact back there to update that time. [They] said 'Okay. We'll try it'. So I call[ed] him, as it's just as clear as you and me talking here, for about four or five minutes, and we had a good conversation. He was standing out in a parking lot by a Safeway grocery store, with this little hand-held radio. And this guy [in the parking lot] comes up to him afterwards and said, 'Who were you talking to?' And he said, 'Well, if I told you, you wouldn't believe me!'" The mission the two friends had spoken about was, for the newly unified Federal Republic of Germany, one of its most important and expensive scientific ventures, costing over half a billion dollars to develop. In order to accomplish the mission, Columbia was again outfitted with a Spacelab module in her payload bay, which had been equipped with no fewer than 88 medical, radiation, materials science and technology experiments.

MINI-MARATHONS

"I'm not fluent in German," said astronaut Jerry Ross, who was on this mission. "Fortunately, most of the international science people work in English anyhow, because you've got all the other languages in Europe that have to find some common language and, fortunately, *that's it!* I tried to learn some German, but I'm not good in foreign languages to start with and trying to do all the other things I was doing, there wasn't time to learn much German."

Designated 'Spacelab-D2' (for 'Deutschland'), it was the second in what originally should have been a series of three missions sponsored by the former West Germany. An earlier flight, Spacelab-D1, had been undertaken with great fanfare and success in the autumn of 1985, only a couple of months before the Challenger disaster. Both it and the D2 mission were expected to pave the way for joint US/West German research on Space Station Freedom sometime in the 1990s. However, unlike the three- to six-month stints that astronauts would undertake on the station, Nagel has described week-long Spacelab missions as scientific mini-marathons.

"You want to load them up as much as you reasonably can," he said, "because you want to get as much for your money, but they're like sprints. You go hard at it for a week or 10 days, and you can't do that on [the] space station. You've got to have some time to back off a little bit, because you can't keep a sprint up for a long time." Nagel was the only astronaut to fly *both* Spacelab-D1 and D2 and had experienced firsthand the hectic dual-shift, around-the-clock pace of such missions.

Such a pace would be entirely inappropriate for space station flights and when NASA flew, in succession, seven of its astronauts to the Russian Mir space station in 1995–1998 to gain experience of long-duration missions, the agency came face-to-face with the immense physical and psychological hurdles involved. Several astronauts returned to Earth totally exhausted after what they described as an "ordeal". It forced a major rethink of NASA's strategy of sending its astronauts – who had previously trained for intensive, fast-paced one- or two-week-long Shuttle missions – into orbit for months at a time and expecting them to perform in the same way.

In the early 1990s, as these space station plans gradually began to crystallise, Spacelab was seen as useful practice for the kind of working environment astronauts would have on a permanent outpost. Indeed, it *was* an excellent research platform, and its dual-shift system to run experiments around-the-clock maximised a crew's productivity in orbit, but it did not cater for other demands that long-duration flights would place on them. Nagel's colleague on Spacelab-D2, Payload Commander Jerry Ross – a veteran of three previous Shuttle missions – would also describe the training as excessively hectic and "not a very viable way to do business".

By the time Columbia reached orbit in the last week of April 1993, Ross and three of his crewmates – Mission Specialist Bernard Harris and two German Payload Specialists, Ulrich Walter and Hans Schlegel – had been training for almost two years and spent the bulk of that time flying backwards and forwards between the United States and Germany. "We spent a long period of time away from family and friends," Ross recalled at a pre-flight press conference in February. "That is not the nicest way to have to do business over extended periods of time."

He added that, during space station training – which would involve cooperation not only with Germany, but with other European countries, together with Canada, Japan and probably Russia – there was a very real risk of burning crews out if they were constantly 'on the road'. "You will have to hire some additional divorce lawyers to keep up with all the families that are splitting up," he said darkly.

AN IMPORTANT MISSION FOR GERMANY

Ross, an astronaut since 1980, knew exactly what he was talking about: not only had he already flown three times before STS-55, but he would also fly *three more* times after the German mission, becoming the first human being to chalk up seven trips into space overall. His expertise would also, after his return from Spacelab-D2, be rewarded with the post of deputy chief of NASA's Astronaut Office. Yet even Ross has described *this* mission as one of the most demanding of his career.

"The Payload Commander is the guy that's responsible basically for interfacing with the payload sponsors and the crew to make sure that [what] the payload sponsors want to happen on orbit is things that the crew can physically do, both from the interfaces to the payloads, the checklists [and] the timeline. I had to do all the coordination, all the dealings – everything – with the safety community, the medical community, the science community. I had to work all that in addition to trying to get three rookies ready to go fly on 90 different, very complex experiments."

It was no mean feat. Nagel had also flown three times, including a previous mission with Ross and also as Pilot on Spacelab-D1. "They [the Germans] were hoping he'd come back and fly again on D2," said Ross. "I was also hoping he may come back and we could fly together again, so I could harass him some more! So he said, yes, the system thought that would be a good thing to do to have that kind of continuity and he [came] back as Commander."

It was becoming common practice on complex research flights like this one to 'carry over' a crew member from one mission to the next to provide expertise and ensure a smooth transition. Similarly, Ken Bowersox – who flew as Pilot on STS-50, the first United States Microgravity Laboratory – would command USML-2. However, as Nagel would say later, the second German Spacelab was somewhat different to the first. "The complement of experiments was more biased in the direction of life sciences. It was [also] going to be a little longer: it was a 10-day mission [whereas] the other one was seven days."

The life sciences bias was further illustrated by the inclusion of a physician, Harris, on the crew. "The Germans had requested a medical doctor Mission Specialist," said Ross, "and since I wasn't one, they were certainly hoping the next one would be. While I didn't personally think that a medical doctor was mandatory, I did think that it was not a bad idea, because probably over 50% of the work we were going to do was life sciences [and] human research type of experiments."

The EDO pallet, which could have extended STS-55's time aloft to a fortnight, was not carried this time, although according to Nagel the Germans *were* offered this option. "They had a tradeoff they could've had there. Could've put [an EDO] pallet in the back with more cryo[genic tanks] for fuel cells and flown even longer, but they chose to have experiments back there on an outside rack instead of additional days in orbit." That 'rack' was a Unique Support Structure (USS) – a bridge in Columbia's payload bay – which carried a set of astronomy, atmospheric physics and materials science experiments.

One of the most important of these USS-mounted experiments was the second Modular Optoelectronic Multispectral Scanner (MOMS-2), an improved version of a device previously carried on Spacelab-D1. It was capable of simultaneously acquiring high-resolution images of Earth's surface for remote-sensing purposes. Although the relatively short length of the D2 mission meant that its full potential could not be achieved, another version was later flown to Mir in 1996, providing better data coverage over longer periods of time.

The instrument's multispectral coverage allowed geologists to better discriminate between different classes of vegetation and rock or soil surface coverings. Columbia's 28.5-degree-inclination orbit allowed photography to take place of terrestrial features between 28.5 degrees North and 28.5 degrees South latitude which, coupled with less-than-ideal solar illumination and cloud cover, initially restricted MOMS-2's studies to North and South America and parts of Africa. However, the 'daytime' ground track drifted westward later in the mission and images were acquired of the Near East, India, Southeast Asia, Indonesia, Australia and the Pacific islands.

NASA's Earth Observations Laboratory at JSC also pushed up the instrument's scientific yield by providing real-time information on individual targets' character-istics from geosynchronous- and polar-orbiting satellite data. This allowed MOMS-2 to take more high-resolution pictures over cloud-free regions. Among the instrument's notable successes were incredibly detailed images of irrigation ditches, rice crops and roads in Vietnam, as well as underwater reef structures just off the Egyptian coastline and high-population-density regions like Delhi in India. It even documented 'outgassing' from Colima Volcano in Mexico, which revealed reddish and pinkish tones under MOMS-2's infrared gaze and distinguished volcanic vegetation from that of nearby fields.

Although Colima had not erupted since 1913, the imagery suggested that it had experienced a number of smaller, 'quieter' upheavals since that time. Columbia's crew also supplemented the MOMS-2 data with their own photographs, taken from hand-held Hasselblad and Linhof large-format cameras, which featured the redistribution of ash from Lascar Volcano in the Altiplano of Chile. This volcano had erupted on 20 April 1993, only six days before the Shuttle lifted off, and the photographs clearly picked up several plumes of wind-blown material.

The chance to photograph a recently erupted volcano was serendipitously presented to the STS-55 crew, who were originally supposed to have been in orbit two months earlier, but were kept on the ground by a string of mechanical problems and a dramatic on-the-pad shutdown of Columbia's three main engines mere seconds before a launch attempt on 22 March. When the flagship Shuttle returned from STS-52 in November 1992, NASA engineers and technicians dove into readying her for the Spacelab-D2 mission, scheduled for launch in late February of the following year.

A TROUBLED ROAD TO SPACE

All went well until the beginning of February. The myriad of experiments arrived in Florida in July 1992 and the laborious process of installing them into the pressurised module began in September. After a slight delay caused by problems latching Columbia's payload bay doors, the 11,340-kg Spacelab-D2 cargo – including its experiment-laden USS pallet – was loaded on board the Shuttle on 11 January 1993 and checks confirmed their electrical and mechanical compatibility. After being transferred to the VAB for attachment to her boosters and tank, the complete stack was rolled out to Pad 39A on 7 February.

The mission's problems really began when Columbia was at the pad. Concerns had been raised since late January that her three main engines might have contained obsolete versions of tip-seal retainers on the blades of their high-pressure liquid oxygen turbopumps. These seals minimised the flow of gas around the tips of the turbine blades and, in doing so, enhanced the turbopumps' performance. Each seal was held in place by its own 'retainer'. The uncertainty was that NASA could not conclusively determine from the STS-55 pre-flight processing paperwork whether the retainers were of the 'old' or 'new' variety.

If they were of the old type, they needed to be checked before each mission; if of the newer type, on the other hand, they could fly several times without inspection. To play things safe, Columbia's main engines were removed, inspected and their retainers checked – they *were* of the newer-specification type. By the time this work had been finished and the engines were back in place, it was the end of February and NASA was forced to reschedule the STS-55 launch for "no earlier than" 14 March.

Despite ridicule heaped on the episode by the media, John Plowdon – the KSC division manager of engine-builder Rocketdyne – stood by the decision to postpone the launch and check the retainers on safety grounds. "We're still confident that was the right decision to make," he told journalists. To accommodate the almost-three-week delay, the film on one USS-mounted astronomical instrument – the Galactic Ultra-wide-angle Schmidt System (GAUSS) – had to be changed and two other Spacelab experiments needed to have their batteries replaced. However, more trouble was afoot.

On 2 March, during a flight readiness test, a 33-cm flex hose in Columbia's aft compartment – which served as part of the hydraulic system to move the main engines, elevons and speed brake – burst and spilt several litres of hydraulic fluid. Although the hose itself was capped within a few seconds of the incident, it was decided to remove and check all 12 hydraulic hoses. Nine were examined and reinstalled and three new ones were fitted. By 9 March the task was complete and Shuttle director Tom Utsman had nothing but praise for the tremendous work done by his team.

"The team has done a great job in addressing and closing issues such as the hydraulic flex hose problem and putting together a new plan for processing activities," he said. He also paid tribute to the US Navy, US Air Force and Hughes Space and Communications Company, who had agreed to postpone the launch of an Atlas rocket carrying a military communications satellite in order to give Columbia another shot in the third week of March. As with all rockets launched from Florida, the ascent trajectories of Shuttle missions are closely monitored and tracked by Eastern Test Range antennas.

Only one rocket can be launched on any one day, thus enabling the range to concentrate its full resources on that mission. The agreement of General Dynamics, who built the Atlas, to offer their 'slot' on 21 March to Columbia provided a much-needed boost, particularly to the Germans, who were reportedly paying a million dollars *every day* just to keep their Spacelab-D2 experiments and ground personnel flight-ready. The mission, with its $560-million pricetag, had already been criticised and the enormous cost of reunification since the Berlin Wall was torn down in 1989 had imposed restrictions on Germany's space ambitions.

The country – or West Germany, at least – had been one of ESA's strongest members before reunification, but now had to commit an increasing share of its finances to extensive public works improvements in the former East Germany. "We all realise the effort that needs to be spent to get [the former East Germany] up to speed, infrastructure-wise," said Spacelab-D2 Project Manager Hauke Dodeck. "We have very tight budgets in all areas, including research and technology." Already, Spacelab-D3 had been cancelled, in favour of limited participation in a general, 'pan-European' Spacelab-E1 mission. Ultimately, that flight, too, never took place.

RED LIGHTS IN THE COCKPIT

Despite a month of delays, Columbia's new launch target was jeopardised when another Eastern Test Range booking – that of a Delta rocket from Cape Canaveral Air Force Station, carrying a US Air Force navigation satellite – was postponed on 18 March due to high winds. In accordance with the range's launch policy to all customers, the US Air Force was given a second chance on the 21st, which pushed STS-55 back to the 22nd (plus its own 'second chance' on the 23rd). This then left the Atlas rocket, with its US Navy satellite, in third place at the end of March.

"Officially," Shuttle Test Director Mike Leinbach told journalists as engineers shifted their preparations to launch Columbia into high gear, "we are on the range for Monday [22nd] and Tuesday [23rd]. Anything after that would have to be negotiated." The attempt on 22 March, which would have led to a liftoff at 2:51 pm, went remarkably smoothly. The seven astronauts – Nagel, Ross, Harris, Walter, Schlegel, Pilot Tom Henricks and Mission Specialist Charlie Precourt – boarded Columbia and moved crisply through their pre-launch checks in the cockpit. With 31 seconds to go, as planned, the Shuttle's onboard computers took command of the countdown.

Just as NASA had done in its previous 53 missions, any problems cropping up in this last half-minute would be dealt with by the computers. During this time, particularly in the last few seconds when the three main engines would be ignited at precise intervals of less than a fifth-of-a-second, it was realised that no human operator could have the same accuracy as the computers. The computers would monitor thousands of separate functions and decide whether or not Columbia was fit to launch. On this day, as circumstances would transpire, she was not.

Six-and-a-half seconds before liftoff, the ignition of the main engines got underway with a low-pitched rumble; within three seconds, however, dramatically, and to gasps from spectators at the press site 5 kilometres away, all three shut down. Subsequent analysis pointed to the incomplete ignition of the Number 3 main engine. It turned out that a liquid oxygen preburner check valve had suffered a leak in the final seconds of the countdown, causing the purge system to be pressurised *above* its maximum-allowable 50 psi. A small piece of rubber, caught in a main engine propellant valve, was identified as the culprit.

This moment, which relived the experiences of two previous Shuttle crews in 1984 and 1985, was one of the most dangerous and unwanted episodes in the entire programme. With unburnt hydrogen possibly hanging underneath the still-hot engines, the risk of an explosion or fire was very real. Although the astronauts had been trained to escape from such an on-the-pad abort and slide to a fortified bunker, the danger of them running through *invisible* hydrogen flames prompted NASA to keep them in the relative safety of Columbia's cockpit.

For 40 tense minutes, Nagel and his crew remained strapped into their now-motionless ship as ground controllers switched off all electronic components that could cause an explosion. Paramedics were rushed to Pad 39A as a precautionary measure. Having watched the clock tick down to the last few seconds and hearing the roar of the engines far below him, coupled with a violent shaking of the cabin,

Columbia during her main engine shutdown on 22 March 1993.

Nagel would say later: "I'd convinced myself and all my crew that we were going to fly."

When the engines died and their noise faded away, he told his three crewmates on the middeck – Ross, Schlegel and Walter – what had happened. For Nagel, Henricks, Harris and flight engineer Precourt on the flight deck, they knew from looking at the instrument panel what was amiss. "I wouldn't call it fear," Nagel said later. "There's a couple of moments wondering what's happened because all you see on board are red lights [on the instrument panel], indicating an engine shutdown. So you know the computers shut down the engines, but you don't know why or exactly what went wrong with them."

The situation was equally tense in the Launch Control Center, next to the VAB, from where Launch Director Bob Sieck, his face drained of colour, watched the proceedings. "Your initial reaction", he said after the abort, "is to make sure there are no fuel leaks or that there's nothing that's broken that's causing a hazardous situation. Really, it was one of those nice boring countdowns until the last few seconds! What *did* work, and worked very well, were the safety systems on board. As a result, the crew is safe and the vehicle is on the pad and safe as well."

Nevertheless, the entire crew emerged from Columbia looking visibly shaken by the incident, particularly Walter, who was experiencing his first countdown. It was becoming increasingly likely that a delay of several more weeks would be needed to remove, replace and retest the three main engines. On 30 March, NASA announced that the *next* Shuttle mission, STS-56 – an important atmospheric-research flight called ATLAS-2, using Discovery – would leapfrog STS-55 into orbit on 6 April. By the time of Columbia's abort, Discovery had already been on adjacent Pad 39B for a week and preparations for *her* flight were proceeding smoothly.

"Flying the missions in this order is the most effective use of all our resources," said Tom Utsman. "The early April launch of ATLAS-2 will give scientists the opportunity to observe changes in Earth's ozone [layer] during the seasonal transition between spring and summer in the northern hemisphere. At the same time, the launch team will be working to get Columbia back to launch configuration for launch on 24 April. NASA is very pleased with the cooperation given by our friends in the German space agency. They have been involved as all possible options were considered. Their willingness to let the STS-56 mission have an early April launch will give the ATLAS folks the chance to collect some very important data on Earth's ozone."

The delay to Spacelab-D2 did, however, require several time-critical experiments to be removed from the pressurised module in Columbia's payload bay, replaced or repaired. Often, the principal investigators for the experiments were present at the launch pad to supervise the operations. After a two-day delay of its own, STS-56 set off on 8 April and flew a highly successful nine-day mission.

'THERE'S ONLY *ONE* JERRY ROSS!'

It was realised that if Columbia set off as planned on the 24th, she would set a new record for the shortest interval between two Shuttle flights – just seven days. This would beat the previous 10-day record jointly held by missions STS-51D and STS-51B in April 1985 and STS-61C and the ill-fated STS-51L (Challenger) in January 1986. "From a programme standpoint, we have no concerns," said Brewster Shaw of the Mission Management Team. "If somebody is not ready, we will wait until they are. But if everyone is comfortable with flying on Saturday [24 April], then we will go fly."

As he prepared for his next attempt, Nagel happily received some advice from fellow astronaut Jim Wetherbee, who had led Columbia's previous mission. "He said he forgot to tell me that when you turn the key, you're supposed to jiggle it to get it started," Nagel joked. "So now I think we'll get going." Unfortunately, fate had one more card to play before letting STS-55 fly. Nine hours before liftoff on 24 April, an intermittent power problem was picked up by technicians in one of the IMUs. The device was replaced and launch rescheduled for the 26th.

"We're disappointed," said Rudolph Teuwsen, a spokesman for DARA, the German space agency, after the 24 April scrub, "but it would have been far worse to have launched and then come back without completing the science." Throughout the two-day postponement, Columbia's crew remained in Florida and finally set off at precisely 2:50 pm on the 26th. The launch did indeed set a new record of nine days between two Shuttle missions and the spacecraft's ascent was trouble-free.

Shortly after they reached orbit, the crew found that there was not one Jerry Ross on board Columbia, but *two*! "Steve [Nagel] had arranged with some of the guys to take a large-sized picture of what he had taken of me looking in one of the rear windows of the orbiter [during a spacewalk] on STS-37. It's my smiling face in my spacesuit, looking in the window, and he's taken a picture of me looking in.

"He'd arranged to have one of those pictures cut out so it fit perfectly into one of those windows, and the window was covered by a cardboard [panel] for launch, so that dirt doesn't get under [it]. He knew I was going to be taking that panel off so I could look out the window to set up cameras for opening the payload bay doors after we got onto orbit. So there I am, pulling this thing off, and I'm looking at this thing in the window and it's *me* looking at myself, and I started laughing! It was hilarious!

"Charlie, Steve and Tom were up in the front [of the flight deck] and they were getting ready for an OMS burn or doing some checklist procedures and they thought Ross had lost it back there! I was just laughing hilariously. Then they turned around and saw the picture up there and Steve had actually forgot[ten] that they were going to put that thing in there. But it was great!"

Under the supervision of a now-calmer Ross, the Spacelab module and USS pallet and their experiments were activated and powered up within hours of reaching orbit. "We've got Jerry hard at work in [the lab]," Tom Henricks told Mission Control as the Payload Commander busied himself with setting the module up for what was intended to be nine days of around-the-clock research. Nagel, Henricks, Ross and Walter comprised the Blue Team, while Precourt, Harris and Schlegel were the Reds for STS-55.

The STS-55 crew in the Spacelab-D2 module. Front row (left to right) are Tom Henricks, Steve Nagel, Ulrich Walter and Charlie Precourt and back row are Bernard Harris, Hans Schlegel and Jerry Ross.

"There was [also] a Crew C, which was the Shuttle crew [as a whole]," Ross said later. "So you basically had *three* different teams that were working. I always tried to have a tag-up at each shift handover so that we could tell the oncoming crew where we were at, what we got done, any problems that were going on, a summary of all the flight notes that had been sent up and everything else, so that we made sure we had a good, clean handover. I think a couple of times during the flight I gave little pep talks.

"I also made sure they'd go to sleep on time so they'd get the right amount of rest, even though I wasn't. I probably didn't get more than five hours of sleep a night for the whole time we were up there, and when I got back on the ground, I was just flat wiped out. Part of it was because I was working extra time in the lab to try to make sure that we kept up with the timelines [and] part was that I don't sleep well in the bunks we had. [They] are like coffins; they're really small. I couldn't even turn over in them. And there's still an ambient noise of people working out there, getting their food and eating, and knocking around and stuff, and my brain was going a million miles an hour. I was thinking about all we'd done that day and what we needed to do the next day. I'm kinda that way! I'm kinda hyper!"

A FULL PLATE OF EXPERIMENTS

Had Challenger not been lost in January 1986, it is likely that Spacelab-D2 would have taken place some time in 1988, about three years after the D1 mission. The delays in getting the Shuttle back into space, however, pushed it back into 1992 and finally 1993, but the enthusiasm of West Germany – and later the unified Federal Republic – remained intact. "It is viewed by the general public, as well as the politicians in our government, as a very important mission," said Heinz Stoewer, a Spacelab-D2 project manager for DARA which, with DLR (the German Aerospace Research Establishment), was financing the flight.

"We definitely have the fullest plate that's ever flown in space," Ross told the preflight press conference in February 1993. That was quite a remarkable statement from the gruff-voiced US Air Force Colonel, whose three previous Shuttle missions had included a total of five satellite deployments and four spacewalks, on one of which he had helped 'save' NASA's Compton Gamma-Ray Observatory by manually unstowing its jammed communications antenna. "We have a large number of experiments. There are constraints on when things can be done and when they can't. We will be very challenged to keep up with the timeline."

In fact, Ross' experience thus far had been in EVA matters and he has admitted to being more interested in building the space station than in a science mission. "Dan Brandenstein, [then] head of the Astronaut Office, called me when I was in quarantine, getting ready to go fly STS-37, and said, 'We want you to be the Payload Commander for STS-55'. I said, 'Dan, you want a scientist for that. You don't want me, an old engineer!' He said, 'No, we want *you*. We want somebody that will work well with the Germans and [get] the flight pulled together.' So, literally, it was less than two weeks after I got down from STS-37 [in April 1991], I was announced as the Payload Commander for STS-55, and I didn't do a lot of the normal post-flight [public relations] activities on STS-37 because I went to Germany probably within three weeks of landing, to start working on STS-55."

In many ways, Spacelab-D2 augmented the microgravity research begun by D1 in the autumn of 1985 with investigations into life sciences, biology, materials processing, fluid physics, technology and – mounted on the USS pallet – astronomy, atmospheric studies and Earth observations. In addition to DLR, DARA and NASA, several of these experiments were also provided with support from ESA and French and Japanese researchers.

"Our scientific methods are governed by the effect of gravity," said Hauke Dodeck before the mission. "Objects fall down, lighter objects float or are carried upwards, heavier ones sink to the bottom. What happens to these processes when there is *no* gravitational force: no sedimentation, no thermal convections, no hydrostatic pressure? What new mixtures, structures and forms are possible? Concrete answers can only be given by space research."

Many of these materials science experiments were housed in three custom-built facilities inside the Spacelab module: the Materials Science Experiment Double Rack for Experiment Modules and Apparatus (MEDEA), Werkstofflabor and the Holographic Optical Laboratory (HOLOP). Others, which required direct exposure to the vacuum of space, were attached to the USS in the Materials Science Autonomous Payload (MAUS) and Atomic Oxygen Exposure Tray (AOET).

These experiments studied the behaviour of 'columns' of fluid under weightless conditions, closely followed the solidification processes of different metallic alloys and undertook crystal growth of various materials. Three separate furnaces were contained within the MEDEA double rack: one that conducted long-duration crystallisation studies, another that processed metallic crystals at extremely high temperatures using the directional solidification technique and a high-precision thermostat that examined the behaviour of metals under carefully controlled temperature conditions.

During the mission, investigator Peter Saum excitedly reported that the astronauts had grown the largest-ever gallium arsenide crystals in MEDEA – measuring some 20 mm in diameter; such materials are used extensively in electronic components, such as light-emitting diodes, semiconducting lasers, photo-detectors and high-speed switching circuits. The second materials science facility, Werkstoff labor, held almost a quarter of Spacelab-D2's 88 experiments and successfully produced a monotectic alloy of bismuth–aluminium, which might one day be used to assemble ball bearings capable of withstanding higher loads and temperatures.

Other research included the growth of crystals of materials that could be used to produce stronger turbine blades and other components for aircraft and spacecraft engines. "If the tests produce the hoped-for results," said Dodeck, "turbine blades can be developed which are strongly resistant to heat and stress, thereby improving the performance and lifetime of aircraft engines." Meanwhile, HOLOP used holography to investigate the processes of heat- and mass-transfer and cooling in transparent materials, which are of importance in metallurgy and casting techniques. The holography, using laser light, made these processes more easily visible.

"HOLOP will transmit video pictures of experiments to the ground while they are being performed," Dodeck said before the mission. "Scientists on Earth can not only watch what happens, but also may intervene in the test sequence, thus demonstrating a concept called 'telescience'." The technique had also been used in conjunction with the USMP-1 experiments on Columbia's previous mission, STS-52. In fact, more than 600 telescience commands were transmitted to HOLOP during the course of Spacelab-D2 from DLR's Microgravity Life Support Center at Cologne-Porz.

To emphasise the German flavour of the mission, like Spacelab-D1, the Payload Operations Control Center (POCC) had shifted from Houston to Oberpfaffenhofen, near Munich. However, during D1, several functions still had to be monitored from Houston because Oberpfaffenhofen's data-transmission capabilities were insufficient to handle all communications traffic. For this second mission, the situation had improved and satellite-transmitted data was received by ground stations on DLR premises and forwarded directly to the control centre. Sitting in the control room as Spacelab-D2's science coordinator was German astronaut Ulf Merbold, who had flown on STS-9 and served as backup Payload Specialist on D1.

'Outside' Columbia, on the USS pallet, two experiments in the MAUS facility investigated complex boiling processes and the diffusion phenomenon of gas bubbles in salt melts. Meanwhile, AOET exposed more than a hundred material samples to the harsh atomic oxygen environment of low-Earth orbit to obtain data on their reaction rates. Such data was deemed important for constructing new materials to help shield Columbus, ESA's laboratory module for the International Space Station, which would have to remain in orbit for many years.

TORTURE CHAMBER

Back inside the Spacelab module, the science crew – Ross, Harris, Walter and Schlegel – also conducted medical experiments using three painful-sounding facilities: Anthrorack, Biolabor and Baroreflex. Televised images from inside the module at various stages of the mission gave viewers the distinct impression of a medieval torture chamber, as the astronauts injected each other with saline as part of studies to replace bodily fluids lost during adaptation to microgravity and repeatedly took blood samples. Fortunately, as the 'flight crew' responsible for managing Columbia's systems, Nagel, Henricks and Precourt were 'immune' to these punishing medical tests.

"[We couldn't] get involved in any invasive experiments," Nagel said with a grin, "which means they can't draw blood [from us]. They can't poke you with needles and things like that. They need you in good shape to bring the [Shuttle] home, which is okay with me!" The Anthrorack facility (from the Greek word 'anthropos', meaning a human being) supported nearly two dozen medical experiments and was capable of performing the first-ever comprehensive, integrated screening of the astronauts' bodies in space. It could make simultaneous measurements of their respiratory, cardiovascular and endocrine systems and even had its own ultrasound device.

"Anthrorack is the most advanced of its type which has flown in space," said Dodeck. The facility enabled the astronauts to examine their cardiovascular, pulmonary and hormonal adaptation to the microgravity environment. Two other studies, funded by NASA, were the Baroreflex Experiment and the Cardiovascular Regulation Experiment. These investigated the relationship between post-flight cardiovascular 'deconditioning' and the baroreflex. Such deconditioning was high-lighted by a characteristic drop in blood pressure when astronauts stood up after landing. It was associated with the decreased workload of the heart while in space and the baroreflex, which maintains appropriate blood pressure throughout the body.

The experiment tested the theory that light-headedness and a reduction in blood pressure after landing may be due to the fact that the baroreflex's ability to regulate it is somewhat reduced after having become adapted to microgravity. In particular, the ability of the body's blood pressure sensors to control the heart rate – known as the 'baroreceptor reflex' – was measured to determine if the predicted impairment *did* occur. To perform the experiment, astronauts wore a silicone rubber neck cuff, through which pulses of pressure and suction were transmitted to baroreceptors. Heart-rate changes were then recorded by an ECG.

All crew members also participated in 'saline loading' – drinking salty water in order to rapidly expand their intravascular volumes – to allow medical scientists to better understand the mechanisms responsible for post-flight cardiovascular deconditioning and orthostatic intolerance. "I think we will find that space adaptation syndrome comes in phases," said medical doctor Harris after the mission. He concluded that acclimatisation to the microgravity environment comes in at least three stages which span undetermined lengths of time and began with dizziness and an upset stomach, gradually leading to a loss of blood volume and a decrease in bone and muscle mass.

ROBOTIC EXPERIMENTS

In readiness for operations on the International Space Station, several experiments focused on the development of new robotics and telescience techniques. Two ESA-provided investigations were intended to validate concepts for the Columbus laboratory: one, the Crew Telesupport Experiment (CTE), was operated by Henricks and successfully achieved two-way communications with Mission Control using a device akin to a child's 'Etch-a-Sketch' toy. The other positioned a variety of accelerometers throughout the Spacelab module to study the effects of crew and Shuttle motions on very sensitive microgravity experiments.

However, the 'toy' that really attracted the media's interest was the Robotic Technology Experiment (ROTEX), a six-jointed robot arm fitted with a variety of tactile and torque sensors, laser range-finders and stereo and fixed cameras. During the mission, it was used to perform a multitude of tasks, including building a small tower of cubes and retrieving, connecting and disconnecting an electrical plug. Operating within its own sealed work cell, ROTEX was used by the crew and – for the very first time, on 1 May – via telescience from the ground.

While the science crew conducted the bulk of the research in the Spacelab module, the flight crew not only monitored Columbia's systems, but also handled the housekeeping chores and unexpected problems that cropped up. One such problem materialised on 28 April, two days into the mission, when the crew reported a puncture in the wall of the waste water tank attached to the toilet, as well as a gaseous nitrogen leak. Video footage was downlinked to Mission Control and Nagel and Henricks took up the panels of Columbia's middeck floor to conduct a visual inspection. They found no sign of a leak, which implied that the tank's 'bellows' remained intact. However, in view of the nitrogen leak, it was decided to use a backup container to hold waste water for the remainder of the mission. "We got some tools with the instructions for the in-flight maintenance and we rerouted the waste water into a bag," said Nagel. "We fly these contingency bags. It's like a canvas bag with a rubber lining in it and we just put the waste water in that bag. Periodically, we'd have to empty the bag [by dumping the water] overboard."

This enjoyable off-duty pursuit delighted the child in Jerry Ross. "Every once in a while, we'd have to sit there and squeeze this bag to shoot the urine back out the side of the orbiter into space. *That* was a delightful thing to do!"

The gaseous nitrogen system for the tank, meanwhile, was turned off, but since it also assisted the crew's drinking water supply, Columbia's galley began dispensing

Hans Schlegel works with ROTEX.

water at a slower-than-expected rate. Late on 28 April, the crew capped a nitrogen line into the now-unused waste water tank, which allowed the nitrogen system to be turned back on and fully pressurise the water supply tanks. The only other problem of note was an overheating freezer in the middeck – used to store biological samples – which obliged the astronauts to transfer them to a backup unit.

"I found out [before the flight] that the freezers we were carrying on a lot of those [Spacelab] flights were failing at a fairly rapid clip," said Ross. "It became apparent to me that probably 40 or 50% of the science was counting on that freezer working and bringing back those samples that we collected over the 10 days on orbit. So I suggested that they carry a second freezer as a backup and forfeit two lockers of space [in the middeck], because if we had a freezer break, we'd be out of luck.

"Well, wouldn't you know it, the freezer failed by the time we got to orbit and never did come back to life, so the other one that was our spare was the one we used throughout the entire flight, and we did all kinds of things to try to nurse it along. We've got these plastic [trash] bags and taped together and ducted cold air coming out of one of the vents in the orbiter and put it right in the intake fan of the freezer so that we could give it additional help to keep itself cool."

AN EXTENDED MISSION

Nagel thinks it was the presence of the time-critical biological samples that influenced NASA's decision to bring Columbia home on 6 May, lest the second freezer also failed. In all other respects, the flight had proved so successful – "We've really got this laboratory cranked-up," Ross had told ground controllers on 27 April – and the crew's use of onboard consumables was so economical that NASA decided to extend it from nine to 10 days.

In fact, both the Americans and the Germans had been working towards the 'extra' day since the start of the mission. Over a three-day period, they had worked with the astronauts to conserve as many of Columbia's consumables as possible and easily accumulated 25 hours' worth of additional margin in the cryogenic fuel cells. On the morning of 3 May, it was official: the Mission Management Team told a happy Nagel that his crew would now return to Earth around 1:00 pm on the 6th.

Early on the morning of 'landing day', the Red Team took charge of deactivating the Spacelab module and storing all of the biological specimens and materials samples from the many experiments. They then bade the German ground controllers "Aufwiedersehen", before shutting the hatch and returning to the middeck, closing the tunnel and packing away the remainder of their equipment in readiness for re-entry. Both payload bay doors were closed and latched without incident by 9:25 am, with the de-orbit burn scheduled for around midday. It was not to happen.

Unacceptable weather at KSC led forecasters to 'wave-off' the crew for one orbit and, when it did not improve, the landing site was changed to Edwards Air Force Base. Nagel was sure Columbia could have stayed aloft an extra day or two to wait for conditions in Florida to improve, but added that this mission was "a big life sciences flight with lots of samples. They [the Germans] were hanging it all on one freezer to save these samples, and I'm sure that's why, when it came time to land, we were weathered out of Florida and went to Edwards.

"If it hadn't been for that, we'd have stayed up an extra day waiting for good weather in Florida, but the Germans were deathly afraid that their second freezer was going to fail and they'd lose all these samples." Nagel performed the almost-three-minute de-orbit burn at 1:29 pm to begin the hour-long glide home. The astronauts landed on Runway 22 at Edwards at 2:29:59 pm, after a mission lasting a few minutes short of 10 full days. During STS-55, Columbia also achieved a personal milestone: becoming the first Shuttle to spend more than a hundred cumulative days in space.

JOHN BLAHA AND THE LONG ROAD TO SLS-2

"John, we're going to fly you one of these days," NASA Launch Director Bob Sieck called over the communications loop as the afternoon of 15 October 1993 wore on. The disappointment of another scrubbed launch attempt was in his tone. "Just hang in there."

"Nice try," responded STS-58 Commander John Blaha from the flight deck, as he and his six crewmates prepared to disembark from Space Shuttle Columbia after two-and-a-half uncomfortable hours on their backs, clad in bulky partial-pressure suits, harnesses and parachutes. It was the second time they had been through this routine to get into space for what would turn out to be NASA's longest-yet Shuttle mission, lasting just over two weeks. For now, however, Columbia was living up to her reputation: an immovable bear difficult to get off the ground, but once in orbit, a beautiful, graceful swan.

More than two months' worth of delays in getting her previous mission – the German-sponsored Spacelab-D2 – into space had already pushed her *next* flight back a month from August until mid-September. That date had also quickly become untenable as another mission, STS-51 on Discovery, experienced 1993's second on-the-pad main engine shutdown and was itself repeatedly delayed. Not until Discovery and her five-man crew were aloft could a firm timeline be set for Columbia's own launch. Eventually, a Flight Readiness Review by top-ranking Shuttle managers scheduled it for 14 October.

Bad weather that day caused a two-hour extension of the launch window and, when it finally cleared, the countdown clock continued ticking towards a scheduled 4:53 pm liftoff. Then a failure in a US Air Force Range Safety command message encoder verifier resulted in the cancellation of the attempt; this system would have been used to destroy the vehicle by remote control should an unthinkable in-flight accident happen. It *had*, in fact, already happened during Challenger's catastrophic ascent in January 1986, when it was employed to destroy the two wayward SRBs as they spiralled out-of-control, potentially threatening populated areas.

The attempt on 15 October was relatively smooth until a problem surfaced with one of two S-band transponders on Columbia. Although it could *transmit* data, the 13.6-kg unit could not *receive* properly. Flight rules stipulated that both transponders – which provide a steady stream of main engine data and voice communications between the ascending Shuttle and Mission Control – must be fully functional for a launch to proceed. The option to sidestep the rule and fly with only one transponder was briefly considered, then a spare unit was rushed to the pad for hasty installation.

Ultimately, however, NASA managers were not happy with engineers replacing it in the short time available before the launch window closed or asking the astronauts to replace it once they reached orbit. "As that scenario unfolded, folks could not quite get comfortable with it. Me included!" said Loren Shriver, the new head of NASA's Mission Management Team and himself a former Shuttle Commander. "*That* scenario would have placed a lot more responsibility and decision-making on the crew. Having been there [myself], there is no substitute for having a little help from the ground."

The STS-58 crew participate in a practice countdown on the launch pad. Front row (left to right) are John Blaha, Marty Fettman and Dave Wolf and back row are Rick Searfoss, Bill McArthur, Shannon Lucid and Rhea Seddon.

Shriver was here referring to the dire predicament Columbia's astronauts would have faced if the one remaining transponder had failed and an in-flight problem had forced Blaha and Pilot Rick Searfoss to perform a hazardous and as-yet-untried emergency landing back at KSC. A failed transponder might have left the crew out of communications with Mission Control for up to 15 critical minutes. However, even if the transponder problem had not arisen, as the weather closed in on the Floridian spaceport it was becoming more likely that Columbia's launch that day probably would not go ahead.

The 'failed' transponder, in fact, had already operated 700 hours *longer* than its 2,600-hour design lifetime, more than proving its worth. It was replaced over the weekend of 16–17 October, in readiness for another attempt the following Monday. It was with characteristic enthusiasm that Blaha led his crew out of the Operations and Checkout Building for what would be 'third time lucky' in getting STS-58 off the ground. Not only would it become the longest Shuttle flight so far, but would also allow Blaha to finally undertake a mission he was originally assigned to fly four years earlier.

An astronaut since 1980, the now-retired US Air Force Colonel waited an unenviable nine years for his first trip into orbit. Although at the time of his selection, Blaha was actually the senior military officer in his class, many regarded him as a steady, dependable astronaut, rather than a mission-hungry 'rising star', and this perhaps contributed to his receiving a flight assignment so late in his career. He was slated to fly as Pilot on STS-61H in the summer of 1986, a flight abruptly cancelled by the Challenger disaster, but finally made it into space on STS-29 in March 1989.

His admirable performance on his first flight led to his rapid reassignment, within a fortnight of landing from STS-29, as Pilot of the first Spacelab Life Sciences mission (SLS-1), scheduled for the summer of the following year. Blaha had only been training for his second flight for a couple of months when tragedy conspired against one of his fellow astronauts and gave him another assignment. Dave Griggs, who was due to fly as Pilot of a top-secret Department of Defense mission in November 1989, was killed in an aircraft crash and Blaha was quickly drafted in to replace him.

After completing Griggs' mission, he went on to command a highly successful nine-day flight in August 1991 which deployed an important communications satellite for NASA. A year later, he at last got his chance to train for a life sciences mission by receiving the command of SLS-2. Joining him were Pilot Searfoss and flight engineer Bill McArthur, together with a four-person science crew: two medical doctors (Payload Commander Rhea Seddon and Mission Specialist Dave Wolf), a biochemist (Mission Specialist Shannon Lucid) and a veterinarian (Payload Specialist Marty Fettman).

A CONTROVERSIAL FLIGHT

The inclusion of a veterinarian on the crew had been on the cards before the SLS-1 mission, because, like its predecessor, the second Spacelab Life Sciences flight would involve extensive physiological examinations of 48 male rats (*Rattus norvegicus*), caged in a pair of Research Animal Holding Facilities (RAHFs). It would also controversially feature the first-ever *in-flight* decapitation and dissection of six of those rats; this had drawn a significant amount of public criticism, but, according to Marty Fettman and Harry Holloway, NASA's Associate Administrator for Life Sciences, it was essential to assess ongoing changes in the rats' body tissues during spaceflight.

"This is really a unique opportunity to collect biological specimens," said Fettman before Columbia's launch. "We believe these tissues will provide some answers to questions that potentially will change our interpretation of past observations." Despite the risk of a public outcry, it was rationalised that examinations of rats brought back after SLS-1 had been unable to conclusively differentiate between the effects of spaceflight and the effects of the rodents' readaptation to terrestrial gravity. The SLS-2 dissections would allow researchers to more precisely assess ongoing tissue changes while in space.

Nevertheless, Holloway had called for an unscheduled pre-flight assessment of the mission's rat research plans; however, he would later deny that he had felt pressurised to do so by either the White House or NASA Administrator Dan Goldin. "I know there were rumours to the contrary, but I did it," he said of the assessment, which was led by Deputy Surgeon-General Robert Whitney of the Department of Health and Human Services. Still, the SLS-2 project management were quick to stress that they were treating the dissection of the rats as delicately as possible.

"We expect some public concern about the animal work we will be doing," NASA's SLS-2 project manager Frank Sulzman said before the flight. "We feel we have done everything that should be done in order to ensure that these animals are treated humanely and that we conform to standard practices for animal care and research." Whitney's probe broadly agreed with Sulzman's comments, describing the agency's animal-care review process as "superb" and commending the use of protocols that used the fewest number of rats necessary to satisfy the needs of more than a hundred experiment investigators.

Another source of controversy, at least within the astronaut corps, surrounded Mission Specialist Dave Wolf, although it would not become public knowledge until after the flight. The story has been told in detail elsewhere, but apparently involved a bizarre FBI sting called 'Operation Lightning Strike' in which Wolf, through his own misfortune, had become entangled. Although the astronaut himself was later exonerated from any blame and had not, as some journalists claimed, accepted bribes, it would harm his career for some years to come.

On 18 October 1993, as John Blaha led Wolf, Searfoss, Seddon, McArthur, Lucid and Fettman out to the launch pad, none of this had apparently surfaced. From Columbia's flight deck, as he tested his communications gear, Blaha told launch controllers "Let's go do it", to which Lockheed engineer Brian Monborne replied, "This time, we're going to send you!" And send them they did: after a slightly extended hold in the countdown at T−5 minutes, when a stray US Navy aircraft entered the launch danger area and had to be shooed out, Columbia rocketed into orbit at 2:53:10 pm.

The ascent itself was normal, although when the two SRBs were recovered from the Atlantic Ocean, it was found that one of their four separation motor covers – used to protect the motors, which separate the boosters from the External Tank – was missing. This had also been seen after several previous launches and an investigation team was established to determine the cause; it would later be suggested that it probably came off during the boosters' descent or water impact. The missing cover did not, however, present a safety-of-flight issue for the crew.

OFF TO WORK

Work on activating SLS-2 – which, like its predecessor, utilised the European-built Spacelab module – began almost immediately after the seven astronauts reached orbit. Even before the hatch to the module had been opened, Lucid and Fettman were busy taking blood draws from each other's arms, while Wolf took their blood pressure in order to gather new data on the astronauts' early adaptation to weightlessness. The findings correlated SLS-1 data by showing a slightly lower central venous pressure than predicted in terrestrial studies, coupled with a slightly larger volume in the heart's left ventricle than would be expected with the lower pressure.

This shed new light on the basic physiology of the heart in microgravity. Immediately after the Spacelab module had been activated, Seddon took ultrasound measurements of Lucid's heart with a state-of-the-art echocardiograph imaging unit. Both Lucid and Fettman had, like Drew Gaffney on SLS-1, travelled into orbit with catheters threaded into veins in their arms, which ran to the tips of their hearts to provide these central venous pressure measurements. Lucid's catheter was removed late on 18 October and Fettman's was taken out the following day.

Data dropouts on the echocardiograph led to the astronauts resorting to the use of a portable unit, known as the American Flight Echocardiograph (AFE). This was a piece of equipment very close to Wolf's heart in more ways than one, because he had helped to design it before becoming an astronaut. "As the person responsible for turning the AFE into flight equipment, you should be very pleased to have it work so well," Alternate Payload Specialist Larry Young told Wolf.

The 9,900-kg SLS-2 payload continued the research begun on SLS-1 two-and-a-half years earlier. Unlike the consistent repeatability of experiments in the physical and chemical sciences, it had long been recognised that physiological studies involved looking at variations in responses from one individual to the next. To reduce the influence of biological variability and improve the quality of the scientific results, it was preferable to collect data on numerous individuals and compare it through statistical analysis. When added to the SLS-1 data, the new results would provide the most detailed and interrelated physiological measurements since Skylab.

This second mission, in which NASA had invested $175 million, in addition to its overhead costs for staging a Shuttle flight, carried 14 major experiments, eight of which used the astronauts themselves as test subjects, while the other six utilised the rats. Body tissues from the latter would be preserved and distributed after Columbia's landing to US, French, Russian and Japanese medical scientists as part of an extensive biospecimen-sharing project. Each of the rats underwent radio-isotope and fluorescent bone-marker injections before launch to measure their blood parameters and bone formation.

The science crew then took frequent blood draws from the rodents' tails during the first part of the mission and performed additional radioisotope and hormone or placebo injections to measure plasma volumes and track their protein metabolism. This was part of a study into how red blood cell mass changes in space. "The mechanisms controlling red blood cell production that are affected in space are the same as those affected in people on the ground with illnesses like leukaemia, chronic renal disease and auto-immune diseases affecting red blood cell production," Fettman told an interviewer during the flight. "If we can show that the mechanisms are the same in the rats as they are in the people in the process of space anaemia, and erythropoietin is an effective counter-measure, this may have long-ranging effects in benefiting both people and animals back on Earth."

Ultimately, the six unlucky rats destined to meet their maker while the Shuttle was in orbit were decapitated by Fettman and Seddon on 30 October using a modified $100 laboratory dispatcher. Pre-flight studies had already opted to decapitate, rather than anaesthetise, the rats because the latter procedure would have caused their neural tissues to deteriorate.

"Things went pretty well," said Fettman after the six-hour dissection procedure. "We're happy to accomplish this. It was a big day for us." Each of the rodent tissues was chemically preserved, refrigerated or frozen in preparation for landing. The mood on board Columbia was one of solemnity, with only Fettman and Seddon working quietly at one stage in the Spacelab module. Even Blaha, who typically would always check up on his crew to see how they were doing and offer his help, only popped his head once or twice over Fettman's shoulder before returning to his other work.

The decapitated rats were part of a series of neurovestibular and musculoskeletal investigations, which examined changes in their gravity-sensing organs and the effect of microgravity exposure on their limb muscles and bones. It was already known that a person's awareness of his or her body orientation, on Earth, is partly due to the detection of gravity by the otolith organ deep within the inner ear. Combined with touch sensors in the skin and, of course, our eyes, this provides a mechanism to sense our relationship with other objects. In space, however, this balanced state is upset somewhat.

Information sent to the brain from the otolith and other sensory organs no longer corresponds to the cues experienced on Earth and some medical scientists have found links between this conflict and instances of disorientation or space motion sickness in astronauts. "Gravity", said Larry Young of Massachusetts Institute of Technology, "is as profound a factor on the evolution and development of biology on Earth as oxygen and water. Yet we know so little about its influence because, until the Space Age, we simply *couldn't get away from it!*"

Research conducted during SLS-1 had led researchers to conclude that gravity sensors in the adult rats adapted to their new environment by changing the number, type and groups of 'synapses' (gaps between their nerve cells across which chemical transmitters propagate). The second mission attempted to uncover the precise nature of this adaptation, as well as structural changes within the rats' inner ears as a response to microgravity exposure. Investigators also used the rats who remained alive throughout the mission to measure the speed of readaptation to terrestrial gravity after Columbia's landing.

It was hoped that such experiments could have clinical applications for motion sickness sufferers or patients with vestibular disorders, which often lead to falls or bouts of dizziness. In addition to the studies of the rats, a series of joint US/ Canadian

Assisted by Rhea Seddon, Marty Fettman goes for a spin in SLS-2's rotating chair.

studies, which were originally carried on Spacelab-1 as well as SLS-1, were reflown as part of investigations into space motion sickness and vestibular changes in humans. For SLS-2, the apparatus included a rotating chair mounted in the Spacelab module's centre aisle, which tested changes in the astronauts' reflexive eye motions.

Seddon, the head of the science crew and a veteran of SLS-1, was first to use the rotating chair on 21 October, as part of studies of the 'vestibulo-ocular reflex' in the eye, "which is what allows us to see while we're moving", according to co-investigator Daniel Merfield of Massachusetts Institute of Technology. "Without this reflex, objects around us would appear blurry as they move."

Lucid manually spun the chair, with Seddon in it, rapidly, stopped it suddenly and then asked her to pitch her head forward. When a person spins around on Earth, his or her eyes tend to move in the opposite direction to stabilise their 'picture' and then snap quickly ahead in a repeating pattern. After about 20 seconds of constant speed, however, both the eye motions and the sense of movement halt. Then, when the actual rotation ceases, the person's motion-sensing organs signal the brain that he or she is moving in the opposite direction and eye motions reflect that change.

However, if the person suddenly leans forward, as Seddon did, gravity sensors overrule the false perception of motion. The astronaut reported that she continued to feel a 'rolling' sensation after tilting her head forward. She would later tell colleagues that her "sense of 'down' had completely gone away".

Other experiments featured a rotating dome, which was placed over the astronaut's head; on the interior was a pattern of dots that appeared to 'rotate' in a direction opposite that of the moving dome itself. The astronaut then used a joystick to indicate their perceived direction and velocity of rotation. Crew members also viewed targets and pointed at them with their eyes closed as part of a study of differences in their perceived relationship between their body and the surroundings.

Seddon was first to try out the dome, on 20 October, and as it began to rotate, her eyes told her that she was spinning. The experiment carefully measured her perception of how quickly she was moving and what kind of eye, head and neck motions she exhibited in response to her perceived movement. "The difference between the data on Earth and the data in space is absolutely amazing," Blaha told Mission Control after Fettman participated in one experiment using the dome.

On several occasions, starting just a few hours after launch on 18 October, Wolf donned a special skull 'cap' fitted with motion sensors. Known as the Acceleration Recording Unit, the device included a pocket tape recorder on which the astronaut reported the time and severity of any symptoms of space motion sickness. Investigators hoped that having crew members wear the skull cap throughout their working day might enable them to correlate instances of sickness with periods of provocative head movement. The science crew also performed vigorous exercise routines and head movements to assess their subjective levels of discomfort.

It was already known that the musculoskeletal systems of both humans and rats are not used as extensively in space as they are on Earth; changes in load-bearing tissues, caused by the absence of gravitational force, therefore lead to reductions in bone and muscle mass, as well as the obvious use of a 'floating', rather than 'walking', motion for locomotion. Furthermore, the system's metabolic state can be altered by dietary intake, exercise levels and space motion sickness. Each of these factors could potentially have a detrimental effect on an astronaut's physical fitness.

Human muscular atrophy in space is characterised by a loss of lean body mass, decreased muscle mass in the calves and decreased muscular strength. During SLS-2, the astronauts ingested amino acids labelled with a non-radioactive isotope of nitrogen which allowed them to track protein metabolism in their bodies. Urine, saliva and blood samples were then periodically taken to determine the rates of protein synthesis and catabolism. Other musculoskeletal experiments looked at the performance of the rats' hindlimbs in microgravity, which showed an almost 40% reduction of their muscle fibres after Columbia's two-week-long mission.

The rats tended to rely more heavily on their forelimbs for bipedal locomotion in space, using their hindlimbs only as grasping aids, and upon their return to Earth exhibited slow movement and an abnormally low body posture. All of these observed characteristics pointed to a weakened muscular state, fatigue and problems with bodily coordination. Similar muscular shrinkage and weakness was also anticipated in humans on long-duration missions on board the International Space Station or on trips to the Moon or Mars.

"It is not surprising that it takes astronauts a few days to recover their pre-flight strength and coordination after flight," said Kenneth Baldwin of the University of California at Irvine, "since their muscles are remodelled by microgravity." Moreover, since muscle protein 'turnover' in rats is much more rapid than in humans, two weeks of microgravity exposure in *them* was roughly equivalent to two *months* in us, which was why they were essential to the study.

In addition to the neurovestibular and musculoskeletal investigations, SLS-2 experiments studied the cardiovascular and regulatory systems of the astronauts' bodies. Already, data from SLS-1 had pointed to increases in heart rate, size and output, which many medical researchers had attributed to the initial increase in central blood volume caused by fluid shifts within the body. Three SLS-2 studies assessed the functional capabilities of this system by monitoring the astronauts' cardiac output, heart rates, arterial and venous blood pressure, blood volume and the amount and distribution of blood and gases in their lungs.

Since the early days of manned spaceflight, cardiovascular 'deconditioning' had been frequently reported by astronauts upon their return to Earth. This was evidenced by a reduction of 'orthostatic tolerance' – light-headedness – and was frequently accompanied by an increased heart rate and decreased pulse pressure. Additionally, measurements of blood and body water highlighted fluid shifts towards the head, which in turn 'fooled' the body into thinking that too much fluid was present and resulted in a reduction of fluid volume. This shifting pattern then influenced cardiac output and arterial and venous pressures.

Moreover, many medical scientists believe that microgravity may affect lung functioning and on SLS-2 sought to investigate the effects of weightlessness on the pulmonary system and particularly on respiration, blood flow and gas exchange. During experiments on SLS-1, cardiac output from all seven astronauts stayed elevated, but total 'peripheral resistance' – the resistance of blood flow throughout their bodies – adapted fairly quickly. Central venous pressures produced unexpected results by *decreasing*, which refuted earlier hypotheses that microgravity-induced fluid shifts should cause it to increase. Despite this, the astronauts' heart sizes increased, as did overall cardiac output.

This finding could point to a general 'opening' of the blood vessels in orbit. In addition to general cardiovascular measurements, central venous pressures were measured using the catheters in Lucid and Fettman's arms. Echocardiograph data was also taken and leg cuffs measured blood flow and volume. Other experiments looked at the unexpected finding from SLS-1 that, far from being more even in space, lung ventilation actually improved only by about half as much as on Earth. The science crew inhaled oxygen, nitrogen and other trace gases, as part of tests to measure blood and gas motions in their pulmonary systems.

To do this, they used the Gas Analyser Mass Spectrometer (GAMS), which measured the composition of both their inhaled and exhaled air. As part of these cardiovascular investigations, the astronauts also employed a bicycle ergometer in the Spacelab aisle to exercise vigorously – so vigorously, in fact, that on 25 October the bike came away from its floor attachment point and had to be hurriedly fixed by Blaha and McArthur.

"One key issue is the ability to maintain blood pressure when standing up after spaceflight," said Alternative Payload Specialist Jay Buckey. "Gravity will have a profound effect on them after they come back from space. Secondly, we want to look at the early adaptation in space. SLS-2 is changing our understanding about how the cardiovascular system adapts to space. Now we're looking forward to their first day back! They've adapted well to space, but now they will have to readjust to 1g, which will have dramatic effects on the cardiovascular system."

Still other investigations focused on the response of the kidneys and endocrine glands to the microgravity environment and their ability to control hormones that regulate blood volume, pressure and essential electrolytes. During SLS-1, these renal/endocrine and haematology studies examined plasma (the fluid component of our blood) and red blood cells. The plasma experiments allowed scientists to identify the types of nutrients circulating throughout the body to determine whether or not the astronauts were well nourished and well hydrated. One notable observation was the decrease in red blood cell production, which it was realised could complicate in-flight illnesses or injuries on long-duration missions.

On Earth, gravity affects bodily fluids by pulling them towards our feet, but in space they are redistributed to the head and torso, which causes changes in the kidneys and fluid-regulating hormones in the cardiovascular and blood systems. During SLS-2, experimenters investigated the theory that the kidneys and endocrine glands adjust the body's fluid-regulating hormones to stimulate an increase in fluid to be excreted. Over longer periods of time, the kidneys and hormones establish new levels of salts, minerals and hormones appropriate for the reduced fluid volume. The plasma shift also affects the blood system through decreases in plasma volume.

In order to determine immediate and long-term changes in their kidney functions, water, salt and mineral balance and fluid shifts, Columbia's astronauts collected urine samples throughout the mission. Additionally, they carefully measured their body weight daily and logged all food, fluids and medication taken while in space. Chemical tracers were also injected into the rats to measure changes in *their* red blood cell and plasma volumes.

A PRODUCTIVE MISSION

Unusually for a Spacelab flight, STS-58 did not operate a dual-shift system, although the entire crew typically put in 14-hour workdays or more. In view of the longer-than-normal duration, and in line with EDO procedures, each astronaut was given several periods of free time to relax and look out of Columbia's windows. "The crew members are an important part of these investigations," said SLS-2 Mission Scientist Howard Schneider. "We want to assure ourselves that we continue to study the physiological effects of *spaceflight* and *not* the physiological effects of fatigue! If the crew is up there and is overly stressed, we don't get good science."

One of the main leisure activities, obviously, was Earth-gazing and Searfoss took a huge pictorial atlas of the world with him among his personal items. "One of the most entertaining things we can do is look out the window, so there will be a lot of nose prints," Seddon said before launch.

During the mission, Blaha spoke for all of them: "We have a beautiful planet. We ought to take care of it and we ought to take care of ourselves. Doing research on this type of laboratory can do a lot to improve the welfare of the five billion people who live on our planet." The mission was also being hailed as a spectacular triumph by the scientists eagerly awaiting their results. "SLS-2 has been an undeniable success," said Mission Manager Lele Newkirk. "I'm proud of the team, the work that the crew has done and the science they have gathered. This mission has provided the scientific community with a great deal of information that will benefit humans on Earth and in space for a long time to come." Howard Schneider agreed, remarking that SLS-2 had "exceeded our expectations" and Payload Operations Director Susan Brand would laud the near-flawless operation of the Spacelab module and the experiment hardware.

Already, STS-58 was due to come close to the duration record set by *another* Columbia crew in 1992, and to keep their flying skills sharp throughout the long days of weightlessness, Blaha and Searfoss took turns on a new computer program called the Portable In-flight Landing Operations Trainer (PILOT). This comprised a high-resolution colour display and hand controller which gave them the 'look' and 'feel' of the Shuttle. Kept in a middeck locker when unused, PILOT was assembled on the console in front of Searfoss' seat and its joystick was affixed to the top of his 'own' hand controller.

The astronauts also participated in now-customary Lower Body Negative Pressure (LBNP) runs to better prepare them for the punishing onset of terrestrial

Rick Searfoss participates in a PILOT simulator session from his seat on Columbia's flight deck.

gravity. Additionally, while the science crew busied themselves with the research in the Spacelab module, Blaha and Searfoss conducted a number of thruster firings in support of the Orbital Acceleration Research Experiment (OARE). This device was designed to make very precise measurements of minute accelerations and jitters and record the levels to which they disturbed particularly sensitive experiments. It had been flown on two previous Spacelab missions in June 1991 and June 1992.

On 26 October, Searfoss sent the device head-over-heels by manoeuvring Columbia through a variety of orientations. He started with the payload bay facing Earthwards and the nose pointing in the direction of travel, then turned the Shuttle end-over-end 360 degrees, put her into a flat spin and gradually rolled her for 20 minutes. He then executed a 'pitch–drag' manoeuvre, whereby he pointed Columbia's nose 'down' 20 degrees and waited for natural atmospheric drag to pull the nose so that it pointed directly Earthwards. This so-called 'gravity-gradient' attitude minimised the need for thruster firings to keep the vehicle stable.

When the time came to close the Spacelab module early on 1 November, in anticipation of landing later that day at Edwards Air Force Base, Columbia's crew had already more-or-less eclipsed the record set by the STS-50 astronauts. Not only that, but Payload Commander Rhea Seddon had set her own personal record. Her husband, Hoot Gibson, was also an astronaut, watching intently on the ground, and

Marty Fettman, Shannon Lucid and Dave Wolf review procedures during a training session in a middeck simulator.

had actually served for the last year as chief of NASA's Astronaut Office. Sometime late in the mission, Seddon surpassed Gibson's total of just over 26 days in space.

"He's still a really good guy [and] I love him a lot," she said in a televised interview, "but I've got more hours in space than he does, so there!" Sticking her tongue out playfully for the television audience and Gibson, she was, however, forced to acknowledge that he had more launches and landings, having flown four times to her three. By the end of STS-58, Seddon had accumulated almost 32 days in space overall.

Unlike the activation of the Spacelab on 18 October, which had been led by Seddon, the *deactivation* of the module was under the supervision of Wolf. Columbia's payload bay doors were closed at 11:30 am and Blaha fired the OMS engines to bring his crew home at 2:05 pm. Wrapping up a perfect flight of 14 days and 12 minutes, he guided the vehicle to a textbook touchdown on Runway 22 at Edwards at 3:05:42 pm. This put STS-58 firmly into fourth place in the United States' list of long-duration missions and made it the longest Shuttle flight so far.

"Congratulations on a very successful life sciences mission," Capcom Curt Brown radioed the crew as Columbia rolled to a stop.

"From the entire crew, we sure appreciated all the help we got from everyone on the ground," replied Blaha from the flight deck. For three of the STS-58 astronauts – Lucid, Wolf and Blaha himself – their future astronaut careers would involve even longer missions of several months apiece on the Russian Mir space station. In fact, Blaha told the author in 1997 that his goal before retiring was to fly a long-duration station mission. He did so, but it brought his long love affair with the astronaut business to an end by turning into the most stressful mission he ever flew.

COOL CREW

When Commander John Casper and Pilot Andy Allen rode Columbia into orbit on the afternoon of 4 March 1994, they officially became the coolest, most chilled-out astronauts around. "We're as cool as a chilled martini at sunset!" Casper had told journalists with a grin before launch. He was referring to a new set of water-cooled long johns – fitted with more than 320 km of tubing – that he and Allen wore under their partial-pressure suits to keep them comfortable during their eight-and-a-half-minute climb into space.

Not only was their ascent comfortable, but the entire countdown had gone exceptionally smoothly. In fact, the only reason Columbia did not launch as scheduled a day earlier, on 3 March, was due to weather forecasters' prediction that conditions would be unacceptable. Citing "predictable weather patterns", they noted surface winds of 30–50 km/h, easily in excess of the maximum-allowable 24 km/h should an emergency landing back at KSC become necessary. Their ability to make such predictions meant that NASA's Mission Management Team could postpone the launch *before* engineers started loading Columbia's External Tank with cryogenic propellants.

The STS-62 crew during training. Left to right are John Casper, Andy Allen, Pierre Thuot, Sam Gemar and Marsha Ivins.

Instead, the countdown clock was held at the $T-11$ hour mark on 3 March and kept at that point until conditions improved; eventually, half an hour after midnight on the 4th, it started ticking again, the giant tank was fuelled without incident and Columbia lifted off precisely on time at 1:53 pm. By now, the Shuttle's ability to set off a cacophony of car alarms with the roar of its engines had been well proved, and this ascent was no exception: Columbia's noise and vibration were more than sufficient to affect vehicles some 10 *kilometres* away from Pad 39B.

The smooth countdown and picture-perfect ascent to orbit – the only deviation in procedures being a slight delay in sending recovery ships to pick up the SRBs, due to high seas – set the stage for what would be one of the quietest and most problem-free missions flown by the Shuttle. At the risk of being dubbed 'boring', STS-62 was a 14-day flight with a payload bay and middeck literally packed to capacity with a wide range of materials processing, space technology, medical, solar physics and robotics experiments, many of which had flown before.

USMP-2: LONG-TERM PAYOFF

Most visible was the second United States Microgravity Payload (USMP-2), which continued the research begun on STS-52 a year-and-a-half earlier by carrying the joint US/French MEPHISTO furnace and the SAMS acceleration monitor, as well as indulging in new scientific studies with three new experiments: the Advanced Automated Directional Solidification Furnace (AADSF), the Isothermal Dendritic Growth Experiment (IDGE) and the Critical Fluid Light Scattering Experiment (nicknamed 'Zeno' by its sponsors). Like USMP-1, the experiments were mounted on two bridge-like MPESS carriers at the midpoint of Columbia's payload bay and operated via 'telescience' by ground controllers.

"There is the potential for a number of important benefits and applications to come out of this flight," Casper had told journalists in a pre-flight news conference, "but the payoff is not always right away, and that's the way it is with this flight." Some of the benefits Casper alluded to were advanced semiconductors, which could be used in computers, calculators and infrared detectors, as well as materials to make stronger turbine blades for aircraft and powerful electronic components. USMP-2's two directional-solidification furnaces – MEPHISTO and AADSF, both attached to the 'front' MPESS – played a pivotal role in this research.

Building on experience from its previous outing on STS-52, MEPHISTO once again conducted studies of the actual process of 'directional solidification', whereby certain materials were melted, then solidified, from end-to-end. The USMP-1 investigations had traced the motion of the 'interface' between the solid part of a sample and its liquid part; on USMP-2, this would expand to look more closely at the precise location, shape and behaviour of the interface. Three rod-shaped samples of a bismuth–tin alloy were again flown and, despite a temperature sensor glitch on 5 March, the furnace was returning analysable data within a couple of days.

During the USMP-1 experiments, investigators had found regular cellular patterns on the alloy's structure close to the point at which its solidification became unstable. This, it was pointed out before the second mission, was an important discovery because it might enable adjustments to be made in order to better manipulate and control the mechanical and electrical properties of resulting materials. Such materials might possibly be used as the basis for stronger and more resilient alloys in the future. The use of telescience allowed ground controllers to adjust MEPHISTO's operating procedures and improve their data.

In fact, cross-sectional analyses of the samples after Columbia's return to Earth would provide the best-yet information on the shape of the solid-to-liquid interface front, as well as refining the directional solidification process. Meanwhile, the AADSF furnace employed directional solidification to process a cylindrical sample of mercury–cadmium telluride, a material that has seen uses in remote-sensing and astronomical detectors. "It is difficult to grow a good [semiconducting] crystal on Earth from the molten state," said the experiment's co-investigator Frank Szofran, "because of convection: the movement of fluids caused by gravity. In microgravity, the substantial reduction of convection makes it possible to grow much better crystals." Directional solidification and relatively rapid cooling of samples had already been identified as two methods of obtaining suitable electronic properties in them, but could not be employed effectively in terrestrial experiments because convection and the 'settling' of molten components introduce defects and imperfections. These, in turn, can then lead to physical flaws in a crystal's internal structure and an uneven distribution of its chemical constituents.

To process its sample during the USMP-2 experiment runs, the AADSF furnace moved the mercury–cadmium telluride through three separate temperature 'zones' – ranging from a 'hot' region of 870 Celsius to a 'cool' region of 340 Celsius – and, in doing so, slowly cooled and solidified them. This use of three temperature zones, and the one-way directional technique of solidification, yielded a 'flatter' solidification front and crystals which scientists could analyse with greater clarity after Columbia returned to Earth. Analysis of data downlinked from AADSF was met with delight by ground-based scientists.

"Our furnace is operating perfectly," said Principal Investigator Sandor Lehoczky during the mission. "In fact, the experiment has gone so well that we were able to use telescience to create a 'demarcation', or reference point, on the crystal at a critical time in our growth process." This marker precisely determined the location and shape of the liquid-to-solid interface in the mercury–cadmium telluride, which moved along the length of the sample at around 0.7 mm per hour. By the time the experiment was completed on 11 March, members of the AADSF science team were overjoyed with their results.

Telescience had already been used with great fanfare on USMP-1 and investigators transmitted hundreds of commands from the ground to adjust their experiments' settings and parameters and make changes as new and unexpected data and changes emerged. The technique was being regarded as essential for operating experiments on board the International Space Station which, at the time, it was thought would operate for long periods without a permanent caretaker crew. "Telescience is the closest thing a scientist can get, without actually being there, to the way he would conduct his experiment on Earth," said USMP-2 Assistant Mission Scientist Don Reiss.

After Columbia's landing, the AADSF sample was carefully polished and etched to enable investigators to better determine the position and shape of the liquid-to-solid front. As the two furnaces continued their work, mounted on the aft MPESS bridge were the IDGE and 'Zeno' experiments: the former of which grew strange, tree-like crystalline structures known as 'dendrites', while the latter investigated the behaviour of xenon as it approached its so-called 'critical point', at which it simultaneously, and temporarily, displayed the characteristics of both a liquid *and* a gas.

Dendrites, the primary focus of the IDGE investigation, are known to develop as materials solidify under certain conditions; their name derives directly from their tree-like shape, which comes from the Greek word for 'tree'. By increasing scientists' understanding of solidification processes, it was hoped to improve the industrial manufacturing of a wide range of different materials, including steel and super-alloys used in applications ranging from making tin foil to cars and jet engines. Each of these materials are formed under conditions that yield dendrites.

During the course of the USMP-2 mission, dendrites were grown and photographed under a variety of pre-programmed experiment parameters on more than 30 occasions. As each growth 'cycle' ended, the tiny tree-like structures were re-melted and another dendrite produced at a different temperature. In total, they were grown at no fewer than 20 different temperature levels and the experiment as a whole performed beyond expectations. "I couldn't be more pleased," IDGE's Principal Investigator Marty Glicksman told journalists on 8 March. "It's working like a charm!"

Throughout the growth processes, a pair of cameras photographed the dendrites for post-flight analysis. About five pictures were taken during each process and slow-scan television images were recorded and downlinked in real time to ground controllers; this enabled them to make adjustments and changes to their operating procedures where necessary. Compared to Earth-processed specimens, the USMP-2 photographs and video showed that space-grown dendrites grew much faster and were larger. This allowed investigators to spend additional time studying their forms and structures.

"These discoveries and their subsequent development simply could not have been accomplished without going into low-Earth orbit, where gravity is reduced up to a million times," said Glicksman. "We are certain that the basis of current theories about dendritic growth are seriously corrupted by convective effects due to gravity. We are confident that the images and data collected during the IDGE experiment will become the 'standard' for the scientific field for some time to come." By the time IDGE operations ended on 13 March, he added that it had delivered "a goldmine of scientific data."

Zeno, meanwhile, examined the behaviour of xenon at its critical point, where its properties change back and forth from a liquid to a gas so rapidly that neither state is fully distinguishable. Such points are difficult to achieve in terrestrial experiments, because the fluid becomes highly compressible or 'elastic'; its *own weight* compressed part of the sample to a density greater than that of the critical density and ultimately caused it to collapse. This meant that, if it *was* possible to reach the critical point, it could not be maintained long enough to conduct detailed studies of it.

In space, however, the virtual 'absence' of gravity and its limitations served to 'widen' the critical zone and provided clearer insights into the phenomenon. Xenon had been chosen as the sample gas for IDGE, because it was known to develop tell-tale patches of 'milky' iridescence as it neared its critical point. A somewhat slower-than-expected cooling of the Zeno unit shortly after activation put the experiment behind its timeline, but by 5 March – after calibration – investigators reported closing in to within 20 millionths of a degree Celsius of xenon's critical point.

Ultimately, according to Zeno Project Scientist Jeff Shaumeyer, the closest point the experiment reached was within 500 *millionths* of a degree Celsius. On 7 March his colleagues reported that they were seeing behaviour in the xenon unlike any ever observed on Earth, indicating that they were nearing the critical point. "We have gone where no-one has gone before," said Principal Investigator Robert Gammon. After pinpointing the critical point, a temperature search and optical laser measurements were performed to obtain the best possible measurements of the region surrounding the phenomenon.

"These light-scattering optical measurements will help to test theories at temperatures closer to the critical point than is possible on Earth," Zeno Project Manager Richard Laurer told journalists on 10 March. As this research went on, astronaut Andy Allen described the purpose and accomplishments of the experiment in a televised downlink. Using a vial filled with freon, he explained to audiences how gases reached their critical points, and added that, knowing for example the critical point of water led to the development of new techniques to decaffeinate coffee.

Similar understanding of other substances, Allen noted, could yield new knowledge about them.

The fifth and final 'experiment' on USMP-2 was the SAMS accelerometer, which included two sensor heads on each MPESS pallet and another deep inside the IDGE unit. In general, its data revealed that no major disturbances were caused to the sensitive microgravity experiments by the movements of Columbia's crew. It did find, however, that larger-scale accelerations, including thruster firings by the vehicle itself, did have a direct impact on the quality of the MEPHISTO and AADSF crystals and was able to precisely measure these effects.

Unlike many flights, which operate scientific payloads alongside one another, that of STS-62 was virtually *two* separate Shuttle missions in one. The USMP-2 operations were scheduled to consume the first 10 or so days, after which they would be shut down and operations on a second payload, sponsored by NASA's Office of Aeronautics and Space Technology and dubbed 'OAST-2', would take precedence. This would also be highlighted by the deliberate lowering of Columbia's orbital altitude early on 14 March to enhance the data-gathering capabilities of the six OAST-2 experiments.

By the time USMP-2 operations drew to a close on the evening of 13 March, virtually all of the scientific personnel involved praised it as a superb success, made possible only through the cooperation of thousands of individuals on the ground and five astronauts in orbit. "One factor that has really contributed to the success of this flight," said USMP-2 Assistant Mission Manager Sherwood Anderson, "is the teamwork among all the investigators to maximise total science return." Added Mission Scientist Peter Curreri: "We have made basic science discoveries that we had not anticipated. It has been a fantastically rich mission."

OAST-2: A TECHNOLOGY TESTBED

Operations with the $30-million OAST-2 payload commenced immediately after USMP-2 deactivation and followed an earlier mission flown on Space Shuttle Discovery in the summer of 1984. Since that flight, the NASA sponsor had been renamed the 'Office of Advanced Concepts and Technology', but nevertheless OAST-2 retained its original designation. It comprised six experiments mounted on a cross-bay Hitchhiker bridge structure and was specifically detailed to conduct advanced-technology experiments – including evaluations of new-specification solar cells and energy-storage systems – and investigate plasma and atomic oxygen interactions with the Shuttle.

All but one of those experiments – the Thermal Energy Storage (TES) – were controlled directly from an avionics unit attached to the Hitchhiker bridge. TES, which was designed to provide new data about the behaviour of thermal energy storage salts in microgravity, was operated by one of the crew members via a laptop computer on Columbia's aft flight deck. As NASA moved into high gear in its plans to develop and build the International Space Station, it was realised that a low-mass, low-cost and high-efficiency electrical power generation and storage system would become increasingly necessary.

The TES experiment consisted of two GAS canisters affixed to the Hitchhiker bridge, one of which contained a 'salt' of lithium fluoride and the other of lithium fluoride and calcium difluoride eutectic. After activation by Mission Specialist Pierre Thuot early on 5 March, TES collected and stored solar energy which was converted into electricity while in Earth's shadow. As the salts in the GAS canisters absorbed thermal energy, they slowly melted and expanded by up to 30%. Then, when cooled, they solidified and shrank, creating 'voids' in the salts which affected their heat-absorption rates.

It was expected that research conducted by TES would assist with the development of future solar dynamic power receivers being considered for the space station. Typically, the experiment underwent a five-hour 'heat-up' period, 10 hours for completing four separate thermal 'cycles' and a lengthy period of cooling. After Columbia's return to Earth, CAT scans were taken of the canisters to provide data on the sizes and distribution of the voids in the salts. Overall, the experiment was highly successful and yielded the first long-duration, high-temperature melting-and-freezing tests with lithium-based heat-storage salts.

Another OAST-2 investigation with its eyes firmly on the future International Space Station was the Solar Array Module Plasma Interaction Experiment (SAMPIE) which exposed several samples of solar array material to the harsh environment of low-Earth orbit to evaluate their performance and degradation over time. Although there is no 'atmosphere', in the normal sense, in the region of space from which the station would operate, there is still the ionosphere high above Earth comprising a plasma of widely spaced ionised atoms of oxygen and nitrogen. These are known to have potentially damaging effects on high-voltage spacecraft surfaces.

Conducting surfaces, whose electrical potential is highly negative with regard to this plasma, are prone to 'arcing' that severely damages solar cells and causes major electromagnetic interference. Intermittent power loss can also be a problem. Before STS-62, engineers had used ground-based plasma chambers to simulate temperatures and conditions in low-Earth orbit, but were unable to replicate them accurately because of differences in pressure, plasma flow, electron temperature and ion species. During the mission, SAMPIE exposed four different types of solar cells to the space environment, with Columbia's payload bay alternately facing the direction of travel (the 'ram side') and away (the 'wake side').

Generating and storing heat energy, and then routeing it to different parts of a spacecraft, were also expected to prove useful as space station construction got underway. The Cryogenic Two Phase (CRYOTP) experiment evaluated the performance of a nitrogen 'heat pipe', the techniques for which had already been tested on several previous Shuttle missions, in cooling electronic components and sensors. Heat pipes are essentially fluid-filled closed tubes which absorb heat at one 'end' and evaporate the liquid into a vapour. This condenses and moves towards the other end of the pipe, from which the heat is released. The condensate then returns to the heated end of the pipe by capillary forces formed in a porous 'wick' and the cycle is repeated. On STS-62, a device called the Brilliant Eyes Thermal Storage Unit (BETSU) took the procedure a step further by absorbing thermal energy from a heater using a 'phase-change' material known as methyl-pentane in a pair of cryo-cooler devices. After activation, the devices cooled the material down to 120 Kelvin and the heater was switched on to melt it. Several melting and freezing cycles were run throughout the mission to better quantify ground-based models.

Shielding astronauts from the harsh radiation environment of low-Earth orbit would also be essential for station operations and another OAST-2 experiment, known as Emulsion Chamber Technology (ECT), sought to collect data on its impact on a series of photographic plates. Highly sensitive X-ray films were interleaved with sheets of lead and the 'tracks' left in them by radiation particles were examined after the mission. It was hoped that these particle patterns would yield more information about how to safe-guard spacecraft from such radiation. Additional objectives included testing the film's sensitivity against deterioration caused by heat, mechanical vibrations and unwanted 'background' radiation.

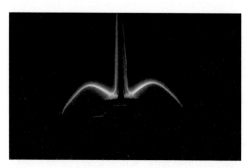

The strange 'Shuttle glow' phenomena envelopes Columbia's OMS pods and tail fin in this view from STS-62.

Not only radiation, but also particles of atomic oxygen and nitrogen in the near-Earth environment, had long been known to potentially have an adverse influence on the Shuttle itself as they ram into it at high velocity. Typically, the impacts of these particles create strange auras of light around the leading or front-facing edges of orbiting spacecraft and can interfere with measurements taken by sensitive optical or other scientific instruments. Two experiments on the OAST-2 payload – the Experimental Investigation of Spacecraft Glow (EISG) and the Spacecraft Kinetic Infrared Test (SKIRT) – were designed to explore these impacts in greater detail.

A set of pressurised nitrogen gas canisters were mounted beneath the sample plate on EISG and used to produce ionising atoms that generated artificial 'spacecraft glow' around Columbia's OMS pods and vertical stabiliser fin. Far-ultraviolet and visible imaging spectrometers then gathered data on a series of samples coated with paints of differing thermal qualities. Infrared sensors also monitored the samples' behaviour. To conduct the experiment, Casper and Allen manoeuvred the Shuttle into seven different elliptical orbits and four circular orbits, one of which took the crew on 16 March to their lowest-yet altitude of just 170 km.

The lowering of the Shuttle's orbit in this way increased the yield of glow-related data. Glow measurements were usually taken during 'nighttime' portions of Columbia's orbit for periods up to about 30 minutes at a time, yielding three-and-a-half hours' worth of data altogether. Additional measurements were taken during scheduled thruster firings and also from Earth's own atmospheric 'airglow'. Meanwhile, the adjoining SKIRT experiment complemented EISG with an infrared-imaging capability. During the course of the mission, the astronauts were required to perform six manoeuvres: four involved nose-down, 360-degree 'rolls' of Columbia and two used the Moon for calibration purposes.

Overall, the experiments were highly successful, although a problem surfaced with EISG's far-ultraviolet spectrometer, which had gathered some good spectral data on 11 March but its capabilities later degraded. Ground-based scientists suspected that it was either operating at a lower gain or was partially blocked. Mission Specialist Marsha Ivins used Columbia's RMS camera at one stage to look at the device as part of troubleshooting efforts. Nevertheless, in general, operations with the two glow experiments ran relatively smoothly and Thuot remarked that the phenomenon was much more pronounced from the lower orbit than when the crew was at a higher altitude.

PREPARING FOR THE SPACE STATION

As well as evaluating materials and technologies needed to build the space station, the astronauts – and Mission Specialist Sam Gemar in particular – assembled and tested a miniaturised *model* of one in Columbia's middeck. Known as the Middeck Zero-Gravity Dynamics Experiment (MODE), it consisted of miniaturised space station 'modules' and components for its girder-like truss structure, which would be used to support electricity-generating solar arrays. The intention of the experiment, which had also been carried on an earlier Shuttle mission in September 1991, was to explore the dynamic behaviour of large deployable structures in the microgravity environment.

In addition to the USMP-2 and OAST-2 operations, a number of other, 'secondary' payloads were accommodated on board the Shuttle. One of these was a nifty extension to the Canadian-built RMS arm called the Dexterous End Effector (DEE). Instead of employing the usual means of capturing targets in space – snaring them with wires that closed around a stubby grappling 'pin' – this experiment used a set of powerful electromagnets to generate an attraction force of 1,450 kg. It was expected that, if such a system achieved 'operational' status, it might increase the robotic arm's dexterity and alignment accuracy.

Furthermore, from the astronauts' points of view, it provided them with a sense of 'touch' with the arm and offered important benefits when building smaller, more compact grapple fixtures for future spacecraft. Although the experiment had been extensively tested on the ground before Columbia carried it into orbit, NASA officials were quick to stress that only a series of tests *in space* could truly demonstrate the electromagnet's capabilities. For STS-62, it comprised four components: a Magnetic End Effector (MEE), a Targeting and Reflective Alignment Concept (TRAC) camera, a Carrier Latch Assembly (CLA) and a Force Torque Sensor (FTS).

During the course of the mission, the new system was rigorously tested. The MEE employed a pair of U-shaped electromagnets to grab and release payloads fitted with a flat, ferrous grapple fixture or 'handle'. This provided a more reliable means of maintaining a good grip on a payload and was a safer method of releasing it in the

Marsha Ivins and Pierre Thuot during DEE tests on Columbia's aft flight deck. Note the joystick-like hand controller for the RMS on the panel in front of them.

event of a problem with the Shuttle or the RMS. If, for example, a mechanical failure occurred while the arm was grappling a large object, the satisfactory closure of Columbia's payload bay doors would be compromised and her return to Earth threatened.

MEE eliminated that risk because its electromagnet could be 'switched off' simply by cutting power to it. It was also a much smaller and lighter device than the standard end effector and had fewer moving parts. Meanwhile, TRAC provided a simpler and faster means of manually or automatically aligning the electromagnet with a 'target' payload by using video images and mirrors. Crew members typically looked through a camera that pointed 'outward' from the centre of DEE at the target until the camera could see its own reflection, then finished the process by lining up a set of cross hairs.

As soon as alignment was achieved, the astronauts 'drove' the arm forward until magnetic forces mated its end effector to the target. The FTS provided feedback on the forces being applied by the arm, while the CLA provided a pair of dummy 'payloads' for use during the demonstration. Beginning on 4 March, within hours of reaching orbit, all three Mission Specialists – Thuot, Gemar and Ivins – took turns evaluating the 315-kg DEE in a series of eight tasks lasting about an hour apiece.

They commented that its ultra-fine guidance would prove useful during delicate space station construction tasks.

Later in the mission, on 15 March, Ivins – who would participate in a major station assembly flight in 2001 – jokingly challenged her colleagues to match the precision with which she had manoeuvred the RMS. Overall, the Mission Specialists returned with good reviews of the new device and used it to insert a series of 'pins' into sockets, with progressively smaller clearance rates from three millimetres to just three-quarters of a millimetre. Additionally, a 30-cm-wide flat beam was inserted into a slot and moved backwards and forwards to correlate readings from the FTS.

SOLAR STUDIES

A number of other, 'secondary', experiments and sensors were also located in and around Columbia's payload bay and middeck. One of these was the Shuttle Solar Backscatter Ultraviolet (SSBUV) instrument, designed to help to calibrate atmospheric ozone measurements being conducted by the Earth-orbiting Total Ozone Mapping Spectrometer (TOMS) and Upper Atmosphere Research Satellite (UARS). It had flown on several previous Shuttle missions as a calibration tool and during STS-62 was used to detect tropospheric sulphur dioxide emissions from several Central American volcanoes, including Colima in Mexico. It also observed sulphur dioxide from industrial byproducts in the troposphere above China and Japan.

The instrument's value lay in its ability to provide very precisely calibrated ozone measurements; it was verified to laboratory standards before launch and checked during and after each mission. This provided a highly precise 'standard' against which data from other Earth-resources satellites – some of which had spent years aloft and whose accuracy may have degraded – could be compared. Specifically, it enabled researchers to update the calibration of another ultraviolet instrument on the National Oceanic and Atmospheric Administration's NOAA-11 satellite, which had been in orbit since 1988. This also allowed calibrations with even earlier TOMS data going back to 1978.

In combination, this provided a highly prized 16-year 'data set' of differing ozone levels in the Earth's atmosphere, which in turn could better inform scientists' theories of climatic changes. Among other disturbing findings, SSBUV data has helped to confirm a 10% ozone depletion in the northern hemisphere at mid-latitudes, resulting from the combined effects of residual aerosols distributed in the upper atmosphere following the eruption of Mount Pinatubo in the Philippines in 1991 and cold stratospheric temperatures during the winter of 1992–1993. Simultaneous observations of the physics and chemistry of the atmosphere by UARS helped to build a more complete picture.

A SUPERB SCIENTIFIC SUCCESS

As part of ongoing research into materials that could withstand the harsh environment of low-Earth orbit and provide potential building blocks for the space station, more than 700 samples in three identical sets were flown as part of the Limited Duration Space Candidate Materials Exposure (LDCE) experiment. Each 'set' was housed in its own container: one remained open throughout the mission, another opened only when the payload bay was facing the direction of travel (the 'ram' position) and a third opened only when the bay faced *opposite* the direction of travel (the 'wake' position).

This allowed station planners to better evaluate the damaging impact of atomic oxygen and nitrogen particles on delicate spacecraft surfaces in different orientations; it also took better advantage of the variety of manoeuvres performed by Columbia's crew during STS-62. In fact, the Shuttle spent its first week aloft with its payload bay facing Earthwards. Then, on the morning of 11 March, Casper and Allen moved their ship so that its tail pointed towards Earth and its payload bay faced the direction of travel in a 'ram' orientation.

Not only did these manoeuvres provide data for the LDCE samples, they also allowed MEPHISTO investigators on USMP-2 to better evaluate the effect of thruster-induced jitters on the directional solidification of their bismuth–tin rods. The IDGE team also saw some interesting changes in their data. "We observed an 8 to 10% change in dendritic growth velocity when the Shuttle changed to a 'tail-down' attitude," said Project Scientist Matthew Koss. "This indicates the exquisite sensitivity of dendritic growth – at very small temperature differences below its freezing point – and our instrument's capability to detect small differences in the microgravity environment."

As OAST-2 operations drew to a close late on 17 March, Columbia's mission had virtually turned into a 'mini-space station' flight in its own right: not only in terms of evaluating technologies that would one day be used to build and maintain the International Space Station, but also in operating a myriad of experiments over a long period of time. Mounted in the back of the Shuttle's payload bay, for the third time, was the EDO pallet, which provided additional cryogenic oxygen and hydrogen reactant tanks to support two weeks aloft.

Also for the third time, Columbia's astronauts participated in readaptation exercises using the Lower Body Negative Pressure (LBNP) apparatus – the cylindrical device which sealed around their waists and drew bodily fluids into their legs – as part of countermeasures to better prepare them for the punishing onset of terrestrial gravity. Throughout the mission, all five STS-62 crew members undertook several 45-minute 'ramp' sessions with the sack-like unit and on 17 March Gemar wore it for four hours. This provided medical researchers with additional data by imitating the sudden return to 'normal' gravitational conditions.

Meanwhile, Casper and Allen, to whom would fall the task of guiding Columbia back through Earth's atmosphere and landing her at KSC, spent part of several of their shifts practising their piloting skills with the PILOT simulator. In recognition of the longer-than-normal nature of STS-62, each astronaut was granted two half-days off during the course of the mission, which gave them ample opportunity to gaze out of the Shuttle's windows at Earth below. Allen also received some happy, and well-deserved, news on 12 March: the US Marine Corps Major was promoted, effective immediately, to the rank of Lieutenant-Colonel.

After a fantastic two weeks in space, a jubilant Casper told journalists during a space-to-ground news conference that the mission showed "that we can do world-class science on the Space Shuttle". The five astronauts then deactivated their multitude of experiments and by 9:30 am on 18 March Columbia's payload bay doors were closed for re-entry. The de-orbit burn followed at 12:16 pm and, streaking through brilliant azure-blue Floridian skies, the spacecraft swooped onto Runway 33 at KSC at 1:09:41 pm. The crew had just missed, by barely an hour, breaking the record set by STS-58 in November 1993.

"It would have been nice to get the record," Casper told journalists after landing, but noted "I think we did a lot of good things." Columbia's landing was not, however, entirely picture-perfect, although many of the details would not be explored in depth until the spring of 2003. Infrared images of the landing showed four fragments of debris falling away from the Shuttle's underside as her nose gear was deployed. Two sources for the debris were subsequently identified: one from around the nose gear's starboard door and another from close to the door's thermal barrier.

Nine years later, during the early stages of Admiral Hal Gehman's presidential inquiry into the cause of Columbia's destruction at the end of the STS-107 mission, the loss of debris during the STS-62 landing was scrutinised in more depth. It revealed no evidence of plasma having entered the nose gear wheel well during that re-entry and determined that the thermal barrier had performed satisfactorily. Nevertheless, the STS-62 incident was serious enough for NASA to spend two days searching the KSC runway for debris. On 1 February 2003, during her 28th atmospheric re-entry, however, Columbia would not be so lucky ...

TWO HISTORIC 'COLUMBIAS'

The name 'Columbia' has long been associated with the people and culture of the United States. It would appear that the name originates from the earliest arrival of European settlers, under the command of Italian-born explorer Christopher Columbus, in the Americas during the autumn of 1492; since then, adjectives such as 'pre-Columbian' and 'post-Columbian' have routinely been applied to the epochs before and after his arrival. Numerous ships have been named 'Columbia', including two that proved instrumental to the United States' fortunes in space.

The achievements of both of these Columbias were appropriately hailed by Capcom Mario Runco on the morning of 16 July 1994, as seven astronauts circled Earth on board Space Shuttle Columbia. "On this day, at this moment, 25 years ago," he told STS-65 Commander Bob Cabana, "three of your predecessors began an epic journey that would change the way we viewed our world. Columbia's journey today, as her namesake did back then, is pushing the frontiers of knowledge and science for all mankind. Thank you, Columbia."

Runco was, of course, referring the exploits of the Apollo 11 crew – Commander Neil Armstrong, Command Module Pilot Mike Collins and Lunar Module Pilot Buzz Aldrin – whose command ship, also dubbed 'Columbia', had ferried them to the Moon to support humanity's first-ever manned lunar landing. A quarter of a century to the day later, Space Shuttle Columbia's crew was midway through a two-week journey that was not destined to travel quite so far, but whose contribution to space science was no less important.

IML-2: AN INTERNATIONAL TRIUMPH

On her last scheduled mission before a planned year-long overhaul, Columbia was again outfitted with the EDO equipment to support her crew for a fortnight aloft. Mounted in her payload bay for the sixth time was the European-built Spacelab module, this time devoted to a variety of life and microgravity science investigations and known as the second International Microgravity Laboratory (IML-2). As its name implies, the mission featured more than 70 experiments provided by 200 scientists from 13 countries, including the multinational ESA, together with Canada, France, Germany, Japan and the United States.

Unlike the IML-1 mission, flown on Space Shuttle Discovery in January 1992, the second flight was twice as long in duration and carried double the number of experiments and research facilities. Several of these were controlled remotely by ground-based scientists, with European and US research teams connected by intercontinental voice, video and data links. Such 'telescience' was seen as useful practice for running long-duration space station experiments. Further, said IML-2 Mission Manager Bob Snyder, "with this amount of science, squeezed into a 14-day mission, it is critical to have both telescience and remote operations".

In another indication of the importance of the experiments, and the need to squeeze as much scientific output from IML-2 as possible, Bob Cabana's crew was split into two 12-hour shifts to operate the laboratory around-the-clock. The Red Team comprised Cabana, Pilot Jim Halsell, Payload Commander Rick Hieb and Japanese Payload Specialist Chiaki Mukai, while their Blue counterparts included Mission Specialists Carl Walz, Leroy Chiao and Don Thomas. The inclusion of Mukai, who became the first female Japanese spacefarer and represented her country's National Space Development Agency (NASDA), highlighted Japan's immense contribution to the IML-2 mission.

The experiments spanned a wide range of the facilities in the Spacelab module, covering both life and microgravity sciences and highlighting the mission's international nature. Germany's Biostack, for example, sandwiched biological specimens between plates of radiation detectors as part of efforts to determine the impact of high-energy cosmic rays passing through the module's outer hull. The specimens were then examined after Columbia's landing to identify the paths and entry points of heavy ions and assess physical changes or damage. The impact of such radiation could, it was recognised, prove detrimental to astronauts on future long-duration space station or Mars trips.

The STS-65 crew gathers in the IML-2 module. Front row (left to right) are Chiaki Mukai, Bob Cabana and Jim Halsell and back row are Rick Hieb, Don Thomas, Carl Walz and Leroy Chiao.

Previous investigations had already confirmed that high-energy particles of solar and galactic origin, and their interaction with Earth's atmosphere, have potentially serious side-effects on living organisms. However, it is not possible to adequately measure these side-effects in ground-based laboratories because our planet's atmosphere – fortunately for us – filters out most of this radiation. During IML-2, three sealed Biostack containers were loaded with shrimp eggs and salad seeds to assess their reactions. Other radiation-measuring experiments were conducted using a Japanese-built monitoring device. Moreover, IML-2 marked the first occasion that such radiation data had been transmitted to Earth in real-time.

Such radiation data was not only important from the points of view of the astronauts themselves, but also the many other biological specimens, animals and fish who called Spacelab their home for the two weeks that Columbia was aloft. One particularly important life sciences facility was ESA's Biorack, which supported a wide range of investigations looking at the side-effects of microgravity and cosmic radiation on genetically modified rapeseed roots, cress seedlings, fruit flies and even human skin cells. Elsewhere in the module was the Aquatic Animal Experiment Unit (AAEU) with its complement of Japanese red-bellied newts, goldfish and Medaka fish.

Obviously, the 'perishable' nature of these living specimens meant that loading them on board Columbia could only take place a few hours before her launch on 8

July 1994. She had returned from STS-62 in excellent condition and routine post-flight maintenance and preparations for IML-2 ran like clockwork. Even the installation of the 9,600-kg, experiment-laden Spacelab module into her payload bay on 9 May, followed by the tunnel adaptor on the 20th, was described as "fairly simple plug-in work" by NASA spokesman George Diller. "There's nothing complicated about it," he added.

FOUR SHUTTLES IN FLORIDA

With the return of Columbia's sister ship Atlantis to Florida in early June, following a lengthy refit in California, all four Shuttles were undergoing simultaneous preparation for missions – a rare occurrence, particularly as KSC only had enough Orbiter Processing Facility bays for three of them! Moreover, continuous early summer rains complicated the issue further, forcing technicians to 'shuttle' the Shuttles backwards and forwards between the three OPF bays and a 'transfer aisle' in the VAB to keep them under cover. Columbia finally entered the VAB for stacking on 8 June and was out at Pad 39A by mid-morning on the 15th.

Once on the pad, the final preparations went relatively smoothly, with the exception of a faulty main engine computer in the last week of June which, although not expected to impact the scheduled 8 July launch date, would mean "we're going to have to work part of the 4 July weekend", according to KSC spokesman Bruce Buckingham. The computer, part of a network designed to detect potential problems and shut down the engines if necessary – as had happened during STS-55's aborted launch attempt a year earlier – was replaced and retested at the pad.

The countdown on the 8th went well until the T−9 minute hold, when concerns were raised over the weather; Air Force meteorologists had predicted a 40% chance of early afternoon thunderstorms at KSC. Their biggest worry was that, should such storms materialise, they might impact visibility at the SLF runway, which Cabana and Halsell would use in the event of an emergency during the first few minutes of ascent. Mission managers restarted the clock and counted down to another hold at T−5 minutes, before stopping again to assess the weather.

Fortunately, none of the holds had to be extended and Columbia rocketed into space at 4:43 pm, precisely at the start of her two-and-a-half-hour window, watched by dozens of Japanese journalists and dignitaries. "People [in Japan] are keeping a close eye on this mission," said NASDA Director Yoshiro Ishizawa. The launch created a blast of such intensity that a corrugated iron roof was torn from a building near Pad 39A. "There was absolutely no way the vehicle or its crew were in danger," Buckingham said later. "The vehicle was several hundred feet off the ground when the roof [came] off."

Columbia's rousing liftoff raised the roof in other ways, too, especially from the point of view of her seven astronauts – four of whom had never ventured into space before. Late on 8 July, as the science crew readied IML-2 for two weeks of intensive research, Cabana showed Mission Control a video recorded from inside the cockpit during the bone-rattling climb to orbit. "As you can see," he commented, matter-offactly, "it vibrates pretty good." The tiny, lipstick-sized camera had been mounted at the rear of Columbia's flight deck and was already being tapped for inclusion in a new IMAX movie.

A BUSY, AROUND-THE-CLOCK MISSION

Cabana and his Red Team crewmates, however, had little time for reflection on 8 July. Shortly after entering space, Walz's Blue Team went to bed for an abbreviated sleep shift, while the Reds oversaw the activation of the Spacelab module and initiation of its first experiments. "We're looking forward to a super two weeks up here," Cabana enthusiastically told Mission Control. Within three hours of reaching orbit, Mukai began the first of what would be 300 hours' worth of operations with the Advanced Protein Crystallisation Facility (APCF) – a device that employed three different methods to grow crystals of valuable proteins.

As well as making crystals of difficult-to-produce and biologically important proteins for pharmaceutical analysis, the facility was also designed to better determine the physical mechanisms that governed the growth process. The three crystal-growth methods used by APCF were vapour diffusion, liquid-to-liquid

Jim Halsell photographs the Earth through Columbia's overhead windows.

diffusion and dialysis. The first suspended proteins at the end of a syringe which was surrounded by material soaked in a concentrated precipitation agent; as water migrated from the protein solution to the precipitation agent, the concentration of protein in the drop increased, supersaturated and the growth process got underway.

The second technique initially separated the proteins, a buffer solution and precipitation agent by a series of 'shutters', which were opened to allow the agent to diffuse into the protein. This caused the protein to become less soluble and initiated crystal growth. Finally, the dialysis method separated the protein and precipitating agent by a thin, semi-permeable membrane which admitted the precipitant and initiated crystal development. Both APCF units were housed in a pair of middeck lockers and, during the course of Columbia's mission, more than 7,000 video images of the crystallisation process were taken.

Meanwhile, in the Spacelab module, Chiao spent part of his first duty shift initiating Biorack experiments. Mukai had already transferred many of the perishable samples from storage lockers on Columbia's middeck into the facility, and late on 8 July Chiao began a five-day experiment to examine the loss of calcium from human bones. A similar investigation had been conducted during IML-1 and its results had suggested that bones typically did not lose a significant amount of calcium, as long as they were exposed to periods of 'compression' – such as vigorous exercise – during flight.

However, it was recognised that more data was needed to counteract the effects of microgravity on the human skeletal system. Biorack was a unique facility in that it could house multiple biological samples, with a total of 19 separate experiments performed during IML-2. Among these was a Norwegian investigation into the growth of genetically modified rapeseed roots and cress seeds. Another study examined the behaviour of fruit flies; its developer, Roberto Marco, was testing his hypothesis that the reason for premature ageing of the insects was due to increased activity as they struggled to move around in the strange microgravity environment.

"Dr Marco tells us the flies in space have been more mobile than their [control] counterparts on the ground at [KSC]," said Project Scientist Enno Brinckmann, the Biorack team's representative at NASA's Marshall Space Flight Center. Two hundred and fifty kilometres above his head, on board Columbia, Rick Hieb was inclined to agree, remarking that the flies were "buzzing around with excellent vitality". By 18 July, 10 days into the mission, however, the astronauts were reporting that the flies' bull-at-a-gate response to microgravity exposure had tailed off and they were acting more like their terrestrial counterparts.

Other Biorack experiments included cultures of human skin fibroblasts – cell-producing connective tissues – and bacterial cells, which had been exposed prior to Columbia's launch to ionised radiation. Don Thomas inserted the samples into Biorack's incubator, to allow them to repair themselves, and from thence into the Spacelab freezer for storage. He also tended samples of baker's yeast, which was being flown as part of Swiss scientist Augusto Cogoli's study into the effects of stirring and mixing on the cultivation of cells in space.

Living and non-living specimens, exposed to varying levels of gravitational influence, were also the focus of research in Germany's Slow Rotating Centrifuge

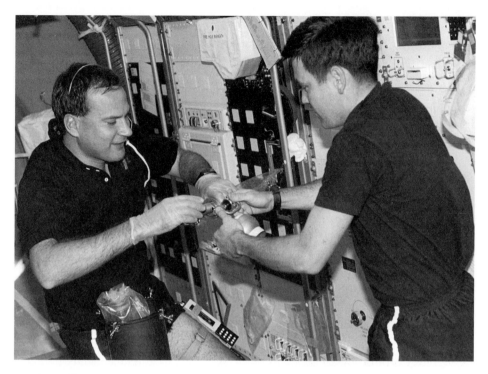

Rick Hieb (left) and Bob Cabana take air samples from the NIZEMI facility in the Spacelab module.

Microscope, dubbed 'NIZEMI'. It was hoped that in space, free of the bulk of Earth's gravitational pull, investigators would be able to determine how certain organisms responded to their unusual new environment and discover more about their internal gravity-sensing mechanisms. During IML-2, the NIZEMI facility imposed gravitational levels ranging from a thousandth of Earth's gravity to around 1.5g on slime mould, *Loxodes*, *Euglena*, jellyfish, cress roots, lymphocytes and a type of green algae called Chara.

In addition to these studies on unicellular and multicellular organisms, samples of non-living matter – including succinonitrile-acetone, a transparent material that solidifies like metal – were also carried in NIZEMI. On 11 July, Chiao inserted samples of the unicellular organism *Loxodes striatus* into the facility to monitor changes in their orientation, velocities and swimming tracks and better determine the stage at which they began to perceive gravitational forces. Investigators believed that such cells functioned similarly to the inner ear of vertebrates, hopefully yielding new insights into the mechanisms by which living organisms sense gravity.

As Chiao worked, Augusto Cogoli – who had another experiment, known as 'Motion', also on board IML-2 – watched video footage of immune system T- and B-cells, which were part of a study into how immune systems operate. Each sample could be observed in unprecedented detail, thanks to the facility's powerful microscope. "Using NIZEMI, we can observe fluid flows and detect the gravity levels at which they begin," said Klaus Leonartz, whose material-solidification experiment used the facility. "We can also determine the effect of the fluid flow on the solid. If we can learn how to make semiconductors or metals more homogeneous, we can improve their properties. By determining gravity thresholds, we can learn how to use other methods, such as electromagnetic forces, to suppress fluid flows during processing on Earth."

Other NIZEMI experiments included cress roots, which Mukai told Principal Investigator Dieter Volkmann of the University of Bonn had all germinated by 13 July. Volkmann was hoping to pinpoint the minimum amount of gravity to which the cress samples would respond, before committing them as a foodstuff on future missions.

Video downlink from the Spacelab module provided him with clear views of how the seeds were behaving. "Thus far, we have observed a difference in gravity sensitivity between the microgravity samples and the 1g samples," Volkmann told journalists on 18 July. "That's a first. The microgravity roots responded in six minutes, while samples grown in the 1g centrifuge took 10 minutes." Meanwhile, Dorothy Spangenberg's study of the development of jellyfish in space was proving "a great success."

PRACTISING FOR THE SPACE STATION

Spangenberg's words not only reflected the exemplary performance of NIZEMI and its experiments, but also the progress of the IML-2 mission overall. As well as providing an excellent testbed for the kind of research that would be conducted on board the space station, Cabana's crew expanded on the experiments performed during IML-1 in the spring of 1992. One of these was a Canadian study of back pain: for many years, two out of every three astronauts had returned to Earth suffering from what was thought to be a result of the lengthening of the spinal column in the microgravity environment.

Canadian astronaut Roberta Bondar had investigated the phenomenon during IML-1 and follow-up research on the second mission focused on determining whether it could indeed be associated with changes in the function of the spinal cord or spinal nerve 'roots' which branch off from it. Concerns had already been raised about the potential impact of spinal column lengthening on changes in the functioning of astronauts' cardiovascular systems and bladders; it was also considered likely that such 'stretching' of the spinal nerves could cause them not to work properly.

During IML-2, the spacing between discs in the spinal vertebrae was measured to determine if that was the reason for the height increases, or if it was actually due to the 'straightening' of the curvature of the back. Typically, IML-1 had illustrated overall height increases of between five and seven centimetres, as well as 'flattening' of the normal spinal contour. Each day throughout their mission, Cabana and his crewmates filled in questionnaires to describe any occurrences of back pain or symptoms of spinal column 'dysfunction', such as a feeling of numbness. Additionally, the astronauts carefully measured their heights daily.

They also took stereo photographs of themselves in different postures to provide further evidence of changes in spinal contour, height and the range of motion of their vertebral columns. This was followed by MRI scans after landing. To stimulate their sensory nerves, they applied tiny electrical impulses to their ankles and measured the time it took for signals to reach their brains using a nerve stimulation and recording device. They also monitored their autonomic nerves by squeezing hand grips, as electrodes gathered blood pressure and heart rate data, and synchronised their breathing on audio tape for another heart rate study.

Meanwhile, a group of Japanese researchers were more than happy with the results of their investigations in the Animal Aquatic Experiment Unit elsewhere in the Spacelab module. This aquarium-type facility provided them with an opportunity to closely examine the spawning, fertilisation, embryonic stages, vestibular functioning and general behaviour of live fish and small amphibians in space. For IML-2, almost 150 pre-fertilised newt eggs, as well as four adult newts, were carried in cassette-sized containers as part of studies of their development throughout the 14-day mission.

Unfortunately, not all of the newts survived, to which researcher Michael Wiederhold told a disappointed Don Thomas that there was always a chance that some might not make it through the flight. In general, the newts developed at about the same rate as ground-based controls. Other AAEU passengers included Akira Takabayashi's goldfish, which were included as part of an investigation into the causes of space motion sickness; on 17 July, live video downlink from the Spacelab module gave him an excellent view as they reacted to light stimulation inside their tank.

Takabayashi later told journalists that, even a few days into the mission, "they [the goldfish] appear to have adapted to the weightlessness of space". Other investigations focused on the adaptation characteristics of Japanese Medaka fish, watched intently by their experiment's Principal Investigator Ken-Ichi Ijiri. The AAEU, which had carried carp on a Japanese-dedicated Spacelab mission two years earlier, performed near-flawlessly during IML-2, providing not only life-support and temperature-control data but also an attachable video system to monitor swimming patterns and fertilisation and embryonic development.

Although life sciences research occupied a significant portion of the crew's time, a large number of facilities were devoted to materials science, processing and fluid physics investigations. One of these was ESA's Critical Point Facility (CPF), which explored the behaviour of certain substances at their so-called 'critical points' – the peculiar stage in which they are technically neither a liquid *nor* a gas, but rather occupy a limbo world in between. More accurately, the material's properties fluctuate backwards and forwards from a liquid to a gaseous state so that its 'bulk' state is indistinguishable.

On Earth, critical point experiments are difficult to perform because, under our gravity, the weight of the material is 'compressed' to a density greater than its own critical density. Scientists were keen to understand the behaviour of different materials at their critical points because such knowledge could provide greater insights into physical problems ranging from phase changes in fluids to magnetisation changes in solids. It had long been recognised that temperature control is extremely important in critical point experiments and the CPF, which was delivered to KSC for IML-2 pre-flight processing in September 1993, included five interchangeable, high-precision thermostats.

One experiment, provided by Dutch scientist Antonius Michels, measured the wave motion of heat within sulphur hexafluoride as it neared its critical point. "The facility functioned flawlessly, especially in providing stability to our sample," he said as his experiment ended on 13 July. Other studies included Richard Farrell's examination of how energy was transported through fluids that had reached their critical points. At one stage in the experiment, a wire was charged inside the test cell to 500 volts to simulate the pressures induced by terrestrial gravity. "The effect was like turning the gravity on and off," Farrell said later.

Elsewhere in the Spacelab module, the Electromagnetic Containerless Processing Facility – known by its German acronym of 'TEMPUS' – provided a levitation melting device for processing metals in an ultra-clean environment. 'Containerless' processing was highly desirable because, on Earth, properties of liquids are known to be affected by the properties of whatever vessel holds them. In microgravity, on the other hand, positioning and control of liquids can be accomplished more accurately and precisely and the amount of power needed for positioning is greatly reduced. This, in turn, reduces motions within the liquid and is less intrusive on the physical phenomena under investigation.

For its first flight on board IML-2, the facility was loaded with 22 spherical specimens, each measuring up to 10 mm in diameter. These were mounted on a storage disk, which rotated until the desired specimen was positioned over a mechanism that transferred it to a processing area where it could be processed in either a vacuum or an ultra-pure helium–argon 'atmosphere'. During processing, TEMPUS provided researchers with the option to manipulate their samples by applying a direct-current magnetic field to change rotation speeds and oscillation rates.

Most of the experiment runs were controlled almost entirely by computer, requiring little in the way of crew interaction, other than starting the facility and shutting it down; TEMPUS was reprogrammed from the ground via telescience, although the astronauts on Columbia could adjust experiment parameters if necessary. In general, the IML-2 operations proved highly successful, although Julian Szekely's study of viscosity, internal friction and surface tension – the forces which keep liquid in a 'drop' – using a 10-mm-diameter sphere of copper had to be halted earlier than planned when the sample inadvertently made contact with its containment cage.

On another occasion, Hieb kept watch over a zirconium–cobalt alloy as the TEMPUS ground team sent commands to levitate, then melt, the small metal sphere inside the facility's processing chamber. It was hoped that such research could aid the production of near-perfect metallic 'glasses' with unique mechanical and physical properties. "The sample looks extremely stable today," Hieb told investigators. In an unexpected moment of serendipity, on 16 July Thomas began processing a niobium–nickel alloy sample and Principal Investigator William Johnson discovered that it had an unknown 'metastable phase'.

In such phases, materials can be quite different from what they are in their 'stable' phases; for example, a diamond is a metastable phase of carbon. "People have been wondering for a long time about the special behaviour of this alloy, but there was no explanation for it," said TEMPUS team member Knut Urban. "The excellent quality of the space images allowed us to detect a phase which had been masked by other forces on Earth." The niobium–nickel sample was solidified and returned to Earth for a detailed microstructural analysis.

Other materials that were melted, levitated and solidified in TEMPUS included an 8-mm-diameter nickel–tin alloy and an aluminium–copper–cobalt sample; all were preserved for analysis after Columbia's landing to determine their structures. Researchers were particularly interested in a newly discovered type of atomic arrangement, known as 'quasi-crystals', which it was hoped could yield materials with high degrees of 'hardness', together with novel electrical and physical properties. This research was of such importance that, during a particularly sensitive TEMPUS run with a sample of pure zirconium on 17 July, Columbia's crew suspended all thruster firings to provide it with an ultra-stable platform.

European participation in the IML-2 mission was represented still further by a facility called RAMSES – a French acronym for Applied Research on Separation Methods Using Space Electrophoresis – which had been developed by the French Space Agency (Centre Nationale d'Etudes Spatiales, or 'CNES') in conjunction with several European industrial partners. Electrophoresis – a process that involves the separation and collection of ultra-pure components of biological substances according to their electrical charges – had been pioneered on several of Columbia's early missions and continues to be an important area of research for the pharmaceutical industry.

Such separation processes are hindered on Earth by our planet's strong gravitational influences, including sedimentation and convection, which have a tendency to 'remix', and thus ruin, the compounds during terrestrial experiments. Chiao activated RAMSES on 13 July, initiating and monitoring a sample of haemoglobin and bovine serum albumin that was being flown to evaluate the degree of protein purification possible in microgravity. "This investigation went better than expected," said Bernard Schoot, whose highly concentrated protein-extract experiment flew on board RAMSES. "Because of the high concentration of protein in this sample, we cannot do this investigation on Earth."

Also provided by Europe was the Bubble, Drop and Particle Unit (BDPU), a multi-user facility designed to study fluid behaviour and interactions, including bubble growth, evaporation, condensation and temperature-induced thermocapillary flows. Previous experiments had revealed the unusual behavioural patterns exhibited by fluids in microgravity – instead of forming teardrop shapes, they become perfectly spherical as their shape is dominated by surface-tension effects rather than the influence of terrestrial gravity. A clear understanding of fluid behaviour in space was considered important for, among other reasons, the design of new and improved spacecraft life-support and fuel-management systems.

Furthermore, the behaviour of fluids was known to have potentially damaging effects on materials-processing techniques, biotechnology and combustion research and surface tension-driven flows also affect crystal growth, welding and flame spreading on liquids. During the IML-2 mission, the BDPU was equipped with a bewildering array of video and still cameras and sensors to carefully scrutinise the behaviour of a variety of fluids in microgravity. One of the most important experiments involved the injection of vapour bubbles into a test cell filled with an alcohol–water solution, after which alternative sides of the cell were heated and cooled.

The results seemed to confirm a theory proposed by the experiment's designer, Antonio Viviani of the Second University of Naples: the bubbles did not always move towards the warmer side of the cell as they might in most materials. "This demonstrates for the first time that, in some fluids of high-technology interest, bubbles can go towards the colder part of the fluid or stop in the middle, due to the particular interaction between temperature and surface tension," he said. It was hoped that such insights could lead to new ways of manufacturing improved glasses, ceramics, composites or alloys.

Later in the mission, as his experiment run concluded on 16 July, Viviani received a round of applause from science teams at NASA's Marshall Space Flight Center – his two-year research and computer-modelling theory seemingly vindicated. "Because of gravity, this experiment could not be done on Earth," he told them, "and it could not be done in space without a good theory, a good facility, outstanding support from the science team and a great crew." Other BDPU investigations included a study of the movement and shape of bubbles and drops in silicone oil and one that involved a three-layer liquid solution.

The Principal Investigator for this experiment, Belgian scientist Jean-Claude Legros, was hoping to learn more about how to control fluid flows within the middle layer of a tri-layered solution of immiscibles; such research could prove beneficial for the production of faster semiconductors and purer crystalline metals. "The rough data we received from our remote support centre in Belgium seems to match our predictions," Legros said on 17 July. On another occasion, Chiao monitored the boiling of a liquid refrigerant, revealing for the first time – and captured by Spacelab's video cameras – the coalescence of two large bubbles, unaffected by buoyancy.

The Japanese, meanwhile, provided their Large Isothermal Furnace (LIF) to heat materials and rapidly cool them to identify relationships between their structures, processing and physical properties. Five samples of ceramic–metallic composites and semiconductor alloys were solidified under varying temperature levels to permit ground-based scientists to refine and improve production practices on Earth. Traditionally, in order to produce lighter, stronger or more temperature-resilient materials, metallurgists combine different metals into alloys, or they may pair dissimilar materials such as metals and ceramics, with suitable physical qualities. However, the key to success is a uniform distribution of various components in the finished product.

This is not always possible, given the complications of terrestrial gravity, which causes substances with dissimilar densities to settle differently as heavier components are pulled downward. The results can diminish the uniformity of a 'new' material's structure, distort its shape and reduce the precision of the casting process needed to produce it. In the microgravity environment, however, where the impact of buoyancy and sedimentation are drastically reduced, it became possible to mix dissimilar materials in spite of great density differences between them.

Usually, LIF followed a pre-programmed heating-and-cooling cycle to process samples – reaching temperatures as high as 1,600 Celsius – while temperature and other data were constantly monitored. As each experiment run ended, helium was injected into the furnace to rapidly cool the sample before removal. The samples were housed in cartridges that the astronauts inserted into the furnace. One of them, provided by Randall German of Pennsylvania State University, examined how gravity changed heavy alloys during a process known as 'liquid phase sintering', whereby two dissimilar metals are combined using heat and pressure, but without reaching the melting point.

A SMOOTH FLIGHT

While the science crew – Hieb, Chiao, Thomas and Mukai – focused on more than six dozen experiments in the Spacelab module during their shifts, Cabana, Halsell and Walz spent their time nursing Columbia's systems as she sailed virtually trouble-free through her 17th mission. One of only a handful of minor problems – a jammed solid-waste compactor piston on the toilet – was quickly fixed by Halsell. The importance of Japanese involvement on the mission, and Mukai's presence, led to a number of calls from dignitaries, including Vice Prime Minister Yohei Kono, NASDA President Masato Yamano and Science and Technology Minister Makiko Tanaka.

It was by now becoming common practice on two-week EDO missions to give each crew member two half-days off duty. The astronauts also participated in 45-minute-long LBNP runs with the cylindrical 'trouser suit' designed to pull fluids into their legs as a potential countermeasure against the punishing onset of return to terrestrial gravity. In general, the device worked well, with the exception of a problem experienced by Mukai when she could not get the apparatus to seal properly around her waist; she later resolved this by wrapping padding around her waist to create a tighter seal.

Cabana and Halsell spent time with the PILOT landing trainer, which helped to sharpen their flying skills after two weeks in the microgravity environment, and they also participated in a series of cognitive tests with the Performance Assessment Workstation (PAWS). This determined their ability to perform operational tasks at different stages of a mission, in order that researchers could better distinguish between the effects of microgravity or fatigue on their performance. It was hoped that this would allow mission planners to reschedule tasks to times when an astronaut's performance was optimum and would be particularly beneficial on long-haul space station flights.

Astronauts, it was recognised, were and still are subjected to a multitude of obvious and hidden stresses during spaceflight, including isolation, lack of privacy, confinement, fatigue and varying work–sleep cycles. Workers placed under similar stresses on Earth are known to suffer from degraded cognitive performance over time and the PAWS tasks were intended to identify these during the course of several long-duration Shuttle missions. On a daily basis, Columbia's crew participated in performance tests on a laptop computer, which monitored the speed and accuracy of their responses to rotated images, letter sequences, calculations, patterns and recollection of numbers.

RECORD-BREAKERS!

Originally scheduled to return to KSC on 22 July, after a mission falling just a few hours short of the Shuttle endurance record set by the STS-58 crew a year earlier, Columbia's crew was obliged to spend an extra day in space due to unacceptable weather in Florida. "Right now, the weather forecast is very good for both Friday [22nd] and Saturday [23rd]," said Flight Director Jeff Bantle early on 22 July, adding prophetically, "but as we all know with the history of KSC weather, sometimes that can change on us."

It did. Both landing opportunities for that day – at 10:47 am and 12:23 pm – were called off due to cloud cover to the east of the SLF runway that Air Force meteorologists were worried could drift over the runway and hamper Cabana and Halsell's visibility on final approach. The weather was fine at Edwards Air Force Base but, with anticipated good conditions in Florida on 23 July, and presumably unwilling to pay the million-dollar fee of ferrying Columbia back to KSC from California on top of a 747, NASA opted to keep the crew aloft for an additional 24 hours.

Immediately after being waved off for 22 July, Cabana and Halsell performed two OMS burns to provide themselves with an additional landing opportunity at Edwards on the 24th if they had to stay aloft even longer. In any case, they were now the Shuttle endurance record-holders: at 4:56 pm on the 22nd, they officially eclipsed the time spent in orbit by John Blaha's STS-58 crew in late 1993. When Capcom Bill McArthur, who had been on board STS-58, congratulated the crew on their achievement, Cabana replied that they could not have done it without Columbia, which he described as "one fine spacecraft".

Seventeen hours later, at 10:38 am on 23 July, on the first of two KSC landing opportunities for that day, Columbia settled gracefully onto Runway 33, wrapping up a spectacular mission of almost 15 days. A subsequent sweep-down of the runway surface also revealed a flattened *fish*, apparently dropped by an osprey or eagle in fright when it heard the sound of Columbia's twin sonic booms! "It's great to be home and we're all feeling great," Cabana told Mission Control as he guided his spacecraft to a halt.

The immense success of the IML-2 research received official acknowledgement from the scientific community in a review meeting at the European Space Operations

Centre (ESOC) in Darmstadt, Germany, in early November 1994, when the consensus from the gathered Principal Investigators was very positive. "This is the important part of the mission, where the scientists get their samples," said Bob Snyder. "In some cases, this is going to take many months, up to possibly several years in the cases where this huge amount of data has to be analysed."

5

Golden age

OVERHAUL

When Columbia disintegrated above Texas on 1 February 2003, killing her seven-member crew, NASA not only lost its flagship Shuttle but also the chance to send her to the International Space Station. For years, Columbia had been restricted to non-station missions because she was the heaviest of the orbiters and, even with major modifications, could not be trimmed down sufficiently to easily carry large components to the orbital complex. Had she landed safely on 1 February after STS-107, however, that restriction would have been lifted and she would have begun processing in earnest for her first trip to the station.

It was not to be. In the aftermath of the tragedy, her payload for the STS-118 flight – a segment of the station's girder-like backbone – was shifted onto one of her sister ships and Columbia's final mission was to provide a research tool for scientists and engineers investigating the development of new thermal protection and other systems for future hypersonic spaceplanes. That she never got to actually fly to the station is a pity, because, during a four-year period from October 1995, she undertook a number of spectacular missions which proved themselves as precursors of the kind of research performed today on board the multinational outpost.

Those 45 months, during which she completed no fewer than nine missions, can arguably be labelled as Columbia's 'golden age'. Eight of them employed the EDO pallet, with its additional reserves of cryogenic oxygen and hydrogen, to support a series of missions that exceeded the 16-day duration mark. Five of them carried the European-built Spacelab module, in which astronauts performed hundreds of investigations in the fields of life and microgravity sciences, biotechnology, fluid physics, protein crystal growth and materials processing. One mission featured the first spacewalks ever staged from Columbia's airlock and another deployed a gigantic X-ray telescope.

On the whole, during her golden age, she performed near-flawlessly, setting a new Shuttle endurance record of almost 18 days in space – which still stands to this day – and carried 48 astronauts into orbit, eleven of them on more than one mission. The length of time it took to prepare her between flights was reduced dramatically, with a processing flow of just 84 days achieved during one turnaround period in 1997. Much of the work to improve her performance was undertaken during a six-month refit at Rockwell International's Palmdale plant in California after the STS-65 mission.

By this time, NASA had already decided that future modifications would be conducted in California. Previously, some of the work had been undertaken in Florida to save the time and money involved in shipping the vehicles across the continent, but it was realised that employing Rockwell staff for the modifications would free up KSC personnel to focus on readying them for flight. "This decision", said Shuttle Director Tom Utsman in March 1994, "will allow the modification effort to be performed by 300 workers at Palmdale, while the 7,000-member KSC team can concentrate on safe and efficient processing."

Originally, Columbia's next modification period was scheduled for the spring of 1995 and would be performed entirely at KSC; this allowed NASA to pencil her in for a 14-day astronomy mission in December 1994 *and* a 16-day microgravity research flight in September of the following year. After the procedural changes, it became increasingly clear that the time needed to remove the hardware from the astronomy mission, fly her out to Palmdale, conduct six months of modifications and inspections and get her ready for an autumn 1995 mission would not permit this.

With this in mind, the astronomy mission – a reflight of the three ultraviolet telescopes from ASTRO-1 – was shifted onto Columbia's sister ship, Endeavour, which gave the workforce a larger window of 14 months to conduct the modifications. Immediately after landing in Florida on 23 July 1994, therefore, the pressure was on to get Columbia out to California well before the end of the year. The IML-2 hardware was removed from her payload bay, her fuel tanks were drained and on 5 October a protective aerodynamic 'tailcone' was installed over her three main engines and two OMS pods.

By 13 October, after a five-day cross-continental trip on top of NASA's 747 carrier aircraft, which involved two stopovers at Huntsville in Alabama and Ellington Field in Texas, she arrived at Palmdale. During the next few months, Columbia underwent 66 modifications to enhance her performance, meet new mission requirements and reduce turnaround times. Among them, she received new wiring to allow her future crews to monitor downlinked data on laptop computers, new filters in her hydrogen flow control valves to reduce the chance of contamination and efforts were made to better control structural corrosion.

Moderate corrosion problems had already been noted, particularly on Columbia's wing leading edges, and improvements were added to rectify the situation. It was speculated that the amount of time she spent on the pad before the installation of a weather-protection system had subjected her airframe to significant amounts of salty spray from the Atlantic. While at Palmdale, she was subjected to more than 460 X-ray and 19 visual inspections. Overall, however, NASA's oldest Shuttle proved to be in excellent condition and, according to Rockwell engineers, was fully capable of completing her estimated operational lifetime of 100 missions.

Aside from the modifications and inspections, two other key differences would characterise Columbia – and her crew – on their next mission. One of these was a set of new-specification main engines, known as 'Block 1'. One of these had been test-flown by Discovery in July 1995, together with two 'old-style' engines, but on her STS-73 mission Columbia would become the first Shuttle to fly into space under the combined thrust of three of the new engines. Central to the system was the so-called Alternate High Pressure Oxidiser Turbopump.

"Completing the flight certification of the [new turbopump] is a major milestone," said main engine deputy manager Otto Goetz of NASA's Marshall Space Flight Center. "The alternate turbopump is now ready for its first flight." Unlike the old-style turbopumps, the newer ones did not need to be removed and inspected after each mission; rather, they only required examination after 10 flights. Their housings were produced through a 'casting' process, which eliminated all but six of the 300 welds that existed in the old turbopump. Moreover, 'harder' ball-bearings, made from silicon nitride, improved reliability and reduced friction during the turbopumps' operations.

The second difference became apparent when the STS-73 crew headed out to the launch pad for the first time: they wore new suits. Since STS-28, each of Columbia's crews had worn David Clark Company partial-pressure ensembles; by 1990, however, work was already underway to develop a full-pressure garment. Known as the Advanced Crew Escape Suit (ACES), its production started in 1993 and the first flight units were delivered to NASA in May 1994. They offered a simple, lightweight and low-bulk suit that astronauts could put on and take off easily, quickly and without help from others.

After her return to Florida on 14 April 1995, Columbia was ensconced in OPF Bay 3 to begin processing for her next mission, STS-73, which was timetabled as a 16-day flight to carry the second United States Microgravity Laboratory (USML-2). In general, its focus was concentrated on the same general areas – fluid physics, materials science, biotechnology, combustion science and commercial space processing – as had been pioneered during the USML-1 mission in the summer of 1992, with many upgraded experiments flying a second time.

A DIFFICULT ROAD TO SPACE

Analysis of data from the first mission had already yielded new insights into the behaviour of fluids under microgravity conditions, led to significant findings for research into developing better spacecraft safety systems, and allowed investigators to develop a clearer understanding of the role of gravity in the production of protein crystals. "The results from these experiments will provide important input into our future technology needs," USML-2 Mission Scientist Marcus Vlasse told journalists on the afternoon of 20 October 1995, as Columbia settled into orbit for the 18th time.

The mission itself would prove an enormous success, but the efforts involved in getting it off the ground would turn into a gargantuan task. On no fewer than *six* occasions would Commander Ken Bowersox and his crew prepare themselves for launch, only to be given the disappointing news of a delay as scrub after scrub plagued their mission. Originally scheduled for launch in late September, Columbia was hauled out to the pad on 28 August. So far, her processing flow had gone smoothly. *That* was about to change.

Already, that summer, a potentially hazardous problem had surfaced concerning erosion in the SRBs' O-ring seals following the STS-71 and STS-70 launches. A leak of hot gas through one of these rubbery seals had led to the loss of Challenger, and Endeavour's STS-69 mission – previously scheduled for 5 August 1995 – was postponed by almost a month until engineers reached a decision that the O-ring damage was within limits. "Depending on how the work goes, there could be as much as a week's threat [to STS-73]," said KSC spokesman George Diller in mid-August. "That would be the worst case."

Fortunately, all went well and Endeavour rocketed into orbit on 7 September, completing a spectacular 11-day mission; her boosters were subsequently checked and "all of the post-flight inspections show that they came out clear", according to Lisa Malone of KSC. Columbia, it seemed, had a green light for her own liftoff on 28 September. However, shortly after technicians began loading her External Tank with 1.9 million litres of cryogenic propellants, the attempt was scrubbed when a sensor detected a hydrogen leak in one of the main engine fuel valves.

The valve was replaced and launch rescheduled for 5 October, but this second attempt to get Columbia off the ground also came to nothing because of fears of high winds, thunderstorms and lightning associated with Hurricane Opal. Weather forecasters expected the hurricane to make landfall near Pensacola on the evening of the 4th, which NASA feared could jeopardise the reloading of the External Tank with propellants and Columbia's launch window the following day. Another try on 6 October was called off when it turned out that fluid had mistakenly been drained from part of the Shuttle's nose-wheel steering hydraulics.

In any event, there was only a 30% chance of acceptable weather conditions, due to the prospect of gusting winds, rain and clouds associated with the hurricane. Columbia's next attempt on the 7th was scrubbed by Launch Director Jim Harrington at $T-20$ seconds following an indication of a fault with one of two Master Events Controllers, which send commands to fire pyrotechnic devices to break the SRBs' hold-down bolts. These controllers are also responsible for separating the spent boosters and External Tank from the Shuttle at the correct instant during ascent.

"We did our best today and we sure hope to see you back here very soon and try all the way down to $T-$zero and a safe launch," Test Director John Guidi told a disappointed Bowersox as preparations to replace the problematic controller got underway. Since the devices were mounted in the spacecraft's aft compartment, engineers firstly had to drain the External Tank of its propellants and the next STS-73 launch attempt was rescheduled for 14 October. This later slipped again to the 15th when additional inspections of the main engines' oxidiser ducts had to be made.

These inspections were mandated following the discovery of a crack in a main engine oxidiser duct which was undergoing tests at NASA's Stennis Space Center in Mississippi. Ultrasonic checks were performed on welds on each of the engine's high-pressure oxidiser turbopump discharge ducts to determine that they were of the correct thickness. Also during the delay period, one of Columbia's five flight deck computers failed and had to be replaced and retested. The 15 October launch target, however, had already come under threat from the weather: it was scrubbed at $T-5$ minutes due to low clouds and rain.

The next launch attempt could not now occur until at least 19 October, because the Eastern Test Range's tracking resources were already supporting another booking: a satellite-carrying Atlas rocket from nearby Cape Canaveral Air Force Station. The Atlas was originally scheduled for 17 October, but was postponed a day because of bad weather, and this in turn pushed Columbia back to the 20th. Even on this seventh attempt, the weather looked grim, prompting KSC spokesman Bruce Buckingham to tell journalists "All you can do is press on and hope for the best."

SEVENTH TIME LUCKY

The repeated launch delays did not appear to have diminished the enthusiasm of Bowersox and his six crewmates – Pilot Kent Rominger, Payload Commander Kathy Thornton, Mission Specialists Cady Coleman and Spanish-born Mike Lopez-Alegria and Payload Specialists Fred Leslie and Al Sacco – as they left the Operations and Checkout Building that morning, wearing back-to-front baseball

STS-73 crew portrait. Front row (left to right) are Al Sacco, Kent Rominger and Mike Lopez-Alegria and back row are Cady Coleman, Ken Bowersox, Fred Leslie and Kathy Thornton.

caps. Sacco would later explain their reasoning: "Our intention was to show that science is *not* for geeks!"

A short, three-minute delay was enforced by a computer glitch when the range command destruct system momentarily experienced a communications dropout, but after so many foiled attempts it seemed that nothing would stop Columbia from spearing into space on 20 October. "Patience and perseverance are a couple of real good virtues to have in this business," said the head of the Mission Management Team, former astronaut Loren Shriver, after the Shuttle's typically rousing 1:53 pm liftoff, "along with a couple of real crack weathermen and a superhuman launch team."

Added astronaut Blaine Hammond, who followed Columbia's climb to orbit from the Capcom's seat in Mission Control, "Good things come to those who wait." Overall, the ascent was normal, with the exception of a minor oil temperature problem with one of three water-spray boilers which required Bowersox and Rominger to shut down an APU sooner than planned. Earlier, as they strapped the crew into their seats, the closeout team at Pad 39B had also been startled by a fire alarm, although fortunately there were no indications of a real blaze having broken out.

Orbiting in a gravity gradient attitude, Columbia's port-side payload bay door is in a 62-degrees-open position to protect its radiators from orbital debris impacts.

In anticipation of the kind of research that the crew would perform during their 16 days aloft, at 3:28 pm, a mere one-and-a-half hours into their mission, the crew opened Columbia's payload bay doors, positioning the port-side one at a 62-degrees-open position, instead of fully open. During STS-73, the Shuttle would operate in a 'gravity gradient' attitude – tail-to-Earth and left-wing-facing-forward – to provide a stable microgravity platform for the sensitive USML-2 experiments. Consequently, the part-open port door helped to minimise the chance of micrometeorites or other orbital debris hitting its delicate radiator panels.

Later, on the afternoon of 25 October, Bowersox briefly opened the port-side door fully for about an hour to enable a waste-water dump from Spacelab's condensate tank to take place. Such dumps had to be performed every few days to get rid of water from the module's dehumidifiers. The payload bay door had to be fully opened during the dump to provide sufficient clearance for the waste water, which was ejected through a nozzle on top of the module's forward endcone. After the completion of the dump, Bowersox returned the door to its 62-degrees-open position.

IMPORTANT GOVERNMENT BUSINESS FOR FRED LESLIE

It was already common practice on research missions of this type for the crew to work in two 12-hour shifts to maintain operations around-the-clock: the Red Team comprised Bowersox, Rominger, Thornton and Sacco, while the all-rookie Blues included Lopez-Alegria, Coleman and Leslie. As with USML-1, which carried two world-class research scientists as Payload Specialists, this second mission proved no exception: Sacco had developed several of the onboard zeolite crystal investigations, while Leslie helped to design a unit called the Geophysical Fluid Flow Cell (GFFC), which was intended to mimic the dynamics of planetary and stellar atmospheres.

"We're primarily researchers and scientists," Leslie said of the role of the Payload Specialist, "and the disadvantage is that often you only fly once." The advantages, however, were that he received a flight assignment much more quickly than 'career' NASA astronauts: "I went into the programme in 1994 and flew in '95. So it was fast!" He also added that, since he was on government business, he received extra pay, but after 'transport' and 'accommodation' deductions, this amounted to just a couple of dollars per day! "You know the government," Leslie grinned. "You can't go anywhere without travel orders!"

During the mission, Leslie had ample opportunity to work with the GFFC, which consisted of two 'hemispheres' – a baseball-sized one, made from stainless steel, mounted inside a larger, transparent one of sapphire – both of which were affixed to a turntable. A thin layer of silicone oil filled the gap between the hemispheres. During operations, the temperatures of both hemispheres, together with the rotation speed of the turntable, were minutely adjusted by the experiment's computer, which also introduced thermally driven motions into the oil. This allowed physicists to simulate and model fluid flows in the atmospheres of rotating stars and planets.

A similar experiment flew on Spacelab-3 in April 1985 and revealed several types of convection difficult to study on Earth, as well as enabling researchers to observe their structures, instabilities and turbulence. During USML-2, the GFFC was employed, among other tasks, to mimic conditions in Earth's 'mantle' – a region of predominantly magnesium-rich silicate rock sandwiched between the core and lithosphere, which undergoes solid-state flow – as well as plasma flows on the Sun. After activation late on 21 October, Principal Investigator John Hart and his team at the University of Colorado at Boulder controlled their experiment remotely from the ground.

Simulating plasma flows very close to the solar surface had been another of the investigations conducted during Spacelab-3 and research on Columbia 10 years later allowed them "to compare USML-2 results with our previous experiment," said Hart. "The instrument seems to be working great." It was hoped that data from GFFC would enable solar physicists to develop better computer models of fluid behaviour on the Sun. "This is all kind of new and exciting stuff," Hart told journalists a few days later, as he watched time-lapse movies of simulated solar flows from the Spacelab module.

The movies were made from a series of still images, snapped every 45 seconds, as the fluids swirled between the unit's two rotating hemispheres. "You can see a lot of the evolution of these solar dynamic flows we've been interested in, and we've seen some surprising turbulence," Hart told Columbia's crew. "We are comparing these results with our computer simulations and other theoretical ideas to understand the extensive turbulence which starts near the polar region and spreads rapidly towards the equator."

Despite minor problems relaying temperature parameters to the GFFC, Leslie was happy with its performance, calling it "a planet in a test tube." Later in the mission, efforts were made to simulate activity in Jupiter's atmosphere, which is of particular interest because this planet radiates significantly more heat than it receives from the Sun, contracting and liberating gravitational potential energy as heat. "These early runs show dramatic changes in flow types, with very small variations in the instrument settings," said University of Colorado team member Scott Kittelman. This focus on mimicking solar and Jovian atmospheric dynamics continued throughout the mission.

"We hope this will give us ideas about the hidden inner atmosphere of Jupiter," said co-investigator Dan Ohlsen on 1 November. However, several experiments had applications that were potentially useful in understanding our own planet. "We can isolate certain phenomena with the experiment," said Leslie during a news conference from orbit on 3 November, "and take out some of the complicating things in terms of climate – things like rainfall and clouds." During each experiment, parameters such as voltage, rotation speed and temperature were adjusted to create unstable and turbulent flows to better explore the dynamics of both oceans and atmospheres.

CRYSTAL GROWTH RESEARCH

Elsewhere in the 10,310-kg USML-2 module were several other experiments modified or improved in the wake of their first flight three years earlier. ESA's Glovebox, for example, had been outfitted with a much larger working area and better lighting for this mission, during which it supported no fewer than seven separate investigations. Experiments on USML-1 had yielded protein crystals of much higher quality than had ever been achieved previously, and by 21 October video from the Spacelab was showing a busy Sacco getting the first zeolite investigations up and running.

Zeolites are used on Earth for the purification of fluids in life-support systems, as well as in the petroleum-refining process and in waste-management and biological fields. In effect, they act as 'molecular sieves' to separate out specific molecules from solutions and might someday allow gasoline, oil and other petroleum products to be refined far less expensively. In addition to operating experiments in the Spacelab module, Sacco tended to his own Zeolite Crystal Growth (ZCG) furnace in Columbia's middeck, which processed 38 sample containers as part of efforts to create large, near-perfect crystals.

Results from USML-1 had already shown that zeolite crystals whose nucleation and growth were carefully controlled from the start of the experiment achieved a much higher level of perfection than Earth-grown ones. "We want to learn more about how zeolites nucleate and grow and learn more about their structure, so we can apply that knowledge to different processes on Earth," said ZCG team member Nurcan Bac of Worcester Polytechnic Institute in Massachusetts. "For instance, one of the zeolites in this experiment is widely used by the petroleum industry to 'crack' heavy oils into gasoline. If we can increase the efficiency of this type of zeolite, we could get more refined petroleum products from the same amount of crude oil." Sacco agreed, pointing out that such experiments could lead to major breakthroughs in the future, thus "pointing us in the right direction for space station research".

Other investigations crammed into lockers in the middeck included a new container known as the Diffusion-controlled Crystallisation Apparatus for Micro-gravity (DCAM), which pioneered a method of autonomously running long-duration protein crystal growth experiments on board the International Space Station. During USML-2, more than 1,500 protein samples were processed, despite a requirement to lower Columbia's cabin temperature early in the mission when one of the experiments' thermoelectric coolers overheated slightly.

GROWING CROPS IN SPACE

Located in the middeck was Astroculture, which had also been on board USML-1 and was flying its final 'test' mission to grow and provide nutrients for potato plants. It grew 10 small tubers to evaluate the extent to which microgravity affected starch accumulation in the plants. Starch, of course, is an important energy-storage compound, but evidence from previous missions had suggested that its accumulation

Cady Coleman, Fred Leslie and Mike Lopez-Alegria bail out of their phonebox-sized sleep stations in the middeck to begin another 12-hour shift in support of USML-2 operations.

was somewhat restricted in microgravity; nevertheless, Astroculture scientists were seeing new growth in their tubers as early as 22 October.

"They look very happy and well, staying very turgid," said Co-Investigator Ted Tibbetts, "which means they have not wilted, so the environment is good for them." Meanwhile, Astroculture's Principal Investigator, Raymond Bulla of the University of Wisconsin at Madison, pointed out leaf patterns in a photograph of the growing potato plants. "The thing that's been exciting to us is that, in the three images that have come down at different times, the leaves are in the same position," he said. "This means that the plants are healthy. If they are having problems, the leaves would have shifted."

By 1 November, when Sacco downlinked video footage of the potatoes, their leaves were starting to wilt, although they continued to perform satisfactorily. This enabled investigators to conclude that Astroculture itself *was* providing the proper nutrients, light, water and humidity. When she landed on 5 November, Columbia brought back five successfully grown potatoes from the experiment. After this final test – Astroculture had previously been carried on USML-1 and two Spacehab missions, during which lighting, humidity, pH, nutrient supply and carbon dioxide and contaminant subsystems were validated – the hardware was made available commercially for sale or lease.

THE 'BIG PICTURE'

Sacco's earlier comment, in relation to his zeolite investigations, of USML-2 pointing scientists in the right direction for space station research, was certainly not the last time Columbia's astronauts would remark on this 'bigger picture', of which their mission was only a small constituent part. "This is a kind of pathfinder for the kind of investigations we'll have on the space station," Kathy Thornton told journalists during a space-to-ground news conference on 2 November. "There are very complicated experiments onboard, but they're working beautifully."

Thornton had been training for this mission for 19 months and took a leading role in activating the USML-2 experiments within hours of Columbia reaching space. As with most Spacelab missions having a microgravity-research emphasis, the module was outfitted with a number of devices to measure the effect of Shuttle accelerations on very sensitive experiments. As Thornton powered-up USML-2 on the afternoon of 20 October, Sacco switched on the mission's two accelerometers, which, said Project Scientist Alex Pline, "allow us to make a judgement as to whether to wait for external movements to settle down before beginning an experiment run".

One of the accelerometers' sensor heads had been deliberately positioned next to a major facility known as the Surface Tension Driven Convection Experiment (STDCE), which had also been on USML-1. This enabled scientists to view in great detail the behaviour of fluid flows in microgravity when there were temperature differences along their interfaces. Such 'thermocapillary' flows occur in many industrial processes on Earth, and when they manifest themselves during melting or resolidification they can create defects in crystals, metals, alloys and ceramics. Gravity-driven convection overshadows the flows in ground-based tests, making them difficult to measure. High above the Earth, in a microgravity environment, on the other hand, it became possible to explore them in greater detail. Fluid physicists expected this knowledge to provide clearer insights into bubble and drop migration, which in turn could aid the development of better fuel-management and life-support systems. As with its previous mission, STDCE used lightweight silicone oil as the test fluid and a laser diode and several cameras measured its transition from a 'steady', two-dimensional flow into an 'oscillatory', three-dimensional one; similar, ground-based tests had shown periodic variations in the fluid's motion speed and temperature.

By comparing conditions for the onset of this oscillation in microgravity and on Earth, it was hoped to identify its cause. During the USML-1 experiment runs, investigators concentrated on steady-state fluid flows, with the results confirming many theoretical predictions from scientists at Case Western Reserve University in Cleveland, Ohio, although no oscillations were actually observed. For the second mission, however, the apparatus had been modified to carry three different-sized experiment containers and several viscosities of oil to create more favourable conditions for oscillation. The facility's imaging system was also upgraded to improve observations of any oscillations.

The new optics, developed by H. Philip Stahl and students from Rose Hulman Institute of Technology in Terre Haute, Indiana, provided precise images of oil surface shapes and flow patterns. "Once we understand when and how oscillations occur," said STDCE Co-Investigator Yasuhiro Kamotani of Case Western, "we should eventually be able to design processes to control them." On 21 October, a day into the mission, Leslie reported the onset of oscillations during the very first

STDCE run, as ground controllers watched three different views simultaneously downlinked by Spacelab's new television system.

Two days later, the astronauts drew down the volume of silicone oil to create a concave surface. As a laser gradually heated the surface, investigators identified the transitional point where oscillations began to occur in each run. "We've never seen this kind of transition before," said Pline of NASA's Lewis Research Center, "because we have no way to create a large curved liquid surface on the ground." After watching four STDCE runs, Principal Investigator Simon Ostrach believed that the onset of oscillations was affected by heat sources, temperature distribution on the fluid's surface and the dimensions of its container.

In subsequent runs, Coleman lowered and increased oil levels to create deeply concave, then convex, surfaces. The 'size' of the laser beam on the oil surface was also adjusted to determine its effect on the direction and nature of fluid flows, as well as lifting its temperature to introduce oscillations. This led to several never-before-seen phenomena. During one test on 28 October, investigators watched erratic flows with no apparent organisation or pattern as Thornton increased oil temperatures beyond the point at which flows began to oscillate. Later, Sacco stepped aside and let ground controllers run the facility remotely via telescience.

THE STRANGE BEHAVIOUR OF FLUIDS IN MICROGRAVITY

Sacco also demonstrated 'surface tension' in space on 2 November, by squeezing orange juice out of a tube. It promptly formed a ball, thus highlighting how surface tension makes fluids assume spherical shapes when they no longer have to contend with gravity. The six-channel, simultaneous video footage from STDCE was only possible thanks to the new High-Packed Digital Television Technical Demonstration, nicknamed 'HI-PAC'. Prior to USML-2, video from Spacelab could only be downlinked on a single channel. As part of the tests of the new system, the crew acquired live images of colleagues sitting at their consoles in Mission Control.

Other fluid physics investigations were conducted in the Drop Physics Module (DPM), another facility carried over from USML-1, which sought to explore future ways of conducting 'containerless' materials processing and encapsulation of living cells for pharmaceutical research. One investigation was the Drop Dynamics Experiment (DDE), provided by Spacelab-3 veteran Taylor Wang and his team from Vanderbilt University in Nashville, Tennessee, which gathered high-quality data in support of such pharmaceutical applications. During USML-2, the experiment investigated the breaking-apart of distorted drops under a fluid of varying viscosity, as well as attempting to encapsulate and create 'compound' drops.

It was hoped that such 'encapsulation' tests would someday allow methods to be routinely exercised to insert living cells for the treatment of hormonal disorders into polymer shells to protect them from immunological attack and provide timed releases. Instances where such techniques might be useful include the treatment of diabetes – perhaps by injecting a pancreatic cell that secretes insulin into the body. Normally, the foreign cell would be attacked by the patient's immune system, but if encapsulated in a shell strong enough to withstand attack, yet porous enough to excrete the insulin, it could provide an effective treatment.

A second DPM experiment was provided by Robert Apfel of Yale University and examined the influence of 'surfactants' – substances which alter a fluid's properties by aiding or inhibiting the way it adheres to, or mixes with, other substances – on the behaviour of drops. On Earth, surfactants are routinely used: soap and water interact in dishwashers, for example, and cosmetics manufacturing, the cleaning-up of oil spills and the dissolution of proteins in synthetic drugs also rely heavily upon them. Apfel's study focused on the oscillation of single drops and the coalescence of several drops with different concentrations of surfactants.

On 22 October, the facility's Project Scientist, Arvid Croonquist, and his team watched live downlinked video of the first liquid drop deployment as Thornton released a 2-cm-diameter glob of water and used precisely controlled soundwaves from a pair of loudspeakers to manipulate its movements. She reported that its performance was as expected. The other members of the science crew also worked with the DPM over the following days, with Sacco commenting "I think it's beautiful" after deploying a 2.5-cm-diameter drop of water on 24 October.

The next day, a cheer rose from Taylor Wang's team when Leslie succeeded in the delicate process of encapsulating an air bubble inside a floating water droplet. "There are actually two factors at work in liquid spherical shells: fluid physics and chemical reactions," Wang told journalists. "With these experiments, we are able to separate them." Sacco later manipulated drops treated with an organic surfactant known as bovine serum albumin for Apfel's study. He deployed two drops and brought them perfectly together until they coalesced into a single blob.

"These are the first and best drop coalescences we've ever had," said DPM Co-Investigator (and USML-1 veteran) Gene Trinh of JPL, adding that "it went very nicely." Cady Coleman, a rookie astronaut, was enjoying not only her first experience of microgravity, but also its unusual effects on fluids: on 31 October, she stretched a 'bridge' of liquid between the facility's two injector tips and later precisely centred a drop of water inside a glob of silicone oil. Moreover, she achieved this while the interfaces of both liquids were moving in opposite directions, in a condition known as 'slosh mode'.

Both Croonquist and Wang praised her efforts, calling them "very important" and having "major significance." From Columbia, Coleman seemed pleased that after a year and a half in the Spacelab simulator practising encapsulating drops, "it's sure nice to see it for real." She also worked with Croonquist, who uplinked instructions to her to use soundwaves to manipulate and finally split apart rotating drops of silicone oil.

Later, Sacco succeeded in forming a polymer membrane between two different chemicals, which coalesced into a single drop. "A picture is worth a thousand words – we'd seen it, now we've done it!" exulted Wang. After deploying the drops, Sacco had allowed them to coalesce and a membrane formed as a consequence of the ensuing chemical reaction – which proved to be an important step in the encapsulation studies.

SEMICONDUCTOR FACTORY IN SPACE

Another facility that had flown on board USML-1 was the Crystal Growth Furnace (CGF), capable of achieving very high operating temperatures in excess of 1,000 Celsius and processing large crystals of semiconducting materials, together with metals and alloys. During USML-2, it was employed to grow samples of cadmium–zinc telluride, mercury–cadmium telluride, gallium arsenide and mercury–zinc telluride. Cadmium–zinc telluride crystals grown on the June 1992 mission had already proved themselves to be the most defect-free specimens ever produced and achieved particular perfection when they did not come into contact with the container walls.

In response to this discovery, the primary CGF sample chamber for USML-2 was outfitted with a spring-loaded piston, which moved to reduce the volume of the cylinder as the material contracted while cooling. This helped to eliminate air voids in the crystal and ensured that it maintained an 'even' contact with the container walls along its entire surface.

Cadmium–zinc telluride is routinely used as a base on which infrared-detecting mercury–cadmium telluride crystals can be grown; the alloyed material – zinc – helps to reduce strain and crystal defects, but makes the resultant alloy relatively soft and easily deformable when produced on Earth. In space, on the other hand, it is possible for investigators to pinpoint the constituents within a crystal that could be altered to improve growth technologies. Already, USML-1 had produced cadmium–zinc telluride crystals *one thousand times* better than any grown on Earth. During USML-2, Principal Investigator David Larson hoped to reproduce this success.

Mercury–cadmium telluride is also used as a construction material for infrared detectors, while gallium arsenide crystals form an integral part of high-speed, low-power digital circuits, optoelectronic-integrated circuits and solid-state lasers. However, during USML-2 scientists were keen to learn more about the processes by which impurities were distributed throughout compounds during their growth. Finally, mercury–zinc telluride crystals were known to have properties which, theoretically at least, make them superior infrared detectors for use in defence, space exploration, medicine and industry.

Moreover, by carefully modifying the proportions of mercury, zinc and tellurium within the alloy, it was hoped to adjust a resultant material's electronic and optical properties for a wide range of applications. Each of the materials was processed successfully by the CGF, in addition to a fifth sample – of germanium, with a trace of gallium – to demonstrate the influence of changes in Columbia's attitude on the crystal growth process. In total, eight semiconducting crystals were grown, as well as a very thin crystal and two crystals that could form the basis for faster and less power-hungry computer chips.

Early on 26 October, the furnace finished melting David Matthiesen's gallium arsenide crystal and slowly resolidified it at just 1.8 mm per hour, operating at its highest-ever temperature, close to 1,250 Celsius! Throughout the resolidification, an electric pulse 'marked' its exact growth rate, as well as its liquid-to-solid 'boundary' for Matthiesen's team. The experiment also involved the addition of a small amount

– known as a 'dopant' – of selenium. "There is less than one part per million of selenium in this sample," said CGF researcher Frank Szofran. "Yet it greatly alters the electrical conductivity of the semiconductors."

Two days later, Thornton and Sacco completed the furnace's shortest-ever processing run: depositing a thin layer of mercury–cadmium telluride on a base material in less than 90 minutes. "We're examining ways to reduce crystal 'birth defects', which are transferred from defects in the substrate material that can't be eliminated," said Principal Investigator Heribert Weidemeier of the Rensselaer Polytechnic Institute. "In the crystal we grew on USML-1, the interface between the substrate and the first layer was much smoother than in crystals produced on the ground. This was totally new, something we had never seen before and had not expected."

On USML-2, Weidemeier's team grew thinner layers to determine the propagation limits of substrate defects in crystals. During similar experiments on Earth, the crystal material firstly forms separate 'islands' on the base, which join to form a complete layer after about two hours of growth. "We deliberately stopped growth in less time than it takes for a layer to form on Earth," Weidemeier said on 29 October. "However, there is a good chance that under microgravity we may get a complete layer in the shorter time period." He hoped this might eventually lead to faster and cheaper terrestrial growth methods.

Other investigations on USML-2 included a record-breaking 1,500 protein samples in both Columbia's middeck and the Spacelab module. In the latter, much

Kathy Thornton works with the Glovebox in the Spacelab module.

of the work was performed in ESA's Glovebox, which Sacco used on 22 October to run an experiment for the Center for Macromolecular Crystallography in Birmingham, Alabama. It included feline calcivirus, akin to a virus responsible for causing digestive problems in humans, and proved so well-formed that it elicited applause from team members. Sacco also set up initial conditions for the growth of a collagen-binding domain protein, which is important in the study of arthritis and joint disease.

Later, Coleman activated a protein known as duck delta crystallin, which is similar to the protein responsible for causing a rare, but deadly, disease in humans. "The reason we do protein crystal growth is to be able to find out what the protein looks like in order to find a cure for a disease," she told journalists during a space-to-ground news conference; the shape of a protein molecule controls its function. The USML-1 mission had proved tremendously successful: of the 30 or so proteins flown, nearly half yielded crystals that were large enough for X-ray analyses of their structures.

A FLAWLESS MISSION

Combustion science was another focus of Glovebox research and, in recognition of his efforts, Leslie received a collective 'high-five' for his "outstanding work" on 2 November, according to team members of the Fibre-Support Droplet Combustion (FSDC) experiment. "Fred gets an A+ in combustion theory!" they exclaimed. He had been busily placing droplets of fuel on a thin fibre, using needles in the experiment module and igniting them with a hot wire. Despite early difficulties with clogged drop accumulators, he performed a number of excellent 'burns', varying the quantities of methanol and methanol–water fuel with each run.

Meanwhile, Columbia herself was performing near-flawlessly on her 18th journey into space. On 23 October, Bowersox told Capcom Tom Jones and Flight Director Rob Kelso that he was impressed by her rock-steady nature. A couple of days later, the crew took time out to tape the ceremonial first pitch for Game Five of the baseball World Series. Uniquely, it marked the first time that the thrower – Bowersox in this case – was not actually in the ballpark for the pitch. The video from Columbia was replayed on the enormous Jacobs Field Jumbotron screen in Cleveland before the game.

Before throwing the slow-spinning pitch, Bowersox wished both the Atlanta Braves and Cleveland Indians good luck; the entire crew later signed their onboard baseballs and gave them to Major League Baseball to be enshrined in the Baseball Hall of Fame in Cooperstown, New York.

As the mission's second week aloft wound down, the crew was asked to place Columbia in several different attitudes to increase temperatures across her belly; flight controllers were concerned that her tyre pressures were lower than acceptable for landing. Thermal conditioning of the belly brought them up from 328 to 334 psi, which were well within the required landing pressure minimum limit of 330 psi. The science crew, meanwhile, finished the remaining USML-2 experiments and the module was deactivated by Lopez-Alegria's Blue Team in the early hours of 5 November.

Right on cue, Bowersox and Rominger fired Columbia's OMS engines at 10:46 am to begin the hour-long descent back to KSC. With double sonic booms echoing across Merritt Island, she touched down at 11:45:22 am, completing a mission of 15 days, 21 hours, 52 minutes and 22 seconds; it was Columbia's longest flight to date and the second-longest in the Shuttle programme. Endeavour's ASTRO-2 mission a few months earlier had already established a new record of more than 16 days, although it would not last for long. In 1996, Columbia would surpass even herself . . .

SATELLITE ON A STRING

On 31 July 1992, Space Shuttle Atlantis headed into orbit with one of the most unusual experimental payloads on record – which STS-46 Mission Specialist Marsha Ivins described as "weird science" – known as the Tethered Satellite System (TSS). Originally conceived by the late Professor Guiseppe Colombo of Padua University, it was intended to demonstrate the 'electrodynamics' of a conducting tether in an electrically charged portion of Earth's atmosphere called the ionosphere. It was envisioned that Colombo's idea might ultimately lead to systems that would use tethers to generate electricity for spacecraft, using Earth's magnetic field as a power source. Furthermore, by reversing the direction of the current in the tether, the force created by its interaction with Earth's magnetic field could potentially put objects in motion, thus boosting a spacecraft's velocity without the need for precious fuel, thereby counter-acting air drag.

Artist's concept of TSS leaving the Shuttle's payload bay.

In the late 1980s and into the 1990s, this was particularly appealing to the designers of the International Space Station, as a means of compensating for the effects of atmospheric drag on the colossal outpost. Additionally, it was optimistically hoped that the concept could lead to the development of devices to trail scientific platforms far below orbital altitudes in difficult-to-study zones, such as the fragile ozone region over the South Pole. Other applications for such tethers included serving as extremely low-frequency antennas, capable of penetrating land and seawater, and perhaps generating artificial gravity or boosting payloads into higher orbits.

During the eight-day STS-46 mission in the summer of 1992 – five years later than planned, due to the Challenger disaster – the satellite underwent its first demonstration in Earth orbit. Unfortunately, it achieved only partial success, when its 20.5-km tether snagged on a bolt in the deployment reel mechanism and refused to unroll more than about 258 m.

Nevertheless, the mission proved the concept sufficiently satisfactorily for a reflight to be proposed in the spring of 1996. A two-week-long flight was required to allow additional time for full deployment and a few days of ionospheric research. Most of the original STS-46 crew were also kept together for the reflight: Andy Allen, who had flown as Pilot of the first mission, was promoted to Commander of its second trip into orbit. He was joined by Payload Commander Franklin Chang-Diaz and Mission Specialists Jeff Hoffman and Claude Nicollier, all of whom had been on STS-46.

Rounding out Allen's crew on STS-75 were a rookie Pilot named Scott 'Doc' Horowitz – the first Shuttle pilot to hold a PhD – and two Italian astronauts: Mission Specialist Maurizio Cheli and Payload Specialist Umberto Guidoni, the latter of whom had also served as an STS-46 backup crew member. Together with Swiss-born Nicollier, this became the first Shuttle mission with three ESA astronauts on board. Unusually, the seven men were split into *three* separate shifts for the TSS deployment operation, reverting to a Spacelab-style dual-shift system after the completion of operations with the tethered satellite.

The Red Team consisted of Horowitz, Cheli and Guidoni, the Blue Team of Chang-Diaz and Nicollier and a unique 'White' Team of Allen and Hoffman; this unusual, staggered shift was added to enable the crew to operate a suite of instruments on the TSS to gather real-world data about how conducting tethers might some day be used to generate electrical power in space. After completion of deployment operations, it was planned for Allen to rejoin the Blue Team and Hoffman the Reds.

The satellite that was the primary focus of these three shifts was a 1.6-m-diameter, Italian-built sphere, weighing 517 kg, with an outer skin of aluminium alloy and coated with an electrically conducting layer of white paint. It was, however, far more than just an oversized metallic football. Piercing its shell were windows for Sun, Earth and charged-particle sensors, a connector for the umbilical tether and doors that provided access to its onboard batteries. Extending from one side of the TSS was a long, fixed instrument boom, while a shorter antenna sprouted from its other side.

To assist with the thermal control of the satellite, the interior of the spherical shell was painted black. If one were to break open the satellite, like an egg, one would see two separate compartments: a Payload Module, housing its scientific instruments, and a Service Module for its subsystems. Additionally, in the centre of the spherical shell was a pressurised nitrogen tank, which provided propellant for the satellite's 12 cold-gas manoeuvring thrusters. If the satellite could be termed an engineering marvel, the 2-mm-thick conducting tether that connected it to a support mast in Columbia's payload bay was no less.

Surrounding its Nomex core was electrically conducting copper wire, insulated with Teflon and coated with ultra-strong braided Kevlar 29 – a similar kind of material to that used in bullet-proof vests – which made it much hardier than it looked. Over the top of the Kevlar was a final, outermost 'jacket' of braided Nomex, which protected it from abrasion and the corrosive effects of atomic oxygen in Earth's rarefied upper atmosphere. During deployment operations, the tether was unreeled from a 2,027-kg mechanism affixed to a Spacelab pallet and the Mission-Peculiar Equipment Support Structure (MPESS) in the payload bay.

Essentially, the mechanism took the form of a four-sided erectable tower, looking like a small broadcasting pylon, which unfolded slowly out of its storage canister using a series of rollers. As the canister rotated, fibreglass batons popped out of their stowed, bent-in-half positions to form cross-members – 'longerons' – that supported the tower's vertical pieces. The tower deployed to a height of 11.8 m above Columbia's payload bay, so that when the satellite was released, there was no risk of it hitting any part of the Shuttle's structure. "The complexity of the experiment is extreme," Andy Allen had said before the flight.

That, however, was the easy part. Deploying the tower and even releasing the TSS itself had already been done by the STS-46 crew nearly four years earlier; what Allen's team planned to do on their 14-day mission in February 1996 was finish the job by getting the satellite and tether to their full 20.5-km length and demonstrating Giuseppe Colombo's concept. In fact, no fewer than 12 separate experiments were planned during the reflight – known as TSS-1R – of which six had been provided by NASA, five by the Italian Space Agency (ASI) and one by the US Air Force.

Several of these experiments were mounted on the MPESS: two were designed to investigate the dynamics of the tether during its deployment phase, another provided theoretical support in the area of electrodynamics, a couple more employed ground-based equipment to measure electromagnetic emissions from the satellite and seven others stimulated or monitored the entire assembly as it reeled its way out of the payload bay. Nearly 22 km of cable was in the deployment mechanism for STS-75, although 'only' 20.5 km of that would actually be unravelled.

BAD SIMULATION RUN

"Arrivederci, au revoir, auf wiedersehen and adios," Allen radioed cheerily from Columbia's flight deck on 22 February 1996, as he and his crewmates lowered their visors and prepared for launch. "We'll see you in a couple of weeks." Without further ado, and after a picture-perfect countdown, they thundered into space precisely on time at 8:18 pm. Unfortunately, the first portion of their ascent did not prove to be quite as perfect as NASA would have liked: a mere four seconds after liftoff, Allen and Horowitz spotted a potentially serious problem on their instrument panel.

One of the Shuttle's three main engines, it seemed, was running at a mere 40% thrust; far lower than the 104% it should have been pumping out in the seconds after launch. After checking with Mission Control, who confirmed that *their* telemetry indicated that all three were indeed performing normally at full power, the crew continued safely into orbit. Of course, if the (ultimately erroneous) instrument readout had been for real, it would have required Allen and Horowitz to perform an emergency landing back at KSC.

"We had a couple of moments there that we got a little adrenaline rush," Allen said later. "I said [to Horowitz], 'This looks like a bad simulation run'." After achieving orbit, the crew divided into their respective shifts, and Blue Team members Chang-Diaz and Nicollier quickly set to work activating the TSS-1R support equipment in readiness for a planned deployment of the satellite early on 24 February. Before turning in at the end of his first shift, Hoffman also tested the reel motor and latching mechanism which secured the spherical satellite to its docking ring on top of the still-folded deployment tower.

By the afternoon of the 23rd, with less than 24 hours to go before deployment, the mission had encountered its first spate of problems: a computer relay, responsible for sending crew-issued commands to the satellite, experienced what appeared to be an electrical overload. Known as a Smart Flexible Multiplexer-Demultiplexer, or 'Smartflex', the relay had to be switched to a backup component; although the backup performed satisfactorily, Mission Control opted to spend several hours evaluating it before giving the go-ahead to begin TSS-1R deployment activities.

Next, a laptop computer on Columbia's aft flight deck encountered difficulties and was exchanged for a spare, which performed sluggishly. Nevertheless, in the early hours of the 24th, the methodical, step-by-step procedure to activate the experiments associated with the tethered satellite got underway. Firstly, the experiments were turned on individually, and then together, in an attempt to isolate the computer problems. By 8:00 am, all experiments were up and running and data was successfully transmitted through the Smartflex to the laptop and from thence to Mission Control in Houston.

LEARNING TO WALK

It was at around this point that managers decided to postpone the TSS-1R deployment until 25 February to gain additional confidence-building and testing time with the Smartflex. Although, so far, it had remained stable, additional time was desirable to allow engineers to better understand its behaviour after several unexpected crashes and restarts. The 24-hour delay also gave them ample opportunity to develop work-around procedures should the Smartflex encounter difficulties during the deployment procedure. It was "like learning to walk before you run," according to Mission Scientist Nobie Stone.

One of these confidence-building tests involved 'mapping' Earth's charged-particle environment, which varied dramatically as Columbia circled the globe in periods of sunlight and darkness every 45 minutes. During this time, Nicollier also activated the Tether Optical Phenomena (TOP) experiment, developed by Stephen Mende of Lockheed Martin's Palo Alto Research Laboratory in California; this employed a hand-held, low-light-level television camera to provide visual data in support of questions concerning tether dynamics and the optical effects created by the satellite itself. Through the overhead flight deck windows, Nicollier acquired stunning views of atmospheric airglow and aurorae over the South Pole using TOP.

"At this point, all of the instruments are working fine and returning data," concluded Adam Drobot of the Science Applications International Corporation in McLean, Virginia, in a live televised interview on 24 February. "I think everybody is excited that things are going very well." Drobot's own experiment, the Theory and Modelling in Support of Tether (TMST), provided theoretical electrodynamic assistance to TSS-1R. Meanwhile, as preparations got underway, other experiments were undergoing tests. One of these was an electron gun called the Shuttle Electrodynamic Tether System (SETS), which was designed to generate an electron beam in support of the science experiments.

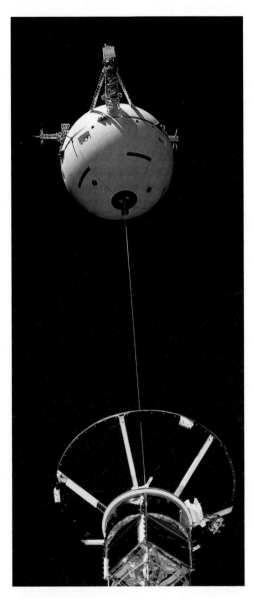

TSS-1R inches away from its mast at the start of deployment operations.

It would also provide voltage and current readings from the tether throughout the deployment process. Multiple test beams were fired from the experiment's electron gun on 24 February to acquire data on Columbia's ionospheric environs. A day later than originally planned, but nevertheless right on cue at 8:45 pm, the deployment got underway and proceeded without a hitch under the watchful eyes of Hoffman and Guidoni. The satellite pushed itself away from its docking ring on top of the mast using its own cold-gas thrusters and was expected to reach a maximum distance of 20.5 km over a five-and-a-half-hour period.

"The satellite is rock-solid," Hoffman excitedly told Mission Control. It was expected that, after reaching its maximum extent, the satellite and tether would remain extended for 22 more hours of studies of electrodynamic phenomena using onboard and payload bay-mounted instruments. A slow 'creep' back towards Columbia, precisely choreographed from Mission Control in Houston, would then have resulted in a docking back onto the mast at around 6:43 pm on 26 February.

Orbital dynamics resulted in TSS-1R initially deploying 'upwards', at a rate of just 15 cm/min, and an angle 40 degrees 'behind' Columbia's path; its motion was carefully controlled by electric motors which reeled out the tether, and by the satellite's cold-gas thrusters. An hour into the deployment, TSS-1R eclipsed the 258-m limit of its predecessor. Gradually, the rate increased to 1.6 km/h, slowing briefly within a couple of kilometres of the Shuttle in order that the 40-degree angle could be reduced to just five degrees. This placed the satellite almost directly 'above' the Shuttle's cabin.

Throughout the deployment, Columbia's RCS thrusters were disabled by Allen and Horowitz to avoid causing unwanted oscillations in the tether. Beginning at a distance of 610 m, the satellite underwent a series of very slow rotations in support of its onboard and payload bay-mounted science investigations. The deployment rate of the tether, meanwhile, increased to a peak speed of 8.1 km/h around 1:00 am on 26 February, when at a distance of 15 km from Columbia. It was shortly after this point that things started to go wrong.

The intention was that, as the tether neared its 20.5-km maximum length, its deployment rate would have been gradually reduced; however, at 1:29:35 am on the 26th, at a tantalisingly close-to-target distance of 19.6 km from the Shuttle, the tether snapped! According to the shocked crew who recorded video footage of the incident, the break appeared to have occurred near the top of the mast in the payload bay. "The tether has broken at the boom!" Hoffman radioed urgently. "It is going away from us", accelerating away at about 670 km during every 90-minute orbit.

By mid-afternoon, it was trailing Columbia by 4,830 km, flying some 50 km 'above' the Shuttle. After winding in the remaining 10 metres or so of tether by 6:58 pm, the astronauts retracted the collapsible mast to its original configuration. Although nearly a day's worth of electrodynamic measurements had been lost, the $154-million reflight was not, however, a total failure. Already, when the satellite was less than 6 km from the mast, Carlo Bonifazi – Principal Investigator of the Deployable Core Equipment (DCORE) investigation – reported his first data.

BLOODIED NOSES

Bonifazi's experiment was mounted in the payload bay on the MPESS structure and its task was to control the flow of electrical current in the tether using two electron guns. Before the break, its first performance run had successfully generated a current of 480 milliamps from the electrical charge that had collected on the satellite's surface: this was 200 times greater than the levels obtained during the first TSS mission in the summer of 1992. Other experiments in Columbia's payload bay continued to operate in support of the satellite and tether until as late as 6 March.

"We did get a lot of good data during the deploy," Hoffman told reporters in a space-to-ground news conference. Currents measured during the deployment procedure were at least three times greater than predicted in analytical models.

Voltages as high as 3,500 volts were developed across the tether, achieving current levels of 480 milliamps. It was also possible for researchers to study the interaction of gas from the satellite's thrusters with Earth's ionosphere. A first-ever direct observation of an ionised shockwave around the satellite itself – impossible to study or model in the laboratory – was also made.

Moreover, as the satellite and its trailing 19 km of tether sped through the ionosphere, it was possible to continue investigations despite the fact that it was no longer physically connected to Columbia. On the evening of 26 February, TSS-1R Mission Manager Robert McBrayer told journalists that a good deal of data was "in the bank", regardless of the mishap, adding that "everyone is happy over what we got and disappointed over what we didn't get". Nobie Stone agreed, saying that "things in the data really pop out at you, that are unexpected or unexplained".

"If you don't ever get your nose bloodied, you're not in the game," said Lead Flight Director Chuck Shaw, paraphrasing Theodore Roosevelt's famous comment. Meanwhile, Capcom Dave Wolf told the disappointed astronauts that it was too early to speculate on the cause of the breakage. "We are really in the data-gathering mode," he told them. "The teams are assembling for a thorough review and we will all have a lot to do with that when you get back." In fact, an investigative board convened that same day, under Kenneth Szalai, the director of NASA's Dryden Flight Research Center in California.

On 27 February, as the free-flying satellite and tether flew above JSC's Electronic Signal Test Lab in Houston, ground controllers transmitted commands to successfully reactivate three of its onboard experiments: the Research on Orbital Plasma Electrodynamics (ROPE), the Magnetic Field Experiment for TSS Missions (TEMAG) and the Research on Electrodynamic Tether Effects (RETE). With a possible two extra days of data-gathering capabilities now re-established before the satellite's batteries were predicted to expire around 1 March, Stone's team scrambled to put together a hasty, last-minute research timetable to squeeze as much as possible from their payload.

The ROPE experiment had been developed by Stone himself and sought to examine the behaviour of charged particles in Earth's ionosphere, as well as those surrounding the satellite and tether, under a variety of conditions. Italian National Research Council physicist Marino Dobrowolny's RETE investigation, meanwhile, measured the electrical potential in the plasma 'sheath' around TSS-1R and identified waves excited by the tether. Meanwhile, the TEMAG study, provided by Franco Mariani of the Second University of Rome, mapped fluctuations in magnetic fields around the satellite.

It was hoped that, even though Columbia and TSS-1R were now physically separated by thousands of kilometres, firing electron beams from guns mounted in the payload bay would still disturb the ionosphere and be detectable by the satellite's instruments. On 28 February, scientists were able to observe a sunlight-induced electrical charge on the satellite's surface as it moved through the daytime and nighttime portions of its orbit. They also succeeded in reactivating and acquiring valuable data from two other satellite-mounted experiments. It was even possible, according to Hoffman, for ground-based observers to see TSS-1R from the southern United States.

Since the tether break, orbital dynamicists had predicted that Columbia would approach to within possible retrieval distance of TSS-1R on 29 February, and such a scenario was briefly considered but ultimately discarded due to insufficient propellant margins on the Shuttle. The retrieval, had it been approved, would have consumed up to six days of the crew's time. In anticipation of its rendezvous with Columbia, the satellite's batteries were placed in a 'low-power' mode from the late evening of 28 February until mid-afternoon on the 29th to keep it alive just long enough.

Right on cue, at 5:15 pm on 1 March, Allen spotted the satellite and its tether, just 75 km away. "Basically, all we really see are pinpoints of light real close together," he said. "It's beautiful!" added Chang-Diaz. By this time, however, TSS-1R's batteries were rapidly failing – very weak signals had been detected through the Merritt Island and Bermuda tracking stations earlier that day – and no further signals were received after 1 March. Nevertheless, it had survived much longer than expected, prompting Stone to joke that "it's like the Energiser bunny – it just kept on going!"

Clearly, a significant amount of electrodynamic data was gathered and Carlo Bonifazi later told reporters, "When you do research, it's not the *amount* of data you collect that's important, but the information you can extract from that data. If we discover, two weeks from now, that we got the information we were looking for to understand the basic behaviour of the system, we could move forward." Added Umberto Guidoni: "We demonstrated that tether dynamic applications *work* and we *can* generate electricity using tethers." Nonetheless, the mood on board Columbia was, for a time, grim.

"Every time I turn around and look through the window and I see this empty bay, it's like a part of myself has left," Maurizio Cheli told journalists. For Allen, who had been on board Atlantis for the previous ill-fated TSS mission, during which a bolt had snagged the tether early in its deployment, it was yet another blow. "Scientists on the ground have lost a lot and we feel for them," he said. "We were looking forward to demonstrating that we could actually retrieve a satellite from 20 km, and we've put an amazing amount of work into it."

Of course, the main scientific breakthrough of the mission was, as Nobie Stone said, the discovery of tether currents *three times higher* than theoretically predicted. It was speculated that this might indicate some degree of ionisation around the satellite, even when its cold-gas thrusters were not switched on. In fact, when the thrusters were activated, the current climbed even higher, to 580 milliamps. Overall, Stone told journalists on 8 March, the day before Columbia came home, that TSS-1R demonstrated that its electrical current-collection and power-generation capabilities were several times higher than predicted.

The EDO pallet and now-empty TSS-1R deployer are visible in Columbia's payload bay during the STS-75 mission.

USMP-3: TORCHBEARER OF THE 'SEMICONDUCTOR AGE'

With the completion of 'direct' satellite operations late on 26 February, the STS-75 astronauts returned to their dual-shift – Red and Blue – system of activities and focused on their other mission objectives. The most important of these was the third United States Microgravity Payload (USMP-3), which included a reflight of the AADSF and MEPHISTO semiconductor-processing furnaces, the Zeno xenon-critical-point experiment and the IDGE dendritic-growth investigation, together with the SAMS and OARE accelerometers. All of these had previously been on board USMP-2 in March 1994 and a few had also ridden on the payload's first flight in October 1992.

Unlike the tethered satellite, USMP-3 performed without problems and, like its two previous outings, was operated more-or-less-autonomously by ground controllers via 'telescience'. Chang-Diaz activated the payload late on 23 February and it was one of Cheli's responsibilities to keep a close eye on it throughout more than two weeks of experiment runs. Within hours of activation, IDGE Principal Investigator Marty Glicksman reported "the best images ever sent down" from his experiment. Elsewhere, Archie Fripp's AADSF grew a large crystal of infrared-detecting lead–tin telluride, while French MEPHISTO scientists monitored the solidification and remelting of three tin–bismuth alloy samples.

"The ages of humanity are known for the type of materials humans used, such as Stone Age [and] Bronze Age," said Fripp. "If we can contribute to the development of the 'Semiconductor Age', then we really have achieved something!" By 2 March his furnace had finished processing its first crystal: it grew 'vertically', pointing away from Earth, with the 'hot' end at the top and the 'cool' end at the bottom, while the second sample grew in the opposite direction. This helped scientists to ascertain the importance of growth in different directions before committing hardware to the International Space Station.

"As the station structure grows, the orientation of its microgravity environment could change and affect the direction of samples growing in an onboard furnace," Fripp explained. "If there's a measurable difference between the uniformity of semiconductor samples grown in different directions on this mission, then we know that this will be significant on the space station." After Columbia's landing, the AADSF samples were compared with Earth-grown ones, in order to shed more light on how to improve ground-based processing methods.

Meanwhile, the ability to command each of the USMP-3 experiments via telescience was proving enormously successful. Commenting on its effectiveness, IDGE Project Scientist Ed Winsa of NASA's Lewis Research Center, said "Time is money. In our limited time in orbit, typical pre-programmed experiments would have missed out on a tremendous amount of science return if we had not been able to adjust [our experiment's] growth cycles." Significantly, on 4 March, the mission made history when IDGE received direct commands from scientists on a US college campus.

"This is the first time a Principal Investigator has commanded a microgravity science experiment on the Space Shuttle from his home institution," said USMP-3

Project Scientist Steve Davison. The investigator concerned, Marty Glicksman of Rensselaer Polytechnic Institute in Troy, New York, was joined for the historic commanding session by one of his team members from Lewis. The instructions were part of a third science phase for the dendritic-growth team, as part of continuing efforts to study how metals solidified in the virtually convection-free environment of low-Earth orbit.

Overall, Glicksman's team collected twice as much data as they needed, which helped to significantly improve post-flight statistical analyses. It was hoped that knowledge from IDGE might aid Earth-based processes such as metal casting, welding and the production of aluminium, steel and copper. The ability to perform real-time materials science in space had already shown experiment investigators that variations in Columbia's microgravity environment did not affect variations in the speed of dendrite growth. "This was an important operational experiment, as well as a significant science experiment," Glicksman told journalists.

"An incredible atomic journey", meanwhile, was going on elsewhere on USMP-3, according to Zeno's Principal Investigator, Robert Gammon. As with its previous run on USMP-2 in March 1994, this experiment sought to precisely identify the 'critical point' of xenon – the peculiar temperature-and-pressure point at which its characteristics rapidly drift between those of a liquid and a gas – as part of efforts to better understand the phenomenon. At such 'supercritical' temperatures, fluids make remarkable solvents as well as having other applications. Carbon dioxide, at its critical point, for example, is routinely used to extract caffeine from coffee beans.

As it approaches its critical point, the normally clear xenon "turns milky-white because it does not know whether it really wants to be like a liquid or like a vapour. It just can't decide," said Gammon. In order to determine exactly where the critical point lay, the Zeno team simultaneously monitored more than 100 channels of data, updated each second, to plot statistical curves that measured the sample's critical temperature. These plots, known as 'correlograms', enabled researchers to show the relationship between temperature and pressure and provide clues as to when the critical point was reached.

MISSION EXTENDED

The USMP-3 operations were running so smoothly, and generating such valuable data, that on 4 March NASA decided to extend the mission to almost 15 days. Andy Allen told Mission Control that his crew was more than willing to support the longer time in orbit. According to USMP-3 Mission Manager Sherwood Anderson, this demonstrated "NASA's commitment to the type of science that will be typical of space station operations". His Mission Scientist, Peter Curreri, went further, adding that having more time to observe Zeno made "the difference between seeing the peak of 'Mount Zeno' and setting foot on the summit".

By 7 March, after 14 days of methodically measuring the temperature and pressure of the xenon sample, the experiment had reached that peak. The transparent xenon sample displayed the unusual critical point condition, with maximum light scattering followed by a sudden increase in cloudiness. The effect was much more distinctive than had been observed during USMP-2 and happened at a much lower temperature than previously thought possible. "We've been to the top, did our flyover of Mount Zeno and have seen the highest count rates from this sample that we've ever seen anywhere," said Gammon.

"There's no other system that lets you approach this kind of intensity of thermal fluctuation – you just can't do this on the ground." He further compared the critical point to the violent reactions in the Sun's core and at the beginning of the Universe when condensed matter expanded and cooled to the states observed today. It was hoped that knowledge from this experiment, which flew again on USMP-4 in late 1997, would be of value to applications such as liquid crystals and semiconductors.

Also on board USMP-3, and flying for the third time, was MEPHISTO, which began its first 32-hour solidification-and-remelting run of three tin–bismuth alloy samples – used as representatives of mixtures employed in semiconductors and metal alloys on Earth – late on 26 February. "Most of the materials we use are formed by solidification," said the experiment's principal investigator, Jean-Jacques Favier. "This is true for semiconductors as well as high-strength alloys." It was anticipated that, in time, the kind of research conducted in Favier's furnace could lead to improvements in the design of metals and faster computers.

The furnace studied the point at which materials changed from a 'melted' into a 'solid' state and did this by selectively heating, then cooling, the samples. At the same time, it also took non-invasive electronic measurements, so as not to interfere with the crystallisation process. At several points during the STS-75 mission, the SAMS accelerometer was used to support MEPHISTO in order to adjust the samples' growth rates in response to Shuttle manoeuvres. This knowledge could help to develop new furnaces which automatically compensate for unavoidable acceleration changes by their carrier spacecraft.

Very little of the research conducted on USMP-3 could have been done without the ability of the ground to command 'their' experiments via telescience and, in total, more than 2,300 instructions were transmitted to Columbia during the two-week mission. "I couldn't be happier. The science we've obtained is fundamental to a lot of processes that are very important to all of us," Curreri concluded as the flight drew to a close.

THE MIDDECK GLOVEBOX

Not all of the experiments were housed on the MPESS-mounted payload bay hardware. A new, and potentially important facility – particularly in anticipation of upcoming International Space Station research – was provided in the Shuttle's middeck. Known as the Middeck Glovebox (MGBX), it was developed by Don Reiss of NASA's Marshall Space Flight Center and was essentially a scaled-down, locker-based version of the unit that had for some years been used in the Spacelab module. Equipped with video and still cameras and foot-activated audio-recording gear, it was used on STS-75 to conduct three combustion experiments.

Jeff Hoffman (left) and Maurizio Cheli work with the Middeck Glovebox.

It was activated late on 22 February, only hours after Columbia reached orbit, and following a test of its cameras and recording equipment, took centre-stage from the 27th. Typically, for each experiment, the crew removed a sample kit from stowage lockers and inserted it through the MGBX door, before tightly sealing the unit. Using gloves, they were then able to manipulate the samples. The three experiments investigated the effect of airflow on flame-spreading characteristics, explored the conditions leading to the ignition of several samples of ashless filter paper and examined the behaviour of soot under microgravity conditions.

On 3 March, Takashi Kashiwagi of the National Institute for Standards and Testing and Principal Investigator of the Radiative Ignition and Transition to Spread Investigation (RITSI) – the filter paper experiment – received a boost in science return thanks to Cheli, who sped up a relatively long process and completed half of a second run ahead of schedule. Early results indicated that flames produced in space differ dramatically from those on Earth. "They were upside-down," exclaimed Kashiwagi, "completely the reverse of flames one usually sees!"

Another experiment, supervised by Horowitz, was the Forced-Flow Flame-Spreading Test (FFFT), in which he ignited a cylindrical sample of ashless paper, commenting on "a blue ball moving down the tube" as he recorded the phenomenon on video. He also informed investigators that he saw "a yellow flame inside the blue flame that is quite spectacular! I've never seen anything like this." All MGBX combustion experiments were completed by 5 March and the crew was instructed to check and evaluate the unit's performance before stowing it away.

"These procedures", said Glovebox Facility Manager David Jex of NASA's Marshall Space Flight Center, "will give the Marshall project team a post-operation look at Glovebox performance following a range of combustion experiments performed over the last few days." Added Peter Curreri: "The mix of remote-controlled experiments in Columbia's [payload bay], with the hands-on Glovebox demonstrations in the Shuttle middeck, have proven the compatibility of the two approaches to conducting science in orbit." In total, 65 samples, ranging from paper to Teflon, were burned and, according to investigators, gathered more than 125% of the planned scientific data.

Knut Sacksteder of NASA's Lewis Research Center, the Principal Investigator for the FFFT experiment, reported 16 flat and cylindrical paper samples successfully burned, yielding significant variations in flame colour, size, growth rate, airflow speed and fuel temperature. Kashiwagi's RITSI team, meanwhile, observed a new combustion phenomenon known as 'tunnelling flames', which moved along a narrow path rather than 'fanning out' from the burn site. Lastly, NASA Lewis' David Urban, developer of the Comparative Soot Diagnostics (CSD) experiment, tested the effectiveness of two different smoke-detection systems.

Such experiments, said Franklin Chang-Diaz, "are going to be valuable if we are going to have a reliable fire-detection system" on the International Space Station. Although Columbia would never visit the orbital complex, her crew *did* succeed in contacting a team of astronauts who in two weeks would fly to another space station: the Russian outpost, Mir. On the morning of 6 March, Allen spoke to Kevin Chilton, who would command the STS-76 mission to Mir later that month. "When you get up here," Allen joked, referring to TSS-1R, "if you see anything we left, bring it home for us."

A BITTERSWEET MISSION

In terms of its microgravity research, STS-75 had been a superb success, hampered only by the tether breakage which lost almost a day of valuable electrodynamic measurements. Nevertheless, the reflight of the unusual satellite did demonstrate the concept of powering spacecraft using conducting tethers and, undoubtedly, will be returned to at some point in the future. Columbia's landing, extended a day to 8 March, was postponed 24 hours due to a forecast of low clouds and the chance of rain and gusty winds. "Conditions are marginal for Friday [8th], worse for Saturday [9th]," was KSC spokesman George Diller's summing-up of the situation.

Although weather was acceptable in California for a landing at Edwards, NASA managers – aware that Columbia had another mission booked for late June – decided to hold out for conditions to improve in Florida. A cold front passed through the spaceport on 7 March and was expected to become stationary over the Caribbean by the 9th, perhaps leading to an upper-level low-pressure system that

In-flight crew portrait. Front row (left to right) are Franklin Chang-Diaz, Andy Allen and Jeff Hoffman and back row are Maurizio Cheli, Claude Nicollier, Scott Horowitz and Umberto Guidoni.

could cause clouds and showers. Fortunately, weather on 9 March turned out to be acceptable and Allen and Horowitz brought Columbia safely onto Runway 33 at 1:58:21 pm.

"All right!" yelled the seven-man crew at the instant of wheels-stop, to which Mission Control responded simply: "We copy your elation!"

Meanwhile, the review panel charged with investigating the cause of the TSS-1R tether breakage was established on 26 February. "Given the public investment in the tethered satellite, it is important that we find out what went wrong," said NASA's Associate Administrator for the Office of Space Flight, Wil Trafton. "To do any less would be a disservice to the American and Italian people." Ultimately, by the time its report appeared in May 1996, the panel blamed a defect in the tether insulation within the Shuttle's payload bay. This had apparently caused a local electrical discharge, which led to the breakage.

SHORT-NOTICE SPACELAB

A few days after Kenneth Szalai's investigative board presented its report on the TSS-1R mishap, Space Shuttle Columbia was being hauled out to Pad 39B to begin final preparations for her 20th trip into orbit. It would be a mission that, perhaps more so than any previous Spacelab flight, would virtually mimic the research that would be conducted on board the International Space Station. Dubbed the Life and Microgravity Spacelab (LMS), the flight, according to its Mission Manager Marc Boudreaux of NASA's Marshall Space Flight Center, had "the key ingredients to take us into the next era of space exploration".

One of these ingredients was the relatively short period of time it took from laying the initial blueprints for the mission to its actual realisation: just 21 months! Historically, Spacelab missions were incredibly complicated affairs, involving hundreds of scientists and engineers from numerous institutions spread across the globe, and took up to four years to prepare from initial conception to getting off the launch pad. In the wake of the IML-2 mission in July 1994, it was recognised that there would be a conspicuous lack of life and microgravity science flights until the space station became operational.

With this in mind, in September of that year NASA announced its intention to stage a one-off 16-day mission in the summer of 1996. Midway through the following May, seven astronauts were assigned to the flight, which would become known as STS-78: Payload Commander Susan Helms, Mission Specialists Rick Linnehan and Chuck Brady, Payload Specialists Jean-Jacques Favier and Bob Thirsk and Alternative Payload Specialists Pedro Duque and Luca Urbani. Five of them would actually fly Columbia for the ambitious mission, while Duque and Urbani would serve as backups and help to coordinate the experiments from NASA's Marshall Space Flight Center.

The LMS mission was intended to be 'international' in its flavour, providing yet another ingredient in preparing for space station activities, and the names of its crew members reflect this: Helms, Linnehan and Brady were American astronauts, Favier was French, Thirsk was Canadian, Duque was a Spaniard and Urbani was Italian. Many of its experiments would be carried over from earlier Spacelab Life Sciences and International Microgravity Laboratory missions, with Principal Investigators based at four European sites and four United States institutions and employing the most extensive use of telescience to date to run them from the ground.

Already, the USMP series had demonstrated the ability of scientists to control their payloads remotely from Earth and on the LMS they would do the same with research facilities in the pressurised Spacelab module. In a similar move to what was planned for International Space Station scientific operations – and as NASA had done on each of its life science-dedicated missions – Columbia's crew would work a single shift, rather than being divided into two 12-hour teams. "It represents kind of

The STS-78 crew. Left to right are Bob Thirsk, Chuck Brady, Tom Henricks, Susan Helms (holding an Olympic torch), Rick Linnehan, Kevin Kregel and Jean-Jacques Favier.

a blueprint, a roadmap, to the space station," said Arnauld Nicogossian, the acting head of NASA's Life and Microgravity Sciences Office.

The benefit of a single-shift mission was that life science experiments, which by their very nature would require a significant amount of crew time and video coverage, but only minimal power and energy, could be performed when all seven astronauts were available as 'test subjects' and operators. The automated and remotely controlled materials science investigations, on the other hand, could then be run via telescience from the ground as the crew slept. Full telecommand and video facilities would still be available, but at the same time the absence of movement from astronauts would not disturb the highly sensitive microgravity investigations.

Developed at a cost of $138 million, the LMS comprised 22 major experiments, supported by NASA and ESA, as well as the Canadian, French and Italian space agencies, together with adjunct research teams from 10 other nations. Specifically, the life science investigations explored the responses of living organisms to the weightless environment, with a particular focus on musculoskeletal physiology. The other, microgravity side of the LMS 'coin' probed subtle influences at work while processing a variety of materials and examining the behaviour of fluids.

In May 1995, around the same time that the first portion of the STS-78 crew was announced, technicians in KSC's Operations and Checkout Building began integrating the LMS hardware into the Spacelab module. By the following April, when the fully experiment-laden, 10,100-kg facility was closed-out and loaded on board Columbia, the crew for the mission had been increased to seven with the addition of Commander Tom Henricks and Pilot Kevin Kregel. The Shuttle arrived in the VAB on 21 May for stacking onto her External Tank and boosters, and was transferred to the launch pad overnight on 29–30 May.

A WORRYING DISCOVERY

Overall, her three-month processing run after STS-75 had gone exceptionally well, further demonstrating the efforts during her last modification period at Palmdale to reduce turnaround times and improve her capabilities. The only problems were the need to replace and retest a Tactical Air Navigation (TACAN) device – a landing aid – and correct a software error with one of Columbia's Master Events Controllers, which fire the pyrotechnics to separate the tank and boosters during ascent. Finally, on 18 June, NASA managers opted to open her aft compartment and X-ray the power drive units for the External Tank access doors in her belly.

The latter were suspected of having loose screws in their circuitry boards (a similar problem had been discovered during routine inspections of Space Shuttle Atlantis), but the X-rays proved them to be secure. Right on time, at 2:49 pm on 20 June, at the opening of a two-and-a-half-hour window for that day, Columbia duly rocketed into orbit for what was already expected to be a record-breaking 17-day Shuttle flight. Officially, it was scheduled for just under 16 days, but NASA managers expected that conservation of electrical power reserves should be sufficient to extend STS-78 by 24 hours or more.

Commander Henricks provided television viewers with a unique perspective of his fourth launch into orbit – and his second on board Columbia – thanks to the presence of a small video camera mounted in the forward section of the flight deck. Its footage began just before the thunderous ignition of the main engines and boosters and proceeded through the separation of the External Tank until the spacecraft settled into her preliminary orbit. The launch itself was certainly a bone-jarring affair, although a subsequent analysis of the recovered boosters revealed worrying damage to their field joints caused by searing hot gases.

It was through such 'blow-by' of one of the boosters' seals that Challenger had been lost in January 1986 and, although the STS-78 ascent did not compromise the astronauts' safety, it was the first time that the new-specification Redesigned Solid Rocket Motor (RSRM) joints had experienced such 'combustion-product penetration'. NASA managers quickly pointed out that there *was* a hot gas 'path' through the motor field joints, but not through the 'capture' joint, and that, therefore, the motors performed within their design requirements. However, the incident did raise questions over an environmentally friendly adhesive that had been added to the boosters.

Even as Columbia circled the Earth, a problem with the adhesive was already

STS-78 blasts off on 20 June 1996 to begin the record-setting, 17-day STS-78 mission.

expected to delay the next Shuttle mission, STS-79. Unfortunately, NASA was unable to revert to its old-style adhesive because environmental regulations had banned its methyl-based material. Developing a new adhesive "could take a while", admitted Shuttle manager Tommy Holloway. The potential damage to the boosters was more serious, however, prompting Holloway to tell reporters: "This is a serious situation until we determine it's not serious." He stressed at the same time that the damage did not affect the rubbery O-rings within Columbia's boosters. "This joint is an order of magnitude more robust than the joint on Challenger." In total, engineers found six joints where hot gas had apparently penetrated their heat shields. Ultimately, the discovery and the repairs that had to be put in place led to the postponement of the next Shuttle mission, STS-79, from 31 July to the middle of September.

This meant that NASA astronaut Shannon Lucid was obliged to spend longer than expected on the Russian Mir space station. "Of course," she said from her Earth-orbital home, "that always happens, no matter where you are." Lucid had been sent to Mir on STS-76 in March 1996 as the second NASA astronaut to spend a period of several months at the station. When she finally returned to Earth on STS-79, she had spent more than six months aloft, setting a new American record.

Aside from the booster concerns, which did not come to light until the end of June, during their post-retrieval disassembly and inspection, Columbia's climb to orbit was nominal. The only other problem was a failure of one of the five GPCs; this unit held a backup set of flight software in case the other four computers became corrupted. One of the four primary computers was loaded with the backup software and the head of NASA's Mission Management Team, Loren Shriver, told journalists that he expected the mission to run to the planned duration.

THE BUSIEST DAY

Once inserted into a 280-km orbit, inclined at 39 degrees to the equator, which enabled the crew to maintain the normal sleep/wake circadian rhythms that they were used to on Earth, the seven astronauts set to work transforming their launch vehicle into a home and laboratory for the next 16 days. By 4:15 pm, less than one-and-a-half hours into the mission, Columbia's payload bay doors were opened and ahead of schedule the Spacelab module and its experiments were activated.

"Today", said LMS Mission Scientist Patton Downey on the evening of 20 June, "was the busiest first shift of activities we've ever had for Spacelab. Virtually every experiment on board either had its equipment activated or checked out." From their stations on Columbia's flight deck, Henricks and Kregel oriented the spacecraft in a 'gravity gradient' attitude, with its tail pointing Earthwards, so that only very few thruster firings were needed and thus did not disrupt the sensitive microgravity experiments.

One of the most important payload packages in the Spacelab module was ESA's Advanced Protein Crystallisation Facility (APCF), which housed 11 separate experiments to study three different growth methods. In total, during the course of the mission, more than 5,000 video images were taken of protein crystals grown in the facility to track their development over time.

The APCF, which had previously been on board Columbia for the IML-2 mission in July 1994, was a relatively late arrival in the LMS payload. It only arrived at KSC from its home base in Europe on 13 June and was installed on board the Spacelab module, in a *vertical* orientation, the day before launch. It was switched on six hours into the mission and the crew typically provided daily reports of the status of its front-panel displays, as the facility had no space-to-ground telemetry capability of its own.

Also provided by ESA was the Advanced Gradient Heating Furnace (AGHF), which was flying for the first time on STS-78 and successfully processed no fewer than 13 samples for one semiconductor and five metallurgical experiments. The furnace was switched on by Favier late on 20 June and operated near-continuously for 16 days, performing better than in ground-based tests. Its objective was to solidify alloys and crystals in a number of experiments designed to understand the conditions in which the structures of freezing materials change in the solidification process.

It was ultimately hoped that AGHF research would increase scientists' knowledge of the physical processes involved in solidification and lead to improvements in ground-based materials research. The first actual experimental run with the furnace got underway when Susan Helms inserted a sample cartridge to examine the transition in solidifying metal mixtures from ordered, column-like grains to unordered, round ones. Processing of AGHF samples was sequential and required the exchange of experiment cartridges and the activation and deactivation by a crew member.

On 22 June, a sample of pure aluminium, reinforced with zirconia particles, was placed into the furnace as part of Doru Stefanescu's investigation into the physics of liquid metals containing ceramic particles as they solidified. The experiment "may lead to more inexpensive ways to make mixtures of metals and ceramics – particularly for the metal casting industry", said Stefanescu, a researcher from the University of Alabama at Tuscaloosa. "We hope to find ways to help manufacturers make composite products with superior quality."

Other experiments included a polycrystalline sample used to gather information on how to combine liquid metal alloy components into precise, well-ordered solid structures. It was anticipated that knowledge from melting and resolidifying such compounds could help manufacturers make higher-quality metal alloys and semiconductors. Later, an aluminium–copper mixture was solidified as part of an investigation designed by Denis Camel of the French Atomic Energy Commission in Grenoble. One sample was solidified at near-constant temperatures, a second at a high-temperature gradient; both were then compared to theoretical models as part of efforts to understand the influence of fluid flows on metal-alloy processing.

Brady and Helms also ran an experiment which sought to control the internal structures of aluminium and indium alloys during solidification, which was expected to have terrestrial benefits in producing new materials for engineering, chemical and electronics applications. On the whole, the facility performed exceptionally well, although on 2 July – after the insertion of a mix of pure aluminium and aluminium–nickel alloy – Marshall Space Flight Center investigators discovered it was not providing data readouts. The problem was quickly resolved, however, when Helms switched off and reset the facility. In total, six individual experiments, many with separate runs, were performed.

FLUID PHYSICS RESEARCH

The third European-provided microgravity facility was the Bubble, Drop and Particle Unit (BDPU), which had previously been on board IML-2, and was used during STS-78 to observe and record the behaviour of fluids under differing temperature levels and concentrations. Early in the mission, on 22 June, it was utilised for one fluid physics investigation which it was hoped could lead to a greater understanding of the processes controlling evaporation and condensation. According to Johannes Straub of the Technical University of Munich, who designed the experiment, the results were expected to refine a long-standing physics study.

In its specially designed test cell within the BDPU, a small heater emitted an electrical charge into liquid freon, supersaturated with gas, which produced a single bubble by boiling. Straub's investigation measured its development and then reversed the process by increasing pressure to cause condensation in which the gas bubble was redissolved into the liquid. "Eventual applications of what is learned from this basic study," said Straub, "will benefit designers of power and chemical plants and of air conditioners, among others."

Despite requiring some minor in-flight maintenance after suffering a blown fuse, the BDPU performed very well and successfully processed all nine of its test containers, including a spare. The blown fuse was repaired on 24 June by Favier and Kregel, using a credit card-sized piece of plastic from the cover of their flight manual between two layers of wires. This insulated the short-circuiting wire from the unit's metallic housing.

One experiment, provided by Antonio Viviani of the Second University at Naples, examined surface tension and interactions of gases and liquids in microgravity. Various sizes of air bubbles were injected into a water-and-alcohol solution with temperature gradients ranging from 'hot' to 'cold'. Another study, devised by R.S. Subramanian of Clarkson University in New York, examined the behaviour of inert gas bubbles within silicone oil. From such studies, it was hoped new insights might be gained into controlling defects in many aspects of materials processing, possibly leading to the production of stronger and more resilient metals, alloys, ceramics and glasses.

A second glitch with the BDPU led to a remarkable repair procedure on 28 June, that had actually been uplinked to the crew in the form of a video from Mission Control; Favier and Kregel were once again able to resuscitate the unit. Their efforts proved successful when they activated the Electrohydrodynamics of Liquid Bridges experiment, which focused on changes occurring in 'bridges' of fluids suspended

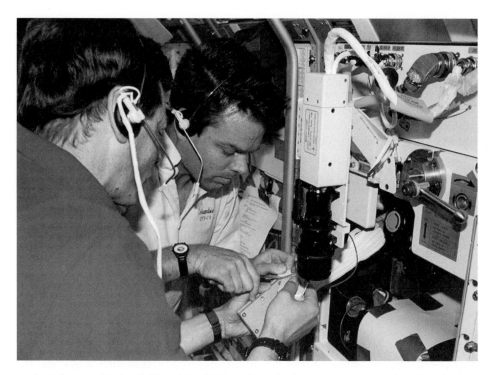

Jean-Jacques Favier (left) and Kevin Kregel repair the Bubble, Drop and Particle Unit (BDPU) in the Spacelab module.

between a pair of electrodes. It was hoped that such research could aid industrial processes in which control of liquid sprays is vitally important, such as inkjet printing and polymer fibre spinning.

Fluids under study in the experiment, which had been designed by Dudley Saville's team from Princeton University in New Jersey, included castor oil, olive oil, eugenol and silicone oil. Another, two-part investigation was conducted by Rodolfo Monti of the University of Naples; its first segment looking at the interaction of moving, pre-formed bubbles and the melting and solidifying 'edge' of a solid, and its second segment examining the ways in which droplets were captured by, or pushed away from, a moving solidification front.

The experiment called for the injection of water droplets of differing diameters into a liquid alloy, in order to study their behaviour during the application of heat. Unfortunately, difficulties were encountered with the injector, which prompted the crew to perform another experiment and defer Monti's study until 4 July. A repair procedure, performed by Kregel and Helms, revealed that the water injector would not retract from its 'deployed' position and the sample cartridge had to be removed. However, ground-based engineers developed a solution to what was conceded to be a minor electrical problem and the experiment was later run satisfactorily.

MEDICAL EXPERIMENTS

Elsewhere in the Spacelab module, the crew – which included two medical doctors (Brady and Thirsk) and a veterinarian (Linnehan) – also concentrated on the second complement of LMS experiments: the life science investigations. These were further categorised under five disciplines: human physiology, musculoskeletal, metabolic, neuroscience and space biology and, according to Victor Schneider of NASA's Life and Microgravity Sciences Office on 5 July, "the information-gathering has just been tremendous. As a scientist, I think the amount of data we have is exciting."

Two hours after launch, Favier and Thirsk kicked-off the human physiology investigations by donning electrodes and sensors to monitor their eye, head and torso movements for Canadian scientist Douglas Watt's Torso Rotation Experiment. It had been known for decades that many astronauts experience motion sickness in space – particularly during their first few days aloft, as their bodies adapt to the strange new environment – and the aim of Watt's study was to identify, and ultimately avoid, movements that contribute to this feeling of illness.

Meanwhile, Linnehan and Brady joined their colleagues in the torso rotation studies and participated in initial tests to evaluate their muscle strength and control. It was already known that muscle fibres became smaller – or 'atrophied' – in microgravity, which resulted in a steady loss of muscular mass. Many of these observations tend to be short-lived, generally vanishing when astronauts return to Earth, but the potential impact of longer-duration, six-month stays on the International Space Station on human muscles and bones remained largely unknown.

Using an ESA-built device known as the Torque Velocity Dynamometer, akin to a piece of exercise equipment found in a gym, the crew were able to take precise measurements and calculate their muscle performance and function, including strength, amounts of force produced and resultant fatigue. Blood samples were taken throughout the mission to enable ground-based physicians to better understand metabolic and biochemical changes in their bodies while in space. Additionally, the Astronaut Lung Function Experiment was used to gauge microgravity's impact on lung performance and respiratory muscles during rest and periods of heavy exercise.

On 22 June, the human physiology studies entered the first of two specialised three-day 'blocks' of time to probe changes in sleep and performance patterns. It marked the first-ever comprehensive study of sleep, circadian rhythms and task performance in space and, according to Principal Investigator Timothy Monk of the University of Pittsburgh, was essential "if we are going to do long-term exploration in space. We have to know what happens when we remove ourselves from real-time cues." As astronauts circle Earth every 90 minutes – experiencing 16 'sunrises' and 'sunsets' in each 24-hour period – their 'normal' timing cues alter significantly.

Investigators hoped that such studies might be beneficial to workers on Earth, whose normal work and sleep schedules change regularly, as well as sufferers of jetlag. Additionally, "ageing and depression are related to the 'clock' going wrong," said Monk. Typically, during the sleep experiments, all four science crew members –

Linnehan, Brady, Favier and Thirsk – wore electrode-laden skullcaps which monitored their brain waves, eye movements and muscular activity. A second block of time for the experiment began on 2 July when they filled in questionnaires at the start and end of their shifts and again wore the skullcaps while asleep.

Following Columbia's landing, the data from both blocks was compared to create a picture of how the astronauts' sleep patterns had changed during the mission. "These were the first integrated studies where we could look at mood, circadian rhythms and sleep," said Victor Schneider. "We need to know how we can help individuals in space and on Earth work different schedules."

Extending this physiological research into the musculoskeletal arena were a battery of experiments to explore the underlying causes of muscular and bone loss, which featured the first-ever collection of muscle tissue biopsy samples before and after the mission. Almost immediately after Columbia touched down at KSC on 7 July, the science crew underwent MRI scans and biopsies for comparison with pre-flight samples. It was expected that this might lead to improved countermeasures to reduce in-flight muscular atrophy. Pre- and post-flight data was also used in support of other investigations looking at changes in the astronauts' muscular activity in orbit.

The crew routinely used a bicycle ergometer in the Shuttle's middeck, which was fitted with a large, weighted flywheel surrounded by a braking band to resist imparting their pedalling motions to the hull, enabling the crew to exercise without disturbing the sensitive microgravity experiments. The Torque Velocity Dynamometer, already mentioned, was used to measure calf-muscle performance during these exercise periods; additionally, Linnehan and Favier wore electrodes on their legs which applied precise electrical stimuli to cause involuntary muscular contractions. Data from these experiments was expected to provide new insights into why muscles lost mass in space.

The dynamometer was also employed for musculoskeletal tests to measure the astronauts' arm and hand-grip strength. On 24 June, they strapped their arms into the machine, curling and extending them as it provided resistance. One of the investigators for the experiment, Pietro di Prampero of the University of Udine in Italy, commented a few days later that "analyses so far have shown smaller changes in the maximal force of leg and arm muscles than expected." Overall, the dynamometer performed near-flawlessly, with the exception of a few mechanical setup problems and software glitches, and operated for 85 hours during the mission.

In light-hearted reference to all of the electrodes worn and blood and tissue samples taken from them, Linnehan, Brady, Favier and Thirsk jokingly called themselves "the rat crew". Even their food and drink intakes were carefully monitored, as part of ongoing studies of metabolic changes and calcium loss during spaceflight. They typically took non-radioactive calcium isotopes at each meal, from 10 days before Columbia's launch until a week into the mission, and by tracking its relationship to food-and-drink intake, scientists were able to distinguish calcium intake and excretion and determine the total amount used by their bones.

Many of the LMS research facilities were cross-disciplinary and Douglas Watt's Torso Rotation Experiment was also employed for some of the neuroscience studies, by providing for the first time an opportunity to bridge the gap between space motion sickness and the causes of disorientation and nausea on Earth. It was hoped that this could help scientists to learn more about the problems associated with postural disorders and vertigo leading to falls and broken bones. The device precisely tracked the positions of the astronauts' eyes, heads and upper bodies as they went about their everyday activities.

The most important observations came shortly after entering orbit and they reported symptoms associated with adaptation to the microgravity environment. Voluntarily fixing the head to the torso with a neck brace has acquired the name 'torso rotation' because the subject has to turn their entire body in order to move their head. On Earth, this gradually leads to motion sickness in an example of deliberate 'egocentric' motor strategy, during which the subject concentrates on a body frame of reference, rather than an 'external world' reference. Similar motor strategies are often adopted by astronauts, exacerbating the onset of space sickness symptoms.

"The findings will make a contribution to a further understanding, counter-measures and rehabilitative programmes for not only astronauts, but also for people in hospitals on Earth," Thirsk told reporters in Toronto during a space-to-ground news conference on 30 June. "With the information we can figure better ways to keep people in space healthier and fight off muscle and bone degeneration and also use the information on Earth."

Another device used as part of the neuroscience research was a piece of headgear for the Canal and Otolith Integration Studies, which investigated the impact of microgravity exposure on the vestibular system of the inner ear and resultant changes in eye–hand–head coordination. Throughout the experiment, astronauts wore high-tech modified ski goggles, which carefully tracked their eye and head motions as they watched a series of illuminated targets. Typically, they remained either in a fixed position on the bicycle ergometer or 'free-floating' in the Spacelab module and the targets displayed themselves across the inner surface of the goggles.

COLUMBIA'S 'OTHER' PASSENGERS

Seven astronauts were not the Shuttle's only living passengers on LMS: also hitching a ride were embryos of the hardy Medaka fish, provided by Debra Wolgemuth's team at Columbia University College of Physicians and Surgeons in New York, which were being flown as part of investigations into gravity's role in animal development. At intervals, an onboard video microscope provided television viewers with pictures of the growth of the transparent embryos. It was recognised, said Wolgemuth, that understanding the impact of microgravity on vertebrate develop-ment would become increasingly important as long-duration space station missions got underway.

A total of 36 embryos of the Medaka – which is known to be especially tolerant of reduced temperatures – developed during the mission. Judging from the video images downlinked from Columbia, "it appears that the samples in orbit are developing at a slower rate than the ground-controlled samples", said LMS Mission Scientist Patton Downey, adding that "gravity is an important factor in the development process. One of the biggest mysteries in science is how you start from one cell and develop different types of structures and tissues. This experiment indicates that gravity may be an important factor in this development."

Other living specimens included 20 loblolly pine seedlings in a special plant-growth facility. When trees growing on Earth bend, then right themselves, they form so-called 'reaction' wood which is structurally inferior. During the LMS mission, biologists carefully examined the cellular structures of the pine seedlings as part of efforts to devise ways to prevent reaction wood formation on Earth, which would prove enormously beneficial to the paper and lumber industries. After several days of growth, Favier and Helms harvested the seedlings, applied a chemical fixative, photographed them and stored them for landing.

Although the bulk of their schedule was spent in the Spacelab, the astronauts had some free time, which first-time spacefarer Thirsk used to look through the windows at the ever-changing Earth 280 km below. "Every morning, within minutes of wakeup," he wrote in an article for the *Toronto Sun* newspaper on 29 June, "we pass over a virtually cloud-free Europe and Asia. I rush unshaven to the window with a camcorder to capture the view one more time. The orbital pass begins over Portugal and Spain, through the Mediterranean Sea, across the boot of Italy and the Peloponnesian peninsula. At eight kilometres per second, we over-fly Cyprus, Israel and the Persian Gulf. Clouds then begin to thicken as we near India and Sri Lanka, which are now in their monsoon season. It is a thrill to recognise features such as the Straits of Gibraltar, the Bay of Naples, Mount Vesuvius, Athens, the Jordan River and the Nile Delta. These are regions of Earth that were the cradle of civilisation millennia ago and even today play a major role in global affairs, [providing] a recap of history and current events in 15 minutes!"

RECORD-BREAKERS

Columbia's marathon mission was expected to get within hours of the record 16-and-a-half days spent aloft by Endeavour's STS-67 crew a year earlier; on 29 June, it became official that STS-78 would snare that record for its own when NASA told the crew that they could remain in space until 7 July. The announcement was accompanied by background music from the movie 'Mission Impossible', but as the astronauts responded with high-fives, Henricks told Mission Control that they were all "willing, able and eagerly anticipating" the extension of their flight to almost 17 days in orbit.

As their mission wore on, it received nothing but praise from scientists. "We have 41 Principal Investigators involved, and all but very few have 100%, if not 200%, of the data they had hoped to collect," said Patton Downey. It had already demonstrated the most complex and successful use of telescience to date, and Kevin Kregel in particular lauded the ability to conduct video-conferences and uplink video-taped repair procedures for the BDPU. "It made fixes a lot easier," he told

Mission Control, "as opposed to sending up the message and trying to interpret the fix on paper."

Right on time, at 11:37 am on 7 July, Henricks and Kregel fired Columbia's OMS engines – during their record-breaking 271st circuit of the globe – to begin the glide back to KSC. An hour later, at 12:37:30 pm, she settled gracefully onto Runway 33, completing a mission of 16 days, 21 hours, 48 minutes and 30 seconds; eclipsing the earlier STS-67 record by about six hours. Television viewers would later be treated to a unique perspective of landing, thanks to the tiny video camera, which was mounted on a bracket on Kregel's glare shield in the Pilot's window.

The lipstick-sized device provided stunning coverage of the last seven minutes of Columbia's 20th descent from space. Immediately after disembarking, Henricks and Kregel participated in an Olympic Torch ceremony and told journalists "it's been a pleasure to stay up this long, and we know it will be a short-lived record". Their words proved prophetic: not only would the STS-78 record be broken again that same year, but it would be broken *by Columbia herself*, on her very next mission. The difference, on *that* flight – STS-80 – was that it would set a Shuttle endurance record that remains unbroken to this day.

SPACE STATION PRACTICE

STS-80 seemed to have it all: two satellite deployments *and* retrievals, a record-smashing 18-day mission, a range of life and microgravity science experiments and the oldest person yet to fly into space. But one thing it did *not* have – and which harked back to Bill Lenoir and Joe Allen's ill-fated experience in late 1982 – was a spacewalk, after a balky airlock handle prevented two STS-80 crew members from becoming the first astronauts to venture outside Columbia in orbit. Nonetheless, it was, as Commander Ken Cockrell put it before liftoff, "a little warm-up for operating the space station".

So it was that, in the early hours of 23 November 1996, less than four days into the mission, no fewer than *three* separate spacecraft trailed one another in a delicate orbital ballet. One was the venerable Columbia herself, while the others were an odd, disk-shaped contraption known as the Wake Shield Facility (WSF) and a German-built facility called the Shuttle Pallet Satellite (SPAS). The latter had flown numerous times since June 1983, and can be thanked for taking the first-ever photograph of a Shuttle in orbit when it photographed Challenger during STS-7.

Originally intended as a means of providing a standardised support structure and resources for research payloads, SPAS flew successfully on three occasions, carrying its first major scientific instrument – an experimental infrared background signature survey for the US Department of Defense – on board Discovery on STS-39 in April 1991. It was later superseded by a heavily upgraded version known as 'ASTRO-SPAS', which first flew in September 1993, transporting an astronomical spectro-meter-and-telescope package called the Orbiting and Retrievable Far and Extreme Ultraviolet Spectrometer (ORFEUS) into space.

The STS-80 crew depart the Operations and Checkout Building, bound for the launch pad. Left to right in pressure suits are Kent Rominger, Tammy Jernigan, Story Musgrave, Tom Jones and Ken Cockrell.

So successful was the maiden voyage of this package that, on STS-80, it flew again. Both ASTRO-SPAS and ORFEUS were the product of a cooperative venture between NASA and the German space agency, DARA. As well as offering support and data-handling services to its scientific package, the satellite provided energy from its powerful lithium–sulphur dioxide battery and achieved ultra-stable attitude control with a three-axis-stabilised cold-gas system, in combination with an onboard star tracker and specially developed GPS receiver. It was planned that, on STS-80, ORFEUS-SPAS-2 would spend up to 14 days in quasi-autonomous flight alongside Columbia.

ORFEUS-SPAS-2: A FREE-FLYING OBSERVATORY

When its spectrometer-and-telescope ORFEUS instrument had been attached, the total payload had a mass of 3,570 kg and, like the Wake Shield Facility, would be deployed and retrieved with the assistance of the Shuttle's RMS arm. ORFEUS itself was devised by several German and US research institutions, with funding provided by NASA and DARA, and its centrepiece was a 1-m-diameter, 2.4-m-long telescope with built-in far-ultraviolet (dubbed 'Echelle') and extreme-ultraviolet spectrographs. An iridium coating on the telescope's primary mirror served as a 'reflection enhancement' to give it ultraviolet sensitivity.

During orbital operations, the two spectrographs were operated alternately, by 'flipping' a mirror into the beam reflected off the instrument's primary mirror. Echelle covered a wavelength range from 90 to 125 nanometres, while its extreme-ultraviolet counterpart encompassed the 40–115 nanometre span. Two reflection gratings then dispersed the incoming light from celestial sources into a spectrum, which was projected onto a two-dimensional microchannel plate detector.

The scientific objectives of the second ORFEUS-SPAS mission were to closely examine the evolution of stars, the structure of distant galaxies and the nature of the 'interstellar medium' – the almost-empty region between stars – in two rarely explored, very-short-wavelength areas of the electromagnetic spectrum. These particular regions are obscured by Earth's atmosphere, thus precluding ground-based observations; nor are they within the Hubble Space Telescope's capabilities. During its maiden flight in the autumn of 1993, ORFEUS-SPAS provided invaluable data on the structure and dynamics of interstellar gas clouds, as well as insights into how molecular hydrogen was created in interstellar space.

It also observed the spectra of a diverse group of important astrophysical objects, including a compact interacting binary star with an enormous magnetic field, three hot white dwarfs and the distant active galaxy PKS2155-304. Despite enormous advances during the Space Age, the formation of stars is still imperfectly understood; nonetheless, the process *is* known to arise in dense clouds of gas and dust. Under gravitational contraction, these clouds can then become dense enough to trigger starbirth. It was hoped that ORFEUS-SPAS-2 data might help to measure the size, distance, density and temperature of such clouds.

This, in turn, was expected to aid astrophysicists' understanding of the circumstances in which interstellar clouds collapse and precisely how new stars are born. When stars form, their evolutionary paths are predominantly governed by their masses: high-mass stars burn their energy reserves more than 100,000 times faster than our Sun, through processes which produce 'bright' ultraviolet emissions

ORFEUS-SPAS-2 flies freely in this scene from STS-80.

and strong winds of hot, ionised material. One of ORFEUS-SPAS-2's main tasks was to observe the surfaces and winds of such objects.

Low-mass stars, like our Sun, burn their energy reserves fairly slowly and do not emit large amounts of ultraviolet radiation. Their surfaces can still become very hot, however, due to turbulent convection which creates shockwaves. ORFEUS-SPAS-2 measured the ultraviolet spectra of such layers in comparatively cool stars to yield insights into the physics of such processes. Many stars ultimately end their lives as 'white dwarfs' which, because they are compact, are very hot and cool very slowly. During this time, they emit much of their radiation at ultraviolet wavelengths, making them among the brightest-known extreme-ultraviolet sources.

Larger stars, on the other hand, often meet their demise by exploding as 'supernovae', thus returning their mass back to the interstellar medium. Another of ORFEUS-SPAS-2's aims was to carefully trace the progress of such stellar remnants. It also explored the 'exchange' of stellar material between the members of binary star systems to investigate how quickly the matter is transferred and discern other characteristics of the hot 'accretion' disks thus formed. Many astrophysicists are certain that similar matter-exchange processes also occur, albeit on a far larger scale, in active galactic nuclei. A large number of these nuclei are suspected to contain massive, hidden black holes, surrounded by accretion disks. For each of these astrophysical objectives, the new data provided by ORFEUS-SPAS-2 was expected to lead to important new insights. Moreover, the second mission of the spectrometer-and-telescope package provided an added bonus: half of its observing time was set aside for the general astronomical community.

Four other scientific payloads were attached to the ASTRO-SPAS satellite in addition to ORFEUS. The Interstellar Medium Absorption Profile Spectrograph (IMAPS) operated for two days during the course of the two-week-long period of free flight, observing numerous extremely bright galactic objects at high resolution. Previously flown on high-atmosphere sounding rockets and on the ORFEUS-SPAS-1 mission, IMAPS was capable of very precisely measuring the motions of interstellar gas clouds to an accuracy of 1.6 km/s.

Meanwhile, the Surface Effects Sample Monitor (SESAM) provided a passive carrier device for evaluating state-of-the-art optics and potential future detectors during various phases of a Shuttle mission. Another experiment helped to develop and validate ground-based simulation facilities for ESA's Automated Transfer Vehicle (ATV) – an unmanned cargo carrier for the International Space Station – as well as evaluating the performance of its onboard software, GPS receivers and optical-rendezvous sensors. Lastly, high-school students from Ottobrunn in Germany provided an electrolysis experiment to investigate various salt solutions.

BOOSTER CONCERNS

When Columbia reached orbit on the evening of 19 November 1996, the ORFEUS-SPAS-2 deployment was the first major objective on the crew's agenda, scheduled to take place seven hours into the STS-80 mission. The launch itself was three weeks overdue. Firstly, problems had surfaced earlier that year with a new-specification adhesive on the SRBs, which engineers suspected was partially to blame for field-joint damage experienced during the STS-78 ascent. *That* had led to the postponement of Atlantis' STS-79 flight and the replacement of her set of SRBs with those previously earmarked for STS-80.

Additionally, preparations at KSC in anticipation of Hurricane Fran had forced managers to delay the launch from 31 October until 8 November and later the 11th. Otherwise, readying Columbia herself for her 21st trip into space was proceeding well: the only minor problem, on 7 October, was the need to replace two of her flight deck windows after an engineering analysis determined that windows with a high number of flights could tend to fracture more easily. One of the windows had been in space eight times previously, the other seven.

Elsewhere, the primary payloads for STS-80 – ORFEUS-SPAS-2 and the Wake Shield Facility – were transferred to Pad 39B on 11 October and installed on board Columbia when she arrived a few days later. Then, on 28 October NASA managers opted to postpone the launch by at least an additional week beyond 11 November to allow engineers time to better understand an issue regarding erosion of one of the SRB nozzles during STS-79's ascent in September.

The concern surrounded insulating material, which had experienced higher-than-normal 'grooving' erosion in the nozzle's 'throat' area. Although this presented no safety-of-flight issue, it was decided to evaluate and clear the issue before committing Columbia to flight. "Our decision to defer setting a launch date allows the team time to ensure the RSRM nozzles are safe to fly," said Shuttle manager Tommy Holloway. "We will take whatever time is necessary to understand the phenomenon seen on the STS-79 boosters before we proceed with the STS-80 launch."

Ultimately, after agreeing on a new target date of 15 November, he told journalists: "Everyone involved with the investigation of this issue has been doing a superb job and it appears that the effort is nearing a point where it will provide us with a good understanding of the phenomenon. However, this additional week will ensure that the final portions of the investigation are not rushed, and it will allow those involved to organise and present their data in the best way possible as they draw their conclusion."

By the end of their investigation, engineers had concluded that the most likely cause for the 'out-of-family' erosion pattern seen on the STS-79 boosters was due to a 'pocketing' erosion effect triggered by slight ply distortions in the ablative material of the nozzle's throat ring and normal variations in other material properties. Ordinarily, the manufacturing of the throat rings is accomplished by wrapping the ablative material in a 'criss-cross' fashion and curing it at elevated temperatures and pressures. However, it was suspected that during the curing procedure, the material near the surface of the insulation shifted slightly, producing the distortions. When hot gas was flowing through the boosters, the distortions significantly raised stresses in the material that could result in the pocketing effect and the ablator wearing away unevenly. Nonetheless, analysis showed that – even with the ply distortion condition in the worst-possible configuration – safety margins could still be maintained. "I am very proud of this Shuttle team and their efforts in reviewing the nozzle issue," said Holloway. "The extra time we took to make sure all of the data was properly reviewed and analysed once again demonstrates that safety remains the Number One priority."

Launch was delayed yet another four days to 19 November, thanks to the postponement of an Atlas rocket from nearby Cape Canaveral Air Force Station and predicted bad weather at KSC. Instead of stopping, and then restarting, the countdown, the clock was brought back from $T-11$ hours to $T-19$ hours and held at that point until the evening of 18 November. Liftoff the following night was postponed at $T-31$ seconds when a slight hydrogen leak was detected in Columbia's aft compartment.

During a two-minute hold, engineers carefully monitored the hydrogen concentrations and concluded that they were at "acceptable levels" and the countdown proceeded without further incident. Among the crew was Mission Specialist Story Musgrave, who not only became the first person to fly a sixth Shuttle mission, but also became the oldest man yet to fly in space: aged 61. "I'm hugely blessed, just hugely blessed," he said before boarding Columbia. In doing so, he also became the only person to have flown on board *all five* Shuttles: two missions on Challenger and one each on Columbia, Discovery, Atlantis and Endeavour.

After a spectacular liftoff at 7:55:47 pm, NASA's long-range tracking cameras spotted something unusual: an apparent *fire* between the two SRBs. However, it was considered so minor that managers did not even mention it during the post-launch press conference, with NASA spokesman Doug Ward explaining that it was nothing unusual "When you get close to booster separation and you get out of Earth's atmosphere, there's no strong airstream to keep the fire from the rocket boosters pointing down," he said, adding that the External Tank was heavily insulated to safeguard against a possible explosion.

Fortunately, when Columbia's astronauts photographed the tank as it tumbled away, eight-and-a-half minutes into the mission, they reported no apparent damage. After establishing themselves in their correct orbit and stowing their pressure suits and seats, they readied ORFEUS-SPAS-2 for its deployment in the early hours of 20 November. With Mission Specialist Tom Jones at the controls of the RMS, the satellite was released at 4:11 am; three hours later, to relief from ground controllers, the ORFEUS telescope's aperture door was reported as having opened satisfactorily. Deployment was an hour later than planned, following a longer-than-expected checkout of the satellite.

Within a day of its release, Columbia led ORFEUS-SPAS-2 by about 53 km and, following the first of several station-keeping manoeuvres, closed in to enable radar data to provide a more precise 'fix' on the satellite. It was hoped this would help Mission Control to better compute future manoeuvres. Over the next few days, until the satellite's retrieval on 3 December, Commander Cockrell and Pilot Kent Rominger performed a series of thruster firings to bring the Shuttle closer to, then further away from, ORFEUS-SPAS-2. This job was complicated somewhat when STS-80's second satellite payload was deployed on 23 November.

WSF-3: AN ORBITING SEMICONDUCTOR FACTORY

The Wake Shield Facility was essentially a disk-shaped experimental 'space factory', measuring 3.6 m wide and made from stainless steel, with the purpose of producing ultra-pure thin films of semiconducting materials that might one day be used as the basis for advanced electronic components. Designed and built by Alex Ignatiev's team at the NASA-funded Space Vacuum Epitaxy Center (SVEC) at the University of Houston, in conjunction with Space Industries Inc., it was making its third journey into space with Columbia's astronauts.

On its maiden mission in February 1994, the facility experienced hardware problems which prevented Discovery's crew from deploying it; nonetheless, a significant amount of valuable data was gathered and one semiconductor film was grown as it drifted through space on the end of the RMS. When it flew again, on Endeavour in September 1995, it *was* successfully deployed and retrieved and made two significant breakthroughs. "It produced a vacuum nearly 100 times better than operating vacuum levels achievable in terrestrial vacuum chambers," said Ignatiev, "and also yielded the purest gallium arsenide and aluminium gallium arsenide thin films ever made."

Essentially, the wake shield exploited the 'moderate' natural vacuum of low-Earth orbit and improved it further by creating an 'ultra-vacuum' behind an object moving boat-like through the ionosphere. As it flew, it pushed residual gas atoms in low-Earth orbit out of its way, leaving few – if any – in its wake. This yielded results on the first two missions of thin films between 100 and 1,000 times purer than any obtained in terrestrial vacuum laboratories, and Ignatiev's team hoped to create not only the next generation of semiconducting films, but also the next generation of devices they will make possible.

"The quality of the materials will surpass what can be done on the ground," Ignatiev said before Columbia lifted off, "so we should be quite competitive." Yet the concept itself dated back to the 1970s, when NASA engineers published papers arguing that a satellite sailing through space would leave an 'ultra-vacuum' in its wake in the same way that a motorboat left a short-lived channel in the water behind it. Unfortunately, in the absence of practical applications, the idea was left

The Wake Shield Facility in free flight during STS-80.

unexplored until Ignatiev's team revived it in the mid-1980s.

In 1986, they joined forces with nine other companies to form a Center for the Commercial Development of Space – one of several industry–academia partnerships sponsored by NASA – and designed and built the WSF using essentially commercial, off-the-shelf technology. Its purely functional shape made it appear like a factory cast-off: a dull, silver-grey disk with a clumsy arrangement of boxes, rods, tubing and angular shapes attached to one side. The forward-facing side (known as the 'ram') was loaded with avionics and monitoring systems, while the back, or 'wake', side was where the ultra-vacuum formed and the semiconductor films produced.

The main objective of the WSF-3 mission on STS-80 was to grow thin 'epitaxial' films which, it was hoped, could have a significant impact on the microelectronics industry. It was recognised that the commercial applications for high-quality semiconductor devices are particularly critical in the consumer technology areas of personal communications, fibre-optic systems, high-speed transistors and processors and optoelectronic devices. Most electronics are made from silicon semiconductors, but 'compound' semiconductors were anticipated to have even higher performance characteristics. Such semiconductors, and the ultra-vacuum needed to grow them, were only possible using the WSF.

Two major components made up the system: a cross-bay carrier in Columbia's payload bay and the 2,090-kg free-flying satellite, which was carried flat, rather than as a vertical 'wafer'. In total, the Wake Shield hardware consumed almost a quarter of the bay. The satellite was equipped with its own set of cold nitrogen gas thrusters to propel itself away from the Shuttle and its own attitude-control system. It also carried a set of silver–zinc batteries to support its thin-film-growth furnaces as well as powering its heaters, controllers and vacuum-measurement gauges and sensors.

A SHORTER-THAN-PLANNED FREE-FLIGHT

Since even a few atoms could have detrimental effects on the factory's ability to grow ultra-pure films, the wake side had to be very carefully cleaned before being installed into the Shuttle's payload bay. Further cleansing was also necessary just prior to its deployment in space. On the afternoon of 22 November, Jones powered-up Columbia's RMS for the second time and grappled WSF-3 at 7:25 pm. Shortly afterwards, Mission Specialist Tammy Jernigan activated the Canadian-built Space Vision System (SVS) to carefully track the satellite's position and Cockrell oriented Columbia in a gravity gradient attitude to minimise disturbances.

Meanwhile, Jones' next task after unberthing the satellite was to 'cleanse' its wake side for two-and-a-half hours with the atomic oxygen prevalent in low-Earth orbit; to do this, he 'hung' it over the port-side payload bay wall with its underside facing into Columbia's direction of travel. Then, at 11:45 pm, he manoeuvred WSF-3 across the bay and positioned it over the starboard wall, this time with its underside facing *away* from the direction of travel, in order to check out its Automatic Data Acquisition and Control System (ADACS) and attitude-control thrusters.

Finally, he oriented the satellite high above the payload bay, again with its underside facing away from Columbia's direction of travel, to await the opening of the first, 41-minute-long deployment 'window' at 1:06 am on 23 November. After a slight delay, he released WSF-3 at 1:38 am as the Shuttle sailed over the western Pacific. Shortly afterwards, the satellite's ground controllers commanded it to fire a tiny nitrogen thruster, which produced 45 grams of thrust and pushed it clear of Columbia. This made it the first 'self-deploying' payload; the Shuttle's own thrusters having been temporarily disabled to reduce possible contamination.

The firing, which lasted for 19 minutes, positioned WSF-3 some 30–50 km away from the Shuttle and set it up for three days of autonomous operations; at this stage, ORFEUS-SPAS-2 was approximately 93 km 'behind' Columbia. By 2:00 pm on 23 November, the first of a planned seven semiconductor-processing runs was underway; however, a potential problem would lead to an earlier-than-scheduled retrieval on the 26th. The crew was awakened early on 24 November to the news that ORFEUS-SPAS-2 seemed to be closing on WSF-3 at a faster-than-expected rate.

Original plans called for Cockrell and Rominger to perform as many as two manoeuvres per day to maintain the proper separation distance from both satellites, and for the Wake Shield to fire one of its thrusters daily. This, it was expected, would keep both some 30–40 km away from the Shuttle, and approximately the same distance from one another. After a lengthy period of tracking analysis and predictions of the satellites' positions, Mission Control decided that their separation distance was well within limits, but Columbia's crew were nevertheless instructed to retrieve WSF-3 three hours ahead of schedule.

The early retrieval did not impact the facility's semiconductor-growth runs; all seven films had been successfully processed on 25 November and WSF-3 was grappled at 12:01 am the following morning. Throughout the rendezvous and retrieval procedure, Columbia drew no closer than 10 km of ORFEUS-SPAS-2, which was exactly as had been predicted by flight controllers and well within safety margins. Yet it was not the end of WSF-3's operations. At 12:06 am on 27 November, Jones again grappled the satellite for what should have been three-and-a-half hours – it turned out to be almost seven hours – of additional experiments.

The work was in support of the facility's Atomic Oxygen Processing investigation, in which it was angled 45 degrees into the Shuttle's direction of travel to explore the usefulness of atomic oxygen in low-Earth orbit to grow aluminium oxide films. So successful were these experiments that scientists were granted more than three additional hours of work, before Jones finally reberthed the satellite on its cross-bay carrier at 6:53 am. During the procedure, Jernigan was also able to gather additional data on WSF-3's position in the payload bay using the SVS.

'BIONIC EYES'?

Applications of the Wake Shield research extend much further than the semiconductor and electronics industries. Ignatiev has pointed to the ability to build replacement rods and cones – which convert light into electrical impulses and transmit them along the optic nerve to the brain, constructing images and enabling us to 'see' – for insertion into the retina, possibly aiding retinitis pigmentosa and macular degeneration, both of which can lead to permanent blindness. "If only we could replace those damaged rods and cones with artificial ones," said Ignatiev, "then a person who is retinally blind might be able to regain some of their sight."

Even as the three WSF missions were underway, SVEC researchers were busy experimenting with thin, photosensitive ceramic films, capable of responding to light in much the same way as rods and cones. Arrays of such films, it is hoped, could be implanted in human eyes to restore lost vision; in effect, creating 'bionic eyes'. Previous tests had been conducted with silicon-based photodetectors, which are toxic to the human body and react unfavourably to fluids in our eyes, whereas ceramic ones are considerably more stable.

"We grew thin oxide films using atomic oxygen in low-Earth orbit as a natural oxidising agent," said Ignatiev. "Those experiments helped us to develop the oxide [ceramic] detectors we're using now for the Bionic Eye project." He pointed out that the ceramic detectors are much like computer chips and can be 'arrayed' in the same way; by arraying them in a hexagonal structure, mimicking the arrangement of the rods and cones they are designed to replace, it was expected that they would not block the flow of nutrients to the eye.

"All of the nutrients feeding the eye flow from the back to the front," Ignatiev explained. "If you plant a large, impervious structure [like silicon detectors] in the eye, nutrients can't flow" and the eye would atrophy. The ceramic detectors, on the other hand, are individual, five-micron-sized units – the exact size of cones – which allow nutrients to flow around them. Artificial retinas constructed at SVEC consisted of 100,000 ceramic detectors, each a twentieth the thickness of a human hair; so small, in fact, that the only way to handle them was to attach them to a one-millimetre-square piece of film.

Within two weeks of insertion into the eyeball, this polymer film simply dissolved and left only the arrays behind; even at the time of the WSF-3 mission, Ignatiev was expecting the first human trials to take place around 2002. "An incision is made in the white portion of the eye and the retina is elevated by injecting fluid underneath," said Charles Garcia of the University of Texas Medical School, the surgeon-in-charge of the procedure. "Within that little 'blister', we place the artificial retina." However, he pointed out that such technology was still in its infancy.

Originally, four WSF missions were planned: three funded by NASA and a fourth by industry. In May 1998 the Spacehab company acquired the rights to market and manage the Wake Shield from the University of Houston. "This is a key first step to reassemble the WSF engineering team and bring the hardware out of storage," Spacehab project manager Mike Chewning said at the time. "We have set a very aggressive goal for the flight system to be ready for a mission in mid-1999 and we are all anxious to get started."

Sadly, WSF-4 – which was to have included solar panels for additional power and a capability to produce as many as 300 semiconductor thin films – never flew and, with the focus of post-Columbia missions exclusively on the International Space Station, in all likelihood never will.

"A LITTLE MORE ELBOW GREASE"

In spite of the success of the mission so far, STS-80 will be remembered as the flight on which both planned spacewalks were cancelled because of a faulty airlock hatch. When Jernigan and Jones were assigned to the crew in January 1996, they eagerly anticipated the chance to perform two excursions, each lasting around six hours, to prepare for the kind of procedures and equipment-handling that would prove so vital when building the International Space Station. Their spacewalks were part of a series of EVA Development Flight Tests (EDFTs), which also served to build space-walking experience among NASA's astronaut corps.

"Of all the space station assembly missions coming up, probably more than 80% of them [require spacewalks]," Jones told journalists before the flight. "They're going to depend on these concepts that we think we've gotten right, but we've got to prove." While Jones and Jernigan were outside in the payload bay – becoming the first astronauts to perform a spacewalk from Columbia – Musgrave would have choreographed their every move from the flight deck, while Rominger would have handled the RMS for several key tests.

During their first spacewalk, planned for 28 November, they were to conduct an 'end-to-end' demonstration to replace an International Space Station battery, using a 1.8-m-tall, 70-kg crane. This was anticipated to take three hours. They would then have tested the crane's ability to lift and move a small 'cable caddy', previously used on the previous EDFT spacewalks in January 1996. Jernigan and Jones' second excursion on 30 November would involve each of them spending two hours working with the battery while standing on a mobile work platform affixed to the RMS.

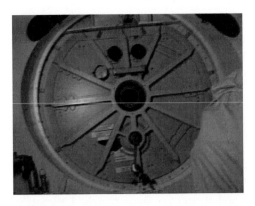

A view inside Columbia's airlock showing the jammed outermost hatch.

Other activities would have included evaluating a Body Restraint Tether (BRT), which offered a kind of stabilising bar for spacewalking astronauts, and a Multi-Use Tether (MUT), capable of fitting square Russian handrails and round American ones. The crane, meanwhile, included a boom capable of telescoping out to between one and five metres in length and could move payloads as massive as 270 kg to various locations on the International Space Station's truss structure.

In anticipation of their first spacewalk, on 21 November, Jernigan and Jones, with assistance from Musgrave, began checking out their suits in Columbia's middeck. Then, late on the 27th, the cabin pressure was lowered from 14.7 psi to 10.2 psi, thus reducing the amount of time that the two spacewalkers would need to pre-breathe pure oxygen before venturing outside. Ironically, as it would turn out, the five astronauts were awakened that evening by Robert Palmer's 'Some Guys Have All The Luck', although Jernigan and Jones' fortunes were bad for that day when the outermost airlock hatch refused to open.

"Initially, I thought we just had a sticky hatch and the fact that Tammy's initial rotation wasn't able to free it up was just an indication that we'd have to put a little more elbow grease into it," Jones told CNN on 2 December. "Certainly, we are feeling some combination of disappointment at the failure of the hatch, but yet pleasure in being part of this mission that's been in every other way very successful," added Jernigan. Both astronauts remarked that spaceflight was a complex business, but remained upbeat, adding: "We'll learn from this experience and go on."

The hatch's handle apparently stopped after about 30 degrees of rotation, making it unable to release a series of latches around its circumference. An engineering team was quickly established to determine the most likely cause of the mishap and Mission Control frantically adjusted the astronauts' schedules in anticipation of a hoped-for second try on 29 November. A minor problem was also noted with a signal conditioner in Jones' spacesuit and it was decided to replace it should the spacewalk go ahead. "We've got high hopes", said Cockrell optimistically, "[for] the rest of the flight."

By the evening of the 29th, however, Mission Operations representative Jeff Bantle was telling journalists that engineering analyses thus far suggested a misalignment of the hatch against the airlock seal. Meanwhile, engineers worked feverishly to assess emergency procedures to open the hatch – which might still be needed if a contingency spacewalk to manually close the payload bay doors became necessary – including warming it by orienting Columbia's topside towards the Sun and having Jernigan and Jones apply pressure from inside the airlock.

"It [takes] fairly light forces overall [to open the hatch]," said Jerry Ross, himself a veteran of four spacewalks. "That's what was a little bit surprising to us." The force needed to open the hatch in normal circumstances, he added, is "certainly not as high as a lug wrench on a bolt". Awakened by David Bowie's song, 'Changes', on 29 November, the crew underwent another day in limbo, working with their many onboard experiments, until finally the Mission Management Team, chaired by Loren Shriver, decided that a conclusive cause could not be pinpointed.

Capcom Dom Gorie relayed the bad news to the five astronauts on 30 November, telling them that both spacewalks would be cancelled. Nevertheless, the following day, Jernigan and Jones conducted a demonstration on Columbia's middeck of a 'pistol-grip' power tool which they would have used during their excursions. They successfully loosened and tightened bolts and screws on floor panels. Ultimately, both astronauts would fly again and perform their 'missed' spacewalks, albeit next time outside the 'real' International Space Station: Jernigan on STS-96 in mid-1999 and Jones on STS-98 in February 2001.

After landing, inspections would reveal a small screw had apparently come loose from an internal assembly and lodged in an 'actuator' – a gearbox-like mechanism which operates linkages that secure the hatch of the airlock – that had subsequently made all of Jernigan and Jones' efforts in vain. When the actuator was replaced, the hatch opened normally.

COLUMBIA BREAKS HER OWN RECORD

A glimmer of good news on the horizon was, however, received on 2 December when the mission was officially extended to almost 17 days, bringing it within hours of the record duration set by Tom Henricks' STS-78 crew a few months earlier. Awakened that day by Jackson Browne's song 'Stay', the decision gave ORFEUS-SPAS-2 an additional day of data-gathering and enabled the crew to press ahead with several of their onboard scientific experiments. One of these was the Space Experiment Module (SEM), provided by NASA's Goddard Space Flight Center and making its first flight. It was carried as part of efforts to increase educational access to space and targeted students from kindergarten to university level, providing research volume in a dustbin-sized GAS canister in Columbia's payload bay. The investigations encompassed studies of gravity and acceleration and observations of bacteria and crystal growth and carried algae, bones, yeast, photographic film and even a variety of children's play items, including crayons, chalk and Silly Putty. Other experiments in the middeck included a pair of medical investigations, studying blood pressure regulation in rats and microgravity's impact on bones at the cellular level.

As the mission drew to a close, Jernigan successfully recaptured ORFEUS-SPAS-2 with the RMS at 8:26 am on 3 December, manoeuvring it through a number of planned exercises to acquire positioning data with the SVS, before reberthing it in the payload bay. Overall, the second flight of the spectrometer-and-telescope package had been a spectacular success, with the instruments yielding data of a much higher quality than had been expected; the quantity, too, was impressive, with more

Columbia is surrounded by servicing vehicles on the runway after completing the longest-ever Shuttle mission.

than 420 observations of around 150 astronomical targets, including the Moon, nearby stars, distant sources in the Milky Way, active galaxies and a quasar.

With a chance of less-than-favourable weather in Florida on 6 December, NASA managers revised their decision to allow the STS-80 astronauts to remain aloft for 17 days and instructed them to begin preparations to come home on the 5th. However, both landing opportunities at KSC were waved off that day, thanks to unacceptable cloud cover moving towards the SLF runway. With a more optimistic outlook for the 6th, Entry Flight Director Wayne Hale opted to forego a landing at Edwards and try again for Florida. Unfortunately, both opportunities *that* day, too, were called off due to low-level fog.

Eventually, following a cold front which moved through KSC on the night of the 6th and produced clear skies, Columbia came home to Runway 33 at 11:49:05 am on 7 December, smashing the record set by the STS-78 crew by around 18 hours. At the instant of touchdown, Ken Cockrell's mission had lasted 17 days, 15 hours, 53 minutes and 18 seconds and covered 11.2 million kilometres. Perhaps partially in recognition of this achievement, Cockrell became chief of NASA's astronaut corps in 1997 and would subsequently command two pivotal Shuttle missions to help build the International Space Station.

To this day, almost a decade later, no other Shuttle crew has surpassed the record set at the end of STS-80; and with the removal of EDO hardware from Endeavour – making none of the surviving vehicles capable of such long-haul feats – it is unlikely that it will be broken before the fleet retires around 2010. A fitting tribute, if ever one were needed, for Columbia.

6

A miniature space station

A BROKEN ANKLE

Columbia's next flight – or rather *flights* – had multiple personalities, but instead of receiving attention from a doctor, it received the undivided focus of thousands of scientists, engineers, technicians, Shuttle processing workers and a close-knit team of seven astronauts. For STS-83 Mission Specialist Don Thomas, 1997 would prove to be one of the high points of his astronaut career: at the end of January, it seemed unlikely that he would be flying *at all* that year, but by the end of July he had flown not once, but *twice!*

Thomas, who had joined the astronaut corps in January 1990, had already flown twice by the time he was named to the STS-83 crew. In many ways, the activities he would be involved with on his third mission closely mimicked the experiments he performed on his IML-2 flight in July 1994. The payload for STS-83 – the Microgravity Science Laboratory (MSL)-1 – was expected to be not only the penultimate voyage of the Spacelab module, but also the last Shuttle flight fully dedicated to materials processing, combustion science and fluid physics research.

All future investigations in these disciplines, it was anticipated, would be conducted by long-duration crews on board the International Space Station, whose inaugural assembly flight was tentatively scheduled for the summer of 1998. In many ways, therefore, STS-83 was to be the last roll of the dice for many physical science investigators, and its importance was highlighted by a very experienced crew: Payload Commander Janice Voss would lead a three-member team – Payload Specialists Roger Crouch and Greg Linteris, together with Thomas – who were responsible for the bulk of the research in the Spacelab module.

Their backgrounds highlighted the nature of the investigations planned on MSL-1: Voss was an electrical engineer, Thomas a materials scientist, Crouch a physicist and Linteris – significantly – a combustion expert. Rounding out Columbia's crew were Commander Jim Halsell, Pilot Susan Still and Mission Specialist Mike Gernhardt, who would supervise the Shuttle's systems throughout a projected 16-day flight. Again highlighting NASA's policy to 'carry over' veteran astronauts on science missions, Halsell had previously accompanied Thomas on IML-2. Still became only the second woman to pilot a Shuttle and, with Gernhardt on board, STS-83 boasted no fewer than *five* PhD scientists on its crew.

Thomas' brush with bad luck came on 29 January 1997, after a year of training for the flight and only nine weeks away from launch: following an emergency egress training exercise at JSC, he slipped down some stairs and broke his ankle. Unlike the Russians, NASA no longer assigned backup crews to its flights, but on this occasion and in light of the mission's importance, it was considered necessary to quickly train another astronaut to stand in for Thomas, if necessary. By mid-February, it was official: astronaut Cady Coleman, who had flown USML-2, would train as his backup.

"We are hopeful that Don will be cleared for flight," said former astronaut Dave Leestma, then-Director of Flight Crew Operations and the man responsible for assigning and managing the training of Shuttle crews. "He is an experienced astronaut with the majority of his required training for this flight already complete. The decision to assign Cady as backup was made to protect all available options." Thomas was also assigned, with Gernhardt, as one of two contingency EVA crew members: Coleman immediately began refresher classes and familiarisation sessions with more than two dozen MSL-1 experiments and research facilities.

"Cady's previous experience makes the amount of training required to bring her up to speed minimal," Leestma added. Thomas, meanwhile, was determined to be ready in time for Columbia's scheduled 3 April liftoff. "I'm in a period of pretty heavy physical therapy right now," he told journalists on 13 March, "spending about five or six hours a day walking in swimming pools, walking with the cast and without the cast, just getting my mobility [and] strength back. We've got three weeks until launch and there's no doubt in my mind or the doctor's mind that I'll be ready in time."

Whichever astronaut had been chosen to fly, few doubted Coleman's capabilities and aptitude, especially Linda Slaker, the Dean of Natural Sciences and Mathematics at the University of Massachusetts, who had known her for some years. "Cady is a very talented experimentalist, both innovative and very reliable," Slaker said. "She has her own speciality – polymeric materials – but she also has to be a quick study, able to learn the fine points of several very expensive experiments from other specialists who will stay on the ground and be able to earn their trust that each experiment will be carried out correctly."

By mid-March, however, Thomas had been cleared as medically fit to fly and confidently awaited Columbia's 22nd launch on 3 April. Preparations for the mission itself had been chugging along relatively smoothly: integration of research hardware into the Spacelab module allocated to MSL-1 had begun in March of the previous year and was finished and loaded on Columbia around the same time that Thomas broke his ankle. The tunnel adaptor was installed to connect the module to the Shuttle's middeck on 14 February 1997 and the vehicle was moved to the VAB for stacking on 5 March.

Not everything went entirely to plan, however: while in the OPF, during planned modifications to existing lines in the bay's hypergolic fuel system, some highly toxic monomethyl hydrazine – the propellant used by Columbia's RCS steering jets – spilled onto two technicians. They and several others received minor treatment for exposure and possible inhalation, but the accident was not expected to impact the scheduled STS-83 launch target. Rollout to Pad 39A, originally set for 10 March, occurred a day later because of a misaligned liquid oxygen umbilical carrier plate and the need to perform additional repairs to the liquid hydrogen tail service mast.

PENULTIMATE SPACELAB

Meanwhile, the seven STS-83 astronauts arrived at KSC and spent some time on board Columbia on the pad, performing their simulated countdown and emergency escape procedures. This gave Still, a native of Augusta, Georgia, the chance to milk her hometown's proud golfing tradition, as she took her best swing to release one of the slidewire escape baskets. "We'll be watching your form," one of her crewmates joked, "and your follow-through!"

It was not only her six colleagues who were watching Still's performance with intense interest: so too was one of her former teachers. "There is one person I'd like to thank by name," she said before the flight, "my junior high school math teacher, Sarah Brown. Due to her teaching ability, I was able to get where I am today. She was hard on the students and really got us to work." In response to Still's tribute, Brown, who retired from teaching in 1976, told journalists "it's a real honour to know what you were doing made a difference".

The STS-83 crew during their Terminal Countdown Demonstration Test (TCDT). Left to right are Jim Halsell, Susan Still, Mike Gernhardt, Don Thomas, Janice Voss, Roger Crouch and Greg Linteris.

Then, on 1 April, and only a few hours after the commencement of the three-day countdown, the Mission Management Team opted to postpone the liftoff until the 4th. It was determined that a water coolant line in Columbia's payload bay had been improperly insulated, and with the possibility that it could freeze during the 16-day mission, work platforms were installed close to the forward bulkhead to enable technicians to reach them. The work to install new thermal insulation blankets was complete by mid-afternoon on 2 April.

"From the standpoint of flight safety and meeting the mission requirements, given enough time to work, I think we would have gotten comfortable with flying 'as is'," said Launch Director Bob Sieck at a news briefing on the 2nd. "But that probably wouldn't have occurred until today or tomorrow and the prudent thing to do was just go in and take care of the concern with these blankets." Rather than stopping and restarting the countdown from scratch, the clock was halted at $T-19$ hours on 1 April, held there and picked up again following the completion of the insulation work.

The prospects of a successful launch, now rescheduled for 7:00 pm on 4 April, seemed good, with the exception of a slight chance of rain showers generated by sea breezes. Mission managers also briefly considered, but later discarded, an option to launch Columbia an hour *earlier* than planned at 6:07 pm, to provide more daylight at the Transoceanic Abort Landing site at Banjul in West Africa and alleviate concerns about delamination of a backup antenna there.

"We're working to get the best science in 16 days in space," Halsell said before the launch. "It's exciting. It's payback time for everybody." It was also somewhat disappointing that, on MSL-1, one of the two Spacelab modules built by ESA would be making its final journey into orbit. The other module was reserved for a medical and behavioural research flight called 'Neurolab', planned for the spring of 1998. "That's currently the last scheduled Spacelab module flight," said STS-83's Lead Flight Director Rob Kelso before Columbia lifted off.

"It's sad that this [Spacelab] era is coming to an end," Mission Scientist Mike Robinson of NASA's Marshall Space Flight Center told a news conference on 18 March. "But it's also exciting to have long durations. Sixteen days isn't enough. We're excited about going to [the] space station." Others echoed his words, with TEMPUS project scientist Egon Egry of the German space agency, DARA, pointing to MSL-1 as an excellent precursor of how future research would be conducted on board the International Space Station.

Following a slight delay in evacuating the closeout crew – the team who strap the astronauts into their seats and conduct final checks at the pad – from the launch danger area, and a problem with excessive concentrations of oxygen in the vehicle's midsection, NASA's 83rd Shuttle mission thundered off to a spectacular start at 7:20:32 pm. It marked the ongoing demonstration of a new Laser Imaging System under development by Naval Research and Development (NRaD). It was hoped that the new system, which was to be provided to the US Air Force's 45th Space Wing, would help to improve launch tracking.

Prior to STS-83, range safety officers monitored Shuttle ascents by optical means, which could be impaired by engine plumes, low-level clouds and fog. By illuminating part of the vehicle with a non-invasive laser beam, it was hoped to acquire clear, defined images even in low-visibility conditions. Ultimately, it was expected this might enable NASA to relax several of its launch constraints. For the first test of the hardware, imaging equipment was mounted at three sites at KSC and Cape Canaveral Air Force Station. These illuminated Columbia's aft section and the SRB 'skirts' at specific points during the countdown and ascent.

"This is a great sunny Florida afternoon here on the Space Coast," said Loren Shriver, the head of NASA's Mission Management Team. "We're happy to have this one off . . . the ascent went very well." Within two-and-a-half hours of reaching orbit, the Blue Team had opened the hatch to the Spacelab module and were activating its research facilities for what Voss described as a mission that covered "the entire spectrum of the sorts of experiments we do on the Shuttle these days and looking forward to [the] space station".

MSL-1: INVESTIGATING PHYSICAL SCIENCE PHENOMENA

For MSL-1, the crew would indeed conduct a wide range of studies to unravel mysteries in three key areas: crystal growth, combustion and the development of techniques to produce stronger, more resilient metals and alloys. Moreover, as with several of Columbia's previous Spacelab missions, STS-83 was expected to evaluate some of the hardware, facilities and procedures that NASA expected to employ on board the International Space Station. It had already become clear from Shuttle research over the years that physical processes ordinarily masked by gravity on Earth were virtually eliminated in space, making it possible for scientists to conduct hitherto impossible experiments.

An important thrust of MSL-1's research was the ability to grow larger and purer crystals of proteins ranging from insulin to HIV-Reverse-Transcriptase and, in so doing, determining their three-dimensional structural 'blueprints'. By unlocking the crystals' structural details in this way, biochemists hoped to better understand how they fit in to the overall biology of the human body, but it has long been recognised that there are more than 300,000 proteins in our bodies, of which barely 1% are fully understood in structural terms.

It was anticipated that MSL-1's growth facilities would be capable of processing more than 1,500 protein crystal samples, perhaps ultimately helping to address the 'social costs' of illnesses and diseases – including cancer, diabetes, alcoholism, Alzheimer's and AIDS – which were estimated in 1997 to top $900 billion *per year* in the United States alone! Many of the proteins associated with these potential killers were the subject of most Shuttle crystal growth experiments, including those on board Halsell's mission.

Additional investigations focused on the differences of the combustion process in the microgravity environment, and its importance was highlighted by Linteris' inclusion on the STS-83 crew. It was hoped that developing a clearer understanding of the peculiarities of different types of fuel, and the fires they produce, could ultimately lead to increased efficiency and reduced emissions in internal-combustion

In the Operations and Checkout Building at KSC, the Spacelab module assigned to STS-83 is prepared for movement to the OPF and installation on board Columbia.

engines. In the United States, in 1997, the annual expenditure on crude oil was estimated by the American Petroleum Institute as close to $200 billion.

"Combustion in general is the major methodology for converting the chemical energy in fuel into useful thermal and mechanical energy," said Fred Dryer of Princeton University. "Combustion is also a major contributor to air pollution, including nitrogen oxides, carbon monoxide, unburned hydrocarbons and particulates. In addition, carbon dioxide – a greenhouse gas – is also produced by the combustion of hydrocarbons. It is relatively easy to produce very highly efficient conversion of the chemical energy in fuels to thermal or mechanical energy, particularly if there are no constraints with regards to emissions. It is much harder to optimise the conversion of thermal energy to useful mechanical energy, particularly with the constraints of emissions. Minimised emissions and best miles-per-gallon require us to carefully control and tailor the combustion process." Prior to MSL-1, this could only be done with sophisticated computers, but the experiments on Columbia provided an opportunity to analyse theoretical predictions and develop new models.

"The pieces of these models, then, can be fed into design codes for things such as engines and engine combustion and can help us improve the efficiency and emissions from those devices," said Dryer. For example, theories held that small fuel droplets should go through three separate 'regimes' during combustion. One of these, known as a 'quasi-steady state', has been frequently studied on the ground: the square of the droplet/flame diameters decreases with time in a linear fashion, eventually extinguishing itself when it became too small to support itself.

The MSL-1 experiments not only provided additional data on that regime in much greater detail than was possible on Earth, but also investigated the *other two* regimes, which had never before been directly observed. "In one of them", said Dryer before the flight, "the flame will decrease and the droplet will actually disappear before the flame structure extinguishes. In the last regime, we will find that the extinction of the droplet is actually defined by radiative loss from the flame. A very large droplet will be formed, burned and then extinguished with much of the mass remaining."

On Earth, mere 1% increase in fuel efficiency – for example, by improving a car's mileage from 25 mpg to 25.25 mpg – would translate into savings of 100 million barrels per year or $5.5 million each day. One of the most important facilities on board Columbia for this research was the Combustion Module (CM)-1, developed by NASA's Lewis Research Center, which contained two experiments: Laminar Soot Processes (LSP) and the Structure of Flame Balls at Low Lewis Number (SOFBALL). Several minor difficulties were experienced during the activation of the facility, when a cable configuration had to be changed by Voss.

UNDERSTANDING COMBUSTION ON EARTH AND IN SPACE

After this procedure was finished, on 5 April Linteris started the LSP investigation, provided by Gerard Faeth of the University of Michigan at Ann Arbor, which was designed to gather data on flame shapes, together with the quantities, temperatures and types of soot produced under various conditions. The experiment supported ongoing efforts to better understand how to control fires and help to limit the number of deaths on Earth from carbon monoxide poisoning associated with soot. A frightening demonstration of how fire behaves in space had already occurred just two months before Columbia's launch.

In February 1997, NASA astronaut Jerry Linenger was entering his fifth week of a planned four-month stay on board the Russian Mir space station, when an accidental fire – triggered by an oxygen-regeneration canister – burned for 10 *minutes*, filled the complex with smoke and forced the entire six-man crew to don breathing masks. Linenger later commented that he was "stunned" at how rapidly the smoke spread. The experience, said Joel Kearns of NASA's microgravity office, "just heightens our concern that we understand this phenomenon both in space and on Earth".

"Soot has a lot of negative attributes and that's why we're concerned about it. It's a pollutant," said Gerard Faeth. "It is harmful to public health. It is the major source of difficulties of unwanted fires in homes. Soot has carbon monoxide associated with it, which is toxic and in that role soot is responsible for the deaths of about 4,000 people a year in the United States and fire injuries of about 25,000."

By 6 April, a mere two days into the mission, the experiment gave him his first glimpse of the concentration and structure of soot from a fire burning in microgravity. "We've hit a home run: it's the first truly steady, non-buoyant flame that's been observed by anybody, anywhere on Earth," Faeth exulted. "It's a real first and the pictures we saw today will probably find their way into textbooks of the future."

Another major study in this area was the Droplet Combustion Apparatus (DCA), which occupied Voss during one of her early shifts on 5 April and housed a variety of experiments to investigate burning drops of different fuels and monitor conditions at the instant of their extinction. A significant amount of the energy produced around the world comes from burning fuels, Vedha Nayagam of Lewis Research Center said, and by studying them in space and comparing their data to theoretical models it was hoped to learn more about their chemical makeup.

Combustion of fuel droplets is an important element in heating materials-processing furnaces, homes and businesses, providing power by gas turbines and for car engines. Ultimately, it was expected that such understanding would enable the development of technologies to burn fuels far more efficiently, minimising pollutant quantities in the process. Inside the DCA, which filled one rack in the MSL-1 module, was the Droplet Combustion Experiment (DCE) which investigated the fundamental combustion aspects of isolated drops under different pressures and oxygen concentrations. Each drop varied between 2 and 5 mm in diameter.

In most practical combustion devices, liquid fuels are mixed with oxidisers and burned in the form of sprays. An essential prerequisite for an understanding of such 'spray combustion' and its application to the design of efficient and clean combustion systems is knowledge of the laws governing droplet combustion. In the absence of buoyancy-induced convection currents, a droplet ignited in microgravity burns with spherical symmetry and yields a simple, one-dimensional system capable of being very precisely modelled. Previous experiments using 'drop towers' at Lewis Research Center could only produce data for droplets up to 3 mm in diameter.

The microgravity environment on board the Spacelab module, on the other hand, provided scientists with an opportunity to better investigate the complicated interactions of physical and chemical processes during droplet combustion. As well as permitting experimentation with varying droplet sizes, it also became possible to study them over longer periods of time. During typical DCE activities, images of the burning droplets and surrounding flames were taken and processed to obtain precise data on their 'burn rates' and other characteristics, including phenomena surrounding the extinction process.

CHANGE OF FORTUNE

"Everything's going great!" exclaimed MSL-1 Mission Manager Teresa Vanhooser from NASA's Marshall Space Flight Center as these and other experiments finally got underway. "We've planned and practised for this mission for a long time and it's wonderful to see it coming together so well," added Mike Robinson. "We're tremendously excited about the opportunities that await us. We can't wait to get down to business – the business of fundamental scientific research in space." That 'business', and the mission's fortunes, were about to dramatically change for the worse.

Since shortly after reaching orbit, Halsell, Still, Gernhardt and Mission Control had been monitoring erratic behaviour from one of Columbia's three $5-million fuel cells. Mounted beneath the payload bay floor, these used a reaction of oxygen and hydrogen to generate electricity to support the Shuttle's systems, run the MSL-1 experiments and, as a byproduct, yield drinking water for the astronauts. Although, technically, one cell provides sufficient electricity to support orbital and landing operations, mission rules dictate that – to provide backup capability during the incredibly dynamic phases of re-entry – *all three* must be working satisfactorily for a flight to continue.

Each cell has three 'stacks', made up of two banks of 16 units each. In one of Fuel Cell 2's stacks, the difference in output voltage between the two banks had been increasing. It had been noted before launch, but cleared for flight. Late on April 5, Halsell and Still adjusted Columbia's electrical configuration to reduce demands on the ailing cell; this allowed mission controllers to stabilise it for ongoing analysis. Overnight, the rate of change in the cell had slowed from five millivolts per hour to around two millivolts, but continued to show a slight upward trend.

"There's always a difference between the two halves of the stack," Mission Operations representative Jeff Bantle told journalists on 5 April, "but we're noticing a changing difference. Actually, that changing difference has levelled off a lot, so the degradation was greater the first 12 hours of the mission."

Bantle's main worry was that if the difference between the two stacks increased to 300 millivolts – and it was touching 250 millivolts by the evening of 5 April – it could force the crew to shut down Fuel Cell 2. "The concern is degradation in a single cell. If it degrades enough, rather than getting power out from the cell, you would have power output *into* that cell. You could actually have crossover and localised heating, exchange of hydrogen and oxygen within the cell and could even have a localised fire. That's the very worst case. That's why we have flight rules that are very conservative to try to avoid and try to shut down and safe a fuel cell before you would ever get to that point."

Early on the morning of the 6th, Halsell and Still performed a manual, 10-minute purge of the cell at Mission Control's request, but as the situation showed no signs of improving that afternoon it was decided to declare a 'Minimum Duration Flight' and bring Columbia home at the earliest opportunity. Capcom Chris Hadfield radioed the disappointing news to the crew.

"The MMT [Mission Management Team] has all players in on the meeting right through from the factory," he told Halsell. "The consensus is they just do not understand the behaviour of Fuel Cell 2. Even though your efforts have done a good job towards stabilising the problem, it's significantly out of family. So we'll shorten the mission."

"That's certainly a disappointment," Halsell replied grimly, "but we know you guys put your best effort forward and you're doing the right thing. We appreciate all the work that's gone into that." Nevertheless, Mission Control retained its collective sense of humour, faxing the crew a tongue-in-cheek list of the Top Ten 'real' reasons why they would be coming home early: lead contenders included running out of 'Columbian' (a play on 'Colombian') coffee, forgetting to record the latest episode of 'Friends', forgetting to do their taxes before launch and – obviously – that the whole thing was a huge April Fool's joke.

Returning to the business at hand, Fuel Cell 2 was shut down at 8:30 pm on 6 April, followed by several other pieces of non-critical equipment on board Columbia to provide additional power to support the MSL-1 experiments for as long as possible. According to Patrick Simpkins, an engineer with the fluid systems division at KSC, the failed cell had flown four times – on STS-75, STS-78, STS-80 and Halsell's mission – and had not shown any indications of problems. Then, 12 hours before the STS-83 liftoff and before it was even *switched on*, it displayed a 500-millivolt discrepancy between its two stacks.

Simpkins added that similarly abnormal behaviour had been noted during two missions by Space Shuttle Atlantis, but that on both of those occasions the cell's discrepancy levelled off to well within safety guidelines shortly after it had been switched on and began taking its full electrical load. With that past experience in mind, engineers activated Columbia's fuel cells and, sure enough, the Number 2 unit settled down to 'normal' levels. The cell apparently performed normally throughout the ascent and the discrepancy materialised again once in orbit.

As a precaution, one of the fuel cells (serial number 116) assigned to Atlantis' upcoming STS-84 mission, scheduled for launch on 15 May, was removed for checks after displaying a similar behavioural 'signature'. "It's not so much that we think 116 is bad," said Simpkins. "It's that we think 116 shows a similar enough signature that we might be able to gather some data. When we shut down Fuel Cell 2 [on Columbia], we probably disturbed the integrity of the system enough that we won't be able to repeat the failure. Or if we *do* find a failure, we won't know if it's attached to what we observed. So we're going to pull 116 on [Atlantis], send it back to the vendor and run some tests. We'll probably put it through a flight profile and see how it behaves." Bantle added that flight controllers had never seen a fuel cell behave in this manner before, adding that "we really don't want to push these limits. We take the fuel cells very seriously."

"This [the Shuttle] is one of the original electric fly-by-wire airplanes," Halsell told journalists during a space-to-ground news conference. "That is, we depend on electricity for this airplane to fly. Therefore, when you lose one-third of your electrical-producing capacity, you have to consider that. So, as a result, after we had to safe Fuel Cell 2, we had a little pow-wow here on the flight deck and we made sure we understood all the emergency procedures as modified by the fact that Fuel Cell 2 had been shut down. So I don't want to convey the opinion that we thought we were in imminent danger. We felt the situation was well in hand [but] we want to be prepared for the next possible failure and we feel like that's where we're at right now." When asked if NASA should have postponed Columbia's launch in light of the failure, Halsell was philosophical: "Hindsight's always going to be 20–20. This is a situation where, with the best information everybody had in hand at the time, the decision was made and I believe it was a correct one at the time."

Nevertheless, the entire crew reacted with "shock and disbelief", according to Don Thomas, at the decision to cut the mission short. "We've been training for this mission for over a year and a half, working hard on the science," he said from orbit. "The people on the ground at Marshall Space Flight Center and the principal investigators all across the country and around the world [have] been working on this mission even longer. So it was a big disappointment, but we've been working as hard as we can, working double-time up here, trying to make up for what we're losing."

The fuel cell failure devastated the carefully planned two weeks of research on board the Spacelab module, as scientists scrambled to reprioritise their schedules to make the most of the one or two days left of their mission before Columbia had to return to Earth. "We're planning for a 16-day flight," said Mike Robinson before the official announcement of a Minimum Duration Flight, "but at the same time, it's better to be safe than sorry, so we've started some planning." He also expressed optimism that, despite the disappointment, he and his colleagues would lobby NASA for a reflight of MSL-1. We would be very anxious to fly again. We have a 16-day mission that's full of science and we'd like to do it all."

A PRODUCTIVE, ALBEIT SHORT, FLIGHT

Certainly, in spite of the difficulties and the tireless efforts of Halsell, Still, Gernhardt and Mission Control to stabilise and save the ailing fuel cell, the research work in the Spacelab module was proceeding in a superb manner, with a significant 'first' in combustion science late on 6 April. "Six burns were successful and, for the first time, we're burning free droplets," said DCE Principal Investigator Forman Williams of the University of California at San Diego. "We can't get this kind of information from ground-based experiments. We have burned at two different atmospheres of oxygen concentration and calculated the burning times of free fuel droplets at each."

Meanwhile, Crouch busied himself with the SOFBALL investigation, designed to explore the conditions under which a 'stable' flameball can exist and if heat loss is responsible, in some way, for its stability while burning. "The two completed runs were successful beyond my wildest dreams," Principal Investigator Paul Ronney of the University of Southern California at Los Angeles said on 7 April.

During the first SOFBALL run, a mixture of hydrogen, oxygen and carbon dioxide burned in the facility for the entire 500-second limit; this was particularly significant because "these are the weakest flames ever burned – lowest temperature, weakest, most diluted mixtures", according to Ronney. "These mixtures will not burn in Earth's gravity. We have known that burning weaker mixtures increases efficiency, but not much is known about the burning limits of these mixtures." As well as providing a clearer understanding of the combustion process, it was anticipated that SOFBALL results would help to improve theoretical models.

"Combustion models give different results for these types of flames," said Ronney. "This is an acid test to show which, if any, current combustion modules should be used." His research could ultimately lead to developments in spacecraft fire-safety systems and had already, by the time of MSL-1, identified anomalous 'flameballs' – essentially stable, stationary, spherically symmetric flames in combustible gas mixtures – from which the experiment received its name. By examining the behaviour of such flameballs in microgravity, including their stability and propagation, more efficient combustion engines, emitting fewer atmospheric pollutants, could be produced.

During typical SOFBALL runs, a 26-litre chamber was filled with a weakly combustible gas (hydrogen and oxygen, highly diluted with an inert gas) and ignited. The resulting flames and their unusual behaviour were then imaged and recorded using a sophisticated battery of video cameras, radiometers, thermocouples and pressure transducers. Unfortunately, the shortened STS-83 mission meant that only two of some 17 planned SOFBALL runs were completed.

Mike Gernhardt shoots photography from Columbia's overhead windows.

Elsewhere in the Spacelab module, other experiments were underway in the area of materials processing, as part of efforts to develop techniques for stronger and more resilient metals, alloys, ceramics and glasses in space. Two facilities used to support these experiments – the Japanese-built Large Isothermal Furnace (LIF) and Germany's Electromagnetic Containerless Processing Facility (TEMPUS) – had already flown on board IML-2 in July 1994. The LIF was activated by Thomas on 5 April and was intended for studies of the diffusion of liquid metals, which cannot be adequately duplicated on Earth because of gravity-induced fluid movements.

One of the first investigations to use the high-temperature furnace was a study of the diffusion of impurities in molten salts. Provided by Tsutomu Yamamura of Tohoku University in Japan, it sought to reveal ideal conditions for the electrolysis of molten salts. Later, Thomas began the Liquid Phase Sintering experiment, which tested theories of how liquefied materials form a mixture without reaching the melting point of the new alloy combination. It was expected that results from this investigation would provide researchers with a clearer understanding of liquid phase sintering in microgravity and compare these findings with theoretical predictions.

Other LIF experiments diffused molten semiconductors – causing two or more types of atoms or molecules to intermingle, in much the same way as a sugar cube would dissolve in a cup of coffee – as part of efforts to explore how uniformly their constituents mixed during the sample-cooling process. Diffusion studies have many terrestrial applications: from very small movements in plasma to massive depletions of the ozone layer. The MSL-1 experiments focused on the usefulness of diffusion processes in support of manufacturing technologies.

The carbonisation of steel and fabrication of transistors have already pointed to the economic and technological importance of the diffusion process; in the semiconductor industry, for example, 'dopants' – small additions of impurities – of antimony, gallium or silicon are routinely added to semiconducting materials such as silicon or germanium to greatly improve the performance of electronic components fabricated from them. Since solid crystals of most semiconductors are grown from their melts, knowledge of diffusion processes in semiconducting melts is important for this industry.

A key problem, however, was that the liquid state of the process was imperfectly understood in terrestrial experiments; however, in the microgravity environment, buoyancy-driven convection is drastically reduced and high-precision diffusion measurements are possible. The MSL-1 experiments, provided by David Matthiesen, explored the diffusion characteristics of gallium, silicon and antimony dopants in samples of germanium. Other investigations focused on the diffusion of liquid lead–tin telluride, which holds great potential as a base material for building infrared lasers and detectors. After Columbia's landing, the sample columns were cut into segments to determine how well their components mixed while cooling.

Meanwhile, the TEMPUS facility was used to study the 'undercooling' and rapid solidification of metals and alloys. Such undercooling typically takes place when a solid is melted into a liquid, then cooled below its normal freezing point without solidifying. The phase change is delayed, and the material is in a *metastable* state. "When the metal or alloy is solidified, it occurs very rapidly, forming new types of materials we cannot manufacture or study in any other way," said the facility's Project Scientist Jan Rogers. A clearer understanding of the process was expected to yield new insights into better casting, welding and soldering techniques and the improved products that will inevitably follow.

Late on 4 April, Crouch activated TEMPUS and configured it for operations. After the decision to bring Columbia home early, William Hoffmeister, the facility's assistant investigator, told journalists that several experiments were pushed up in the timeline to acquire as much scientific data as possible in the short time available. The crew was able to activate, observe and complete an experiment run by melting a zirconium metal sample and levitating it, as part of efforts to study the relationship between internal flows in liquids and the amount of undercooling that a sample can tolerate before it solidifies.

"To understand this experiment," said Robert Bayuzick of Vanderbilt University in Nashville, Tennessee, "imagine if you cooled a glass of distilled water. The temperature could go below freezing without the water actually becoming ice. That is undercooling. However, if the glass were tapped or disturbed, then the water would freeze very quickly. This process may have many benefits to industry. New, enhanced properties in never-before-seen materials could become possible."

As with its first flight on IML-2, TEMPUS provided the means for physically manipulating samples, controlling rotations and oscillations and even 'squeezing' them through the application of an electromagnetic field. The experiments involving zirconium, in particular, were expected to help determine the behaviour of this strong, ductile refractory metal, which has found applications in nuclear reactors and chemical processing equipment.

With the impending landing, a mere quarter of the way through the planned 16-day mission, the TEMPUS team had good prior knowledge of how to reprioritise their research schedules to make the best of unexpected situations. On the facility's maiden flight in July 1994, a misaligned coil forced the investigators to shorten and replan several of their key experiments. "It's a smart group of people," Mike Robinson said of the team before Columbia lifted off on 4 April. "I have a lot of faith in them."

Other areas of research on MSL-1 included plant growth, which could lead to the development of life-saving medicines and other important compounds. "A fundamental objective of this research", said investigator Gerard Heyenga of NASA's Ames Research Center, "is to evaluate whether microgravity may be used to alter specific metabolic pathways in plants, and ultimately apply this technology for Earth-based benefits." He hypothesised that extended exposure of plants to the microgravity environment could reduce their expenditure of energy on structural components, thereby increasing flow through other metabolic pathways, many of which yield materials of important medicinal value.

Of even greater significance, potentially, was developing a clearer understanding of how such pathways are controlled at genetic levels. "Such knowledge would allow us to manipulate or genetically engineer plants with desired metabolic traits," Heyenga added. "For example, this information could be applied to the lumber industry in the production of trees with a low lignin content, greatly reducing the cost of paper production both economically and environmentally." Conversely, it could also be applied to improving timber quality in fast-growing softwoods, thus reducing the need to harvest slow-growing hardwoods.

"If this hypothesis is correct and achievable," he said, "it obviously represents the basis for a multi-billion-dollar industry and certainly highlights the value of space-related research and such facilities as the space station." To conduct the experiments, Columbia carried an advanced piece of hardware developed by BioServe – one of NASA's Commercial Space Centers, situated at the University of Colorado at Boulder – known as the Plant Generic Bioprocessing Apparatus (PGBA). Previously flown on the STS-77 mission in May 1996, it "produced a particularly high quality of plant material, which provided a good basis for further research", said Heyenga.

As well as offering a highly controlled environment with lighting, temperature and gas-exchange functions, the unit was fitted with a 'nutrient pack' to supply water and other nutrients to nine different plant species on MSL-1. "This technology will open an entirely new area of space plant physiology, allowing the study of issues not previously possible," said Heyenga. "It is likely to lead to some very exciting results." For Columbia's mission, the plant species included a member of the black pepper family, selected through a cooperative project with a Brazilian research group.

Also making its second Shuttle flight was the Middeck Glovebox (MGBX), which supported a variety of fluid physics, materials research and combustion experiments. One study, provided by Peter Voorhees of Northwestern University in Evanston, Illinois, investigated coarsening in metallic mixtures at high temperatures. In

Don Thomas works at the Glovebox inside the Spacelab module.

ground-based research, during coarsening, small particles shrink by losing atoms to larger particles, resulting in a lack of uniform particle distribution and thus weakening the resultant materials and shortening their lifespans. Findings from such research could lead to improved manufacturing processes and stronger, longer-lasting materials.

Other Glovebox investigations included the Fibre-Supported Droplet Combustion experiment: a study of fundamental phenomena associated with liquid-fuel-droplet combustion in air – such as how atmospheric pollutants are formed – and had previously flown on USML-2 in late 1995. On that mission, it had demonstrated that data for fuel droplets as large as 5 mm in diameter are useful for testing droplet-burning theories and verified predictions that methanol droplets would be extinguished at diameters that increase with increasing droplet diameter. However, USML-2 data overpredicted the extinction diameter, leading to the addition of radiometers for MSL-1 to monitor radiation emissions.

The behavioural characteristics of bubbles and droplets in response to ultrasonic radiation pressure – perhaps leading to new ways of preventing complications caused during materials processing – were carefully examined in other Glovebox studies, as scientists assessed their ability to control bubbles' locations, manipulate 'double bubbles' and maximise their shapes. Another experiment explored the performance, including the mechanisms leading to unstable operations and failure, of specialised heat-transfer devices in microgravity.

Such technologies are especially important as a means of providing passive thermal control of various spacecraft components, whose electronics are packed closely together and generate heat which can often limit their performance. Mounted in Columbia's payload bay was the Cryogenic Flexible Diode (CRYOFD), attached to a Hitchhiker mounting plate, which tested a pair of experimental heat pipes. One, provided by NASA's Goddard Space Flight Center in conjunction with the US Air Force's Phillips Laboratory at Kirtland Air Force Base in New Mexico, evaluated a unique flexible 'wick' to permit easier integration into future spacecraft and even help point instruments.

REFLIGHT

With such a multitude of investigations packed on board Columbia, it was a pity that the mission should have been curtailed and, even before Halsell's crew returned home, informal plans were already afoot to refly MSL-1 later in the year. Ironically, delays in beginning construction of the International Space Station – the very project for which this mission was providing a technological testbed – made the reflight possible! Originally, Space Shuttle Endeavour was slated to take the first station components into orbit on STS-88 in December 1997, but delays of a crucial Russian service module postponed it until the summer of 1998, at least.

By pushing Endeavour's mission back until the following year, NASA had more leeway to turn Columbia around to fly MSL-1 again in the summer of 1997 and complete another, previously scheduled 16-day flight (STS-87) before the year's end. As Endeavour vacated the December launch slot, ample time was provided for Columbia to fly on two more occasions in 1997, hopefully causing as little disruption to the Shuttle manifest as possible. However, despite significant advances in the space agency's ability to process Space Shuttles more quickly and with fewer problems than ever before, meeting an early July launch date would be challenging.

So it was that, as he climbed down the airport-style steps from Columbia on 8 April, after a picture-perfect 6:33 pm touchdown on KSC's Runway 15, Halsell shook hands with Director Roy Bridges and received the news that "we're going to try to give you an oil change and send you back". All seven astronauts, but particularly the members of STS-83's science crew, welcomed the possibility. "In some ways, this could make for a more meaningful flight in the long run," admitted Crouch, "but, certainly, *this* one was a bummer."

NASA managers discussed the plan over the next couple of weeks, juggling the launch schedule for the remainder of the year in a bid to slot Columbia in again somewhere around 1 July. Atlantis' two trips to the Mir space station, scheduled for mid-May and mid-September, remained on target, while Discovery's 11-day mid-July mission was postponed until the second week of August. Assuming Columbia was ready for the reflight – a mission originally named 'STS-83R', but later STS-94 – on 1 July or shortly thereafter, she should be more than capable of being ready in time for a late November STS-87 launch.

"We're ready to go fly," said Halsell as the initial plans were being wrung out. "If it were up to me, I'd like to give the guys a week or two off to let them decompress from this flight and then we'll come back and start ramping up again for the next flight. I think we should make it clear that this is still under study by everybody; it's not a done deal yet. But certainly, this crew is ready to support with relatively minor additional training."

The schedule was incredibly tight, but according to Shuttle manager Tommy Holloway, who made the official announcement on 25 April, "we now are in a position to do everything possible to complete the mission with minimal impact on downstream flights. Also, it provides us with a unique opportunity to demonstrate our ability to respond to challenges such as this one." From a technical standpoint, the three-month turnaround *was* feasible, particularly since Columbia would carry the same payload and would not need to go through the lengthy removal of her 'old' payload and installation and checkout of a 'new' one.

MAKING LEMONADE OUT OF LEMONS

Costs of reflying MSL-1 would clearly be considerably reduced: NASA normally spent around $500 million per mission, although much of that went into hardware testing, processing, training, planning and simulations that did not need repeating. Holloway quoted $50–60 million, although Joel Kearns pointed out that already-budgeted 'reserve' funds could bring that sum down even more. Additionally, said Holloway, "it would be a very good test of a capability we should have in place for the station – to bring an element of the station back, for whatever reason, and turn it around in as reasonable time as practical".

True to predictions, the reflight *was* cheaper: $55 million for the actual processing of Columbia, according to KSC officials, plus an extra $8.6 million for expenses related to the MSL-1 turnaround. "Our approach", said Lead Flight Director Rob Kelso, "has been to treat this flight as a launch delay. The crew is exactly the same, the flight directors are all the same and the flight control team is almost identical . . . it's a mirror-image flight in many respects." Even the embroidered patch, worn by Halsell's crew was exactly the same, just with a different-coloured border: red for STS-83, blue for STS-94.

Naturally, the scientists were ecstatic at the chance of flying again so soon. "From a payloads standpoint, we could be ready," said Mike Robinson. "All the science teams say they could be ready. It's going to be tight, but again, the majority of the [experiment] samples were not processed, so they don't have to be turned around. There would be no facility upgrades or changes." After Columbia had been towed from the KSC runway into the OPF, the Spacelab module remained in her payload bay, although the tunnel adaptor was removed to provide better access to it.

Ordinarily, between flights, the modules were transferred to the Operations and Checkout Building, but during the short MSL-1 turnaround period technicians overcame the Shuttle's cramped living conditions and successfully completed many critical tasks, including replenishing fluids for the combustion experiments. "This is the first time that a payload has remained in an orbiter between flights," said KSC payload manager Scott Higginbotham. "We are excited about having accomplished something that has never been tried before."

The Shuttle processing team, too, had its own challenges. Normally, they supervised a spacecraft for around 85 days, but the reflight required a turnaround of only 56 days in the OPF. To accommodate this short timescale, and ensure that all necessary work, such as the replacement of two APUs and several RCS jets in Columbia's nose, was completed, they deferred certain structural inspections until her next mission. Overall, it was down to planning. "Once the plan was in place," said Columbia's flow director, Grant Cates, "the team approached this challenge in much the same way that they approach every flow."

Fuel Cells 1 and 2 were both removed and returned to their vendor – Connecticut-based International Fuel Cells – for analysis; although the exact cause was not identified, it was believed to be an isolated incident and engineers took steps to develop monitors for the cells to provide better performance data. Meanwhile, as Columbia rolled into the VAB for stacking, in record time, on 4 June, and from there to Pad 39A on the 11th, the only problems being tracked were efforts to replace three dozen possibly cracked tiles with thicker versions on her forward RCS pod.

Undoubtedly helping the early July target date for STS-94, Columbia was fitted with three 'borrowed' main engines and two 'borrowed' SRBs: the engines were those previously assigned to Atlantis' September flight and the booster set was originally slated to propel Discovery into orbit in August. "The NASA contractor workforce has put forth an outstanding effort in getting the MSL-1 mission ready to fly again," said JSC Director George Abbey. "The quick turnaround in Columbia's processing for launch will allow the crew and the international team of investigators the opportunity to finish the important work they began earlier this year."

For Halsell, it was "a marvellous, once-in-a-career opportunity and something that we all feel very lucky to be a part of", he told a news conference in late June. "I was talking to one of our Capcoms, [fellow astronaut] Bill McArthur, a few days ago and he said this whole reflight business is making lemonade out of lemons. That's kind of the way we feel about it. The people here at the Cape, as soon as our wheels stopped, were working very hard to make a record-breaking turnaround on Columbia. Meanwhile, the scientific investigators had a chance to look at their data results and we're going to change some of our procedures a little bit . . . and maybe take advantage of this reflight in the sense that we'll be able to get even better science, better data for them than we would have the first time around. So it's a great opportunity. Our hats are off to the people at the Cape for making this whole thing possible."

As forecasters continued to assess the Florida weather in the last couple of days of June, the prospects of achieving an on-time liftoff seemed grim, with thunderstorms expected on 1 July and only marginal improvements on the 2nd and 3rd. "With the dynamic [Florida weather], you can never be sure," said NASA's Doug Lyons, who supervised Columbia's countdown. "They are only forecasts." As part of efforts to avoid the thunderstorms, on 30 June managers opted to bring the launch *forward* by 47 minutes, opening the 1 July window at 5:50 pm instead of 6:37 pm.

This removed one end-of-mission daylight landing opportunity at Edwards Air Force Base, if required, but nevertheless enabled *another two* opportunities at KSC.

After a 12-minute delay due to initially unacceptable conditions at the SLF, the crew was advised of the decision to proceed with the launch attempt. "Columbia, it looks like we've got the weather lined up with us, so we're going to get you guys out of here," launch controllers told Halsell. "Have a good flight." The launch itself, at precisely 6:02 pm, was exceptional, with the only minor glitch being a faulty hydraulic sensor during ascent.

"It looks like Columbia's performing like a champ," Capcom Dom Gorie radioed shortly after the Shuttle reached space.

"Roger, Houston, we copy," replied Still from the Pilot's seat. "And thanks [to] the whole ascent team for getting us to a safe orbit."

COMBUSTION SCIENCE EXPERIMENTS

Within four hours, Janice Voss and Roger Crouch entered the once-again-pristine Spacelab module to begin what they hoped would be 16 days of around-the-clock research. "It's good to be back," Voss radioed to Mission Control, as Crouch cartwheeled and backflipped behind her in the voluminous lab. Among the most important unfinished business from the April flight was the completion of 144 scheduled tests in the CM-1; in the event, more than 200 were actually conducted during STS-94.

Among the payloads getting a reflight was SOFBALL, whose off-the-shelf gas chromatograph performed flawlessly throughout the mission, successfully verifying the composition of a variety of premixed gases prior to combustion and determining the remaining reactant and other combustion products. Unfortunately, in light of the constraints imposed upon the chromatograph by the CM-1 hardware, it was not possible to measure the gases to the required accuracy. Improvements were made to the unit after Columbia's landing and another flight of the hardware would later be pencilled-in for a subsequent mission.

Nonetheless, SOFBALL's Principal Investigator, Paul Ronney, was overjoyed with the results, which included the weakest flames ever burned. "Ecstatic would be an understatement," he said as the mission approached its midpoint. "The results so far are just beyond our wildest dreams. We're getting more data than we know what to do with!" The science crew mixed a variety of gases, including tiny quantities of hydrogen and oxygen, which – although too small to be flammable on Earth – burned for more than eight minutes in CM-1. "We saw some flame shapes that nobody had seen before," Crouch said later in the mission.

During the inaugural SOFBALL run on 8 July, supervised by Linteris, the fuel mixture was so weak that it only produced what researcher Karen Weiland referred to as "flame kernels". A later test, using a richer mixture of hydrogen, oxygen and sulphur hexafluoride, proved much more successful, burning for 500 seconds and allowing investigators to gain insights into how different concentrations of fuel and oxidiser affected the flameballs' stability and existence. "In pre-mixed gases used for combustion on Earth, we simply don't understand the mechanisms of flame extinction, what stabilises it or what keeps it going," said Paul Ronney.

Janice Voss works with the Combustion Module.

"These [SOFBALL] stationary spherical flameballs are the simplest to study, learn from and then apply to Earth applications." Such knowledge could, he added, help engineers to develop improved fire safety equipment for mineshafts, chemical plants or on board future spacecraft. Many of the flames produced during MSL-1 were so weak that they equalled only one watt of energy, compared to approximately 50 watts normally produced by a single candle on a birthday cake.

The peculiar sense of *déjà vu* was not lost on Columbia's crew. "This is the first chance to refly a payload and a crew altogether in the same group so quickly," said Voss during a space-to-ground news conference on 3 July. "It's been much easier to get back into the swing of things, and all the experiments are going great, and we all feel extremely comfortable and well-prepared because we've done this so recently."

However, she added that communication between the two shifts (Red and Blue) was difficult at times. "There's a lot of issues, like where everything is stowed, and people always find their favourite places. On a single-shift flight, where we all sleep at the same time, it's a little bit easier to negotiate, because everyone's awake and you can ask them where they put something. But on a dual-shift flight you have to be very careful to work together as a team across those few hours when you're both awake."

Watching the conduct of the MSL-1 experiments from the ground, LSP Principal Investigator Gerard Faeth expressed surprise as he saw weightless flames almost twice the size of any produced on Earth. "It does highlight the importance of us finding out something about what fires are like in spacecraft environments," he told journalists. Faeth's investigation of soot behaviour, it was hoped, could lead to a better understanding of how it forms and perhaps aid efforts to more effectively tackle forest fires and develop 'cleaner'-burning fuels.

"We have a lot of breakthroughs we're working on here," Don Thomas explained in an interview on 13 July, adding that "it might be a few years before they find their way into everyday life down on Earth, but this is basic research and we're pioneering out here". Nevertheless, Thomas' work with SOFBALL received the distinction of having a newly discovered combustion 'effect' named after him! One surprising finding from the experiment involved what happened when two small fuel droplets burn in close proximity: they initially moved away, then approached each other in a phenomenon dubbed the 'Thomas Twin Effect'.

Meanwhile, Faeth's LSP experiment also progressed well, using ethylene gas as part of research into more environmentally friendly, fuel-burning engines. Already, the April mission had given him an important 'first look' at how soot particles formed in microgravity and the shortened flight enabled his team to enhance LSP for STS-94. "We've learned that as we increase pressure, the amount of soot in the flame and the amount of soot it emits increases," he said. "It's probably one of the reasons that high-pressure combustion processes – such as what goes on in a bus engine – tend to emit a lot of soot."

One LSP run on 3 July involved a propane-fuelled study of soot, producing a "beautiful and steady flame", according to Linteris. "We will use the first few experiment runs to set parameters for the remaining runs," said Faeth. "Determining the settings for future runs will improve the efficiency of the experiment operations, which are designed to determine under what conditions soot is produced by flames and what the composition of soot is." The benefit of flying for 16 days, said Voss, was that "you get faster. By the end of the flight, we were getting very efficient running these experiments."

As well as using different fuels – propane, then ethylene – the experiment also burned them under different atmospheric pressures, "because soot is very sensitive to pressure", explained Faeth. "The higher the pressure, the more soot produced. Different fuel types also make a big difference. Natural gas tends to make little soot, propane produces more soot and ethylene, used in diesel engines, produces even more." By the time LSP operations concluded, the experiment had conducted 17 tests "and yielded good data", according to Faeth.

Other, related research focused once again on Vedha Nayagam's Droplet Combustion Experiment (DCE), which involved several 'phases' of observations of the burning characteristics of heptane fuel droplets under a range of atmospheric pressures. "In each phase," Nayagam said, "we are keeping the pressure the same and slowly reducing the oxygen to see if the fuels can still burn and, if so, *how* they burn. On Earth, we encounter low combustion scenarios – for instance, in gas turbines – so it's important to know what happens when pressure is reduced."

The three 'phases' were 'normal' atmospheric pressure – the same as that found at sea-level on Earth – as well as half, then one-quarter, of Earth's atmospheric pressure. Characterising the results from the experiment, Principal Investigator Forman Williams told journalists that "the crew had a tougher time igniting the droplets at this lowered pressure. We expected that. But when the fuel droplets *did* ignite, they burned stronger and more vigorously than we expected."

"In one atmospheric pressure, the flame burned out, leaving a residue of fuel," added Nayagam. "At the lower pressure, the flame is larger so the same results were expected, but the flame collapsed back on the droplet, completely consuming it. This tells us something about the extinction mechanisms." Typically, fuel droplets were formed by squirting heptane through a pair of injectors on opposing sides of a test platform within the chamber. When the drop formed, the injectors were retracted and the fuel was sparked by two hot-wire igniters; the entire procedure was carefully recorded by video cameras and high-resolution photography.

On the whole, the DCE apparatus performed admirably, with the minor exception of three brief malfunctions of the computer responsible for overseeing all of the experiments on board the Spacelab module; it was quickly rebooted and normal operations resumed. This prompted Assistant Mission Scientist Patton Downey to praise "great teamwork by the crew and science teams on the ground [which] has worked around some anomalies, enabling us to collect very valuable data". Another problem with one of the experiment's imaging cameras was soon resolved.

By the time the DCE hardware was finally shut down by Voss and its control computer removed on the evening of 14 July, heptane droplets ranging in diameter from 2 to 4 mm were successfully burned. "The large droplet extinguished early in its burn time, due to loss of significant combustion energy by radiation," said Fred Dryer. "The small droplet burned to near-completion because less energy is lost by radiation as the droplet's initial diameter is decreased. We would have received opposite results if we'd conducted this same experiment on Earth."

SPACE STATION HARDWARE DEMONSTRATIONS

Elsewhere on Columbia, payloads were being moved from the middeck to the Spacelab module and back, thanks to a revolutionary new storage rack known as 'EXPRESS' – an acronym for Expedite the Processing of Experiments for the Space Station – which was being pioneered on MSL-1 as a potential way of getting experiments and equipment quickly to and from the International Space Station. "That's why they call it 'EXPRESS'," said Boeing engineer Greg Day, "to get scientists' experiments up there faster than we have in the past."

The idea behind the new facility was to enable researchers to essentially plug their hardware, electrical power and data connections into the EXPRESS rack, giving them an easier, more generic interface and the chance to get experiments into space just 10–12 months after conception, rather than the three-and-a-half years ordinarily demanded for Spacelab missions. One possible advantage was that graduate students seeking to complete master's or doctoral degrees in a few years could get their research projects into space in a reasonable time.

"The rack itself is just the facility," said Bill Telesco, an engineer with Teledyne Brown Engineering in Huntsville, Alabama, "and you just plug in your equipment", sliding it into predefined locations, tightening screws and hooking up utilities such as computer interfaces, video and electrical power and water cooling. For MSL-1, the facility was used to support the PGBA plant-growth device and the Physics of Hard Spheres (PHaSE), both of which were transferred from middeck storage lockers to EXPRESS when Columbia reached orbit.

The plant growth unit investigated the use of specimens of direct relevance to the treatment of cancer and malaria, as well as those that could be employed in the lumber and paper industries. Peter Wong, a researcher from the University of Colorado, provided wheat in the hope that the plants could produce nodules that might convert nitrogen from the air for self-fertilisation. "Congress wants to know when the space programme can be self-supportive," said Wong. "They are spending lots of money and want to know how we can use space for commercial purposes."

Unfortunately, plans to watch the plants growing on a live video feed to the University of Colorado hit a snag when the lens clouded up. During the transfer of the PGBA from Columbia's middeck to the EXPRESS rack, the crew also encountered brief difficulties removing a long-handled Allen key from a connecting bolt. When operational, however, the experiment successfully explored the production of lignin, energy-generating secondary metabolites and changes in sugars and starches in plants as part of efforts to improve growth and production on Earth.

Other plants under study included a species of sage native to Southeast Asia, also used as an important anti-malarial medicinal drug, and *Catharanthus roseus*, which produces vinca alkaloids, used as part of the chemotherapy treatment of cancer. "Things are going great," said lead scientist Louis Stodieck of the University of Colorado at Boulder. "We've gotten data showing that the plants are very healthy. I'm very enthused and looking forward to the Shuttle landing and the opportunity to study the plants."

Meanwhile, PHaSE studied the transitions involved in the formation of colloidal crystals – collections of fine particles suspended in liquids, such as milk or ink – in the microgravity environment. In space, it was recognised that colloidal particles freely interact without the complications of settling, which normally occur on Earth. If the particle interactions between these 'colloidal suspensions' could be precisely and accurately modelled, they could provide the key to understanding fundamental problems in matter physics, perhaps ultimately leading to the development of new 'designer' materials. Industries which produce semiconductors, electro-optics, ceramics and composites were expected to benefit from the PHaSE results.

MATERIALS RESEARCH

In the materials-processing arena, major focal points for the STS-94 crew were again the Japanese-built LIF and German-provided TEMPUS facilities, both of which were activated shortly after Columbia entered orbit. After switching on LIF early on 2 July, Thomas kicked off a series of investigations to analyse diffusion in liquid tin telluride, developed by Misako Uchida of Ishikawajima-Harima Heavy Industries in Tokyo. Diffusion – whereby liquid metals mix without the need for stirring – of liquid tin telluride in microgravity was expected to lead to advances in making improved infrared detectors and lasers for terrestrial applications.

Minor teething troubles hit the facility on 6 July, when it was found that LIF was using up helium from its built-in purge – used to rapidly cool samples after heating – faster than predicted, but this was not expected to impact its ability to gather data. Diffusion of liquid metals is an imperfect procedure on Earth, because gravity-driven movements in the fluids cause complications. As on the original MSL-1 mission in April, Tsutomu Yamamura's study of the ideal conditions needed for electrolysis of molten salts was again conducted as part of research into refining basic science and engineering processes.

Other experiments included the Liquid Phase Sintering study, initiated by Crouch during his 3 July shift. This subjected tungsten, nickel, iron and copper to intense 1,500 Celsius heating to create solid–liquid mixtures. "Sintering is thermal heating that causes particles to bond together," said Randall German of Pennsylvania State University. "On Earth, sintering distorts the material. We are trying to learn the rules of why things distort on Earth. And *we are!*" In industry, sintering is routinely employed to form very hard, very dense solids, which can then be used to make cutting tools, car transmission gears and radiation shields.

Sintering normally involves heating metallic or ceramic 'powders' which, when placed under high pressures and temperatures, liquefy and bond to form strong materials. However, on Earth, gravity affects the dispersal of the powders and causes the resultant solid to be considerably less 'uniform' in nature. "How to make materials of the right shape and size", said German, "with no distortion is fundamental to powder metallurgy, so we are learning why distortion occurs on Earth."

Although technically a member of the three-person 'orbiter' crew, and responsible for monitoring Columbia's systems during his shift, Mike Gernhardt participated in numerous experiment sessions with LIF, performing an experiment on 8 July to explore the diffusion characteristics of 'dopants' – or impurities – in samples of melted germanium, which is often used as a semiconductor, as well as an alloying agent. "It's a fundamental scientific study," said David Matthiesen of Case Western Reserve University in Ohio. "We're trying to measure the fundamental thermo-physical properties of this semiconductor." Such experiments could eventually lead to faster-performing electronic components, including transistors and integrated circuits.

THE 'REVOLUTIONARY' TEMPUS

With four times as much time aloft as had been possible on STS-83, the scientific teams associated with both LIF and TEMPUS were gathering so much data that many were confidently expecting to spend at least a year analysing it all. The unique TEMPUS electromagnetic-levitation facility yielded the first measurements of specific heat and thermal expansion of glass-forming metallic alloys and, in so doing, obtained the highest temperature (2,000 Celsius) and largest undercooling ever achieved in a space-borne furnace. Moreover, significant progress was made in learning how to control and position liquid drops, using electromagnetic means.

So revolutionary was the facility, which flew first on IML-2 in July 1994, that NASA guaranteed its German manufacturers a free-of-charge flight on board the Shuttle in exchange for American participation in using it. In total, 20 investigations were conducted and almost a quarter of a million commands transmitted during the 16-day mission. On 10 July, materials scientist Robert Bayuzick of Vanderbilt University expressed delight at how well his zirconium samples – found in ceramic and refractory compounds – turned out. "We're doing fundamental science, so like always it's a long stretch between the fundamental science and the applications."

The shortened flight in April had actually proved beneficial to several investigators using TEMPUS, since they were able to examine the characteristics of the facility and their samples and make adjustments and improvements where necessary. On the whole, it performed admirably, with the exception of a problem with one of its television cameras, although a few experiments had to be terminated when their samples inadvertently came into contact with the side of the container. "Before the sample stuck to the coils, we were able to get some good data," said Principal Investigator Hans Fecht of the Technical University in Berlin.

Other studies of the heat capacity, thermal conductivity, nucleation rates, surface tension, viscosity and thermal expansion of a variety of glass-forming alloys proved enormously successful. During one experiment run, Thomas measured the heat of undercooled liquid metals to determine the ability of certain zirconium-based alloys to form glass. "This is the first time that these particular alloys could be prepared and measured in this way," said Fecht. "Undercooling to 200 Celsius is a lot! We have excellent and very important data."

Such fundamental discoveries and measurements of these alloys, some of which were being observed for the first time, could, said Project Scientist Jan Rogers, be used to improve sporting goods, including golf clubs, due to their excellent elastic properties. Mixtures of gold-coated aluminium–copper–iron and aluminium–copper–cobalt, in balls the size of blobs of chewing gum, were also extensively undercooled in TEMPUS. "The gold is protecting the sample from oxidation," explained D.M. Herlach of the German Aerospace Research Establishment in Cologne. "If the sample were to oxidise, it could not be undercooled."

Throughout the high-temperature heating and undercooling processes, electromagnetic 'pulses' were imparted by TEMPUS to squeeze, then release, the samples. "By measuring the [resulting] oscillations," researcher Bob Myers of Massachusetts Institute of Technology said on 10 July, after watching Linteris work with a specimen of palladium–silicon, "we can determine the surface tension and viscosity, or resistance to flow. We're using a new technique which is allowing us to measure the viscosity of these samples for the first time, measurements which can't be obtained on Earth. And even some of the surface tension measurements are a first. Preliminary results look very promising."

SUCCESS, SECOND TIME AROUND

Myers' words were summing-up the entire mission, which was turning out to be far more productive than scientists had ever imagined. In total, 206 fires were set, more than 700 protein crystals were grown and a variety of fluid physics and combustion experiments were conducted in the Middeck Glovebox (MGBX). A record 35,000 telecommands were relayed to the Spacelab research facilities from NASA's Marshall Space Flight Center, prompting Mission Scientist Mike Robinson to tell journalists that "if you walk around the science operations area, there are a *lot* of smiles!"

Although Columbia had enough consumables for a 17-day mission, no formal request to extend STS-94 was made, and early on 17 July Halsell and Still performed their de-orbit burn to bring her home. Leaving an orange streak across the clear Floridian skies, she settled onto Runway 33 at 10:46:36 am, wrapping up a mission just eight hours short of a full 16 days. The crew were elated at their success – "As much as the other one was a bummer, this is a *real* trip!" Crouch told journalists – but glad to be home after two weeks away from loved ones.

"The opportunity is great," said Jim Halsell, "and it's the opportunity to appreciate some of the things you've missed [and] some of the people that you love and you look forward to seeing again in the near future." He paid particular tribute to Columbia, which "performed absolutely flawlessly" for his crew. "Days have gone by", he told journalists during a space-to-ground news conference, "without having to do an 'error log reset', which is our way of saying there have just been no problems whatsoever. Our flavour for this flight is that it has done what [it] set out to do."

LEONID'S LONG WAIT

When STS-87 Commander Kevin Kregel arrived at KSC with his five crewmates in their T-38 jets on 15 November 1997, his words to the assembled journalists proved prophetic. "The crew is looking forward to launching on the fourth United States Microgravity Payload mission," he told them. "It's going to be a real exciting one, we're doing a lot of great science. Winston and Takao are going to do a spacewalk; Takao is going to be the first Japanese to do a spacewalk. Leonid will be the first Ukrainian to fly on the Shuttle. It's really going to be a super mission."

Kregel's third flight into space, and his first time in the Commander's seat, would indeed be exciting and would accomplish all of the tasks he mentioned, but the STS-87 crew would return to Earth with mixed emotions: elation at having safely and successfully conducted two weeks of orbital research, tempered with disappointment at the botched deployment of an important scientific satellite. The crew, which included two veteran astronauts and four rookies, was international in nature, with representatives of four different nations on board.

One of the rookies, Payload Specialist Leonid Kadenyuk, was named in May as the first representative of independent Ukraine to fly into space. Although he blended seamlessly into Kregel's crew, other STS-87 astronauts commented on the language barrier early in their training. "[Leonid] took English lessons and improved, so it wasn't a problem," said Mission Specialist Kalpana Chawla. "Plus, we [took] 80-hour Russian courses so we could speak a little bit of Russian. Not much, but enough that if a word was too big for him, you could open up your dictionary and show him what you were saying."

Interestingly, the 46-year-old Kadenyuk had been an astronaut – or, rather, *cosmonaut* – for much longer than any of his STS-87 crewmates. Born in Chemivtsi, he graduated from the Chemihiv Higher Aviation School, the State Scientific Research Institute of the Russian Air Forces and the famed Yuri Gagarin Cosmonaut Training Centre, earning a master's degree from the Moscow Aviation Institute. A test pilot, he flew 57 different types of aircraft, and from 1976 trained to command both the Soyuz spacecraft and the ill-fated Buran Shuttle. With such an impressive résumé, it is surprising that Kadenyuk's wait to get into space was so long.

"I believe that every person has his destiny," he wryly told journalists before lifting off on STS-87, "and my destiny has been to wait for a long time." In fact, it was not until after Ukraine gained its independence from the former Soviet Union in 1991 that the first hint of a chance to fly would arise. In the winter of 1994, during a state visit by the newly elected President Leonid Kuchma to the United States, a Bilateral Civil Space Agreement was signed, one of the provisions of which was an option to fly a Ukrainian cosmonaut on the Shuttle.

Two years later, Kadenyuk and another trainee, Yaroslav Pustovyi, were chosen by the National Space Agency of Ukraine (NSAU) to train for the position. Despite the historic significance of the venture for his nation, Kadenyuk felt he was not, technically, the 'first' Ukrainian in space. "I think that the first Ukrainian in space was our legendary Pavel Romanovich Popovich, who was Cosmonaut Number Four in the Soviet Union [on board Vostok 4 in August 1962]," he told a NASA interviewer. "But now, of course, since Ukraine has become independent, this will be the first flight of a Ukrainian. I believe the first flight of any cosmonaut, of any government, is a very important event in the life of that country. I am very proud that it has fallen to me to play this role and I will do everything I can to be worthy of this honour." Kadenyuk also added that he saw his mission on Columbia not as a 'one-off' publicity stunt, but as the start of a vigorous programme of international cooperation in space for Ukraine. He pointed out that his country was already responsible for building several important modern launch vehicles, including the Zenit.

FIRST SPACEWALK FROM COLUMBIA

The other five STS-87 crew members had already been training for six months by the time Kadenyuk joined them in May 1997. Commander Kregel, who flew a year earlier on STS-78 and previously in July 1995, would be joined by Pilot Steve Lindsey and Mission Specialists Kalpana Chawla, Winston Scott and Takao Doi. Of these, only Scott – one of very few black members of NASA's astronaut corps, and one of only a handful to have completed a spacewalk – had flown before; his selection for STS-87 was partly to make what would become the historic first spacewalk ever performed from Columbia's airlock.

That excursion, which would be devoted to simulating tasks in support of International Space Station construction, was a repeat of the work planned for Tammy Jernigan and Tom Jones in late 1996; they had been forced to call off both STS-80 spacewalks because of a stuck airlock handle. As well as proving a milestone for Columbia, the STS-87 outing would also make history by featuring the first-ever Japanese spacewalker, Takao Doi. Not surprisingly, *he* was excited by the opportunity but, despite months of training, was aware that the real thing might be very different.

"In order to move from one place to another," he explained to an interviewer, "you always have to grasp something to prevent yourself from floating away. But if your grip is *too tight*, your body will move in the *other* direction! You must softly touch the handrails and move along. That's the basic point of moving in space. It's impossible to make a complete simulation of the space environment during training. Even the training in the [water tank at JSC] is different, because the water generate[s] resistance and therefore body movements will be more stable and predictable than in space. The only way to master moving in space is to actually do this *in space!*" Doi got his chance on 25 November 1997, five days into the mission, although his tasks that night proved somewhat different from what he had trained for.

A FREE-FLYING SOLAR OBSERVATORY

One of Columbia's payloads was an important research satellite with a name even NASA's best acronym-makers could be proud of: 'SPARTAN', or the 'Shuttle-Pointed Autonomous Research Tool for Astronomy'. Weighing 1,350 kg, it was a cube-shaped box loaded with two instruments and a set of adjunct experiments and was supposed to fly freely for two days.

In effect, it was a carrier – a 'tool', as its name implied – for a variety of payloads, only a few of which had anything to do with astronomy. It had been used in January 1996, for example, to carry advanced technology experiments and on another mission in May to demonstrate an inflatable communications antenna. For Kregel's mission, it was designated 'Spartan-201' and would carry a White Light Coronagraph (WLC) and an Ultraviolet Coronal Spectrometer (UCS) to explore the outermost layers of the Sun's atmosphere (its 'corona'). The two instruments had flown before on board Spartan, on no fewer than *three* occasions.

Developed by the High Altitude Laboratory, the WLC was a specialised telescope that produced artificial 'eclipses' of the Sun to conduct detailed coronal observations. Meanwhile, the UCS, which had been provided by the Smithsonian Astrophysical Observatory of Harvard University, took spectroscopic measurements of the primary light emitted by highly ionised atoms, as part of efforts to determine velocities, temperatures and densities of the coronal plasma. Adjoining the

Firmly held by the RMS, the Spartan-201 satellite awaits deployment.

instruments were three technology experiments that would support their observations. The Technology Experiment Augmenting Spartan (TEXAS) would provide a real-time communications link with the WLC, as well as adjusting its alignment for optimal performance.

Spartan-201's first flight had taken place in April 1993, when it acquired more than 20 hours of data on the solar atmosphere, investigating specifically the north and south polar coronal 'holes', a southeast 'helmet' streamer and a very active region above the Sun's western limb. Eighteen months later it flew again, this time conducting observations and proton-temperature measurements in concert with the joint NASA/ESA Ulysses solar probe, which was passing 'under' our parent star's south pole at the time. Lastly, almost exactly a year later in September 1995, it performed similar work as Ulysses hurtled over the Sun's north pole.

For its fourth trip, Spartan-201 would be investigating the mechanisms responsible for heating the corona and accelerating the anomalous 'solar wind', which originates there. It was also supposed to conduct joint observations with another scientific satellite, the Solar and Heliospheric Observatory (SOHO), which had been launched in December 1995, as part of efforts to expand physicists' understanding of conditions within the corona.

"The corona has three important properties," said NASA manager George Withbroe. "It's very hot, about three million Fahrenheit, it flows away from the Sun at very high speeds – about a million mph – and it's very gusty. Spartan-201 is designed to address three questions: Why is the corona so hot? How does it get accelerated to such high speeds? And why is it so gusty?" Insights into these mysteries were expected to allow scientists to better comprehend how stars 'seed' neighbouring regions of interstellar space with matter. Moreover, said Withbroe, "when the [solar] gusts hit the Earth, they cause a variety of effects. One spectacular example are the aurorae. When you get strong gusts, they can occasionally pump up the Van Allen radiation belts and the particles in these belts can affect spacecraft. For example, during a strong gust in April [1997], a communications satellite lost 15% of its power. A third effect of these gusts is when they hit [Earth's] magnetic field, they *move it around!* When you move a magnetic field and you have long conductors, like pipelines and electrical power lines, you generate currents. Most of the time, these are just a nuisance, but occasionally, when you get a really strong gust, you can damage power transformers in electric power stations. In 1989, in a particularly spectacular event, Hydro-Quebec was knocked out for nine hours, affecting six million people."

According to pre-flight plans, Spartan-201 would be deployed by Chawla, using the RMS, on the second day of the mission and would be retrieved about 50 hours later. For its ride into orbit, the satellite was affixed to a Mission-Peculiar Equipment Support Structure (MPESS) in Columbia's payload bay. It was an important responsibility for Chawla, who became the first Indian-born woman to fly into space, but as circumstances would transpire, it would turn into her nemesis. Bad luck could not have seemed further away, however, when Columbia lifted off precisely on time at 7:46 pm on 19 November, beginning 1997's final Shuttle mission.

DOING THE TWIST

The ascent, though picture-perfect, was unusual, as it featured the first use of a novel 'heads-up' manoeuvre about six minutes after launch. In effect, the onboard GPCs commanded the main engines, over a 20-second period, to roll Columbia 180 degrees from a 'belly-up' to a 'belly-down' orientation, enabling the crew to communicate with Mission Control through the TDRS network two-and-a-half minutes sooner than had previously been possible. As well as eliminating the need for a ground-tracking station at Bermuda – the use of which had required the Shuttle to ascend belly-up – the move saved NASA $5 million per year.

The computers pulled off the manoeuvre effortlessly and Columbia rolled left at 5 degrees per second, resulting in a minimal 15-second communications loss with Mission Control. "In the roll, we transitioned from ground station telemetry to TDRS telemetry," said Shuttle launch integration manager Don McMonagle, himself a former astronaut. "That transition occurred in less than 15 seconds, which is precisely as we expected. We're very satisfied that that unique aspect of this ascent trajectory went as planned." The manoeuvre, which was to be used on future flights on an easterly heading, allowed Columbia's cockpit antennas to 'lock-on' to the overhead TDRS.

"Nice roll," radioed Capcom Scott Horowitz from Mission Control.

"Copy and concur," Kregel replied simply. The entire manoeuvre was controlled by the onboard computers, and the crew did not even know if they would roll to the left or the right until it actually happened; *that* decision was made in real-time, based on incoming velocity and orientation data. Telemetry indicated the manoeuvre, which took place 480 km downrange of KSC and at an altitude of about 110 km, went smoothly, although unfortunately it was not visible on long-range tracking cameras. "We do not characterise this as a risky manoeuvre," McMonagle told journalists. "This was analysed so we would make this roll to heads-up *after* we're outside the atmosphere, so the concerns about aerodynamic loads on the vehicle are not a problem. This occur[red] six minutes into the launch, so we're well out of the atmosphere and the roll to heads-up is completely unaffected by air. To quote Kevin Kregel, he expects it to be an E-ticket ride, but it's predictable and completely certified."

"We had to do a fair amount of analysis to ensure that we weren't doing something dumb," said Flight Director Wayne Hale. "One thing we didn't want to perturb was the [RTLS] abort mode," which Kregel would perform in the event of an emergency early in the ascent. "That [the RTLS] is a fairly intricate manoeuvre, it's been analysed to death and we spent a lot of money making sure it would work if we ever had to do that and it is based on a heads-down trajectory. So we picked a time [for the roll] that was *after* negative return [to KSC]." The analysis concluded that, even with electrical failures or two main engines shutting down, "it never goes what you would call 'out of control' ".

"It'll be pretty neat to see from a piloting perspective," said Horowitz before the launch, "because here you are, upside-down, and then you're going to roll to right-side-up. It may actually make the ride a little bit more comfortable." To be safe, the manoeuvre was carried out before Columbia was out of range of the Air Force's Eastern Test Range tracking radars, to enable Mission Control to maintain contact with the crew throughout, until TDRS-supported communications commenced.

"The whole sequence is very fast," Chawla said of her first Shuttle launch. "I was doing the photography [of the jettisoned External Tank] and was disoriented because I didn't know where my legs were! You only have two-and-a-half minutes [to photograph it] because after that it is not going to be visible. During the time it was in view, I took 40 pictures. The people on the ground were happy because it showed them the sequence for how the venting was taking place. They hadn't seen it on previous flights. NASA did a lot of analysis to see if this could become a problem on another day and they concluded that it wasn't."

SPARTAN IN A SPIN

Immediately after reaching orbit and setting up Columbia for two weeks of research, Chawla checked out the RMS in readiness for the planned Spartan deployment a little over a day into the mission. It was at this point that problems emerged. One of its scientific objectives was to conduct simultaneous observations of the solar corona with SOHO, as part of efforts to recalibrate the latter's instruments and ensure that its sensitivity had not degraded after two years in space. However, on the very day that STS-87 was launched, SOHO suffered a voltage surge that temporarily disrupted its pointing control system.

The billion-dollar satellite is in solar orbit, 1.6 million km closer than Earth's own orbital course, and it was hoped that the joint experiment with Spartan could allow both spacecraft to gather data from their different vantage points.

Shortly before going to bed on the evening of 19 November, Kregel's crew was advised of SOHO's problem and informed of the decision to postpone Spartan's deployment until the 21st, by which time it was expected that SOHO would be back up and running. "We plan to execute our [observing] programme as planned," said astronomer Richard Fisher, one of Spartan's principal investigators. "It will just start one day later." By the afternoon of 20 November, SOHO was back on-line, although Mission Operations representative Lee Briscoe explained that even in the case of further problems, the deployment would have gone ahead regardless.

"There's both a science programme and a *joint* science programme," added Fisher. "I can't exactly quantify for you what percentages [between them] are, but we would go ahead and complete the Spartan-specific science programme and we'd

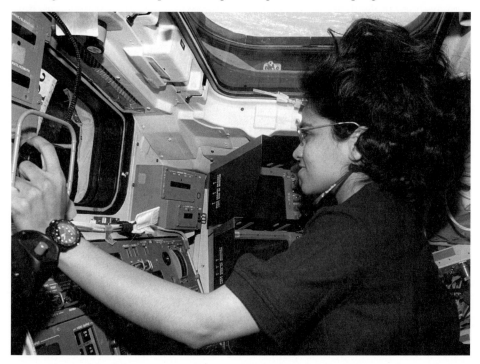

Kalpana Chawla controls the RMS from her station on Columbia's aft flight deck during Spartan-201 proximity operations.

expect there would be no disruption of that. And that would give us a considerable yield. We would be very happy with that." Late the following night, Chawla duly grappled Spartan to begin the deployment procedure and released the satellite at 9:04 pm, as Columbia flew over the Pacific Ocean. What happened next, however, would dog her career.

"We pick up the Spartan and take it to its deploy location," she had told an interviewer before launch. "It's critical that the location be right, because it's supposed to be looking at the Sun at that time. At that time we let it go, and we watch it for about two minutes when it does a pirouette. That tells us that all the automated tasks that are supposed to go on have started on time. If that does not happen, then we'll take it back and put it in the bay for another try on another day."

After releasing the satellite, the crew was surprised when it failed to perform its pirouette, implying a problem with its attitude-control system which provided fine pointing towards solar targets. Had Chawla, responsible for supervising Spartan's pre-deployment checkout, missed issuing a critical command? It was a question that would be revisited in the coming weeks.

"The command that we didn't receive onboard the spacecraft would have powered down some of the systems that were up in a standby health-check mode," said Spartan Mission Manager Craig Tooley. "We basically power up a bunch of subsystems and check their status and then bring them back down into a quiescent state to await deploy[ment], where the sequence is then picked up again and the flight programme kicks in. *That* step – that brought those down into a quiescent state and had them waiting for deployment – wasn't received by the spacecraft.

"Therefore, the attitude-control system was still there warming up, but not configured to deal with the error signals from the sensors or fire some pyrotechnic gas valves that would have turned on the thrusters. So it was basically 'idle' and it never got out of 'idle' mode." Chawla duly moved in to regrapple the satellite, but did not receive a firm 'capture' indication on her control panel in Columbia's aft flight deck and backed the arm away. In doing so, she accidentally bumped Spartan and imparted a two-degrees-per-second rotational spin on it.

"Houston, we have a tip-off rate. It's rolling around," Kregel told Mission Control. "We will go ahead and match it and regrapple." For much of the next hour, he used Columbia's RCS jets to manoeuvre his ship in such a way that its rotation matched that of Spartan, so that Chawla could try again to regrapple it. The concern was one of timing: a second deployment attempt had to occur within an hour of the first to prevent a timing device from automatically closing the fuel valves to the satellite's pointing system.

RESCUING SPARTAN

However, his efforts exhausted the Shuttle's fuel budget for the operation and Flight Director Bill Reeves called off the attempt at 10 pm, instructing Kregel to manoeuvre to a separation distance 64 km 'behind' Spartan; already, by this time, flight controllers had in the backs of their minds a plan to re-rendezvous with it on 24 November and send Scott and Doi out on a spacewalk to manually grab it. "We feel very confident this is going to work [and] the crew is quite capable of doing this," senior flight director Bob Castle told journalists on the 22nd.

"The flying tasks of bringing the orbiter up close to the tumbling spacecraft I think are very doable." Although the Spartan rescue was only expected to require a couple of hours, the decision was taken to run the spacewalk to the planned six hours and have Scott and Doi complete most of the other tasks they had originally trained for. After checking out their equipment, the duo began donning their spacesuits, with Chawla's assistance, around 9 pm on 24 November, as Kregel and Lindsey continued a series of thruster firings to gradually close the gap between Columbia and Spartan.

Winston Scott and Takao Doi in water-cooled underwear prior to donning their spacesuits for the first EVA in Columbia's career.

The most critical manoeuvre, known as the 'Terminal Initiation' burn, came at 10:51 pm and placed the Shuttle on an intercepting path to arrive directly beneath the slowly rotating satellite. A couple of minutes past midnight, the duo became the first astronauts to vacate Columbia's airlock in space – prompting Doi to remark on a "nice view over the Earth" – and quickly moved to their positions at either end of the MPESS carrier as Spartan tumbled slowly 210 m away. When Kregel had positioned Columbia within reach, he radioed the two spacewalkers, giving them a 'go' to grab the satellite.

"Winston," he told Scott, "I think if we're patient, it looks like the telescope is rotating, the top end of the longitudinal axis looks like it will come down to you and the bottom one right up to Takao. If we just wait, it'll come right to us."

"Okay, copy that, Kevin," Scott replied. "We'll just be patient and see what happens." A few minutes later, the elation was clear in his voice as the satellite tumbled closer. "It's perfect: the telescope is right between us! It's maybe four feet above my head." According to Kregel, Spartan appeared to be relatively stable throughout the rendezvous, which suggested its backup attitude-control system had activated at some point after Columbia's departure on 21 November, damping out the two-degrees-per-second rotation to leave a much more tumbling motion.

Right on cue, at 2:09 am, as they flew over the southern Pacific Ocean, the two spacewalkers reached out and grabbed the satellite. "Okay," said Scott, "now that we've got it, Mr Doi, let's decide what we're going to do with it." They did have some difficulty in manually latching it back onto its MPESS carrier, but Chawla, operating the RMS from Columbia's aft flight deck, succeeded in driving it onto its retention latches. "KC's got it!" radioed Kregel. "You guys can let go." The entire procedure was over by 3:23 am, with Spartan snared in the payload bay.

"OUR LAST CHANCE TO LOOK AT THIS"

Yet the spacewalk had not even reached its halfway mark, for Scott and Doi would remain outside for a total of 7 hours and 43 minutes, making this one of the longest excursions on record. Following the failure of Columbia's hatch to open in November 1996, which prevented Jernigan and Jones from testing a series of prototype International Space Station tools, most of their tasks were offloaded onto the STS-87 spacewalkers. Scott and Doi had trained to work with a small crane, which they would use to move a dummy station battery, and a variety of different safety tethers.

It was hoped that the almost 2-m-tall crane would prove useful in transporting Orbital Replacement Units (ORUs), such as batteries, with masses as great as 270 kg from translation carts on the station's external truss to various work sites. "We had a very successful test of the crane," Scott said later. "What we did not do was to get a chance to exercise all the options. The crane actually operates very, very smoothly. It's meeting most of our expectations. I say 'most' because there's a little bit of flexibility in the boom that we didn't anticipate."

Winston Scott (left) and Takao Doi get themselves into position to grab Spartan-201 with their gloved hands.

After the crane had been installed into its socket on the port-side of Columbia's payload bay, Doi evaluated its operating characteristics, while Scott removed a large 225-kg dummy battery and its carrier device from the starboard wall. These were attached to the end of the crane to assess its ability to move large masses. The crane's boom was extended by turning a ratchet fitting with a power tool or, as a backup, by using a manually operated hand crank.

The only problem the two men encountered was a jammed torque multiplier, used to tighten the mock battery onto its mounting location in the payload bay. However, Scott finally freed it by following instructions radioed to him by Capcom Chris Hadfield in Mission Control. Also in the control centre, watching a live televised downlink of the excursion, were the spacewalkers' wives.

"I just wanted to tell you that in the viewing room here, Marilyn and Hitomi are wearing their EVA shirts and proudly watching you fellows at work," Hadfield told them.

"All right! Tell Marilyn and Hitomi both 'hello' from Winston," Scott replied. "We're glad they're there."

"And hello from Takao," Doi chimed in. "Thanks for watching us."

"And for Marilyn," Scott added, with a grin, "I just had to stop by and pick up a satellite. I'll be home by suppertime!"

After stowing the crane and mock battery, the spacewalkers cleared away their tools, before returning to Columbia's cabin; the excursion officially ended at 7:45 am, when they repressurised the airlock and switched their spacesuits off internal battery power. "On behalf of all of us here on the ground, congratulations!" Hadfield radioed from Houston. "That was a long and challenging spacewalk. It was the first one from Columbia and Winston, you've now been outside for 10 times around the world. And Takao's definitely got more EVA time than any Japanese citizen in history!"

One of the tasks assigned to each of the astronauts performing these practical evaluations of potential space station tools was to fill in a questionnaire to capture their initial thoughts. In general, Scott and Doi gave most of the activities they performed 'As' and 'Bs', which helped engineers to finalise the designs before committing them to actual construction missions. The responses from the STS-87 crew were particularly important, because theirs was the last scheduled mission to involve a spacewalk before the first station components were due to be hauled into orbit in the summer of 1998.

"This is our last chance to look at this before we employ it somewhere down the line on [the] space station," said Greg Harbaugh, a former astronaut and then-acting head of the EVA Projects Office. "We have to recognise that EVA is very much fundamental to the success of [the] station. It is going to be an almost everyday occurrence for the next several years, so we'd better get used to it. If we have any misgivings about EVA, it's time to get those behind us. Starting with STS-88 [the first station-assembly mission], we are into what we call the 'EVA wall' and we better be ready to step up to it. It's a gigantic undertaking, something like three times the amount of EVA work that we've ever done in the history of the American space programme. But I think we're up to the challenge. I think we're going to be ready for it and we're going to succeed. But the best way to do that is to mitigate the risk as you go along, to gather what information you can as you go along."

AERCAM/SPRINT

Although their spacewalk had lasted far longer than the six hours originally planned, and Scott and Doi had accomplished a significant amount of their objectives, there were a number of tasks that had been omitted. One of these was an odd, beachball-like contraption known as 'AERCam/Sprint' – or the Autonomous EVA Robotic Camera/Sprint – which was supposed to demonstrate a free-flying camera that could be remotely controlled by an astronaut on board the Shuttle to inspect the exterior of the International Space Station for possible damage.

The concept had attracted particular interest from NASA after the collision of an unmanned Progress resupply freighter into the Mir space station in June 1997, which forced the evacuation and depressurisation of one of its scientific research modules. Had a free-flying camera of AERCam/Sprint's capabilities been available, it might have been possible to investigate and acquire high-quality photographs of the damage, without risking human lives on a spacewalk. "They're [Mir's crew] limited on the views they can get right now because of where the [station's] windows are located," project manager Cliff Hess said before Columbia's launch. "That's exactly what AERCam is meant for: to give you some other view that you can't get otherwise."

Future applications could include sophisticated sensors to 'sniff' out coolant leaks or mechanical arms to hold spacewalkers' tools. "One thing you could do is maybe go out periodically and do a scan of the [International Space Station's] solar arrays, to look maybe for meteoroids that may have hit it," speculated Hess. "You could automate something like this free-flyer to do a scan somewhat like a crop-duster." Multiple AERCams could be mounted on the station's exterior to carry out routine inspections.

"I think it's very useful," said Steve Lindsey, who tested AERCam for its first outing. "There are other applications for it, [such as] moving large masses from one part of the station or another. It's a really neat project. The potential of this technology is limitless." The focus of all this attention looked innocuous enough: a 36-cm-diameter, 16-kg aluminium sphere, coated with soft Nomex felt. This, and onboard control mechanisms that constrain its speed to no more than a few centimetres per second, allow it to be employed safely in close proximity to spacewalking astronauts. If its velocity somehow doubled that, explained Hess, AERCam would automatically shut itself down.

"We don't want anything to inadvertently hit a person or any part of [the Shuttle]," added Kregel before the flight. "The likelihood of damage is very minimal, but our plan is to not let that happen. If it is really out of control, we would just [fly Columbia] out of the way." The free-flyer was designed to move itself using a system of 12 tiny cold-gas thrusters, operated by Lindsey via a joystick, antenna and two laptop computers from the aft flight deck.

Housed in the airlock, the original plan called for Scott to hand-release it for a half-hour joyride. If time permitted, Lindsey hoped to 'fly' it up to 50 m 'above' Columbia, take a series of video images, then guide it back into Scott's gloved hands. Two miniature colour television cameras, with 6-mm and 12-mm lenses, would provide stunning views for the crew and Mission Control. "It's unlike anything I've ever flown before," said Lindsey before the mission. "In some ways, the closest thing you could compare it to is flying a radio-controlled airplane."

REFINING SPACEWALKING TECHNIQUES

Although, in comparison to the crane and battery tests, the AERCam/Sprint run was a low-priority objective, it was nevertheless with disappointment that Columbia's crew realised they would not get an opportunity to test-fly it. "It was supposed to be one [spacewalk planned]," said Chawla after the mission, "but we knew from our timeline that if we were going to fly Sprint, we wouldn't have time to do it. In all of our training runs, the tasks that we were supposed to do took longer than the timeline called for."

It was with pleasant surprise that mission managers decided on 1 December to add a second EVA to the mission, although its primary focus was not to fly AERCam/Sprint, but rather to repeat and gather additional data on the crane tests from Scott and Doi's first spacewalk. According to Greg Harbaugh, planners had been unable to acquire all the data they needed from the spacewalk on 25 November. Specifically, Scott encountered difficulties connecting a mock ORU to the crane when its boom proved more flexible than anticipated.

"This crane, and the ORUs that are handled by it at the end of the boom, are fundamental to the success of the station," said Harbaugh, himself a seasoned spacewalker. "They are used over and over again on the station. From our assessment, Winston had some difficulty mating the large ORU – the big battery – to the end of the boom. That was not entirely unexpected, but now we have an opportunity to go back and refine our techniques based on what we observed the first time.

"Winston did a lot of work. He tried it several times, with Takao assisting, with the technique we thought would work going in, which was basically Winston placing it down from above on top of the crane. When that didn't work, Winston pushed down and Takao tried to pitch up with the crane, a sort of 'push-up/pull-down'-type sequence. And *that* didn't work. The technique that did work was where Winston got a little bit sideways and grabbed the crane from below and the box from above and 'squeezed' them together in a sort of sandwich technique. The concern with that technique is we are not confident that we're going to have the luxury at all the work sites on [the] space station of being in position to grab those two pieces and squeezing them together like that. We want to take advantage of the opportunity to really wring this out while we've got it on this mission."

Scott and Doi's first focus was working with a mock station component, somewhat smaller than the battery they had handled on their 25 November excursion, known as a 'cable caddy'; this was another task originally planned for STS-80. The caddy, which was about a sixth as massive as the battery, was envisioned as a means of holding up to 20 m of replacement electrical cable. Scott began the task of mounting it onto the crane in the same way as he had done with the battery: pushing it down onto the crane's latches, then 'shoehorning' it into place.

He also rotated the assembly 90 degrees, using a tether to help to anchor it in place, then forced the caddy straight 'down'. His efforts produced no apparent problems. "The first thing I'm going to do is try to dock the cable caddy to the grid facing upward out of the [payload] bay," he radioed for his crewmates' and Mission Control's benefit. "Let's see how the soft-dock mechanism works ... I have a successful soft-dock! There is compliance in the crane; it moves a little bit, but not very much – I'd estimate maybe six or eight inches before I got soft-dock."

Overall, he gave the crane an 'A' with the smaller cable caddy. Flight controllers asked him if the 'shoehorning' technique of mounting the caddy onto the crane might also work with larger components like the battery. "Well," Scott replied, "that's a possibility that [it] would have been successful had we thought about trying it. The problem with the large mass is controlling it. This one is so small, I can precisely position [it], then rock it forward. But the answer is 'yes', it is possible I might have been able to do it with the large ORU by rocking it."

However, he cautioned against forcing large components into place on the crane. "I think you'd get in trouble trying to use fast impulses with that large ORU," he explained. "Remember, [the battery] is 500–600 lb and I deliberately kept my inputs low. I think if you put large inputs, or thought of slamming it home if you will, if it docked, you're going to have this large mass swinging back up at you because of the compliance in the boom. If it doesn't dock, you've got this large mass with a lot of rates. I would firmly not recommend that."

Next, Scott tried locking the cable caddy onto the crane using a semi-rigid tether to keep the boom relatively steady. "This tether technique is working pretty well," he said. "It did prevent crane movement aft. If I add more force, it's nice and stable. I can even push against it if I need to. I really think that this could be a possibility with the large ORU, [but] the problem might be controlling the large ORU. I can restrain the crane from fore-and-aft movement with my tether, but part of the problem with the large ORU is up-and-down movement."

"We need to come out of this mission with a clear understanding of the technique we expect to employ and confidence the system will be tolerant to that technique so we will be assured that we can move these boxes around," said Harbaugh. "The implications for [the] space station are that we expect to do a lot of moving around of boxes. We're going to be moving them from one place to another – from the payload bay to various points on the station – and we expect this crane to help us greatly. If we can't move these boxes around and use them to replace boxes that have failed out there, then we have a fundamental design flaw in our thinking – our methodology – for how we're going to put this station together."

Two different safety tethers were also used by Scott and Doi. One, the Body Restraint Tether (BRT), was designed to hold a spacewalker steady while clamped to a handrail; this had the benefit of freeing the astronaut's hands for working. Scott had already tested it during his previous spacewalk on STS-72 in January 1996 and Doi evaluated it on STS-87. The second tether was the Multi-Use Tether (MUT) which, although similar to BRT, had the capability of performing a wider range of tasks. For example, different 'end effectors' could be attached to it to grip ORUs, tools or handrails.

FREE-FLYING CAMERA

Near the end of the spacewalk, Scott and Lindsey at last got the chance to test-fly AERCam/Sprint. "What I'm going to do is retrieve the Sprint from the airlock and I'll make my way up to the foot restraint [and] mount myself in the foot restraint," Scott told an interviewer before Columbia lifted off. "Once I've done that, I'll power it on and look for a sequence of five flashing lights that tells me the power-up went well and everything's working properly. Then, when Steve gives me the 'go', I'll rotate it about one or more axes to perform another self-test. When we get the indication that the self-test is complete, I'll just stand-by. Steve will go through a series of tests and checks from his console [on the flight deck]. When he gives me the word, I'll simply release it, sit back and watch and catch it once the flight is over. He will pilot it back to my location, I'll reach out and grab it, put a tether on it, power it down and that will be the end."

Scott releases AERCam/Sprint.

The first flight of AERCam went like clockwork and, after Lindsey had tested its tiny nitrogen-gas thrusters – which Scott described as a sensation of "very tiny thumps" against his spacesuit glove – the $3-million robot beachball was released into the payload bay at 12:15 pm. Spectacular images of Columbia and the blue-and-white Earth were transmitted to Mission Control from its two tiny television cameras and the robot's handling characteristics were so good that Lindsey was granted additional time to evaluate it.

"Steve's just raving about the handling qualities," Kregel said at one point. "He says it's flying just like the simulator."

"Hey, pretty cool," replied Capcom Bill McArthur, as he watched stunning images on his monitor in Mission Control, then suggested, "Maybe a view a little bit higher looking down at the orbiter could play real well down here."

After more than an hour of tests, Lindsey gently manoeuvred AERCam/Sprint back towards Scott's gloved hands at 1:27 pm. "Keep it coming . . . another couple of inches," he told his joystick-wielding colleague. "I can feel the thrusters on my glove. Okay, I've got the free-flier!" Lindsey had nothing but praise for the robot's designers and Harbaugh, too, was more than happy. "Having this kind of 'floating eyeball' in our hip pocket, available when we need it for space station, is going to be worth its weight in gold over the next few years."

Cliff Hess was elated, telling journalists that AERCam only used 65% of its onboard nitrogen supply and a mere 8% of the power from its lithium batteries, reaching a maximum distance of about 12 m from Columbia. "The team is on 'cloud nine' after the wonderful performance that we had with the Sprint," he said. "We think there are some improvements that could be made, as we thought there would be. We need to add more autonomy, to free up the pilot workload, so there's less [work] to actually fly it: more intelligence in the software. We can put on additional sensing in the future, depending on what we use it for, and then I think we'd like to have the capability to deploy and retrieve it without requiring an EVA crew member. We could fly it into something like a [GAS] can that could be mounted on the Shuttle or a station truss. I think it shows what can be done." Scott and Doi stowed the robot back in Columbia's airlock and wrapped up the final few tasks of their spacewalk, which officially ended at 4:08 pm.

USMP-4: "A GREAT MISSION"

The two spacewalks and the Spartan fiasco had so far distracted many from the highly prized research that was going on elsewhere on a battery of experiments that formed the fourth United States Microgravity Payload (USMP-4), which was flying its final mission. The payload was activated, under Chawla's supervision, late on 19 November. Despite having to make adjustments to their science-gathering schedules to accommodate the initial one-day postponement of Spartan, Assistant Mission Manager Jimmie Johnson told journalists on 20 November that from preliminary results, "it looks like we're going to have a great mission".

Three of the USMP-4 experiments were already familiar from earlier flights: the

AADSF and MEPHISTO solidification furnaces and the IDGE dendritic-growth facility. Another, the Confined Helium Experiment (CHeX), was making its first sojourn into space to test theories of the influences of boundaries on matter by measuring the heat capacity of helium while confined to two dimensions. A clearer understanding of the effects of 'miniaturisation' on material properties was expected to lead to smaller and more efficient and cheaper electronic devices, including high-speed computers.

Almost daily, CHeX's Principal Investigator John Lipa pointed out, the semiconductor industry reduces the size of devices such as computer chips in a bid to improve transmission speeds and power consumption and reduce costs. However, electrical performance is also affected by size, thickness, surface irregularities and changes in a given material's physical properties. As these changes take effect, they often result in defects. Lipa's team from Stanford University sought to discover what caused the changes, using a sample of liquid helium, which conducts heat a thousand times more efficiently than any other material.

"We are studying the novel properties of matter when it becomes very thin," said co-investigator Talso Chui of NASA's Jet Propulsion Laboratory. "Hopefully, we'll learn some secrets of nature."

CHeX consisted of a refrigerated dewar, coated with a magnetic shield and holding 392 crystal silicon disks, each just two-thousandths-of-an-inch thick, which forced the liquid helium into very thin layers. Through these layers, Lipa's team took precise temperature measurements of within a tenth of a *billionth* of a Kelvin. "The temperature resolution", said USMP-4 Mission Scientist Peter Curreri, "is equivalent to the ability to measure the distance between Los Angeles and New York to the thickness of a fingernail! The information from this experiment can be applied to future generations of microprocessors, where we'll approach the finite size limit."

It was a pity that the media focused so intently on the Spartan fiasco, because the USMP-4 investigators were quietly gathering their own treasure trove of scientific data. The AADSF processed two different alloys – lead–tin telluride and mercury–cadmium telluride which could form the basis of advanced infrared detectors or lasers. "We completed processing one sample in both the low-temperature and high-temperature modes of operation," said Archie Fripp of NASA's Langley Research Center, one of AADSF's principal investigators. "The low-temperature sample looked at thermal convection, or flows, and the high-temperature sample studied the changes caused by manipulating the growth rate of the crystal."

Unfortunately, operations with the furnace ended earlier than planned on 30 November, when the science team noted unexpected readings from several temperature sensors used to control the solidification process. This prevented investigators from completing the second growth of three lead–tin telluride samples. "Prior to this mission, the furnace was modified to allow the exchange of samples," said Johnson. "Although we were unable to complete the third run, the two we've completed will potentially yield more science from this mission than the previous two combined."

In particular, a single, unique electrical crystal with exceptional uniformity of composition was grown. "We could only get this uniformity in the microgravity of space," said Marshall investigator Don Gillies. Added lead researcher Sandor Lehoczky: "One of our objectives on this flight was to get benchmarks: near-perfect materials that can be used to compare with and judge materials made on Earth." After Columbia's landing, the crystals were polished and etched in the hope that they could be used to detect previously undetectable infrared energy levels. "We're working on the materials for the future," said Marshall's Dale Watring.

Meanwhile, the French-built MEPHISTO furnace processed three identical bismuth–tin alloy samples, while sensors monitored temperature fluctuations and the position of the solid-to-liquid border, 'marking' them with electric pulses. Its data was correlated with measurements from the SAMS device to determine the impact of Shuttle motions on the solid-to-liquid interface. Once again, the furnace performed like a champ. "We are very happy with the MEPHISTO hardware," said Gerard Cambon of the French space agency, CNES. "The French team has participated in this experiment for the last 18 months and we have worked in a cooperative effort to enhance the science gains."

On its two previous flights, the IDGE device had used ultra-pure succinonitrile (SCN), an organic crystal that mimicked the solidification of so-called 'body-centred-cubic' materials, such as ferrous metals. For USMP-4, however, it employed a different test material: pivalic acid (PVA), a 'face-centred-cubic' material that solidified like many non-ferrous metals. Its transparent nature and low melting point made it an ideal sample for fundamental 'benchmark' observations. According to IDGE Principal Investigator Marty Glicksman on 20 November, the information gathered "is providing extremely valuable insight[s] into the solidification behaviour of [PVA]. It will doubtless lead to new scientific findings about dendrite-growth dynamics."

Whether melting and resolidifying bismuth–tin, lead–tin telluride or mercury–cadmium telluride, monitoring the heat characteristics of liquid helium or growing PVA dendrites, the presence of the two acceleration monitors – SAMS and the Orbital Acceleration Research Experiment (OARE) – helped to identify the impact of vibrations and tremors, some of which could be caused by crew movements on their middeck ergometer. "If we can identify what's causing a disturbance to an experiment," said Melissa Rogers of NASA's Lewis Research Center, "a science team can then decide whether to ask for that interfering fan, pump or whatever to be turned off."

Still other USMP-4 investigations were conducted in Columbia's cabin, using the versatile Middeck Glovebox facility, which Chawla used to investigate ways of creating uniform mixtures of liquids to form specific metal alloys, for possible use as a basis of future semiconductors. Liquids that do not mix well – called 'immiscibles' – were deliberately chosen as part of this experiment to pinpoint the causes of their separation. Another study, provided by Doru Stefanescu of the University of Alabama at Tuscaloosa, examined the solidification of liquid–metal alloys. Lastly, a combustion experiment to investigate laminar gas flows was carried out.

Such flows are a key phenomenon in combustion processes, such as those used to power aircraft engines, natural-gas power plants and modern ship. The MGBX experiment was expected to provide clues of why jet engines occasionally 'flame-out' when fuel injected into a moving airflow moves out of the combustion chamber and extinguishes. "We hypothesised that flames would remain stable at a higher forced airflow in the microgravity environment, and that's exactly what we saw," said Principal Investigator Dennis Stocker of NASA's Lewis Research Center.

"This research will have a direct bearing on engine safety and furnace efficiency," added researcher Lea-Der Chen of the University of Iowa. "This data will also allow us to critically test our computer simulations by comparing them to the data we received from the experiment."

A REDEPLOYMENT OF SPARTAN?

The success of USMP-4 had, however, been tempered by the botched deployment of Spartan-201, although flight controllers hoped that all might not be lost. On 29 November, Kregel performed a status check of the satellite to gather data on its health in anticipation of a possible redeployment towards the end of the mission. Six to 20 hours of free-flying solar observations might still be possible, as opposed to the 50 originally planned, but ultimately the option was called off because of concerns about available propellant reserves in Columbia's RCS tanks.

Propellant for only one rendezvous – not two – had been budgeted by mission planners when they calculated the amount of manoeuvring fuel the Shuttle would require to support the Spartan deployment and retrieval activities. Unfortunately, most of that propellant had been used by Kregel during the manual rescue on 24 November and the Mission Management Team determined that the crew would not be able to conserve enough fuel in Columbia's tanks to ensure a successful recapture. Moreover, if a similar problem was encountered while attempting to grapple the satellite, or additional station-keeping time was needed, propellant margins would run dangerously low.

"Right now," said Lee Briscoe of Mission Operations on 28 November, "we have about 115–120 lb [of propellant] in the forward [tanks]. We're looking for something like 200 lb to be able to go do [a second rendezvous]. We're short a little bit. If we can [conserve] propellant and things go very well, then we can see what we've got. In the next two days, we'll be watching that, see what our propellant situation will be in the forward [tanks], before we will be able to tell managers what we have the ability to do."

However, during all satellite-deployment missions, flight controllers ensure that enough fuel is on board to guarantee success, even if critical systems fail during the rendezvous and force the Commander to expend more propellant than intended. For example, if Columbia's Ku-band radar had failed during the final approach to Spartan, Kregel would not have been able to make a high-precision rendezvous and would have had to use more fuel. "In computing our margins for this case," said Lee Briscoe of Mission Operations, "we just didn't have that kind of propellant. The things we were looking at here were doing strictly a 'mean' rendezvous: we were not protecting the Ku-band failure case [and] then we were having to allow for some bit of flyaround and station-keeping gas. That's not necessarily an easy thing to do and, if you were deploying a brand-new, fresh spacecraft, you wouldn't do it in those circumstances." This was a pity, particularly as Kregel performed a status check on Spartan on 29 November and found it to be in excellent health. This forced managers into a quandary and they deferred their final decision until the 30th.

Two possible options were explored: releasing the satellite on 1 December and picking it up the following day, or setting it free on the 2nd and retrieving it about 20 hours later. Both would require a one-day extension of the mission to 17 days. Ultimately, the risks proved just too high. Not only could Columbia probably not support a full rendezvous, taking into account the potential failures, but there was a chance the $10-million Spartan might be lost and the complex deployment and retrieval procedures would adversely impact the high-priority last few days of USMP-4 operations.

"Here's a case where we have the Spartan in the [payload] bay. We have it: it's a *healthy spacecraft* [and] we can bring it back," Briscoe rationalised. "If you were to deploy it under these kinds of propellant margins, you could stand a 40 or 50% chance of not bringing it back if you had dispersions or failures as you tried to re-rendezvous with it. Based on that, the management team decided we would go ahead and forgo another deploy and retrieval. The flight control team wasn't comfortable with the amount of propellant that we had."

Nor could Spartan conduct its solar observations while attached to the end of the RMS; although Columbia *could* be oriented with great precision, it was not good enough to aim the instruments with sufficient accuracy. By 30 November, it was official: there would be no redeployment, the crew would return to Earth as planned on 5 December and Spartan would be impounded as part of an investigation into what went wrong. Already, some members of the media were pointing fingers at Chawla for having botched the deployment and many were not afraid to question the crew during space-to-ground news conferences.

BITTERSWEET VOYAGE

On 2 December, Kregel told a journalist from the *Orlando Sentinel* that Chawla had taken the events in her stride, refusing to discuss blame. "We'd be very foolish if we tried to second-guess or tried to figure out what the actual turn of events were without having all the information," he said. "We're six folks up here, we know what happened on our side, we'll get together with the folks on the ground and we'll put the whole story together and make sure it never happens again. Sure, we're always a bit disappointed if we don't get the full mission accomplished, but we *did* retrieve the satellite, and so the important thing is we're bringing Spartan back down to Earth and it'll get to fly another day." Floating at Kregel's shoulder, Scott agreed: "We think it's more important to get the spacecraft back, refurbish it and fly it again. After all, the Sun will be there and we don't want to risk losing the satellite altogether."

Whatever the exact cause of the failure was, Flight Director John Shannon confirmed on 4 December that NASA had formed an investigative team. "We are planning to impound the Spartan after landing and the crew will be part of the investigation process," he said. Some engineers at KSC believed that the fiasco was due to an oversight by Chawla, although Lee Briscoe came to her defence by speculating that a Payload General Support Computer (PGSC) could have malfunctioned and prevented her inputted commands from reaching the satellite properly.

"The guys have gone back and looked at the data, they've dumped the PGSC log files and when they look at [those], they don't see some of the indications there that the command was issued to the spacecraft that puts it in the mode to be ready to deploy," Briscoe said. "Did the crew miss that step? That's a possibility. Was there something in the PGSC software that didn't issue it? That's a possibility."

As discussions, and private re-criminations, got underway on the ground, the crew pressed on with the last few days of what had otherwise proved to be a highly successful mission. In addition to USMP-4 operations and Scott and Doi's two spacewalks, a number of other scientific investigations were carried out. One of these was the Collaborative Ukrainian Experiment (CUE), under Kadenyuk's supervision, which explored the impact of the space environment on the pollination and fertilisation of *Brassica rapa* (turnip) and soybean seedlings.

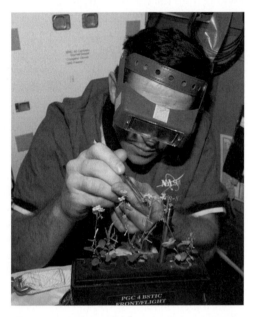

Leonid Kadenyuk tends seedlings as part of the Collaborative Ukrainian Experiment (CUE) in Columbia's middeck.

Watching the experiment, and participating in identical 'control' studies on the ground, were around half a million students and teachers in the United States and over 20,000 others in Ukraine. The underlying objective of CUE was to develop an understanding of how to grow food plants in microgravity, which will prove highly desirable in future years when long-duration missions to the Moon or Mars become a reality, and key focal points included comparing ultrastructural changes, biochemical composition and functional differences in space.

The experiment was approved and signed by then-NASA Administrator Dan Goldin and his counterpart Alexander Negoda of the National Space Agency of Ukraine in 1995; later that same year, then-President Bill Clinton announced that a cosmonaut would accompany CUE on its Shuttle flight. Two years later, Leonid Kadenyuk was assigned to join Kregel's STS-87 crew, with Yaroslav Pustovyi backing him up and performing control experiments in parallel on the ground. In general, the plants returned to Earth were healthy, but some were smaller than expected, averaging 4 or 5 cm long rather than the 12 cm that was hoped for.

Other experiments mounted in Columbia's payload bay included a passive heat pipe being flown to validate technologies for upcoming space missions and an evaluation of a new sodium–sulphur battery cell for future geosynchronous- and low-Earth-orbiting satellites. Elsewhere, housed in a GAS canister, was a study of the characteristics of transitional and turbulent gas-jet diffusion flames; such research could not be conducted on Earth because gravity-driven buoyancy introduced flow instabilities. Two final sensors – the Shuttle Ozone Limb Sounding Experiment (SOLSE) and Limb Ozone Retrieval Experiment (LORE) – investigated the distribution of ozone and measured changes in atmospheric composition in high resolution.

NO BLAME FOR CHAWLA

Despite their heavy workload, and several unsympathetic questions from the media, the rookies nevertheless managed to enjoy their first experience of spaceflight. "On the tenth or eleventh day, I wanted to do one full [orbit] and sit and watch the Earth," recalled Chawla. "Doing so was mind-boggling. It really instilled this huge sense of how small Earth is. An hour and a half and I could go around it! I could do all of the math and logic for why that was, but in the big picture the thing that stayed with me is this place is *very small!* I felt every person needs to experience this, because maybe we would take better care of this place. This planet below you is our campsite and you know of no other camp ground. I didn't think this view would be something so philosophical – I thought I would go around and see the continents and the oceans. But it was *much* more than that." It was a memory that would stay with her long after Kregel brought Columbia smoothly to land on Runway 33 at KSC at 12:20 pm on 5 December, wrapping up a bittersweet, two-and-a-half-week mission.

"We had a very interesting and eventful 16 days," he told the assembled journalists as the crew posed on the runway. "We had a lot of successes. We had a little bit of downtrodden times there, but together as a team, I think we ended up with a very super mission." Columbia, however, had returned home having endured significant damage to her thermal-protection coating: light spots were easily detectable on many tiles and her overall condition was "not normal". Ordinarily, around 40 tiles' worth of damage was expected; on STS-87, more than *a hundred tiles* were found to be irreparable!

Such damage, as the STS-107 disaster has tragically taught us, is mainly caused by fragments of ice or foam falling from the External Tank during ascent, but the impacts on Kregel's flight did not follow aerodynamic expectations and were far higher in number. No fewer than 308 'hits' were counted during post-flight inspections, over a hundred of which were greater than 2.5 cm across and some penetrated 75% of the total depth of the heat-shielding tiles. The chief suspects were new, environmentally friendly products added to the External Tank, including a new foam insulation, of which large quantities had apparently hit Columbia.

It was ultimately determined, through a combination of theoretical modelling and still photography, that foam fell from the massive fuel tank and was carried by

The STS-87 crew in front of Columbia on the runway. Left to right are Winston Scott, Takao Doi, Kevin Kregel, Leonid Kadenyuk, Kalpana Chawla and Steve Lindsey.

aerodynamic flows to 'pelt' the Shuttle's nose and fuselage during her climb to orbit. Possible causes included a defective coating of primer or the new foam insulation on the External Tank. The worrying STS-87 incident would take centre stage in the Columbia Accident Investigation Board's findings less than six years later ...

Meanwhile, another investigatory board – chaired by former astronaut Rick Hieb – had already got its teeth into identifying the cause of the Spartan failure. Initially, it blamed 'crew error', but later absolved Chawla, acknowledging that instructions given to the astronauts might have been unclear. "They [the investigative team] decided on things that were missing from our training," she said later. "For example, we did not simulate lighting conditions. Also, we don't have established rules for how to match [rotation] rates. Later, it became apparent that we couldn't match that rate, so we were just wasting propellant. The biggest thing against the crew was crew resource management, where if more people had been watching what was going on, it could have been prevented. But then again, NASA said there was no margin for error with the Spartan deployment."

Kregel added that the failure gave them a chance to demonstrate their ability to develop alternate procedures in orbit, which would undoubtedly be needed to work around problems encountered during International Space Station operations. "A big success was, in a short timeframe, all the folks [got] together [and] came up with a plan to retrieve a very valuable asset. If that doesn't show the ability of humans to adapt to changing situations in space, then I don't know what is."

TOM CRUISE'S STUNTMAN

Before joining NASA, Scott Altman's claim to fame was as a stand-in for Tom Cruise during the making of the movie 'Top Gun'. In 1986, he was attached to Fighter Squadron 51 at Naval Air Station (NAS) Miramar, during which time he completed two deployments to the Western Pacific and Indian Oceans, flying the F-14A Tomcat. It was in this capacity that he performed several stunts for Cruise's blockbuster. Little did Altman know that 12 years later he would strap into a far bigger and more powerful vehicle: as Pilot of Space Shuttle Columbia, for his first trip into orbit.

In doing so, the newly promoted US Navy Commander would, at over 6 feet, also form part of the tallest crew ever launched into space. Five of Columbia's seven astronauts for the 16-day STS-90 mission exceeded 6 feet in height, with another coming close; in fact, only diminutive Mission Specialist Kay Hire, who became the first KSC employee to fly into space, missed the mark by a considerable margin. Joining Altman and Hire were four science crew members – Payload Commander Rick Linnehan, Mission Specialist Dave Williams and Payload Specialists Jay Buckey and Jim Pawelczyk – and veteran Commander Rick Searfoss.

The qualifications of the science crew – Linnehan was a veterinarian, Williams and Buckey were physicians and Pawelczyk a physiologist – and the mission's name, 'Neurolab', implied that STS-90 was a medical and biological research flight. In fact, it was also the final mission of the versatile Spacelab module and the last dedicated life sciences flight on the Shuttle's agenda before anticipated long-duration research commenced on board the International Space Station in the summer of 1999. With this last throw of the dice for the life sciences community, Searfoss was happy to participate in as many experiments as possible.

"EVERYONE IS A MISSION SPECIALIST"

In the earlier days of the space programme, pilot astronauts typically avoided medical matters as if they were a plague of locusts; all were keenly aware of the old adage that there were only two ways they could walk out of a doctor's office – *fine* or *grounded!* Searfoss, however, had no concerns about getting involved in Neurolab's medical investigations. "This is research," he told an interviewer before Columbia lifted off. "No-one's out there to ground you. No-one's out there to find something that's going to keep you on deck forever."

Searfoss, a US Air Force Colonel who had also flown on the SLS-2 medical research mission five years before, lived by another saying. "Test pilots comprise about a third of the astronaut corps," he said after the mission. "I think that diversity in every respect, both technical background and nationality, is very healthy and great for the programme. There is a saying among astronauts that once you get into orbit, *everyone* is a Mission Specialist! Certainly, the test pilot skills are an absolute must for the launch and entry phases, but in orbit you're there to do a job.

"Our job on the Neurolab mission was to do as much out-of-this-world science as we could. As a test pilot, engineer and physics person, there is no way I could have brought all of that science home. There were four brilliant docs on board to make that happen. Quite frankly, I'm a little disgusted with some Commanders who have not stepped up to the plate in their leadership role to get as much science out of the mission as they should have. Certainly, it *has* been an evolution. Early in the Shuttle programme, we had to focus on maturing the programme.

"As that maturity has come, I believe the best Commanders have been the ones who realise we have an important role in making sure they exercise the proper kind of leadership in order to get the science done. We'll get better at this as we progress and now we've got the International Space Station operational, we should really do some good science up there."

The science on Neurolab was among the most complex ever attempted, focusing on the least-understood part of the human body – the nervous system – and investigating the effects of microgravity on blood pressure, balance, movement and sleep-regulation. "Everyone has the perception this is all Buck Rogers or Buzz Lightyear," Linnehan said before the flight, "but on science missions, *everyone* is really a subject. I freely accept that and, maybe, in a perverse sense, enjoy it!"

Nor was the mission exclusively focused on its human subjects, for the crew would also transport over 2,000 animals into orbit and, controversially, euthanise and dissect some of them to perform complex surgical procedures. However, Linnehan insisted that all of Columbia's passengers were treated respectfully. "I can guarantee the animals are well-fed, well-housed and well cared for," he said. "It's my duty to check [them] every day to make sure everything looks good as far as their food, water and general health. I have absolute authority on-orbit, if I need to, to stop an experiment if an animal becomes sick."

NEUROLAB AND THE 'DECADE OF THE BRAIN'

The $99-million Neurolab came about following then-President George H.W. Bush's declaration of the 1990s as the 'Decade of the Brain', in recognition of advancements made in humanity's understanding of its basic structure and function. The 26 investigations in the Spacelab module for the two-week mission were broadly grouped into eight sets: four of which used Columbia's crew as test subjects, while the others focused on mice, snails, crickets and fish. The mission attracted participation from the United States' National Institutes of Health, NASA and several international sponsors, including ESA and the Canadian, German, French and Japanese space agencies.

Key questions still required answers, despite 37 years of experience of sending humans into space, including the mystery of how astronauts adapt so quickly to the weightless environment, even though *all* of our basic movements were learned, since birth, in the presence of gravity. Moreover, physiologists were intrigued to learn how gravity-sensitive organs, such as the inner ear, cardiovascular system and muscles, cope in space and how astronauts' sleep patterns and biological clocks were affected as the Sun 'rose' and 'set' 16 times in each 24-hour period.

It was as part of these studies that Columbia's animal passengers entered the equation. How, researchers wondered, would *their* inner ear mechanisms – so vital for balance and movement – change in microgravity and, furthermore, what of animals actually *born* in space? Would their gravity-sensitive organs develop differently to their Earth-born counterparts and, if so, in what ways? The ultimate question was: *Must* gravity necessarily be present at the point in an animal's life when basic locomotion skills such as walking and climbing are learned? The STS-90 crew would also participate in this research, through studies before, during and after the flight.

"By involving a broad group of national and international scientific research agencies, Neurolab has drawn together outstanding specialist-investigators to study the brain," said William Heetderks of the National Institute of Neurological Disorders and Stroke, one of the mission's project managers. "This is an ambitious and unprecedented undertaking that will allow us to greatly expand our knowledge of how the nervous system develops in, and adapts to, microgravity. We will be able to observe important changes on a cellular level in muscles and the nervous system that will improve our understanding of muscle development and motor behaviour. Prior to Neurolab, these questions were unexplored and unanswered because they require spaceflight and weightlessness as a component of study design." Arnauld Nicogossian, NASA's head of life sciences, added that this information could then be applied to future missions – with upcoming six-month increments on the International Space Station being the most immediate concern – and help to improve the lives of people on the ground. Knowledge gained from space missions proved particularly useful in the latter case, because many of the physical obstacles confronted by astronauts mimic the kind of ailments associated with ageing and major diseases on Earth.

ABSEILING INTO THE SPACELAB

Preparing Columbia for this marathon mission, which would also become her last long-duration flight for almost five years, was as complicated an affair as the Neurolab experiments themselves. After having been transferred to Pad 39B on 23 March 1998, in anticipation of a 16 April launch, many of her mammalian and aquatic passengers had to wait until the very final hours of the countdown before they could be loaded into their cages in the Spacelab module. This proved an interesting event, worthy of comment, particularly as the Shuttle was positioned *vertically*, on her tail.

Working from Columbia's middeck, two technicians were lowered, one at a time, in sling-like seats down the 6-m-long tunnel into Neurolab in the payload bay. One technician waited in a shelf-like 'bend' in the tunnel, while the other entered the module itself to await the animal cages and aquariums that were then lowered on separate slings.

"The late stow is a 13-hour activity that will conclude about 3:30 am [on the 16th]," said Payload Operations Manager Scott Higginbotham before the intricate procedure got underway. "The stow itself is made up of three 'waves' of activity. The first wave [is] when we will install some passive items into the Spacelab, we'll activate the life science laboratory equipment refrigerator and, once it's chilled down, we'll install a number of chemical sets into it. And then we'll activate the Research Animal Holding Facilities in Racks 3 and 7. Wave Two is basically devoted to Rack 10: the vestibular function experiment unit. We'll activate that unit and install four fish packages. Each of these fish packets contains [a] single oyster toadfish." The final step was to load 24 cages, holding a total of 132 mice, into the lab. "Then, we'll begin four hours of module closeout activities," he concluded.

The delicate procedure was completed without a hitch, although telemetry from a couple of fish subjects – whose heads had been 'wired' with tiny transmitters – proved spotty, which Higginbotham attributed to "the fish not cooperating". When a fish 'hid' in the corner of its tank, for example, the antenna system used to gather telemetry data occasionally did not work properly. It was expected that accurate telemetry from all of the fish would be available as soon as Columbia reached orbit.

Otherwise, readying NASA's oldest Shuttle for her 25th trip into space was proceeding as 'swimmingly' as the fish themselves, with US Air Force meteorologists predicting on 13 April an 80% probability of acceptable weather for an on-time launch. Searfoss and his six crewmates, who arrived at KSC in their fleet of four T-38 jets that evening, were equally enthusiastic and were even treated to a visit by President Bill Clinton the following day.

"I hope you find out a lot of things about the human neurological system to help *me*," he told Searfoss with a grin, "because I'm moving into those years where I'm getting dizzy and I'm having all these problems, and I expect you to come back with all the answers!"

"Well, thank you, Mr President," replied Searfoss. "We'll take that on board as one of the challenges we'll try to meet."

"INCREDIBLE!"

The answers Clinton required would be another day in coming, because Columbia's 16 April launch was postponed 24 hours to allow technicians to replace a faulty Network Signal Processor (NSP) – one of two devices that relay voice and data communications between the astronauts and Mission Control. According to Shuttle integration manager Bob Sieck, the device was over a decade old. "Although the other unit on board was functioning fine, should it fail, your only ability to communicate with the vehicle would be via UHF radio and you would not have any ability to uplink or update [Columbia's] computers from the ground. So we went by our Launch Commit Criteria – our rules – that say this redundancy has to be in place prior to the mission. And when we tried every capability of this box to restore this uplink capability and all those failed, we had no choice but to delay the launch and change out this hardware." A replacement NSP was 'borrowed' from Columbia's sister ship, Endeavour, and took most of 16 April to fit in place. Otherwise, said Sieck, no issues were being tracked by his team. "It's just unfortunate that the

After a one-day delay, Columbia heads for orbit on 17 April 1998.

hardware picked *this time* to fail! I wouldn't say it's wearing out. We all know solid-state hardware usually fails when you turn it on or the previous time when you turn it off, and that looks to be the case here. The word 'bummer' comes to mind. Hardware does this to you at times and you have to accept this as part of the business. There's always the opportunity to do more training and the weather's great [in Florida]. It would be a good day to spend at the beach!"

By the evening of the 16th, the replacement unit had been installed in its correct place in an avionics bay behind a row of lockers in the middeck and satisfactorily tested. Meanwhile, during the day, the animal-holding lockers – carrying 18 pregnant mice and 1,514 crickets – were removed for maintenance and later replaced. Another delay on 17 April was expected to lead to a four-day postponement because many more of the animals would have to be replaced. Fortunately, launch that day went without a hitch and Columbia speared for the heavens at 6:19 pm.

"Ascent was incredible!" wrote Canadian astronaut Williams, making his first spaceflight, in his diary. For launch, he rode into orbit on the flight deck, giving him a spectacular all-round view of the dynamic eight-and-a-half-minute climb. "The view was very impressive as we completed the roll programme [manoeuvre, 10 seconds after liftoff] and progressed to SRB separation. Before SRB separation, there was a lot of vibration but the ride quickly became very smooth as we climbed to our final orbital altitude." As with the STS-87 ascent, Columbia had a never-before-tried surprise in store for Williams and his crewmates.

AN EXTRA PUSH ALONG THE WAY

Ten seconds after the separation of the boosters, two-and-a-half minutes into the climb, the OMS engines fired in a novel engineering test. Normally only ignited in space, they were used this time while still in the 'sensible' atmosphere as part of evaluating ways to enable the vehicle to carry 230 kg of extra cargo into orbit. "The goal", said Altman of the so-called 'OMS Assist' exercise, which lasted 1 minute and 42 seconds, "is to help to increase the Shuttle's performance margin, to get more mass to orbit by using some of that performance of the OMS engines."

In total, 1,810 kg of propellant was expended. Four minutes later, as STS-87 did the previous November, Columbia's computers rolled the spacecraft 180 degrees from a 'heads-down' to a 'heads-up' orientation to improve space-to-ground communications through the TDRS network. All in all, it was an awesome ride. "The physical sensation", recalled Searfoss, who was doing this for the third time, "starts out with a tremendous jolt when you first release off the pad and then there is an awful lot of vibration because as you go through the atmosphere you are affected by the upper-level winds and turbulence. Also [along] the way, the [SRBs] produce a lot of vibration. When the boosters drop off, the ride becomes very, very smooth. The last six-and-a-half minutes is a very smooth ride, but you get progressively pressed further and further into your seat as the acceleration picks up, so in the last couple of minutes you're feeling about 3 gs, so your body weighs three times its normal weight. It's not particularly uncomfortable. The seat supports you very well. You can't move around very much, but you don't want to move around anyway. You just want to enjoy the ride. When the main engines stop, and you get to zero gravity, the transition is instant. You go from 3 gs, pressed into your seat, to floating up against your straps. Various things start to float up. It's a very sudden and amazing transition to a whole different world."

A SEVERAL-THOUSAND-STRONG CREW

The seven astronauts and 2,000 mammalian and aquatic passengers on board Columbia may have been joined, albeit *very* briefly, by another as the final seconds ticked away before launch: a wayward *bat*, which had apparently attached itself to the External Tank. "We did take his body temperature [with an infrared camera]," said Launch Director Dave King. "He was 68 degrees [and] the tank surface was 62 degrees, so we've decided he was just trying to cool off. Some have said he may have heard the crickets in Neurolab, [but] it was his choice whether to hang around when we started the engines or not!" Hopefully, for the bat's sake, he took flight well before Columbia's main engines roared to life! Its sensitive ears would not have survived the acoustic shock.

It was the first time that crickets – more than 1,500 of them, in fact – had flown into space. However, there was no possibility of any chirping from Neurolab: crickets 'sing' by rubbing their wings and those on board Columbia were not yet old enough to have the wings needed to serenade Searfoss' crew. "If they start to sing, then we have a result; an unexpected one," laughed neurobiologist and cricket expert Eberhard Horn of the University of Ulm, near Munich. "The crickets have an external gravity sensor, so you can see immediately what happens in space with such an animal." Some researchers had suspected that these sensors would not develop normally in weightlessness, particularly in the case of the younger crickets. Consequently, almost half of them were 'spun' in a gravity-simulating centrifuge, while the others were exposed to 'ordinary' microgravity. After Columbia's return to Earth, they were frozen and dissected by Horn's team as part of investigations into how their bodies changed in space.

As well as humans, rodents and crickets, Neurolab also carried several hundred snails and fish into orbit, whose gravity-sensing mechanisms are either similar to our own or very simple and easy to analyse. "Gravity is always present on Earth, so it's been hard to explore its role in development and in controlling movement," said Christopher Platt of the National Science Foundation's division of integrative biology and neuroscience. "Neurolab allows unique tests that will shed light on how gravitational sensors work. These studies may tell us how exposure to lack of gravity may lead to abnormalities in the otolith organs – relevant to long-term spaceflight and to certain kinds of posture and balance problems in people on Earth." Furthermore, the results of the aquatic experiments could lead to advances in the development of electrodes as connections to the nervous systems of people with deafness caused by hair-cell damage.

A MAGNIFICENT MISSION

For the astronauts and thousands of scientists, the opportunity to finally reach orbit was met with jubilation. "To say the payload team is ecstatic would be an understatement," said Higginbotham late on 17 April. "Some of us have been working on this mission for years and to finally see it take flight is a feeling that's

The STS-90 crew assembles inside Neurolab for an in-flight portrait. Top row (left to right) are Jay Buckey, Jim Pawelczyk, Rick Linnehan and Dave Williams and bottom row are Scott Altman, Rick Searfoss and Kay Hire.

awfully hard to describe. We're looking forward to 16 days of magnificent science on board this last planned Spacelab mission." By that time, Linnehan and Williams were already in the pressurised module, setting up equipment for an intensive research programme.

As with previous life science flights, Neurolab would operate a single-shift system, with all seven astronauts waking up at the same time, in order to preserve their circadian rhythms and support the labour-intensive investigations. For five of them, it was their first experience of space flight, and all were determined to enjoy it, despite their heavy workload. "I feel a tremendous pride to be a Polish-American," said Pawelczyk, who took a small Polish flag with him, "and I wanted to provide some way to share that enthusiasm and pay a little bit of tribute back to the people of Poland."

Williams, meanwhile, carried his wife's flying wings into space with him. "I thought that if I'm going to have a chance to fly off the planet and orbit around the Earth, I would like to take her wings with me," he told an interviewer before launch.

Kay Hire, who had been employed as an engineer at KSC from 1989 until her selection as an astronaut candidate in December 1994, said that helping to prepare the Shuttle for launch and now actually *flying* one "kind of completes the circle for me".

After a busy first day establishing their orbital research laboratory, the crew was awakened early on 18 April by Aretha Franklin's 'Think', which provided an appropriate start of their two weeks of neurological research. "Good morning, Columbia; time to get those neurons into action," Capcom Chris Hadfield told them, to which Searfoss replied that his crew was rousing themselves for "a fine Neurolab day of work here". Not only were the seven astronauts working feverishly; so too were the staff in Mission Control, as evidenced by an entry in Canadian astronaut Hadfield's journal.

"For tomorrow, Flight Day Three," he wrote, "we have added several small things: a maintenance procedure for a broken pump on a Japanese experiment, a contingency procedure for an Animal Enclosure Module (capping an air port) and changing the orbiter's attitude to keep some thruster jets from getting too cold in the shade. Columbia is very healthy, so things are nice and easy for us. We'll send today's Execute Package to the printer on board Columbia just after the crew wakes up at 8:39 am Eastern [Time]. We have 11 messages to send, totalling about 30 pages. After they receive them, the crew will cut and paste the pages into their checklists using scissors and tape. We've spent the night writing all the messages and reviewing them for errors and crew-friendliness. Yesterday, I woke up the crew with the song 'Think', by Aretha Franklin (since it's Neurolab, I went with the brain theme). Today, they'll hear 'Take Me Out to the Ball Game', sung badly by Harry Carey, the late-great Chicago Cubs announcer. It's for Scott Altman, an Illinois native, who's doing a [public affairs] event with Chicago radio today."

Such messages were typical for each Shuttle mission and highlighted, among other things, the close-knit working relationship and personal bond between the astronauts in orbit and their counterparts on the ground. Overall, Columbia was performing admirably, with the first problem of significant note arising just before bedtime on 24 April, when a valve malfunctioned in the Regenerative Carbon Dioxide Removal System (RCRS), triggering an alarm in the cabin and seriously threatening an early return to Earth.

RCRS PROBLEMS

Prior to the first EDO mission in June 1992, exhaled carbon dioxide from each of Columbia's crew members was scrubbed from the cabin and Spacelab atmospheres using lithium hydroxide canisters, which had to be changed four times daily. The new RCRS has already been described, but in case of failures a series of 28 lithium hydroxide canisters were kept in reserve, offering five extra days of science-gathering and two 'contingency days' – in case of one or two 'waved-off' landing attempts due to bad weather – before the Shuttle would be forced to come home.

Kay Hire had reported hearing a strange noise about the time the device failed. "It was a noise I did not recognise," she told flight controllers, refuting suggestions that it might have been a normal sound from the RCRS' compressor. Already running two hours behind schedule, the crew initially thought the trouble involved one of two electronic controllers; they switched to the backup and restarted the device, but within 10 minutes it shut down again. As a precautionary measure, Mission Control asked them to load two lithium hydroxide canisters, providing ample time to evaluate and recover from the problem.

"The crew was never in any danger," insisted NASA spokesman James Hartsfield on the morning of 25 April. "Carbon dioxide levels remained normal the whole time. A rise in carbon dioxide occurs very slowly [and] the crew was never anywhere near having anything like that occur." Meanwhile, as part of troubleshooting efforts, Searfoss and Altman opened the RCRS, removed a hose clamp and used aluminium tape to bypass a suspected faulty check valve, which was apparently allowing cabin pressure to leak into the system and throw off its electronics controllers. The fix worked.

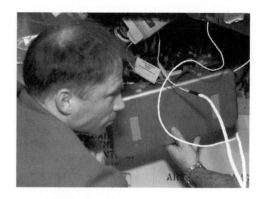

Rick Searfoss and Scott Altman (out of frame) repair the Regenerative Carbon Dioxide Removal System (RCRS).

"We have some good news for you," said fellow astronaut Mike Gernhardt, on his shift in the Capcom's seat in Mission Control. "It seems to be working as expected. It looks like we have headed off the possibility of a shortened mission."

RODENT EXPERIMENTS

Meanwhile, the research in the Spacelab module was proceeding well, including the controversial dissection of several mice in the General Purpose Work Station (GPWS) by Williams and Buckey on 18 April. Four adults were decapitated in order to recover tissue samples from their brains and inner ears, which were needed by members of Neurolab's 'neuronal plasticity' team, who were investigating how nerve cells 'rewire' themselves under the stress of a new and unfamiliar environment. It was hoped that this work could ultimately lead to new insights into neurological ailments on Earth.

"There are different kinds of neurologic assaults – stroke, Parkinson's disease, some kinds of balance disorders – that trigger the brain to reorganise in these ways," said researcher Gay Holstein. "So we can use this microgravity experiment to gain some more insight into what's the best way to treat these neurologic patients." Jim Pawelczyk, who participated in several dissections, added that the surgery had to be conducted quickly to avoid the onset of degradation in certain nerve fibres. In response to questions of whether the poking-and-prodding medical experiments had spoiled his first day in space, he replied simply "That's why we're here!"

Although it was not the first time that animal dissections had been conducted on the Shuttle – that had been pioneered by the SLS-2 crew in October 1993 – many scientists considered the previous work to be messy and time-consuming. Neurolab, on the other hand, demonstrated the ability of trained medical specialists to conduct incredibly intricate surgery in a weightless environment. "We're thrilled!" said Holstein, of the Mount Sinai School of Medicine. "I can't wait to obtain the tissue and start to do my experiments." Later, embryos were taken from euthanised pregnant mice as part of studies into gravity's role in their early development.

The results of such studies will prove crucial in determining whether or not humans or animals could someday be *born* in space, thus enabling colonies to be established and long-duration trips to the stars, which will undoubtedly take place over many generations. Before dissecting them, the astronauts injected the pregnant mice with cell 'markers' to label the brain cells in their embryos, track the development and migration of the cells and compare the results with data gathered from Earth-born mice. "The dissections are proceeding real well," Pawelczyk said during the operation.

Typically, the pea-sized foetuses ranged between 10 and 14 days old and, after removal from their mothers, were preserved whole until the end of the mission. Other mice, with 'hyperdrive' units fitted to their heads connected to the hippocampus area of their brains, were placed in the GPWS on two different, maze-like 'racetracks' as part of the Escher Staircase Behaviour Testing experiment. The hippocampus helps develop spatial maps to help the mice to navigate from place to place and the investigation sought to explain the disorientation frequently experienced by Alzheimer's sufferers.

The hyperdrive units did not create unpleasant side-effects for the mice because their brains had no pain receptors. "It's actually crucial to us that our animals are quite happy, not suffering in any way," said psychologist Bruce McNaughton of the University of Arizona, a memory and ageing expert. "If they are, they won't perform the tasks we ask them to. So it's extremely important in our line of research [that] we make extremely good friends with our animals."

Many healthy astronauts frequently report unpleasant sensations of disorientation when they first leave terrestrial gravity, giving them a temporary kinship with Alzheimer's patients. "We're trying to reverse-engineer the most complex structure in the known Universe," said McNaughton. "Having that knowledge is a critical prerequisite to any kind of intervention. This structure we're looking at is actually the highest level-of-association cortex in our brains and it's the first thing that goes wrong during the breakdown that begins to occur in early Alzheimer's disease. By understanding the basic biology of how this system works, it will give us the understanding we need to be able to tell what is going wrong in this system when it breaks down."

The mice had been taught to use the racetracks – one dubbed the 'Escher staircase' because of its twisting, three-dimensional resemblance to a well-known lithograph by Dutch artist M.C. Escher, the other known as the 'magic carpet' because astronauts flipped and turned it once the mice were on board – before Columbia lifted off, receiving rewards of sweetened condensed milk after completing a certain number of right turns. In space, however, the rules were changed and investigators monitored neural activity as the mice struggled to recognise 'home base' after making fewer turns than they had been taught on the ground.

"In the future, obviously you would like to have some sort of family life on the Moon or Mars," said neuroscientist Kerry Walton of New York University Medical Center, and by studying critical periods of development in the mice, "we would

As the Sun peeps through the aft flight deck windows, Neurolab is illuminated in Columbia's payload bay.

know how long a child of a particular age could be in an environment other than Earth. This way, maybe people could grow up to be 'dual-adapted' so that they could switch from an Earth environment to low-gravity in the same way that children can switch between one language and another."

Several of these mice exercised their 'space legs' for the first time on 22 April, using their forelimbs to scoot around a small jungle gym, but hardly using their hindlimbs, according to the crew. As they moved, the astronauts videotaped them and marked their joints with permanent black ink, so as to allow each motion to be meticulously analysed after Columbia's landing. "We think it's really neat because these animals – their nervous systems – are going to develop and essentially they're going to think they were born in zero-gravity," said Walton, but added that she expected them to quickly readapt to terrestrial conditions.

"It's amazing to watch these animals behave in orbit," said Linnehan, who had worked with hippos, rhinos, giraffes and bears at Baltimore Zoo before joining the US Navy's marine mammals project and later becoming an astronaut. "They act just we do. They learn very fast how to get around their cages and get to food and water. Just last night [24 April], we were checking on some of the young rats, watching them eat. It was kind of akin to the way *we* eat. We float around and hold our food to feed ourselves. One of the rats was holding onto a piece of food [with] his front paws, munching on it leisurely, letting [it] go and going over to drink some water, coming back and grabbing the food again."

Later, on 30 April, the first survivable surgical procedure was conducted on six mice: Buckey and Williams anaesthetised and injected them with dye markers in a muscle-development study. Although the procedure itself was relatively straightforward, NASA's chief veterinarian Joe Bielitzky noted that it paved the way for more elaborate research on the space station.

"Everything went well; all six have recovered," Buckey reported jubilantly at the completion of the procedure.

Elsewhere, a minor problem cropped up with the four oyster toadfish on board Columbia. They possess gravity-sensing organs reminiscent of human inner ears, and although each had been fitted with a small transmitter before launch to monitor neuronal activity, the signals were found to be intermittent, possibly due to an electrical fault or problems with the oxygen-circulation pumps in their aquariums. Nonetheless, Stephen Highstein of the Washington University School of Medicine said, "We think we can achieve 100% of our science goals even with intermittent data."

NEONATAL DEATHS

Other, more serious problems were afoot, however. On 27 April, NASA revealed that dozens of young (neonatal) mice in the Spacelab module had died unexpectedly; others were in such bad shape that they had to be put to sleep by the astronauts. "This hit literally out of the blue," said Bielitzky. The crew managed to resuscitate a few others with bottle feedings and Linnehan later thanked his crewmates for helping to hand-feed and care for the mice, checking on them each night before bedtime.

At first, Linnehan speculated that the deaths might have been caused by the design of the rodents' cages. "It may be a function of the way the cages were built and the way the mothers moved around in the cages," he said. "Some of the animals were not able to maintain a hold or get to a mother at certain times they needed to. We saw them free-floating and they appeared to be doing well, but I just don't think they could figure out the environment well enough to get to the mother and nurse; so they became dehydrated. As you know, when you become dehydrated, you become depressed and that is a cyclical thing that builds and builds and you get to the point where some of them just succumb to that." In addition to providing fluids for the mice, Linnehan and his colleagues hand-fed the rodents nutritional and subcutaneous water supplements and antibiotics to stem the losses.

Back on Earth, the immense death toll – more than 50% of all the neonatals – had drawn condemnation from animal-rights groups, including People for the Ethical Treatment of Animals (PETA), who charged NASA with abuse. Bielitzky denied the allegation, promising a full investigation, but PETA went further and pressed Congress to ban further animal experiments on the Shuttle. A higher-than-normal death count during a previous rodent experiment on STS-72 in January 1996 should, they argued, have provided a red-flag warning sign to the space agency. "Certainly," Bielitzky acquiesced, "you would not refly this if you did not have great confidence the mortality rate would fall into the 10–15% range."

This disaster could not have happened at a worse time for the neonatals, who were particularly vulnerable and just learning to walk and seek out solid food. The poor state of some of them forced Linnehan to cancel one of the experiments, telling Mission Control that "there is no meaningful data to be gained with these animals at this point". In total, more than 55 of the mice were lost. "It was an unforeseeable event," Linnehan said after the flight, "and it's regrettable that it happened. However, we still got back most of the primary science."

It later became apparent that the ill-fated neonatals were victims of a higher-than-normal rate of maternal neglect, who had themselves become dehydrated and unable to lactate adequately. Consequently, as the youngest mice grew anaemic, the mothers stopped feeding and grooming them. Linnehan drew praise from animal-rights groups on 1 May, after defying an instruction to destroy a rodent when ultra-fine electrodes implanted in its brain broke loose. "Based on his expertise and professional opinion, he determined the animal was not in any danger and determined it was appropriate to return that animal to its housing," said mission scientist Jerry Homick.

A FAST PACE ON BOARD 'BLURROLAB'

The intense pace of the mission, and the additional free time given up to care for the sick rodents – which meant most of the astronauts averaged three or four hours of sleep each night – had turned it into something of a blur for Linnehan. "I started calling it 'Blurrolab', instead of 'Neurolab'," he said, "because I just could never remember what day it was or where we were, we just kept on going." Nevertheless, despite the neonatal fatalities, the mission was shaping up to be a superb scientific success.

"We went into this mission facing a number of challenges," said Homick. "We knew we had a difficult timeline to work with, we knew that we had a number of complex hardware systems ... to acquire the data and we knew we were going to be implementing a number of very difficult experimental procedures using cutting-edge technologies. With all of that in mind, we did expect to achieve a great deal of success with this mission and I'm pleased to report that I think we've exceeded our expectations."

Although the astronauts had averaged only a few hours of sleep each night during the rodent crisis, Dave Williams at least was able to sleep fitfully, which aided another of Neurolab's investigations: a series of comprehensive measurements of brainwaves, eye movement, respiration, heart rates, internal body temperatures and snoring while at rest. As well as helping to improve the performance of astronauts in space, such experiments were also expected to improve the sleep–wake cycles of people on Earth. "We are really a sleep-deprived society," said Charles Czeisler of Harvard Medical School, "in which people burn the candle at both ends."

Not only does the hectic workload of astronauts affect their performance, but they also experience a 'sunrise' and 'sunset' during every 90-minute orbit of Earth and regularly take sleeping medication during missions. "Twenty percent of the

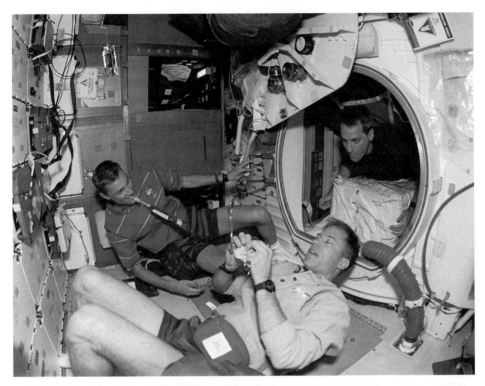

Unusual 'fisheye' view of the cramped middeck. Rick Linnehan pokes his head through the airlock from the tunnel leading into Neurolab as Jim Pawelczyk (left) and Rick Searfoss (lower) enjoy off-duty time. An unidentified crew member (top) is floating through the hatch to the flight deck. Above and slightly to the right of Pawelczyk are the crew's sleep stations and on the far left are a row of storage lockers.

crews on single-shift missions like this one take some sort of sleeping pill while they're in the space environment," said Czeisler. "So they do have a problem. That's three to eight times the rate of the general population." During STS-90, Linnehan, Williams, Buckey and Pawelczyk also donned body suits and sensor-laden skullcaps before going to bed to monitor their sleep patterns.

"Over four or five days," continued Czeisler, "if you lose two or three hours of sleep per night, it's the equivalent of losing a full night of sleep, and that causes a lot of detriments in our ability to perform effectively. It impairs the ability to consolidate short-term memory, causes a slowing of reaction time and it increases the probability of lapses of attention that may occur when you're carrying out a routine, highly over-learned task, such as on Earth driving a car. NASA is concerned about the potential impact of this cumulative sleep deprivation on mission safety and success."

During the mission, although Williams admitted that his sleep had been reduced, he was nevertheless still able to rest adequately. He added, interestingly, that although he was weightless and had no 'pressure points' – such as a bed to lie on – "I still turn over in my sleep. I don't know why I do this, but it still happens." Referring to what was essentially a mini-sleep laboratory that he donned before bedtime, he felt its potential spinoffs on Earth could lead to patients doing studies of sleep phenomena at home, rather than in hospital.

Williams' inclusion on the STS-90 crew was fortuitous, as two of the medical experiments were Canadian. The Visuo-Motor Coordination Facility (VCF) studied changes in movement during weightlessness that affect astronauts' pointing and grasping abilities, while the Role of Visual Cues in Spatial Orientation – shortened to 'Visual Cues' – created 'fake gravity' by applying pressure to the soles of the feet to find out if it 'overrode' visual cues and allowed them to readapt to a terrestrial-type environment. This was expected to aid research into the causes and treatments for motion sickness.

Developed by the Canadian Space Agency, the VCF was a particularly interesting device that projected visual targets onto a screen; as these appeared, one of the science crew members grasped and pointed at them, and tracked them as they moved, using a special, instrumented glove. The motor skills thus demonstrated, at various stages throughout the 16-day mission, were used to record changes in the astronauts' nervous systems as they adapted to the microgravity environment. Other associated experiments on board Neurolab could, it was hoped, one day enable engineers to build advanced robots capable of performing incredibly intricate tasks.

Tasks such as catching a ball are second-nature to most of us, but to train a *robot* to do it involves sophisticated stereo vision, a powerful computer to calculate the

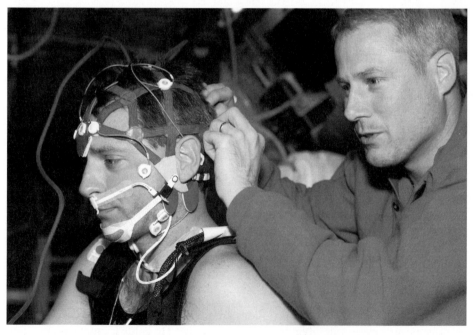

Jay Buckey 'wires up' Rick Linnehan with a special skullcap for a sleep study.

object's flight path, mass and estimate its time of arrival and, of course, position itself appropriately to capture it without causing damage. "During the few hundred milliseconds when the ball first appears in the field-of-view of the subject, the brain apparently computes the velocity and acceleration of the ball and also, from its surface appearance and texture and volume, it infers a mass," said investigator Alain Berthoz.

"The third thing the brain does – which is what is of interest to us here – is that it applies an internal representation of the effect of gravity on such a mass and therefore predicts the effect of gravity." Berthoz's team was also intrigued by the question of *how* the brain computes the distance to the incoming ball, although some psychologists have speculated that our brain actually works out the time-to-impact based on how fast the ball's image on our retina changes as it gets closer. In humans, this is all learned at a very early age.

"Probably during the first year of life, the brain constructs these internal representations of the laws of mechanics – of Newtonian mechanics," said Berthoz. "This is done not only for our limbs. The brain has to know not only the impact of the ball, but also it has to somehow know in advance the *properties* of our limbs. This apparently is done in the first year of life." To understand the motor-sensory processes that control these actions, Berthoz's team flew a spring-propelled ball on board Neurolab, which the astronauts – their arm, heads and bodies outfitted with sensors – were obliged to catch.

"The ball will go with a constant velocity; it will not be accelerated as it is on Earth," explained Berthoz. "So the muscular contractions that the brains of the astronauts will produce, if they use the internal model of gravity, will be non-functional. Therefore, in flight, the brain has to recognise this very basic, fundamental prediction and organisation. That's the idea." Elsewhere, other experiments used a rotating chair, mounted in the centre aisle of the Spacelab module, to stimulate the astronauts' vestibular systems with spinning and tilting sensations.

Known as the Visual and Vestibular Integration System (VVIS), the chair was expected to yield data on how the nervous system 'rewires' itself to account for the lack of 'normal' gravity. Performed by each of the science crew members – Searfoss, Altman and Hire, as the orbiter crew, were exempt from the tests – the experiment was done six times during the 16-day mission. As each VVIS run started, the astronauts' eyes were shielded from external stimuli, giving their nervous systems no visual references, and a video camera was trained on their faces to capture their reactions to the spinning motion.

The orbiter crew did have a number of scientific tasks to perform, however. One of Hire's responsibilities was a device in the middeck known as the Bioreactor Demonstration Experiment (BDE) which, on STS-90, grew cultures of renal tissue and bone marrow. Both of these were widely anticipated to yield substances that might be of use for kidney disease, AIDS and other immune-system ailments, as well as for the chemotherapy treatment of cancer sufferers.

As well as proving somewhat dizzying, a few Neurolab investigations were potentially 'stinging' for the crew: one demonstrated an innovative technique known as 'microneurography', whereby a very fine needle – about the same size as an acupuncture needle – was inserted into a nerve just below the knee. This allowed nerve signals travelling from the brain to the blood vessels to be measured directly, while the crew's cardiovascular systems were monitored using the Lower Body Negative Pressure (LBNP) device. All of this data allowed investigators to carefully track how well the nervous system controlled cardiovascular function in the astronauts' bodies.

Results were expected to aid sufferers of autonomic blood pressure disorders, particularly 'orthostatic intolerance' – an inability to maintain proper blood pressure while standing for long periods. "This is a multifaceted problem," said Pawelczyk, "but it shares many features similar to those experienced by astronauts after flight. The experiments we'll be conducting really focus at the crux of the matter: how it is that nerve signals travel from the brain to blood vessels and cause the process called 'basal constriction'. Basal constriction, the narrowing of blood vessels in the lower body, has the effect of boosting pressure elsewhere in the system.

"The way to think of this is like thinking of a garden hose. If you want to get more pressure out of a garden hose, you can do one of two things. You can either turn up the faucet and increase the flow or you can put your thumb over the end of the hose and increase the pressure that way. *That* process of placing your thumb over the end of the hose is just like the process of basal constriction that occurs in vasculature. All our blood vessels are surrounded by muscle that's fed by 'sympathetic nerves'.

"What we'll actually be doing in flight is recording from those nerves in flight. This will be the first time we've ever done neural recording in humans in flight. We'll be looking for specific fibres that feed blood vessels that travel to skeletal muscles. Skeletal muscle is about half our body mass. Changes that we can elicit there in skeletal muscle have a profound impact on blood pressure regulation. We'll be asking the questions whether or not the signals sent from the brain to blood vessels to cause basal constriction are appropriate for the kinds of stresses we'll be applying."

To prepare for the microneurography procedure, Linnehan, Williams, Buckey and Pawelczyk spent two months at Vanderbilt University training with the hardware. The insertion of the needle, which all four typically achieved within about 40 minutes during several runs in orbit, was both delicate and difficult. "Finding a nerve", said Principal Investigator David Robertson, "is a lot more difficult than finding a vein. This is a very difficult thing to do on Earth, and the idea that it can be done in space is a little bit astounding to many people."

As the mission drew towards its conclusion, on 29 April a problem arose with a blockage in Columbia's waste water dump line, which flight controllers speculated may have been caused by a clogged filter. An initial attempt by Searfoss and Altman to bypass the line by routeing a hose through a spare filter and venting waste water overboard did not resolve the problem. Meanwhile, the following day the Mission Management Team opted not to extend STS-90 to 17 days, after the Neurolab scientists indicated that they already had enough data and additional time in space would not be necessary.

This was a pity, because the conservation efforts of the crew had added enough consumables for an additional day-and-a-half of research work. "We looked at the mission extension and it was tempting," said science manager Lele Newkirk, "but because we have Sunday [3 May] and Monday [4 May] to land at Kennedy, obviously we chose to land on Sunday. The KSC landing is important primarily to the animal projects, that's where we want to receive all the animals and they do all the post-flight data collection down there."

Although weather in Florida was expected to remain acceptable on 4 May, it was not *as good*, which prompted Newkirk to opt for an on-time landing, rather than risking a wave-off to land at Edwards Air Force Base in California. Moreover, the KSC weather was anticipated to become more doubtful late on the 4th. "There is a front that we expect to start to affect the Cape on Monday [4th]," said Entry Flight Director John Shannon. "It will bring rain showers and low cloud ceilings." Summing up, he felt that the decision to come home on time was correct.

In the meantime, as his crew readied their ship for its return to Earth, Searfoss successfully transferred more than 30 kg of waste water into a rubber contingency bag on 2 May. This provided enough volume in the tank to accommodate the seven astronauts' needs even if bad weather kept them in orbit longer than planned. The procedure was successful and Searfoss reported no unpleasant odours from the bag or the tank, joking "We'd like to join the plumbers' union!" to mission controllers. A day later, with pinpoint precision, Columbia swept smoothly and perfectly onto Runway 33 at KSC.

Sixteen days in the closed microgravity environment on board Columbia did not, Searfoss later told an interviewer, induce a claustrophobic reaction among any of the seven astronauts. "The Shuttle appears to be pretty small for seven people to live in for a couple of weeks, but when you go to space, it seems like you have a lot more room," he said. "Try this thought experiment: on Earth, all the people in a room have their bodies at approximately the same level, sharing the same area. In space, those people would utilise *all of the room's volume!*

"They'd be hanging out up around the ceiling and drifting freely. In space, you don't get any feeling of claustrophobia. It's a great experience to be up there and work in that kind of environment."

Even before Columbia lifted off, there was a possibility that their great experience would be repeated. At a press conference on 15 April, Shuttle manager Tommy Holloway told journalists that engineers were looking at the feasibility of flying Neurolab again in the late summer of 1998. The reason: more International Space Station delays had opened up a three-month gap in the Shuttle launch schedule. "There are two or three things we need to think about," Holloway explained. "First of all, the value of Neurolab and what we can learn from it. Secondly, the overall workload across the system, in terms of what [NASA] wants to do." Ultimately, partly to protect an anticipated early September launch of the first station components on the already-long-delayed STS-88, the decision was taken not to refly Neurolab, but the results from STS-90 proved more than enough to keep neuroscientists and biologists busy for several years to come.

"For the most part," said Mary Anne Frey, NASA's chief scientist for the mission, "our scientists received more data than they anticipated." When the inaugural results were published in April 1999, they offered promising insights into the neurological mechanisms responsible for Alzheimer's disease and epilepsy and the astronauts' cognitive work with the spring-loaded ball could ultimately lead to new diagnostic and rehabilitation tools for sufferers of brain injuries. Neurolab's sleep studies, too, could find Earth-bound applications to aid shift workers, the elderly, jetlag sufferers and insomniacs.

Summing up the scientific accomplishments of their mission, the payload crew members were both philosophical – and jocular. "If we don't figure out how to stem nerve-muscle degeneration," said Linnehan, "we're not going to be able to travel to other planets or live in space stations. We would face the risk of fractures when returning to Earth." On the other hand, said Buckey, who later returned to his previous job as a professor of medicine at Dartmouth Medical School, at least in space "you can put your pants on both legs at the same time!"

STRICTLY NON-STATION MISSIONS ONLY

"It's great to be back in zero-g again," said STS-93 Commander Eileen Collins early on 23 July 1999, as she and her four crewmates set about preparing Columbia for five days of orbital activities, but added darkly: "A few things to work on ascent kept it interesting." Those 'things', within seconds of blasting off, almost forced Collins – the first woman to lead a space mission – and Pilot Jeff Ashby to perform a hair-raising and never-before-tried emergency return to KSC and grounded the Shuttle fleet for almost six months.

Columbia's 26th flight was a long time coming. More than a year had elapsed since her Neurolab mission but the cause of the delay had nothing to do with NASA's venerable old workhorse herself. Originally scheduled for launch at the end of August 1998, it was postponed until just before Christmas, then January 1999, then April and ultimately midsummer by a chain of technical problems with her primary payload for the STS-93 mission – NASA's $1.5-billion Chandra X-ray Observatory (CXO) – and the Inertial Upper Stage (IUS) booster that would propel it into its unusual orbital slot.

Given the delays, it is a pity that Shuttle managers did not opt to refly Neurolab in the summer of 1998 because, with only five missions achieved overall that year – as opposed to the normal seven or eight – the manifest for the reusable spacecraft was remarkably light. That was not to imply that no missions awaited the fleet: more than 40 flights to assemble the International Space Station were waiting in the wings, as was most of the already-built hardware they would truck into orbit, but repeated delays to crucial Russian components had temporarily stalled the construction effort.

By the time the first assembly mission, STS-88, had finally carried a US-built connecting node called Unity into orbit in December 1998 and bolted it onto the Russian control module, the project was already running billions of dollars over budget and years behind schedule. More trouble was in store: the habitat module, again Russian-built, would require another 18 months before it finally reached space, delaying the arrival of the station's first full-time crew until October 2000.

Columbia, sadly, would be excluded from this mammoth construction effort; as the oldest Shuttle, and despite numerous weight-saving measures during her numerous overhauls since 1984, she remained considerably heavier than Discovery, Atlantis and Endeavour and thus could not transport large pieces of station hardware into orbit. It was even more saddening and bitterly ironic that this decision was later partially reversed and, had Columbia not been lost on 1 February 2003, her *next* flight in November of that year would have been her inaugural trip to the International Space Station.

In the meantime, she was restricted to exclusively non-station flights: such as the deployment of Chandra, repair and servicing calls to the Hubble Space Telescope and a series of ill-defined 'research missions' that would, it seemed, utilise the Spacehab company's pressurised module rather than ESA's now-retired Spacelab. Clearly, although she would not be granted a significant piece of the station pie, Columbia was still a valuable asset to NASA and the Chandra mission demonstrated this: she was the *only* member of the Shuttle fleet capable of transporting the gigantic observatory into orbit.

The reason for this was Chandra's sheer size. When bolted onto its IUS and mounted on a supporting 'tilt table' in the payload bay, the observatory took up about 17.4 m of Columbia's 18.3-m-long storage capacity and weighed more than 22,600 kg. In anticipation of their back-to-back station assembly missions, the other three Shuttles had already had their airlocks structurally removed from *inside* the middeck and mounted inside the docking system in the forward quarter of the payload bay. The result: there was simply not enough room on Discovery, Atlantis or Endeavour to house Chandra.

Even Columbia needed to lose 3,200 kg of additional mass before she could take the observatory. To achieve some of these savings, engineers used older, lighter main engines, which lacked the newer, more rugged high-speed fuel pumps and combustion chambers. "We put Columbia on a strict diet to get to this mission," said processing manager Grant Cates. "That work actually began [in 1996] with the identification of this mission and the weight reduction that would be required." Nevertheless, her weight would still creep above NASA's normal safety limit if an emergency landing were needed shortly after liftoff.

A HEAVY PAYLOAD

In such a dire eventuality, Columbia would tip the scales at 113,000 kg, some 590 kg heavier than safety rules determined to be the maximum-allowable landing weight. In the case of STS-93, a one-time-only waiver was granted to this rule, based on a detailed analysis of the payload, the Shuttle's centre-of-gravity constraints and a host of other interrelated factors. For Eileen Collins, the challenge of possibly having to perform a heavier-than-normal emergency landing did not faze her: "We would land at 205 knots, which is very close to the maximum certification [of] around 214. There are some challenges there, [but] I feel very confident we've looked at the abort landings and they're well within the safe limits

The silvery, cigar-shaped Chandra observatory, attached to its Inertial Upper Stage (IUS), is raised to its deployment angle a few hours into the STS-93 mission.

of landing the Shuttle." Collins' confidence in her abilities and those of her crew would come very close to being tested. Joining her and Ashby on board Columbia were Mission Specialists Cady Coleman, the notoriously launch-delay-prone Steve Hawley and Frenchman Michel Tognini. All but the Pilot had flown into space before: interestingly, Ashby had originally been assigned to another mission in the summer of 1997 but had resigned that post in order to care for his cancer-stricken wife.

In some ways, STS-93 harked back to the early days of the Shuttle programme. Not only would it last a mere five days – making it NASA's shortest-planned mission for almost eight years – but it would also feature the deployment of a payload on top of the Boeing-built IUS. This two-stage, solid-fuel booster had been used on several occasions to deliver a series of TDRS satellites and a number of military payloads into high orbits; additionally, it had supported the launches of the Magellan and Galileo space probes to Venus and Jupiter, respectively, and the Ulysses solar polar explorer.

Nor was it exclusively a booster that could only operate from the Shuttle; many of its missions had utilised expendable launch vehicles, including the US Air Force's Titan IV rocket. Built by Boeing, under contract to the Air Force, the IUS for the Chandra mission was purchased by NASA's Marshall Space Flight Center. As its name implies, the 14,740-kg booster is inertially guided by its own onboard navigation equipment, and measures 5 m long by 3 m in diameter.

Its first stage ordinarily carried up to 9,800 kg of propellant and produced a total thrust of 20,000 kg; in fact, when it was built, it was advertised as capable of firing for up to 150 seconds, making it the longest-burning solid-fuel rocket motor ever developed for use in space. For the Chandra mission, it would carry considerably *less* propellant in its first stage – mainly due to the Shuttle weight issue – but to counteract this the motor's thrust was increased to 21,000 kg and its 'burn time' established at 125 seconds.

Meanwhile, the second stage typically carried around 2,700 kg of propellant and produced some 8,200 kg of thrust. Also housed in the second stage was an avionics system to provide guidance, navigation, control, telemetry, electrical power, command and data-management services to the booster, as well as supporting the payload itself. For its journey into orbit on the Shuttle, it was attached to a so-called 'Airborne Support Equipment' assembly, which resembled a giant mechanical doughnut with a central, circular opening for the booster's massive first-stage nozzle and 'tilted' the Chandra/IUS stack to the required angle for deployment.

IUS PROBLEMS

The booster's first Shuttle use in April 1983 had been less-than-spectacular, delivering the first TDRS payload into a lower-than-intended orbit. This embarrassing failure led to a number of other missions being either cancelled or repeatedly postponed. Although the system certainly matured with age, it was still prone to technical problems; one of which conspired to push back the already-delayed Chandra launch. On 9 April 1999, a Titan IV rocket carried a top-secret missile early warning satellite into orbit; the first and second stages of its attached IUS failed to separate properly and the $250 million payload was lost.

Ordinarily, NASA would have watched the resultant investigation with interest and incorporated its findings into its own plans to prepare Chandra's IUS for launch. That was complicated, however, by the fact that the investigation itself was top-secret. Moreover, said Scott Higginbotham, the KSC engineer overseeing Chandra's pre-launch processing, the IUS assigned to STS-93 was impounded as part of the investigative process! Columbia's launch, by then scheduled for 9 July, came under review and the impounded IUS had a domino-like effect on the training of the crew and ground staff.

"We were planning a two-day-long sim[ulation], starting 14 April, that would involve all the different control centres; a joint integrated simulation with everybody," said Lead Flight Director Bryan Austin. "That was going to be a big deal. That has been postponed because the Sunnyvale [IUS control centre in California] Air Force folks and Boeing IUS people were going to be taken away initially to be part of the investigation. That exercise has been put on hold. That kind of put a wrench in things in terms of our sim schedule."

The Chandra launch was also particularly critical because it was the Shuttle's first IUS launch since July 1995 and, pointed out Austin, "there's been a lot of change in the expertise level, collectively. For the most part, IUS deploy procedures are the same, but something that to me has been a struggle for this flight with some of our IUS friends is to get them to realise that the payload on the other end of the IUS is *not* the typical thing the Shuttle has been doing."

He was referring to the fact that, as soon as the Chandra/IUS transitioned to internal battery power, just minutes before deployment, the observatory would be on its own. On previous missions, if something went awry at the last moment, the IUS and its payload could be retracted for another try or brought back to Earth. However, mainly due to power and temperature constraints, mission controllers had just one shot at a successful Chandra deployment. "It's either going to be orbiting space trash," said Austin, "or it's going to go out and do its mission."

MAKING SUPERMAN JEALOUS

In fact, Chandra was the third of a quartet of 'Great Observatories' that NASA had been planning for more than two decades to explore the Universe using sensors that covered virtually the entire electromagnetic spectrum. The first two observatories – Hubble and the Compton Gamma-Ray Observatory – had been orbited by two Shuttle crews in the early 1990s and focused on visible and ultraviolet studies, as well as measurements of high-energy gamma rays. Two others would then cover X-ray (Chandra) and infrared (the 2003-launched Spitzer Space Telescope) wavelengths.

"Hubble revealed the visible side of the Universe," theorist Michael Turner of the University of Chicago said, "but most of the Universe *does not emit visible light!* It's only visible by other means, in particular the X-rays. Chandra will give us the same clarity of vision as Hubble does, but for the 'dark side' of the Universe [that] we know the least about." Comparing astronomers' capabilities before and after Chandra, Steve Hawley – an astronomer – likened it to the difference between the small reflecting telescope he used as a child and the 5-m telescope on Mount Palomar in California.

"We can make Superman jealous with *our* X-ray vision!" chimed Ken Ledbetter, NASA's director of mission development for Chandra. The observatory, which received Congressional approval in 1987, was equipped with a high-resolution camera, an imaging spectrometer and the most precisely figured X-ray mirrors yet built. It was hoped to equal, or perhaps even surpass, Hubble by studying some of the most exotic phenomena in the Universe, including black holes, quasars and white dwarfs.

"The observatory is a major improvement in X-ray astronomy over anything that has been done before and is to come," said Martin Weisskopf, Chandra's lead astronomer, before launch. "The key features that make it so special are [firstly] the angular resolution: it's ability to distinguish and pick out objects. The resolution of Chandra is such that you can read a newspaper at a distance of half a mile. Another feature is its ability to concentrate the X-rays within the 'core' of that image and *that* results from the X-ray telescope being so smooth. How smooth is it? If you took the state of Colorado and made it as smooth as these optics, the largest feature would be less than an inch high!"

"Eager is not the word," Weisskopf added, after spending two decades preparing for the mission. "Slavering at the mouth comes to mind!" The spacecraft at the centre of this praise looked like a tapering, 13.8-m-long metallic cigar with two solar panels at its base to provide electrical power. At the opposite end, mounted in the telescope's primary focus, were its two scientific instruments: the High Resolution Camera and the CCD Imaging Spectrometer. The telescope's cylindrical mirrors were coated with reflective iridium, giving Chandra 10 times the resolution of existing X-ray astronomical detectors and 50 times the sensitivity.

"It is a *very* substantial step forward in terms of sensitivity," said Harvey Tananbaum, the director of the Chandra Science Center in Cambridge, Massachu-setts. "Factors of 10 [are] the equivalent on the ground of building a telescope three times larger. We're gaining, not because it's the biggest telescope, [but] we're gaining because the resolution is so great that we're focusing the signal from a point source onto a very small piece of the detector."

ASTRONOMICAL TARGETS

Of the many potential scientific targets for the observatory, black holes were sure to seize the interest of the public, even though so little is known about them. "We don't really know what happens very close to a black hole and how space and time are warped," said Tananbaum. "Einstein's equations make some predictions and, so far, the data available seem to support that. We are pretty convinced that black holes actually *do* exist." By definition, however, they are 'unobservable', but can be studied indirectly by analysing radiation emitted by material being sucked into them.

As gas and dust is accelerated, for example, it collides with increased energy levels and emits X-rays before vanishing. By examining such emissions in unprecedented detail, Weisskopf explained, astrophysicists hoped "to really nail down black hole signatures". Additionally, supernova explosions thought to lead to black holes came under Chandra's X-ray gaze: in fact, one of its first celestial targets was the remnant of a massive star in the Large Magellanic Cloud that was seen to explode in 1989.

"It's thought all sorts of shocks will start to occur in the next few years when the expanding cloud of debris starts hitting all the other stuff that was thrown off from the outer envelope of the star before it exploded," said Tananbaum. "We should be doing assaying of the abundances, the temperature and density and pressure states of that different material. It should look like a Christmas tree." Further, as the expanding shockwave from the supernova reaches material thrown off earlier, energetic collisions should create heavier elements and generate radiation to yield clues about the star's original chemical composition.

Tananbaum even hoped, optimistically perhaps, that if Chandra's planned lifespan of "at least five years" could be extended to a full decade, as the material expands, it might be possible to slightly separate any central source, "which could be a pulsar or it could be just a cooling neutron star. We may actually be able to see the surface of the neutron star as it's cooling just a few years after the explosion."

On a far larger scale, Chandra would also focus on the amount of 'dark matter' present in the Universe by carefully examining galactic clusters. Such galaxies that make up these clusters are deeply embedded in huge clouds of hot, X-ray-emitting gas and are held in place by gravity generated by all the components of the cluster. "The only thing that can be holding it is gravity," said Tananbaum, "and then you can say to yourself: I see the gas [and] the stars and the galaxies; *they* generate a certain amount of gravity. Is that enough to hold onto the hot gas? The answer is always 'no' by factors of 10, or in some cases more." Chandra's observations were expected to enable astronomers to refine their numbers of how much 'normal' matter is present in a given cluster and thus how much 'dark matter' must be present to generate gravity needed to hold it together.

"It may not be able to tell exclusively *what* the dark matter is," he added, "but we should be able to see how much is there with greater precision." Analyses of supernovae shockwaves and neutron star emissions were also expected to provide insights into the physics behind nuclear fusion. "We'll be able to dissect the energy spectrum and really get a handle on the nitty-gritty details of the emission mechanisms in a laboratory that you just don't have on Earth," said Tananbaum. "That's going to have some incredibly profound consequences down the road."

GENESIS OF CHANDRA

Chandra was originally conceived in the 1970s when NASA envisaged a Large Orbiting X-ray Telescope, which Weisskopf described as "a one-metre-class, arc-second-class X-ray observatory. That got descoped and became the Einstein Observatory." The success of the latter prompted the agency to include an X-ray telescope on its wish list for a four-spacecraft flotilla of Great Observatories. Early plans called for it to be launched into a low-Earth orbit and periodically serviced by spacewalking Shuttle astronauts for a projected 15-year lifespan, but in 1991 escalating costs and technical problems forced NASA to rethink the mission.

"What [we] did," said Weisskopf, "was to take the first complement of instruments, which was very powerful, and break it up into two missions." One would transport a high-resolution camera and imaging X-ray spectrometer into a highly elliptical orbit which, although out of reach of Shuttle repair crews, would be able to gather data for no fewer than 55 hours of each 64-hour circuit of Earth. Meanwhile, the second mission, fitted with a super-cooled X-ray spectrometer, would be launched into a lower orbit; this was cancelled by NASA in 1993 in another round of cost-cutting.

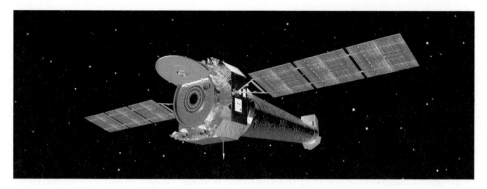

A computer-generated graphic of Chandra when operational.

Even with the cancellation of the second mission, Chandra – known at the time as the Advanced X-ray Astrophysics Facility (AXAF) – was still expected to cost the better part of $3 billion during its first eight years of operation, including the spacecraft itself, its Shuttle and IUS launch costs, annual mission-control and data-analysis fees, use of NASA's TDRS communications network and a one-time charge for the use of an X-ray mirror-testing facility in Alabama. Nevertheless, even this steep pricetag was $4 billion cheaper than the originally planned, all-in-one AXAF in low-Earth orbit.

"The alternative we were looking at was Chandra in this orbit, or no Chandra at all," Weisskopf said, "and so we took the risk to reduce the cost because the option scientifically was infinitely better to have a Chandra in orbit [than] no Chandra at all. We saved billions of dollars with that decision." The decision to launch the observatory into an orbit that was beyond the capabilities of Shuttle repair crews was, however, a tough sell. Already, Hubble had been sent into orbit with a flawed primary mirror and only the skills of spacewalking Shuttle astronauts had saved it.

Still, the unusual orbit – which would bring Chandra as close as 10,000 km and as far as 140,000 km, or a third of the way to the Moon, from Earth – offered several important advantages. "The orbit", said Tananbaum, "is a lot more benign in the sense that you're away from Earth most of the time, so you're not cycling from light to dark, warm to cold, every hour and a half. We don't have to deal with the temperature cycling, which tends to wear out both electrical and mechanical components."

Additionally, the observatory would spend 85% of its time above the radiation belts, allowing it a large, uninterrupted portion of each orbit for celestial observations, and avoiding interference from energetic particles that could otherwise overwhelm its sensitive instruments. "Once it's up there and working, there's reason to be hopeful that it will work for 10 to 15 years anyway in that orbit," said Tananbaum, "[but] we've got to get it right the first time."

THE ROCKY ROAD TO SPACE

As with the other Great Observatories, it was always intended that Chandra would be named after an eminent astrophysicist whose own research had helped to pave the way for the work it would perform. Hence, the Hubble Space Telescope was named in honour of Edwin Hubble, the Compton Gamma-Ray Observatory for Arthur Holly Compton and, in 2004, the Spitzer Space Telescope for Lyman Spitzer. Chandra honoured the Indian-born scientist Subrahmanyan Chandrasekhar, affectionately known as 'Chandra' to his friends, who has been widely labelled as the 'father' of modern astrophysics.

Chandra, who died in 1995 at the age of 85, was not only a Nobel Prize-winner for his contributions to astronomy, but also conducted valuable theoretical work on stellar evolution that established a basis for the existence of neutron stars and black holes – the very objects his mechanical namesake would be observing from its high orbit. "Chandra thought black holes were the most beautiful objects in the Universe," said Lalitha Chandrasekhar, his 88-year-old widow, who attended Columbia's launch. "I hear a lot of people say they are bizarre, they are exotic, but to Chandra they were just beautiful."

It was not, however, only the redesign of the observatory and the troublesome IUS that kept the mechanical Chandra on the ground, but also technical glitches with the spacecraft itself. Problems completing its construction at prime contractor TRW's Redondo Beach facility in California had pushed the launch date from August to December 1998, then January of the following year, and ultimately April when a number of computer software errors arose. On 20 January, only days before Chandra was due to be shipped from California to Florida, NASA announced yet another delay.

It appeared, the agency stated in a hurriedly composed press release, that TRW needed to evaluate and correct a potential problem with several printed circuit boards in the observatory's command-and-data-management system. Several other, TRW-built satellites had turned up faulty copper circuitry, similar to that installed on board Chandra. Aware of its high-priority status and that its orbit would render it irreparable, the decision was taken to remove and replace the boards. "This potentially was very critical to our mission," said Ken Ledbetter. "We're very pleased we found this problem before we got any further along."

Although the replacement of the circuit boards delayed Chandra's delivery to Florida by only a week, it pushed STS-93's launch back by five weeks to late May, because of a need to conduct weeks of testing of the newly fitted units at KSC. This target launch date, however, conflicted with that of STS-96, the first Shuttle mission to dock with the International Space Station and the decision was taken to slip Columbia until the second week of July. Then, with only two weeks to go, Chandra engineers were alerted to *another* problem: this time, with 20 electricity-storing capacitors on board the observatory.

Fortunately, at the eleventh hour, the capacitors were given the all-clear. It was another heart-stopping moment in the often tumultuous genesis of Chandra, but scientists remained confident and excited. "I think we've done everything humanly possible to ensure a successful launching," said Ed Weiler, NASA's Associate Administrator for Space Science. "Let me remind everybody this is *not* a trip to Grandma's on a Sunday afternoon! It's high-tech stuff. There are always risks."

"CUTOFF! GIVE CUTOFF!"

It would not be the last potential show-stopper, however, for Columbia's launch was twice postponed, before she finally set off on her third attempt. To circumvent the possibility that Mission Specialist Steve Hawley – now with four spaceflights and around a dozen launch delays under his belt – was not to blame, he wore a brown paper bag over his head to go to the launch pad. The Hawley gremlin seemed to have struck again on 20 July, when the STS-93 countdown was halted seven seconds before launch, after a high level of hydrogen gas was detected in Columbia's aft compartment.

It was a particularly dangerous time, coming half-a-second before main engine ignition; if the halt had been called after the engines were alight, the result would have been an on-the-pad abort and probably a month-long delay in getting the Shuttle ready for another attempt. The cause of the 20 July problem seemed to be a hydrogen 'spike', which a sharp-eyed launch controller spotted briefly peaking at 640 parts per million – double the maximum-allowable 'safe' limit.

During the momentary crisis, the mood in the Launch Control Center at KSC was tense, as indicated by the voices on the communications loop. Sixteen seconds before launch, it seemed, one of two hazardous-gas-detection systems indicated the 640 ppm hydrogen concentration and even though the second device showed a more normal level of 110–115 ppm, launch controller Ozzie Fish radioed his colleague at the Ground Launch Sequencer (GLS) to manually stop the countdown. To the assembled spectators at KSC, listening to spokesman Bruce Buckingham's commentary over the loudspeaker, all was normal, however.

"T minus 15 seconds," announced Buckingham, then "T minus 12 ... ten ... nine ... "

Inside the Launch Control Center, Fish urgently radioed: "GLS, give cutoff!"

" ... eight, seven ... " continued Buckingham.

"Cutoff! Give cutoff!" interjected NASA Test Director Doug Lyons.

"Cutoff is given," replied the controller at the GLS console.

"We have hydrogen in the aft [compartment]," Fish reported, "at 640 ppm."

By now well past what would have been a 'normal' ignition of the main engines, Buckingham announced the disappointing news to the assembled crowds. Back in the Launch Control Center, the hydrogen concentration was already decreasing back to normal levels. Lyons polled his team, asking them if any emergency safing procedures were needed, such as evacuating the crew from Columbia, and was told that this would not be necessary. Within 10 seconds of the call for cutoff, the indication of high levels had dropped to 115 ppm. Engineers would later blame the problem on faulty instrumentation and flawed telemetry.

DOING HER JOB

Although disappointing, the abort had – at least, indirectly – shown that NASA was not making special provisions to get STS-93 away on time for the sake of several high-level spectators in the audience. Collins' presence on the crew as the first woman to command a space mission had, for the past year-and-a-half, dominated the news. Sitting in the VIP area was none other than First Lady Hillary Clinton, who had formally announced Collins' assignment as Commander of STS-93 on 5 March 1998 at a press conference in the Roosevelt Room at the White House.

For Collins herself, who had worked her way through her 'apprenticeship' just like the male pilots in the Astronaut Office, having a woman lead a mission was simply long overdue, but to Clinton it provided a public-relations boon. In fact, some observers commented at the time that the naming of a Shuttle Commander from the White House was rare, if not unprecedented. However, White House spokesman Barry Toiv defended it as "an important occasion to mark the advancement of science, the advancement of the space programme and the advancement of women in our society".

"Eileen's just trying to do her job," said Coleman. "At the same time, I'm actually very excited about the historical significance – not for Eileen, not for me – but for the little girls out there. It's really bringing home the fact to them that if Eileen can do

The STS-93 crew in Columbia's middeck. At the front are Eileen Collins (left) and Michel Tognini, while in the background are Steve Hawley, Jeff Ashby and Cady Coleman.

what she has set out to do, if all of us can be astronauts, it will help them to realise that the world can be theirs." Added Ashby: "Eileen has made me feel very comfortable and treated me not like a rookie, but as somebody who has flown before and I respect her for that."

The scrubbed launch attempt on 20 July had, therefore, demonstrated that NASA was not prepared to compromise safety to get Columbia off the ground, even with the president's wife in attendance. Fortunately, the fact that the three main engines had not yet been ignited meant that another try could be scheduled for the 22nd; a third opportunity was also possibly available on the 23rd, but after that a three-week delay until the middle of August was on the cards, because the US Air Force was starting a major upgrade of its Eastern Test Range's tracking network.

In eager anticipation for another attempt early on 22 July, eight middeck payloads were removed and serviced and hydrogen sensors in the aft compartment were recalibrated; additionally, the hydrogen igniters – the system that produces what looks like a shower of sparks to clear dangerous unburned hydrogen from beneath the Shuttle's main engines seconds before they roar to life – had to be replaced. The second effort to get Columbia away was also, it seemed, hit by the Steve Hawley gremlin, when lightning strikes were detected just 8.5 miles from the launch pad. According to flight rules, no lightning must be present within a 20-mile radius. The countdown was held at $T-5$ minutes, in the hope that conditions might improve, but when they did not, the launch was scrubbed.

"Due to the storm that's slowly moving to the south, we need to scrub for the day," said Launch Director Ralph Roe. His gloomy news was relayed by Test Director Doug Lyons to the disappointed crew: "Eileen, we gave it our best shot with this storm today, but it didn't agree with us, so our best bet is to give it another try another day."

"Okay, CDR [Commander] copies," replied Collins. "And we thought you guys did just a great job tonight. We're proud of the work and the crew will be ready to go at the next opportunity."

Later that night, Grant Cates told journalists that he was unsure when another attempt could be made, but luckily NASA twisted Boeing's arm to postpone its scheduled launch of a Delta II rocket with four communications satellites and persuaded the Air Force to begin its refurbishment of ground-based tracking facilities a little later than planned. Columbia was granted one more try early on 23 July.

A safe launch was paramount and as Don McMonagle, the head of the Mission Management Team, explained, the mission would be rescheduled for 18 August if its third attempt was also scrubbed. In the early hours of the 23rd, Collins and her crewmates clambered inside Columbia and were quickly strapped into their seats. "Thanks for all the great work," she told launch controllers as the countdown wore on, "and we'll see you in five days." Right on time, and true to form, STS-93 sprang from Pad 39B at 4:31 am, turning night into day across the marshy Florida landscape.

NEAR-DISASTER

However, the commentator's excitement-tinged announcement – "We have ignition and liftoff of Columbia, reaching new heights for women and X-ray astronomy" – masked a serious problem brewing in the Shuttle's main engines. It came to the attention of Collins and Ashby five seconds after leaving the pad, when they noted a voltage drop on one of their ship's electrical buses, which caused one of two backup controllers on two of the three engines to abruptly shut down. The third engine was unaffected by the glitch and, luckily, all three performed normally, propelling Columbia into a 246-km-high, 28.45-degree-inclination orbit.

Nonetheless, the scare was significant: on no other mission had a Shuttle crew come so close to having to perform a never-before-tried abort landing back at KSC. Had the primary controllers, which immediately took over, also quit, an engine failure was likely and *that* would have required Collins to wait for SRB separation, flip Columbia over, fly 'backwards' at more than 10 times the speed of sound in order to slow down and head back west, then jettison the ET and guide her ship down to the SLF in the darkness.

"We were prepared for that," she said later. "We were listening for the engine-performance calls [from Mission Control] on ascent. This crew would have been ready to do whatever was needed." Columbia, luckily, was safely in space, but moving 4.5 m/s slower than expected; although that was an almost-imperceptible discrepancy, in view of her 28,000 km/h orbital velocity, it was enough for puzzled engineers to question whether there might have been a 1,800-kg shortfall in the 545,000 kg of liquid oxygen pumped into the External Tank before launch.

NASA confirmed on 24 July that the loading of propellants had been done properly, although *that*, in addition to the electrical short that knocked out the electrical bus, would become the subject of an investigation. "Since it [the short] was very localised in its effect," said Randy Stone, NASA's director of mission operations, "it could be a short at the main engine controller itself, but this is just speculation. Clearly, we must understand it before we fly again." High above Earth, although acknowledging the mishap, Collins downplayed it during a space-to-ground interview and praised her crew for handling it so well.

Analysis of still and video images taken during the launch also revealed another problem: what appeared to be a leakage of hydrogen gas from one of the main engines throughout most of Columbia's ascent. The images, particularly those from cameras mounted on Pad 39B, revealed a narrow, bright area inside the nozzle of the right-hand engine, possibly indicative of a weld-seam breach in one of more than 1,000 stainless steel hydrogen-circulation tubes. However, Wayne Hale, Columbia's mission operations director, was reluctant to comment on the cause of the leak, at least until the troublesome engine was back on Earth.

"Once we get the engine back," he told journalists, "we will look at the nozzle and see if we really had a leak, then we will turn the metallurgists loose. It could be very simple and a quick fix or something that takes longer. I'm not sure I would call it a constraint to [future] flight yet." He suggested, however, that the leak might explain the otherwise-inexplicable 'missing' liquid oxygen: as hydrogen was lost at a rate of about 1 kg/s, the main engine controllers compensated by gobbling oxygen at a faster pace.

CHANDRA READY TO OPEN ITS X-RAY EYES

It had been one of the most potentially hazardous ascents yet completed and would lead directly to the grounding of the rest of the Shuttle fleet for almost six months. For Eileen Collins' crew, savouring their inaugural moments of weightlessness on the mission, they had other business to take care of: getting their $1.5-billion X-ray observatory ready for deployment. The first few hours were spent checking the health of both Chandra and its IUS, primarily under the direction of Coleman and Tognini, before the stack was tilted up to its deployment angle of 58 degrees.

"Michel and I work as a team," Coleman had told an interviewer before launch. "I put my finger on a switch, he verifies it's the right switch and that is very, very helpful to me. We also have a third person in the background – Steve Hawley – whose job is the big picture of the deploy. It's very human to make a mistake and we cannot afford that. So we're doing everything we can to prevent that."

If the crew had missed their first deployment window for Chandra, matters would have been complicated somewhat. "If we have to keep [the observatory and IUS] in the bay overnight, it really constrains things," said Bryan Austin. "If they lose any power to the heaters that keep the [propellant] lines from freezing, it really gets kinda dicey in terms of being able to still possibly even support a mission, because we cannot put them in a warm-enough attitude to keep everything warm without hurting [Chandra]."

Fortunately, all went well on the first attempt. After a critical "Go/No-Go" decision by flight controllers in Houston and at the Chandra Operations Control Center (COCC) in Cambridge, Massachusetts, the IUS was switched to internal power and cables routeing electricity to the observatory were severed, which transferred it to its internal batteries. At 12:47 pm, about eight hours after launch, as the Shuttle flew high above Indonesia, Coleman commanded the stack to be spring-ejected from its cradle in the payload bay.

"Houston, we have a good deploy," radioed Collins. "Chandra is ready to open the eyes of X-ray astronomy to the world." Added a happy Coleman: "There are five big smiles in here!" After the mission, Coleman would tell an interviewer that she was so taken aback by the beauty of the observatory disappearing into the inky blackness on its way to its highly elliptical orbit that she was "almost too excited to video!" Shortly after the deployment, Collins and Ashby manoeuvred Columbia into a so-called 'window-protection' orientation, whereby the Shuttle's belly was pointed towards the IUS' nozzle.

An hour later, at 1:47 pm, with Columbia about 50 km 'behind' the Chandra/IUS combination, the first-stage engine fired for just over two minutes. Approximately 60 seconds later, it was jettisoned and the second-stage took over for another couple of minutes. The booster's next job was to keep the observatory properly oriented as its two solar panels unfurled. Shortly before the second-stage separation, after insertion into a preliminary elliptical orbit, at 2:22 pm Chandra's solar panels unfolded perfectly. The separation of the second stage went without incident at 2:49 pm.

By this point, Chandra was in an orbit with an apogee of almost 74,000 km and a perigee of 325 km; this was subsequently adjusted using the observatory's own thrusters over a period of three weeks to achieve a final elliptical path with a high point of 140,000 km and a low point of 10,000 km. After being informed of the successful first firing at 2:16 am on 25 July, Collins replied "That's good news. There are a lot of cheering people up here on board Columbia."

The IUS team, needless to say, were also elated. "We were extremely confident in placing Chandra in its orbit," said NASA's IUS representative Rob Kelso after the deployment. "In addition, this mission culminated in more than three years of training for the IUS flight team at the [US Air Force's] Onizuka Air Station. We couldn't be more pleased." Also pleasing – but in a different way – was the discovery of what Wayne Hale called "our smoking gun" for the mishap during ascent: Collins found a tripped circuit breaker in Columbia's cockpit for the centre main engine controller.

The tripped breaker persuaded mission managers more than ever before that the controller of the centre engine, or at least its wiring, was responsible for the electrical short. "Every time you launch, you take a significant risk," said Hale, brushing off suggestions that Columbia had flirted with disaster. "Our job is to manage the risk, to make sure the equipment works the very best it can and that the crew and flight control team are trained to deal with all of the problems."

MIDDECK EXPERIMENTS

With the Chandra deployment behind them, the five astronauts focused the last few days of their mission on a variety of scientific experiments in the middeck. One of these, tended by Hawley, was the Southwest Ultraviolet Imaging System (SWUIS), a telescope and ultraviolet-sensitive CCD device which he used to observe Mercury, Venus, Jupiter and the Moon. Its chief value was its ability to study objects much closer to the Sun than could be observed using the Hubble Space Telescope: such as a hypothetical belt of debris known as the 'Vulcanoids' inside the orbital path of Mercury.

First flown in August 1997, SWUIS had previously taken nearly half a million high-resolution images of Comet Hale-Bopp, providing important insights into its water- and dust-production rates as it receded from the Sun on its return trip to the Oort Cloud far beyond Pluto. Coupled with its unusually wide field of view, SWUIS proved useful on STS-93 by examining faint emissions from Jupiter's upper atmosphere – supporting simultaneous observations by the Galileo spacecraft, in orbit around the giant planet at the time – as well as mapping the Moon and imaging the clouds of Venus.

During typical operations, Hawley operated the telescope through the small window in the middeck access hatch, while Collins and Ashby periodically adjusted Columbia's attitude to support its observations. Elsewhere, Coleman monitored a number of protein crystal growth investigations and Tognini tended a biological cell culture experiment. Other tests included a number of OMS and RCS pulses to provide data for the military Midcourse Space Experiment (MSX) satellite, which had been placed into orbit in 1996 to collect ultraviolet, visible and infrared data from thruster firings.

After a comparatively short – for Columbia at least – five days in space, it was time for the crew to return home. "In a way, it's going to be hard to come home, because I do like being in space so much," Collins told an interviewer on 26 July. "It's a little difficult the first couple of days adjusting to zero-g and just getting used to being here again. I guess you could say it's like riding on rollerskates; it's a little bit different. But once you adapt, after three or four days, it feels almost natural to be up here."

Weather forecasts for 28 July indicated clear skies, with a possibility of thunderstorms inside a 45-km radius of the SLF. Collins fired the OMS engines at 2:19 am to begin the descent through the atmosphere. Passing over Baja California and northwest Mexico, bisecting Texas from west to east and crossing southern Lousiana, Columbia swept into a darkened KSC an hour later with trademark double sonic booms. Little did the STS-93 crew know at the time, but what was found during post-flight inspections of the main engines would effectively ground the Shuttle fleet until the end of the year.

7

Uncertain future

A SHIP WITHOUT A MISSION

For a while, in the spring of 2000, Columbia was a ship without a mission. Then, just like the archetypal British bus, not one, but *two* suddenly turned up. One was a 16-day scientific research mission using a brand-new laboratory module, the other a much-needed servicing of NASA's coveted Hubble Space Telescope. After that, the missions for the venerable old workhorse dried up. It seemed that Columbia – once the pride of the Shuttle fleet, a veteran of 26 spaceflights – would end her days ignominiously in a hangar with a whimper rather than a bang, to be cannibalised for parts.

Tragically, as circumstances would transpire, the opposite was true.

The reasons were partly financial, partly political and partly due to Columbia's hefty weight. Already, by the turn of the millennium, NASA was seriously considering putting its flagship orbiter into storage as a means of dealing with rising Shuttle operating costs over the coming years. "It's on the table. It's an option," spokesman Dwayne Brown had told journalists. "We'll see what happens." Other options including scaling back or cancelling planned fleetwide upgrades – including the development of advanced APUs – and even risking staff layoffs by closing Shuttle test facilities across the United States.

Columbia's weight had already caused headaches in terms of the multi-billion-dollar International Space Station, construction of which commenced in November 1998. At more than 90,000 kg, and even after numerous improvements to shave unnecessary weight from her, she was still too heavy to haul large segments of station hardware into orbit. This was a pity, because her 26 missions between April 1981 and July 1999 had more than demonstrated her capabilities: 185 million kilometres flown, 274 days spent in orbit and 115 people carried into space, including citizens of the United States, Germany, Canada, Japan, Italy, Switzerland, Ukraine and France.

Of those 115 astronauts, two dozen had ridden Columbia into space on more than one occasion – and a handful as many as three times – and helped deploy or retrieve over a dozen communications, military, technological and scientific satellites and operate medical, biological, materials processing, fluid physics, combustion and astronomical experiments on a dozen Spacelab research flights. Her ability to provide a stable platform for sensitive, high-priority experiments had also, by Congressional mandate, assured her a 16-day research mission called STS-107 and it was optimistically hoped by the scientific community that similar flights would follow.

At one point, Shuttle manifests called for one long-duration flight by Columbia every year, devoted exclusively to science, during the first phase of building the International Space Station, as a means of ensuring that the United States' worldwide 'lead' in the microgravity research arena did not suffer. It was expected, eventually, that investigations would be conducted on a permanent basis on board the station, but until then Columbia could readily accomplish scientists' research needs. Still other options for the ageing Shuttle included drop-testing the X-38 emergency crew-return vehicle for the station and perhaps resuming military flights for the Department of Defense.

"Those discussions are ongoing," Shuttle manager Ron Dittemore said of the Pentagon talks, "and whether they're going to conclude in a month or six months or a year, I can't say." Nevertheless, a presidential mandate enforced in September 1986, within months of the Challenger disaster, had all but banned commercial flights on the Shuttle and no requests from the Pentagon were forthcoming. Other possibilities included modifying Columbia for station trips in a 'visiting' – as opposed to 'construction' – capacity or using her to evaluate new technologies before committing them to her sister ships.

By the autumn of 2000, therefore, when astronauts began training for the research and Hubble missions, Columbia's future was very uncertain and it seemed likely that after these two flights she would be decommissioned, at least temporarily, until future opportunities crystallised. Moreover, as International Space Station assembly resumed that autumn, having itself being stalled for over a year, Columbia was far from ready to fly. She was three-quarters of the way through a protracted, $164 million series of modifications and structural improvements that were expected – ironically, perhaps – to make her the most advanced and capable Shuttle in NASA's fleet.

A NEAR-DISASTROUS ASCENT

Had the STS-93 Chandra-deployment mission flown, as planned, in the winter of 1998, Columbia's modification period should have been completed in time to enable her to stage a high-priority repair of the Hubble Space Telescope in June 2000, featuring no fewer than *six* spacewalks – a new Shuttle record. It never happened. Delays in getting Chandra and its IUS booster ready pushed STS-93 into mid-1999; then an electrical short four seconds after liftoff, which knocked out two critical main engine controllers, leaving them running on the backups, forced a lengthy series of inspections of Columbia's 375 km of wiring.

Not only that, but Columbia also suffered a small leak of liquid hydrogen from her right-hand main engine, causing all three to shut down four seconds early. The leak was minor, but significant enough for the main engines to consume liquid oxygen at a higher-than-normal rate to maintain thrust during the remainder of ascent, resulting in the engines shutting down earlier than expected when the oxygen tank ran dry. This, in effect, left the STS-93 crew in an orbit 11 km lower than intended. Fortunately, the mission otherwise ran perfectly.

Detailed analysis after Columbia's landing traced the cause of the leak to several impact-damaged coolant tubes lining the interior of the affected main engine nozzle. This damage, engineers later concluded after extensive borescope inspections, was probably caused by a 2.5-cm-long steel-alloy pin which came loose and hurtled through the combustion chamber, holing three of the 1,080 coolant tubes. Indeed, photographs taken 15 seconds after STS-93's liftoff revealed an unusual bright 'streak' coming from the suspect engine's bell, strongly indicative of leaking hydrogen.

"It appears pretty conclusively that we were impacted by some type of debris," said Shuttle planning manager Bill Gerstenmeier in early August 1999. "What you see is evidence of an impact that scraped away some of the metal and thinned it out enough that the tube could no longer hold pressure. The tubes ruptured, creating the three holes." The question of whether NASA's decision to fly older engines on STS-93 – part of efforts to shave weight off Columbia to house Chandra – contributed to the problem was not overlooked. In fact, the leaking engine had been used 18 times since 1983.

"We very carefully review these engines prior to flight," said Gerstenmeier. "These were perfectly acceptable to use and more middle-of-the-road even than some of the other engines we have flown."

The working-loose of the pin was a one-off incident and had no immediate implications for the fleet. However, two failed engine controllers proved much more serious and far-reaching. By mid-August, NASA gave the controllers themselves a clean bill of health and focused its attention on damaged wiring insulation beneath Columbia's payload bay floor. "From looking at [the wiring] with the electron microscope," said spokesman James Hartsfield, "the indentations in the wire were a tell-tale sign that it was really an impact type of damage," adding that the damage could have developed at any point since the orbiter had been built.

Worse, if it had occurred during routine maintenance, the other orbiters might be similarly affected. Endeavour, Discovery and Atlantis, which were partway through processing flows for their own missions later in 1999, were inspected and instances of damaged wiring were repaired or replaced. Steps were also taken to install flexible plastic tubing over some wiring, as well as smoothing and coating rough edges nearby and fitting protective shields. Columbia, meanwhile, had her wiring inspections added on to her already-scheduled modification period at the Palmdale plant in California.

LEAN, MEAN MACHINE

In fact, by the time Columbia finally returned to KSC in early March 2001, she had become, in the words of Ron Dittemore, "a leaner, meaner machine [that] is fit to fly for many more years". More than 500 kg of weight had been removed and she was also wired-up to accept an International Space Station-specification airlock and docking adaptor in her payload bay. Overall, she underwent 133 structural inspections, wiring repairs and upgrades, including – most notably – a state-of-the-art 'glass cockpit' that was more advanced, lighter and used less electricity than the previous design.

For Duane 'Digger' Carey, who would fly as Pilot of Columbia's Hubble servicing mission, the new cockpit was by far the most striking and important difference for him and STS-109 Commander Scott Altman. "There's a tremendous amount of work that's been done that you *can't see* in Columbia," he told a NASA interviewer shortly before blasting into space, "but the big one for Scott and me is the cockpit." To give it its official name, the Multifunction Electronic Display Subsystem (MEDS), the cockpit boasted 11 flat-panel, full-colour displays to replace 32 old-style mechanical gauges.

"We now have an up-to-date instrument display suite in the orbiter," continued Carey. "Columbia's the second orbiter [after Atlantis] to get this particular upgrade. The nice thing about [it] is that, as we get smarter about what types of displays we

During Columbia's debut with the new Multifunction Electronic Display Subsystem (MEDS), Digger Carey (left) and Nancy Currie occupy the forward flight deck during rendezvous operations with the Hubble Space Telescope.

want to see – to help us to have 'situational awareness' in the orbiter – we can actually quite easily upgrade our displays. Right now, the displays just kind of emulate whatever our old displays had in the orbiter, but in future we have a lot of flexibility in what we can do with those displays."

Ultimately, it was hoped that MEDS would lead to the development of a 'smart cockpit' that would provide Commanders and Pilots with more flexibility to conduct mission operations. However, the new cockpit was only the most visible of the upgrades. Most of the modification period was overshadowed by the need to thoroughly inspect and protect her wiring from the kind of damage that had caused the short-circuit on STS-93. Additionally, thermal blankets replaced heat-resistant tiles in some areas, further reducing the ship's weight, radiator valves were added to better isolate leaks and the middeck floor was beefed-up to withstand loads of up to 20g.

AN ASTRONOMICAL ICON

Beginning in the autumn of 2000, two Shuttle crews began training for Columbia's first missions after her modification period: the 16-day STS-107 research flight and the Hubble Space Telescope servicing. For a time, it was unclear as to precisely which order the missions would be flown, but eventually Altman's STS-109 flight took precedence. As NASA's scientific showpiece, it was essential that Hubble was kept operational until a next-generation telescope could replace it in 2010, and a number of important upgrades and new instruments were waiting in the wings to be installed in the observatory.

Sent aloft in April 1990, Hubble has become perhaps the most widely known and best-loved icon of astronomical technology. "I think the most important thing Hubble has done is answer questions we didn't even know *how* to ask in 1990," said Ed Weiler, NASA's Associate Administrator for Space Science, before Columbia lifted off on STS-109. "Hubble keeps telling us that our textbooks aren't being followed by the actual Universe." For at least one of the STS-109 astronauts, however, visiting Hubble to conduct repairs was more of a pilgrimage or a Holy Grail than 'just' a Shuttle mission.

"Servicing [Hubble] is by far and away the most meaningful thing I've ever done in my life," said John Grunsfeld, a professional astrophysicist before becoming an astronaut. In addition to STS-109, he flew a previous Hubble servicing in December 1999. "People have to decide for themselves what kind of things they want to do [in life] and what they're worth risking and, for me, the Hubble is worth risking my life. It's *that* important. It's just an incredible worldwide resource. It's teaching us so much about our world, the Universe, who we are and our place in the cosmos."

The telescope that received such praise dates from the 1970s and has been billed as nothing short of a cosmic time machine, capable of unravelling fundamental mysteries about the origin, evolution and fate of the Universe. "That was the culmination of decades of work by hundreds of astronomers [and] thousands of engineers," said Weiler. "It was probably the best day of our lives [when Hubble launched]. We all felt on top of Mount Everest that day." Already, Hubble had waited patiently for four years as NASA recovered from the loss of Challenger. Sadly, more trouble was afoot.

In June 1990, six weeks after it was placed into orbit, scientists realised with horror that the telescope's primary mirror had been polished to the wrong specification and was suffering from a complaint called 'spherical aberration'. In effect, its 2.4-m-diameter mirror had been ground *too flat* by only the tiniest of measures – a fiftieth the thickness of a single human hair – but more than enough to have a detrimental impact on Hubble's observations, blurring all of its images of stars, galaxies and other celestial objects and rendering it the butt of jokes rather than an icon to be revered.

Weiler has referred to that dismal time as like a journey from the lofty summit of Mount Everest down not to the surface of Death Valley, "but perhaps six feet *under*! We had to explain to the American public that Hubble had a major flaw in its main mirror that would affect *all* the science we had to do." Nevertheless, a servicing crew was launched in December 1993 and resolved the aberration by fitting new corrective optics. Starting the following spring, the rejuvenated Hubble began a decade of work that has significantly changed humanity's perception of the Universe.

Among its notable discoveries is the possibility that the cosmos is younger than previously thought – around 13 billion years old, as opposed to earlier estimates of 15–20 billion years – and its observations have lent weight to theories that, rather than collapsing in a cataclysmic 'Big Crunch', it will continue expanding forever. Other research has yielded persuasive evidence for the existence of supermassive black holes at the centres of many galaxies, planets encircling distant stars and views of stellar nurseries and graveyards. Quite simply, no other telescope has done so much to rewrite textbooks in such a short space of time.

"It's Americana," said Rick Linnehan, who would accompany Grunsfeld on three spacewalks to repair the orbiting observatory during STS-109, becoming one of only a handful of astronauts to have flown Columbia three times. "It's like cheeseburgers and Clint Eastwood. I mean, there was a Pearl Jam CD cover a year or two back that had one of the Hubble images on it. It just pervades everything now. That's really impressive because this is *raw science* – photographs coming down that not only are important to science, but they're beautiful. They're considered *art!*"

The photographic artwork flowing from Hubble has also included views of Martian dust storms and the once-in-a-lifetime collision of Comet Shoemaker–Levy 9 into Jupiter in July 1994. In terms of expense, the joint US/European project is estimated to have cost around six billion dollars in the quarter of a century from conception to the present day, including a decade and a half of orbital operations and four servicing missions by the Shuttle. From the outset, it was always planned that in order to have a fully operational, world-class, long-term orbiting observatory, there *had* to be an option to upgrade it in space.

By the late 1970s, the Shuttle was nearing completion and Hubble was accordingly equipped with grapple fixtures to allow the RMS arm to retrieve and anchor it into its cavernous payload bay for servicing. Seventy-six yellow handrails were attached to its hull to enable future spacewalking astronauts to crawl easily along its 13 m length to remove and replace scientific instruments. To ensure that it operated high enough to be outside the 'sensible' atmosphere, but low enough to be reached by the Shuttle, it was inserted into a 530-km orbit above the Earth.

Original plans called for the return to Earth, refurbishment and relaunching of the telescope every five years, with a servicing mission every couple of years to remove and replace worn-out scientific instruments. By the early 1980s, the possibility of contamination and extreme structural loads on the Shuttle during re-entry would severely hamper such a plan. Ultimately, NASA settled on repair missions at three-year intervals. Four were completed: an inaugural one to replace Hubble's solar arrays and fit corrective optics; a second to add new instruments; a third to replace the telescope's 'brain'; and a fourth – STS-109 – to conduct further upgrades.

To understand more fully the observatory that Scott Altman's crew finally saw as a dot in space on the morning of 3 March 2002, it is important to summarise the first three servicing missions, since Hubble had changed a great deal both internally and externally since April 1990. The first repair mission, in December 1993, fitted corrective optics, new solar panels and replaced the first Wide Field Planetary Camera (WFPC) with an improved model capable of operating with the new optics.

Three years later, in February 1997, another Shuttle crew removed two instruments and installed two new ones: a Space Telescope Imaging Spectrograph (STIS) to investigate the chemical composition of various celestial sources and a Near-Infrared Camera and Multi-Object Spectrometer (NICMOS) to provide Hubble with the capability to conduct infrared imaging and spectroscopy. Unfortunately, the latter relied on cryogenic nitrogen to keep its detectors cool enough to function, and this became exhausted earlier than planned in the spring of 1999. That was a pity, because NICMOS had already produced impressive infrared results, exploring in particular the atmospheres of Uranus and Neptune.

HST SM-3: 'THE BIGGIE'

The third servicing mission, to which four spacewalking astronauts were assigned in the summer of 1998, was projected for launch in June 2000 and should have been 'the biggie' with no fewer than six excursions to change the telescope's large solar array wings for smaller, but heavier and more rigid, electricity-generating panels, replace its computer 'brain', install gyroscopes to 'steer' Hubble and fit a new $75-million instrument called the Advanced Camera for Surveys (ACS). The latter would provide it with a wide-field survey capability from visible to near-infrared wavelengths, as well as imaging from near-ultraviolet to near-infrared.

All that changed in early 1999, when NASA became increasingly concerned by a rash of failures of the telescope's existing gyroscopes. It was thought that the failures had been triggered by wiring corrosion. Fitted with six of the devices for redundancy, at least three are needed to provide stability and pointing accuracy. In March, when the space agency hurriedly split the third Shuttle servicing mission into two halves – one in October 1999 to carry out the most critical repairs, the other some time the following year to replace the solar arrays and fit the ACS – Hubble was operating on only three gyroscopes.

After several more months of fleetwide wiring inspections in the wake of the STS-93 scare, Discovery and her seven-man crew finally set off just before Christmas and successfully staged three spacewalks to replace all six gyroscopes with new models and fit the new computer 'brain'. Hubble was, quite literally, brought back from the brink of disaster when the third servicing crew reached it: a fourth gyroscope had failed in November 1999, forcing NASA to put the telescope in 'safe mode' and suspending all astronomical observations until urgent repairs could be made.

By the time Discovery's crew returned to Earth, all that remained to be done from the 'original' third servicing mission was the replacement of the solar panels and the installation of the ACS. Additionally, it was intended that the next repair crew would fit a new 'cryocooler' to enable NICMOS to resume work. Although it would officially be the fourth Shuttle visit to the observatory, Altman's flight was actually labelled 'HST SM-3B' – or Hubble Space Telescope Servicing Mission-3B – because it represented the other 'half' of what should have been the third servicing mission.

Entirely appropriately, John Grunsfeld, who had been on board the December 1999 ('3A') mission and was one of the spacewalkers originally trained to fly 'the biggie', was reassigned as Payload Commander of 3B. He would be joined by Linnehan for three spacewalks, while physicist Jim Newman – a veteran astronaut who helped to assemble the first segments of the International Space Station in December 1998 – would work with rookie spacefarer Mike Massimino on two other excursions. By the time Columbia returned to Earth at the end of the SM-3B mission, she would have significantly improved Hubble's observing power.

For the astronauts, however, their personal Hippocratic Oath was to 'Do No Harm'. "I think of [Hubble] as a big old beast and I'm doing surgery on it," said Linnehan, a veterinarian by training who previously flew two life sciences missions on Columbia. "The one cardinal rule about surgery is you *don't* want to make it worse! You don't want to leave any tools or sponges inside the patient. You want to make sure everything works at least as well as it did before you went in, and hopefully a lot better."

NASA's then-Administrator, Dan Goldin, was also keenly aware of the telescope's importance – not only to the astronomical community, but as an icon of space science for the general public. In 2001, recalled Newman, he called the four SM-3B spacewalkers to his Washington office and "let us know that, on spaceflights, there's always risk and sometimes failure happens. [But] then he looked us in the eyes and said: 'That's *not* an option for you!' Hubble is very, very near and dear to a number of people, so we have been impressed with that responsibility."

In order of priority, the most important task was replacing the observatory's central Power Control Unit (PCU) – essentially its 'heart' – followed by the installation of the ACS and a cryocooler to revive NICMOS. However, the order of the spacewalks was not dictated by priority or technical complexity, but rather by the need to create a 'best fit'. "It's like a puzzle," said STS-109 Lead Flight Director Bryan Austin. "In putting the puzzle together, the board is only so big – it fits five EVAs – and the puzzle pieces unfortunately are big and they only fit really in one way."

The PCU replacement was most critical and caused the greatest deal of worry in the minds of both mission planners and the astronauts themselves. "It's kinda like open-heart surgery," said Austin, "except we don't have [Hubble] on a heart–lung machine." Many scientists associated with the telescope admitted they would be on the edge of their seats. "It violates a long-standing policy in the space business that if something's working well, you don't turn it *off* and just hope it comes back *on*," said Ed Weiler. "We're not doing that cavalierly. We fully anticipate that everything will work just fine."

If the procedure failed, added Linnehan, "Hubble is dead." Nevertheless, replacing the troublesome unit was essential: a series of potentially devastating failures after 12 years of orbital wear and tear had rendered it a health risk to the telescope. Its job was to route electrical power from the solar arrays to the scientific instruments, control systems and batteries. The risk was that, in order to replace it, the spacewalkers had to *totally power down* Hubble to protect themselves from electrocution while they worked. They would then have barely eight hours to complete the delicate operation before thermal extremes inflicted permanent damage.

By the autumn of 2001, the plan was to replace the telescope's two solar arrays on the first and second spacewalks. This would enable the new arrays' attached diode controller boxes to better manage the electrical system while Hubble was fully shut down for the PCU changeout; then, on the third spacewalk, Grunsfeld and Linnehan would conduct the critical operation. Unfortunately for them, the 73-kg device had not been designed to be replaced and was wired into the electrical system by 36 closely spaced, hard-to-reach connectors that had to be removed one at a time by the two fully suited men.

It would be busy, with very little margin or time for error. "We're book-keeping just a couple of minutes for connector mate [and] connector de-mate," said Austin before the flight. "So multiply that times 36: that's a lot of time, and if you're off by 30 seconds – times 36 of those – that's about 15 minutes or so! Things can add up real easily if you start to run into problems."

Of course, there would be no point in fitting the new camera and NICMOS cryocooler if Hubble did not come back to life after its heart surgery; consequently they were shifted onto the fourth and fifth spacewalks. For a year and a half before Columbia finally lifted off, Grunsfeld, Linnehan, Newman and Massimino meticulously practised each of their tasks in the water tank at JSC, wearing training versions of the spacesuits they would use and working on full-scale Hubble mockups in full-size payload bays. For every hour of their 30 hours of spacewalks, they spent at least 12 hours underwater . . .

"Do we think we can do it all?" asked Anne Kinney, NASA's Director of Astronomy and Physics, before the launch. "Yes, we do. We wouldn't plan it if we didn't think that. But it is very ambitious. It's *not* easy. We'll worry all the way."

The flight would run for 11 days and its target launch date had already slipped to November 2001 because of the longer-than-expected time needed to fix Columbia's wiring at Palmdale. Despite the achievements of earlier servicing missions, many within NASA looked at SM-3B as the most complicated of the four and few could

Scott Altman (left) addresses his crew before a pre-flight cake-cutting ceremony. Seated from left to right are John Grunsfeld, Mike Massimino, Digger Carey, Nancy Currie, Jim Newman and Rick Linnehan.

have felt the pressure more than Scott Altman. A seasoned Shuttle pilot with two missions under his belt – including Neurolab in the spring of 1998 – he was now saddled not only with the burden of command, but also with the most exacting flying challenge of his career.

"As the Commander," he told a CNN interviewer shortly before launch, "going through the whole training flow, your focus really widens out and you are responsible for much more [than as a Pilot]. I guess it's the mantle of command. You think about how training is going, how the crew is doing together, what our performance is like in the sim[ulator]s and then try to coordinate all that and produce a polished product."

Joining Altman, Carey, Grunsfeld, Linnehan, Newman and Massimino to round out the seven-member crew was petite Mission Specialist Nancy Currie, whose duties were divided between acting as flight engineer during ascent and re-entry and deftly operating the RMS. Before the launch, she described herself as having "a quarterback attitude with a magic touch." It was her job to help Altman and Carey to monitor the Shuttle's displays and subsystems. "I'm kinda 'quarterbacking' to make sure everybody's in the right procedure, on the right page, adjusting or helping with any switch throws as necessary," she told an interviewer before launch.

"I'm flying with some very large guys on this crew, so my seating height is quite a bit lower than them. There are actually some things I can see in the cockpit that they can't see. I can look up and verify switch positions and actually direct their hand to certain switches." As one of the few active US Army officers in the astronaut corps, Currie was making her fourth Shuttle flight; she had also been on board the first International Space Station assembly mission, manipulating the RMS to precisely attach the Unity node onto the Russian-built Zarya control module.

Her experience with the mechanical arm was beneficial on STS-109, as she was required to not only use it to pull Hubble out of space and anchor it into Columbia's payload bay – and redeploy it later in the mission – but also to move her colleagues around as they undertook nearly 30 hours' worth of gruelling spacewalks. "In terms of masses, the Russian [module] was slightly more massive [than Hubble]," she said before the launch, "so I'm familiar with moving large masses; and essentially the slower you go [with the RMS], the better off you are. That's kinda my trademark and I take a lot of grief [in the Astronaut Office] for going very slowly. It's kind of a joke of who can fly the slowest, and definitely I think I win that prize! But what I found is the slower you move, then the less problems you have, say, with oscillations in the arm [and] the controllability of the arm. When you say 'Stop', the payload stops, and because of the upgrades we've made to the arm [since 1996] and really tremendously increased the capability, [it] flies exceptionally smoothly, even with a massive payload like [Hubble]."

A LONG ROAD TO SPACE

It was with great joy that Currie and her crewmates sighted the telescope as a steadily brightening star in the early hours of 3 March 2002. Not only were they closing in, slowly but surely, on the quarry they had studied and prepared themselves to repair for more than a year, but they also felt joy and fulfilment that their own mission was now well and truly out of the woods. Unforeseen circumstances two days earlier had overshadowed an already-troubled and long-delayed mission.

Originally scheduled for launch in November 2001, additional time needed to prepare Columbia for her first trip into orbit in more than two years pushed the launch of Altman's team into the spring of the following year. A target date of 14 February was postponed by two weeks when one of four reaction wheels on board the orbiting Hubble suffered a temporary, seven-minute-long hiccup. Although the wheel, which helps to point the telescope and hold its position while making observations, later began rotating correctly, the scare raised eyebrows over its reliability and NASA opted to replace it.

This was easier said than done, particularly for the crew, who spent the last year precisely choreographing every move of their five already-complex spacewalks. Now their work would be upset by having to add the reaction wheel changeout. "It's certainly been an added issue," said Altman in late January. "The whole set of EVAs are really interlocked with each other to make sure that [the choreography] was the most efficient. Now [we had to] take this giant jigsaw puzzle that [we've] put together and throw another piece on the table and say 'Put that in and make it look perfect!' It's taken a lot of time and effort on everybody's part – the training team, the flight controllers and the crew – to work together and come up with a plan where we can take what were already five very full EVAs and add an additional task and still feel relatively confident that we can get everything done in five spacewalks."

That confidence was reflected in the faces of all seven astronauts when they arrived at KSC in a fleet of T-38 jets on the evening of 25 February. Weather forecasters were already predicting a 70% chance of acceptable conditions for launch in three days' time. "To fly down was really a treat tonight, to fly over the orbiter and take a look at it sitting there on the pad; it looks ready to go," Altman told the small crowd of journalists. He also paid tribute to ongoing 40th anniversary celebrations that week of America's first orbital spaceflight by John Glenn.

Unfortunately, the crew's confidence was not, in fact, mirrored by the weather: by the evening of the 26th, US Air Force meteorologists were tracking an approaching cold front, which threatened to come close to the low-temperature safety limit put in place for Shuttle launches. Freezing conditions were already expected to hit the city of Melbourne, just to the south of KSC, and on the morning of 27 February the decision was made to postpone Columbia's liftoff until 1 March.

Even the rollout to Pad 39A a month earlier had not gone entirely smoothly. Originally scheduled to take place on 23 January, the six-and-a-half-hour trek was postponed when the gigantic crawler transporter literally broke down in the VAB's doorway. "I've got to admit it: we're actually stuck!" NASA spokesman Bruce Buckingham told journalists. The cause of the problem was traced to a fault with the crawler's steering mechanism, which was promptly repaired. Dismal weather then kept the STS-109 stack under cover for two more days.

Fortunately, said Buckingham, plenty of contingency time had been purposely built into the rollout schedule and Columbia was securely on the pad by the evening of 28 January. For the second time since the terrorist attacks on New York and Washington, the rollout of the $2-billion national asset was accompanied by strict security as F-15 fighters patrolled an extensive 'no-fly' zone around the KSC area. Final preparations for the previous Shuttle mission, STS-108 in December 2001, had followed a similar protocol.

Shortly after Columbia arrived on the pad, concerns were raised about the flight-readiness of a number of bearings in the orbiter's main landing gear wheels. During examinations at Wright-Patterson Air Force Base in Ohio, several bearings were heat-tested at temperatures of 300 Celsius and performed satisfactorily, but during subsequent checks at 500 Celsius one of them suffered minute cracks. It was found that the cracked bearing was not sufficiently heat-treated to withstand the tremendous thermal loads placed on the wheels during touchdown. More worrying: eight similar bearings had ended up in Columbia's landing gear.

"GETTING OUTTA TOWN IN A HURRY!"

By the evening of 28 February, however, and in light of the fact that the failure of the bearing at 500 Celsius was a relatively minor concern, the Mission Management Team cleared STS-109 for launch early the following morning. Altman led his crew out of the Operations and Checkout Building at 7:45 am for what was essentially a

Spectacular early-morning liftoff for STS-109 on 1 March 2002.

picture-perfect countdown, the only concerns being a possibility that a broken deck of clouds could form below 2,400 m and violate launch safety rules.

"The launch count was just about as smooth as it could be," Grunsfeld wrote later in one of his NASA-authorised journal entries for the mission. "Lying on our backs, fully dressed in our orange suits and parachutes, the wait for liftoff was not particularly comfortable. As the count progressed, it was clear to us that any concerns about clouds or winds were not going to stop our trip up to Hubble. On board Columbia, we worked through our procedures down to the last couple of minutes. Then it was up to the computers on Columbia, the main engines and finally the [SRBs].

"In the last six seconds, the main engines announced that they were ready to rock and roll. Then, in an instant, the [SRBs] lit and we *knew* we were getting outta town in a hurry! Incredibly, for a few seconds, time seemed to slow down. We were clear of the tower, executing a slow roll to the heads-down attitude, when out of the front windows I saw a mass of billowy white clouds in our path. The light illuminating the clouds was coming from the fiery exhaust of our rocket engines."

On the ground, it was 11:22:02 am GMT when Columbia thundered aloft, precisely on the opening of a 62-minute 'window' to rendezvous with Hubble in two days' time. In fact, directly overhead – some 580 km above Sarasota – the telescope itself orbited, patiently waiting for its next group of human visitors. The effect for spectators of Columbia's ascent was one of lighting up the pre-dawn darkness and knifing straight through a thin layer of clouds. Passing through them helped Grunsfeld to reset his sense of time back to normal, "as I was able to grasp the velocity at which we were travelling. The clouds went zipping by and we were again heading upward into a black sky. A few minutes later, the Sun burst into the cabin as we ascended into daylight, and we rolled heads-up for the remainder of the ascent. In the final couple of minutes, we were accelerating at three times the force of gravity! Duane Carey commented on how hard it was to talk, although I saw he had no trouble reaching for items against the pull of the engines. Eight-and-a-half minutes after we had started, the engines shut off and we were in orbit."

Back on *terra firma*, NASA Administrator Sean O'Keefe, attending his first Shuttle launch after being in the job for just over two months, was flabbergasted. "It was something that was a sight to behold – absolutely extraordinary!" he said. Added NASA Launch Director Mike Leinbach: "It was just a beautiful launch."

NOT OUT OF THE WOODS YET

Reaching orbit after their long wait – and the exhilaration of finally being on their way towards a 3 March retrieval of Hubble – was soon tempered by the worrying possibility that the crew might have to return immediately to Earth. Eighty-five minutes into the mission, the payload bay doors were opened and the radiators lining their interior faces began work dumping waste heat into space. However, it became clear that one of two freon loops used to keep the spacecraft cool, particularly during the thermal extremes of atmospheric re-entry, was behaving sluggishly.

Initial speculation was that the loop might have been blocked by some kind of foreign debris: perhaps a small piece of 'welding slag', or solder, from Columbia's overhaul in California had broken loose during ascent and lodged itself in the coolant line. The dilemma was that at least one coolant loop had to be fully functional to cool the ship during re-entry. According to proscriptive mission rules, if only one loop was operating, a Minimum Duration Flight would be declared and Altman's crew would return to Earth at the earliest opportunity.

This did not bode well for Hubble or the mission. "We on the ground are looking at the degraded flow in freon loop 1 and at present we are pressing on with the nominal timeline," Capcom Steve MacLean told Altman on the afternoon of 1 March. "However, there is a [Mission Management Team meeting this evening] when they will discuss the details and impacts of the degraded flow. So first thing in the morning, we'll give you the outcome of their discussions and we're all hoping that we can rendezvous with the Hubble and fix it."

Gradually, however, the degraded loop began to stabilise and Shuttle manager Ron Dittemore, after extensive analysis, doubted that the second loop was affected, citing Columbia's violent climb to orbit as proof. "The shake, rattle and roll of ascent really is a very dynamic test," he said. "Not only do you get a lot of vibration, but you get a lot of acoustic vibration. If there *was* anything in loop 2, it would have broken loose by now. That's why we believe that we saw the debris in cooling loop 1. [The launch] is the time it was going to break loose. We believe it's going to remain stable and support the remainder of the flight." Strictly speaking, keeping Columbia in orbit with only one fully functional loop infringed flight rules, but Dittemore was quick to stress that the second loop was "rock-solid" and even the contaminated first one should be able to hold its own during re-entry. By the afternoon of 2 March, his concluding prognosis was optimistic: "The flow-rate we see on cooling loop 1 is large enough that it would be able to support a full nominal entry if called upon to do it all on its own."

The good news was relayed to the astronauts by Capcom Mario Runco, who pointed out that although managers were still looking very carefully at loop 1, the chances of Columbia remaining in orbit for the full 11 days seemed brighter. "I appreciate you putting those words together for us," Altman told Runco. "We will stand by for the final decision." Privately, and despite their concerns, the crew had been far too busy executing critical thruster firings and conducting spacesuit checkouts in anticipation of the Hubble retrieval to dwell too much on the coolant loop issue.

Altman and Carey had their hands full with a series of carefully executed thruster burns to gradually close the gap between themselves and the telescope. Meanwhile, at the rear of Columbia's flight deck, Currie and Massimino activated and checked out the RMS arm, as their colleagues set up laptop computers and reduced the cabin air pressure from 14.7 psi to 10.2 psi in readiness for five days of back-to-back spacewalks. This last move helped to rid nitrogen from their bloodstreams before they donned their spacesuits, in order to prevent attacks of the 'bends'.

Back on Earth, Hubble's ground control team closed the telescope's aperture door by remote control. They also retracted its two high-gain Ku-band antennas and disengaged the brakes on its two solar arrays in anticipation of their scheduled replacement during the first and second spacewalks of the mission.

"We've been extremely busy up here [and] we haven't had a lot of time to ponder that," Grunsfeld said of the freon loop glitch on 2 March. "Of course, we know it's a serious problem back there [with loop 1] and one that shouldn't be ignored and so we're letting the smart folks on the ground really worry for us. If we had to come back, we know that Hubble's important enough that Hubble would be well taken care of." With the benefit of hindsight, and NASA's changing attitude towards the welfare of the telescope following the STS-107 disaster, Grunsfeld's words today seem bitterly ironic.

By 3 March, as the Hubble rendezvous drew closer, it seemed they were finally out of the woods in terms of the freon problem and Mission Manager Phil Engelauf defended the decision to slightly 'bend' the flight rules to keep Columbia in orbit. He pointed out that many of the rules were written before the first Shuttle flight and were overly conservative. "We've got 107 flights of experience under our belts now," he told a press conference. "We've continued to evolve the math model [and] we've got performance history of the flight hardware that we didn't have before."

The result was a more accurate understanding of how each Shuttle actually performed in space, how much heat it really generated and how much freon flow was truly needed to support its systems during re-entry. "[If we] aborted the mission for a very small violation of the rule," said Engelauf, "without going and looking at the real data to see if we couldn't do better than that, you guys would be all over us for not doing our jobs," adding "it would be a foolish waste of NASA's resources for us to arbitrarily come home out of over-conservatism."

HUBBLE CAPTURED

As engineers on the ground wrestled with their data and rule books, up on Columbia's middeck, Altman wrestled with a balky airlock handle on the afternoon on 1 March. For an instant – and bringing back unpleasant memories of STS-80 – it seemed that the mission might again be compromised and the spacewalks curtailed or even cancelled when the hatch refused to unlatch properly. Fortunately, Newman lent a hand and the pair quickly reported success. "I fixed it!" Newman radioed jubilantly to Mission Control. "I kept the handle hard and flush against the airlock and the handle is now rotating effortlessly."

"That's excellent news," replied Steve MacLean from Houston.

The rendezvous with Hubble was nothing short of an incredibly delicate, two-day orbital ballet. "From launch until rendezvous," said Altman before launch, "everything is planned and watched over by people on the ground and us in orbit as we do different burns, modifying our altitude so we can kinda catch up to Hubble by going faster in a lower orbit and then bringing us up to the point where Hubble and I are both flying in formation ... approaching very gently apparently until the point where it's hanging motionless over our payload bay and Nancy can go grab it.

"A lot of people think the Shuttle is always flown by computers and it's not as much of a hands-on flying task. In rendezvous, that's certainly not true. The computers get us close, but then we rely on the pilots to take us the rest of the way home until we can rendezvous and just be there in the same spot of space." That 'same spot' was reached on the mission's third day – 'Rendezvous Day' – and Altman's crew knew from more than a year of training that it would be a hive of activity in Columbia's cramped flight deck.

"It's kind of a trick [to] make sure we distribute the duties evenly, but that we don't overcrowd the flight deck also, because a couple of us will be manually flying," said Currie before the mission. "Scott will be manually flying the Shuttle and I'll be manually flying the [RMS] arm at the final phase. So you actually don't want too many people up there all at one time. We've rehearsed who moves where, and at one point we all kinda get up and switch seats."

The crew was awakened at 1:52 am on 3 March for Rendezvous Day, to the theme tune from the 1960s television show 'Mission Impossible'. "Columbia, Houston," radioed Capcom Dan Burbank. "Your mission, if you choose to accept it, is to rendezvous and grapple the Hubble Space Telescope and then spend five days massively re-outfitting and upgrading the telescope. Hubble is 1,400 miles ahead of you and you're closing at 600 miles an hour." Then, with a grin, Burbank added, "This tape will self-destruct in five seconds!"

"All right!" exulted Altman, eagerly taking up the challenge. His first major task was a thruster firing to directly intercept the telescope, which placed Columbia into position just under a kilometre 'below' its quarry some 90 minutes later. He then took manual control of his ship, easing it slowly to within 10 m of Hubble – close enough for Currie to reach out with the RMS and grapple it.

In his journal entry for Rendezvous Day, Grunsfeld wrote: "Scott slowly brought the Shuttle in close, with Duane backing him up. Jim Newman was on the laptop computer providing situational awareness calls from a program that displays our trajectory on the screen. Rick manned the hand-held laser [and] Nancy and Mike prepared the robotic arm to reach out and grab HST. My job was to work the Hubble communication procedures, which also allowed me to take pictures of HST on approach."

The time was 9:31 am and the two now-joined spacecraft flew silently over the Pacific Ocean, southwest of the coast of Mexico.

"Houston, we have Hubble on the arm," Altman radioed excitedly.

"Copy, Scooter [Altman's nickname in the Astronaut Office]. Outstanding work!" replied Capcom Mario Runco. "There's a big sigh of relief we heard from Goddard [Space Flight Center] all the way here."

"I think it echoed up here as well," said Altman.

After Currie had anchored Hubble onto the Flight Support System (FSS) platform at the rear end of Columbia's payload bay, the next step, in readiness for the first spacewalk on 4 March, was to retract both of the telescope's rectangular solar arrays. Although their drive motors had not been used since they were first unfurled during the SM-1 mission in December 1993, they performed flawlessly and after commands were transmitted from Goddard Space Flight Center it only took

Deftly manipulated by Nancy Currie, Columbia's RMS mechanical arm has Hubble firmly in its grasp.

about five minutes to fold up each 12-m-long array, in a manner not dissimilar to a kitchen roller blind, into its storage canister.

The old arrays each weighed 154 kg and originally generated 4,600 watts of electrical power from silicon solar cells; the new models, scheduled to be fitted during the first two STS-109 spacewalks, were considerably heavier at 290 kg apiece, but far smaller at just 7 m long. However, based on the design presently used in Iridium communications satellites, they provided 30% more power, generating 5,270 watts with their gallium arsenide solar cells. This translated into the capability of astronomical teams, for the first time, to operate *all four* of the telescope's scientific instruments – WFPC-2, STIS, NICMOS and the ACS – simultaneously.

The new arrays, framed by stronger aluminium–lithium alloy, were also rigid and less susceptible to 'wobbling' than the older, flexible panels, thus reducing the risk of vibrations affecting Hubble's astronomical observations. Interestingly, rather than rolling up like a blind, the new arrays folded in half like a book and offered greater reliability than the older set, which by March 2002 provided only 63% of their original power output, thanks to eight years of temperature extremes and punishing radiation, coupled with structural problems and shorted circuitry.

"We've got a winner on both sides," observed Capcom Steve MacLean as the second array folded away without a hitch. "It's a good start to five more great EVAs." In terms of the condition of the telescope itself, which had not been seen with human eyes for more than two years, it seemed to have fared extraordinarily well, with photographic surveys revealing it to be in excellent shape. "Our initial examination does not indicate any significant or noticeable changes since December 1999," said Hubble's manager Preston Burch of NASA's Goddard Space Flight Center.

"Prior to STS-103, [the SM-3A] mission, we had a number of briefings from the folks at Goddard that we should expect Hubble to be in rough shape," Grunsfeld said before Columbia lifted off. "We were led to expect possibly that the insulation on the outside ... would be very degraded and peeling off [due to the effects of atomic oxygen and solar particles]. They found a little of that on [SM-2 in 1997], so it was expected it would be much worse on 103. We were told that the solar arrays might be warped a bit more, so we went up on STS-103 armed with tools to cover handrails and repair insulation. What we found was that it was in almost the same shape as it was on STS-82: all of the damage that occurred had pretty much happened during its infancy and that ... the outside of the telescope was still pretty nice. The patches they put in place on [SM-2] were still in place and the handrails were uniformly in good shape." Grunsfeld's expectation to find Hubble in similarly good condition on STS-109 had paid off.

"MASTERFUL" PERFORMANCES

An astonishing sight greeted Grunsfeld and Linnehan as they floated out of Columbia's airlock at 6:37 am on 4 March to begin their first spacewalk: an unrestricted view of 'seas' of Saharan sand, together with stony plains, rock-strewn plateaux and interspersed with dark oases of life. The Shuttle was flying 580 km over a slumbering North Africa. "Oh, wow!" exulted Grunsfeld, now well into his element on his fourth Shuttle flight and – counting his SM-3A experience – his third spacewalk in total. "Beautiful view!"

Next, turning to the glistening silver cylinder standing on end at the rear of the payload bay, he introduced himself to 'Mr Hubble', announcing that Columbia's crew had come to "give you more power to see the planets, stars and the Universe". Without further ado, Grunsfeld set to work removing the old starboard solar array from the telescope and Linnehan stowed it in a carrier on the FSS pallet. Next, Grunsfeld installed the diode box to ensure that power from the soon-to-be-fitted new arrays flowed into the telescope's batteries, and not vice-versa.

Meanwhile, after some difficulty removing the new, $9.5-million array from its storage canister, Linnehan – his feet anchored into a 'cherry picker' on the end of the RMS – carefully brought it into position. Grunsfeld, keeping himself steady by hanging onto one of Hubble's handrails, then connected it, cranked its two halves open and began wiring it to the diode box. Next, in readiness for the PCU replacement during the third spacewalk on 6 March, they placed thermal covers over vulnerable components to ensure that the telescope's sensitive scientific instruments would not freeze or overheat.

"You guys did a superb job today! We enjoyed watching you work," Capcom Mario Runco told the entire crew after Grunsfeld and Linnehan returned to the safety of Columbia's middeck. "You made it look easy once again." The first spacewalk had lasted a minute over seven hours and its only real problem had been the failure of a telemetry sensor in Grunsfeld's suit, although the flight surgeon was still able to monitor his biomedical data. The problem was quickly corrected by resetting power to the suit's built-in communications system.

A similar procedure, albeit for Hubble's port-side array, was the order of business for Newman and Massimino early on 5 March. They were also charged with carrying out the replacement of the troublesome reaction wheel – which had actually, in the last few months, eclipsed the PCU changeout as the most critical task on STS-109 – during a spacewalk which lasted 7 hours and 16 minutes. As with so many other first-time spacewalkers, Massimino, a former professor of mechanical engineering before joining NASA's astronaut corps in 1996, was gripped with childlike wonder as he floated out of Columbia's airlock.

"This is incredible!" were his first words.

"Welcome to the wonderful world of spacewalking," replied Grunsfeld from his choreography station on the aft flight deck.

"Thank you, John," said Massimino. "Let's start with my first task." Most of his time was spent anchored to the end of the RMS, while veteran colleague Newman – making his fourth Shuttle flight and participating in his fourth spacewalk overall – scaled hand-over-hand up and down the side of the telescope. Using a pistol-grip power socket wrench, Newman made short work of unbolting the old solar array and quickly handed them to Massimino, who stored them on the FSS pallet for return to Earth. Next, he hauled the new array and its relay box up to Hubble for Newman to install.

At times, not surprisingly, Massimino seemed to need reassurance from his veteran colleagues, particularly when moving the new array up to the telescope. "You're gonna be fine," Newman told him. "Keep it coming. Looking good. Very nice. Take a big breath and relax. You're doing a great job."

After the pair had wired the port-side array into its diode box, Massimino removed the troublesome reaction wheel and carried it down to the payload bay, where Newman handed him a replacement which he promptly fitted. Initial validation tests performed by the Space Telescope Operations Control Center in Greenbelt, Maryland, indicated that both the new array and new reaction wheel were functioning perfectly. To wrap up their marathon excursion, the two men set up foot restraints and removed a thermal cover in anticipation of Grunsfeld and Linnehan's challenging PCU operation the next day.

The most challenging spacewalk of the mission began with a water leak from John Grunsfeld's $12-million suit; after first pondering whether to postpone the excursion for 24 hours, mission managers directed him to use one of the other, drier ensembles. It was subsequently decided, said EVA coordinator Dana Weigel, that the leaky suit would not be used again during the mission. The suit swap meant that he and Linnehan actually departed Columbia's airlock two hours later than intended at 8:28 am on 6 March and quickly set to work setting up their tools and tethers.

"A kind of late, but hopefully powerful start," Grunsfeld said as he ventured outside. His first task to fit thermal covers on the diode boxes of the new solar arrays and deploy the thermal shields set up by Newman and Massimino the previous day. Linnehan, his feet again fixed to the end of the RMS, then began the process of disconnecting Hubble's six batteries, effectively cutting off power to the telescope at 9:37 am. This gave the two men, at most, eight hours to replace the refrigerator-sized PCU.

Complicating matters was that it was encircled by bulky bundles of cables, which made the 36 electrical connectors incredibly difficult to reach. In fact, they were spaced so tightly that Grunsfeld and Linnehan had to use a wrench, rather than their gloved hands, to remove them. "What makes it difficult is, as you're facing the PCU, those connectors are on the left-hand side: they're not staring right at you," said Bryan Austin. "That's on the side that [the equipment] bay door is hinged. For the suited crewman to reach his hand in there, he's pretty much reaching in there blind. I kind of equate it to changing sparkplugs on your car. There's always those sparkplugs down there where you can't see real well. You've just gotta go down and *feel* and make sure you're oriented such that you're unscrewing it without a lot of offset force." The astronauts quickly set to work opening and latching the doors to the Number 4 equipment bay and Linnehan started at the top and worked down through the double rows of cables.

"The cables are pretty thick, John," he said as he peered into the open equipment bay, located a third of the way up the four-storey telescope.

"Yeah, I can see that," replied Grunsfeld. Undeterred, they methodically disengaged the connectors one at a time and fastened them temporarily to a cable caddy. Several rubber loops on the caddy broke, forcing them to tuck a few connectors behind the nearest wire harnesses, while still others were stiff and difficult to unfasten. Fortunately, they managed to keep close to the timeline and it took them barely four hours to unplug each of the connectors, remove the old unit and substitute it for a new version.

"These are not our typical, fully EVA-friendly connectors with the big 'wing tabs' on them that make it convenient for a suited crewman to manipulate," said Austin. "We've worked really hard to have a special tool that we're going to use to get a good grip on these and get through this." Had Grunsfeld and Linnehan been unable to complete the full procedure, Hubble could have been left safely overnight, as long as at least 11 connectors to the PCU were in place. Fortunately, the operation proceeded without a hitch and by 2:02 pm the telescope came back to life.

"You did it, buddy! You did it!" radioed a jubilant Linnehan with a laugh. "Good job." Partial cutting of power had been done during SM-1 to install

As Nancy Currie operates the RMS, two of her colleagues labour outside in the payload bay on the towering Hubble Space Telescope.

corrective optics and again on SM-3A to fit the new brain, but neither went so far as the SM-3B requirement to have the *entire telescope* shut down for a period of several hours. Mario Runco radioed the welcome news that all seven astronauts were waiting for: "Columbia, Houston, with a post-operative report. We have a heartbeat!" At a press conference later that day, Anne Kinney called their work a "dramatic and masterful performance".

It was a triumphant end to a series of problems with the old PCU that spanned nine years. In fact, a simple loose screw had impeded its ability to satisfactorily route electricity from the solar arrays to the instruments, control systems and batteries. Without the surgery, it would have been impossible for Hubble to run more than one of its instruments simultaneously. "As with any beloved relative, you're worried about sending them in for bypass surgery or even a heart transplant," said the telescope's chief scientist David Leckrone, "but you realise the risk of *not* doing it is severe."

For John Grunsfeld, who half-joked before the mission that failure would mean he could never show his face at meetings of the American Astronomical Society again, it was perhaps the pinnacle of his astronaut career. The following year, within months of the STS-107 disaster, he was transferred to NASA's Washington headquarters to become the agency's Chief Scientist. His journal notes of the critical PCU replacement highlight the intense levels of "concentration, patience and a little bit of skill" that he and Linnehan demonstrated during their 6-hour 48-minute spacewalk.

"On board Columbia," Grunsfeld wrote of that evening's activities within the safe confines of the Shuttle's middeck, "I went to sleep satisfied I did an honest day's work, and very tired."

A MORE CAPABLE HUBBLE

"Beautiful day for a spacewalk," chirped Jim Newman as he entered the payload bay on the morning of 7 March to begin upgrading the observatory in earnest. He and Massimino quickly set to work opening an equipment bay at Hubble's base and pulled out the Faint Object Camera (FOC), the last 'original' instrument still in use since April 1990. It had been out of action since 1997 and was the last instrument to rely upon the corrective optics 'bench' fitted during SM-1. That bench had already been stowed and was due for removal during the SM-4 mission in April 2004.

Sadly, at the time of writing, it seems unlikely that SM-4 will ever take place ...

Filling the gap left by the FOC – and promising to significantly enhance Hubble's capabilities – was the 395-kg, telephone-box-sized ACS. Riding the cherry picker on the RMS, Newman first removed the old camera and temporarily attached it to Columbia's payload bay wall. Nearby, Massimino partially entered the telescope to install a cooling system cable harness that would later be hooked up the new NICMOS cryocooler. This had forced Grunsfeld to admit some good-natured jealousy of his rookie colleague before the mission.

"His head will basically be at the soul of Hubble because this is where all the light from the telescope comes through into the scientific instruments," he told an interviewer. "The sad thing is, the cover on the telescope will be closed. For me, as an astronomer, *that* would be a very exciting place to be!"

Massimino, however, had little time to ponder where he was and quickly rejoined Newman to slot the ACS into its new berth.

"Okay, you're right over the lip of the telescope by about four inches," he radioed to Newman. "Keep it coming. Come just a little left, a half an inch to the left. That's good."

"I think we're going in," replied Newman, holding the new camera by a pair of handrails affixed to its outermost edge.

"Looks real good," came the guiding words of Massimino. "Nice and smooth. You've got about one inch to go, maybe two inches. There you go. You got it." The two men then bolted the ACS firmly into place with a power socket wrench. Not surprisingly, the camera team was overjoyed. "What the crew did was put the turbocharger on this telescope," ACS Deputy Principal Investigator Garth Illingworth of the University of California at Santa Cruz said of the new camera, which was expected to cover twice as much sky with twice the clarity and four times the speed of WFPC-2.

Its sensitivity was so great, said Hubble scientist Holland Ford of the Johns Hopkins University before the flight, that it would be able to discern two fireflies positioned 2 metres apart in Tokyo – from a vantage point in *Washington*!

Among the most exotic applications of ACS is the photographic detection of planets *outside* our Solar System. Although so-called 'extrasolar' planets had been identified through indirect means, research scientist David Golimowski of the Johns Hopkins University expressed optimism that they could be detected around several nearby stars. "It's a long shot," he admitted, "[but] if all the planets align just right and if we have very good fortune in the telescope performance and also in the way we plan the observations, there is a possibility that we could image a planet."

Golimowski's team was particularly hunting for reasonably large, Jupiter-sized worlds, which he estimated to be around a billion times fainter than their host stars. In order to pick out such dim objects, the ACS was equipped with a coronagraph to blot out most of the starlight so that it could record much less intense reflected light from nearby objects. Candidate hosts included nearby Alpha Centauri, for which at least 400 minutes of observing time was already booked for planet-hunting before the ACS had even been launched, and a brown dwarf called Gliese 229B.

If the latter, which orbits a 'normal' star, does indeed harbour a planetary world, Golimowski expected it to 'wobble' slightly under the gravitational tug of its smaller companion. In addition to searching for planets outside our Solar System, Garth Illingworth added that the ACS would help astrophysicists to investigate the history of the cosmos by exploring the fringes of the 'dark age' in the Universe.

"In the next decade, *that's* the place where astronomers are going to be looking to see when the first light of the Universe was actually happening," said Wendy Freedman of the Carnegie Observatories in Pasadena. "When did the Universe light up with stars? When did the galaxies actually form? With the advanced camera, I think we'll get our first glimpse of some early galaxies." Added Holland Ford: "We will be able to enter the 'twilight zone', when galaxies were just beginning to form out of the blackness following the cooling of the Universe from the Big Bang."

After its 10-week checkout, it was anticipated that the ACS would detect more faint objects during its first 18 months than had been picked up by *all* previous Hubble instruments. Ford cited the vast amount of data from the 'Deep Field' observations made by the telescope in December 1995: for 10 days that month, it had been aimed at a patch of sky free of nearby galaxies – roughly the size of a grain of rice held at arm's length – and WFPC-2 shot 342 images which allowed astronomers to identify 1,500 galaxies dating to within a billion years of the Big Bang.

With the ACS, it was expected to do similar 'Deep Field' observations in a fifth of the time. Together with the revived NICMOS, it was hoped that the camera could play a part in identifying exploded stars known as 'type 1A supernovae', the light from which strongly suggests that the Universe's rate of expansion is *increasing* rather than decreasing. The nature of the mysterious force powering that acceleration, known as 'dark energy', has been described by David Leckrone as possibly "the most important question in the physical sciences today."

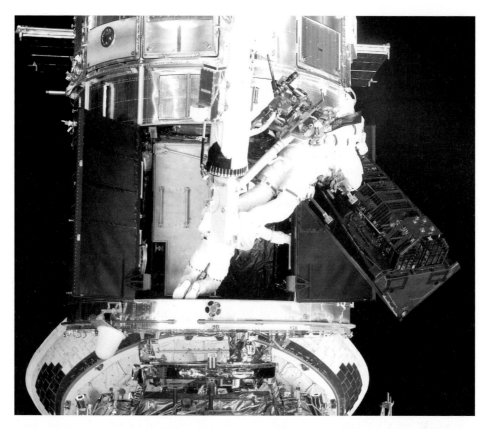

Jim Newman removes the Faint Object Camera (FOC) from Hubble to make room for the soon-to-be-fitted Advanced Camera for Surveys (ACS). In the background, Mike Massimino waits inside the telescope's equipment bay.

Celestial objects of particular fascination for ACS researchers include intrinsically bright, but very distant quasars at the very edge of the detectable Universe, which are thought to have supermassive, 'feeding' black holes at their centres. With the ability of the coronagraph to block out the bright cores of the quasars, it was hoped to gain more insightful views of the physical conditions in their outermost reaches. "We're looking forward to taking images of quasars," said Holland Ford, "and seeing the structures that surround [them] much better with ACS' higher resolution and higher sensitivity."

In its first three years of operations, the ACS has captured a stunning collision between a pair of spiral galaxies known as 'the Mice' – which may illustrate what could happen if our Milky Way collides with the Andromeda galaxy several billion years into the future – and revealed vivid colours and glowing 'ridges' of gas in a stellar nursery within the Swan Nebula. Closer to home, it also observed a craggy-looking 'mountaintop' of cold gas and dust called the Cone Nebula.

On 7 March 2002, however, after successfully installing the ACS, Newman and

Massimino had scarcely time to respond to the congratulations from their colleagues on Columbia's flight deck, before plunging straight into their next task: preparing for the fifth spacewalk by installing an electronics module to power the NICMOS cryocooler. This was duly mounted on the 'floor' of Hubble's aft equipment bay. After finally closing the doors to the bay, the two men removed thermal and light shields that had been used earlier to support the PCU replacement.

Eighty percent of STS-109's spacewalks had been completed and yet, said Digger Carey, there was no opportunity for the crew to rest on their laurels. "These are five long days," he had told an interviewer before Columbia lifted off. "One of the main challenges on the mission is to keep that mental focus: to wake up each morning and say, 'Okay, this is the big day. Got to get this done today. Got to get through the day and do everything perfect.' And then waking up the next morning and having that same attitude."

There was simply no option, Carey added, to get through the last EVA by coasting; nor was the hard part of the mission 'over' as soon as Hubble was redeployed, or as soon as the crew got through re-entry or even after wheels-stop on the runway. He remembered some advice from Nancy Currie, who had been through three missions before STS-109: "She said, 'The only time I feel like I can relax is when I'm outside the orbiter and I'm in that astronaut bus to go back to crew quarters. *Then* you can relax!'"

It was, therefore, with renewed vigour that the crew prepared for Grunsfeld and Linnehan's spacewalk on 8 March to install the cryocooler and, hopefully, reactivate the $110-million NICMOS. This instrument had been sorely missed by the astronomical community since January 1999, "especially by those of us who study protoplanetary and debris disks around young stars", said Ray Jayawardhana of the University of California at Berkeley. Although ground-based telescopes were catching up to Hubble's resolution, thanks to new technologies like 'adaptive optics', the return of NICMOS was still awaited with great excitement.

Mark McCaughrean of the Astrophysical Institute in Potsdam, Germany, for example, had used the instrument before it shut down and felt it "definitely has a few tricks up its sleeves", including the capability to observe celestial objects in certain portions of the infrared band. Cool stars and young giant planets, he said, contain steam in their atmospheres, which has proved difficult to study from the ground because of the intervention of our atmosphere. It was the ability of NICMOS to conduct infrared observations from *above* the atmosphere that made it so valuable.

The problem with NICMOS was its cooling system – essentially a giant thermos flask containing a block of solid nitrogen – which had sublimated away prematurely when an internal short brought it into contact with the surrounding structure. The instrument required cooling to remove its own excess heat and thus prevent this heat from interfering with the sensitive infrared observations. Columbia's crew carried a new mechanical chiller, developed jointly between NASA and the US Air Force, into orbit to circulate ammonia-cooled neon through NICMOS' imager. This was expected to keep its infrared detectors at an acceptable temperature of 60 Kelvin.

"Kind of like installing an external air conditioner in a house for the first time," was how Grunsfeld described fitting the tricky cryocooler. The device itself was experimental in nature, meaning that – although NICMOS' revival was highly desirable from a scientific standpoint – it was at the bottom of the list of priorities for STS-109. It comprised a series of ultra-high-speed microturbines, spinning at nearly half a million revolutions per minute, which had been successfully demonstrated four years earlier as part of the Hubble Optical Systems Test (HOST) experiment on STS-95. This was to be an operational test of the unit.

The installation of the $21-million cryocooler got underway at 8:46 am on 8 March, with Linnehan riding the RMS to firstly open the aft shroud doors on the telescope and then pick up the new device from its storage location in Columbia's payload bay. "I felt like we were opening the doors to a sacred shrine," Grunsfeld later wrote in his journal, "going inside the area where the scientific instruments on Hubble live. In our training, we were taught to have the utmost respect for the delicacy of the instruments and to treat them with kid gloves. Inside the aft shroud, I tried to move as carefully as I could, even though I was in a bulky and clumsy spacesuit. The interior of the telescope is as clean and pristine as it was when launched. After removing a serpentine vent hose that was originally used for venting gas from the old solid nitrogen cooler on NICMOS, we installed the cryogenic cooler itself. It looked a bit odd to take the pristine, almost spartan interior of Hubble and add a box covered with cables and hoses and valves."

The men laboured through orbital daylight and darkness to fit it and connect cables from the electronics support module – installed the previous day by Newman and Massimino – to the new 136-kg cryocooler. Their next step was to attach a 4-m-long radiator to the outside of the telescope to expel NICMOS' waste heat into space when fully operational. At first, this proved problematic. "It wouldn't go on!" recalled Grunsfeld. "We pushed and pushed, but try as we could, it didn't seem to line up. Rick and I took it off Hubble and realigned it by moving the latches a bit. This time, the alignment looked good, but still we couldn't push it on. With some significant effort, Rick got his handrail on and I took one latch at a time and pressed it into the telescope as hard as I could. After a couple of tries at my maximum effort, I squeezed the first, then second latch on the handrails. A small success!" Linnehan, stationed 'underneath' Hubble, then fed tethered wires from the radiator through a hole in the base of the telescope to Grunsfeld, who hooked them up to NICMOS itself.

"I felt as if I was ice-fishing," wrote Grunsfeld, "although I could actually see Rick through the hole. He caught the tether, attached it to a long set of cables and cooling lines and I pulled it through the hole right into where the science instruments live. It was like a giant boa constrictor and much stiffer than we had seen in training. Rick joined me and we began the process of hooking up electronics connectors and the ammonia cooling line to the [cryocooler]." Their task completed, the men set about cleaning up Columbia's payload bay and stowing their tools.

"Good job, you guys," enthused Currie from the aft flight deck.

The final minutes of the spacewalk also gave Grunsfeld an opportunity to ride the RMS, which had been deftly manipulated from within the flight deck by Currie and Altman over the past five days. "What a view I had," the veteran astronaut exulted, "[with] Columbia below, the bright blue Earth above and the Hubble Space

Telescope on my side. It seemed as if time was standing still: it was so touching a moment for me. My last activity was to remove a protective cover from an antenna on the bottom of the telescope.

"At the end of five spacewalks to improve the telescope," he continued sadly, "I gave Hubble a final small tap goodbye and wished it well on its journey of discovery. It is likely I will never see the Hubble Space Telescope again, but I have been touched by its magic and changed forever." The five outings by Grunsfeld, Linnehan, Newman and Massimino had set a new single-flight Shuttle record of 35 hours and 55 minutes overall. Across the board, in all four servicing missions since 1993, astronauts had made 18 spacewalks and spent more than 129 hours repairing the telescope.

HOMEWARD BOUND

"Good luck, Mr Hubble," came the call from Columbia's flight deck at 10:04 am on 9 March, as seven pairs of eyes – some a little teary – watched their companion of the last week depart to continue its journey of astronomical discovery. Shortly after Currie released the silvery observatory from the RMS, Altman and Carey fired the OMS engines to create a safe separation distance between the two spacecraft.

As the telescope drifted serenely away into the inky blackness, high above the Atlantic Ocean and the French Guadeloupe islands, Capcom Mario Runco – himself a veteran of three spaceflights – was overwhelmed; so much, in fact, that he had donned a black-tie tuxedo for the occasion. Echoing the words of the crew, he wished Hubble well. At the time, it was confidently expected that one further servicing mission in April 2004 should be enough to keep it fully operational until the next-generation James Webb Space Telescope (JWST) enters service around 2011.

Circumstances in the next year, of course, changed Hubble's prospects quite considerably and by 2004 NASA had already taken the decision not to risk any more Shuttle crews trekking up to the vaunted observatory. In the wake of the STS-107 tragedy, it was decided instead to devote their resources to completing the International Space Station and retiring the three remaining Shuttles by 2010. The abandonment of Hubble has been greeted by strong criticism from Congress, astronomers and the general public across the globe; clearly, the telescope has opened people's eyes and minds to the Universe around them as never before.

At the time of writing, Hubble continues to operate well and should remain to do so for the near future. Yet, judging from past problems with gyroscopes, reaction wheels and the likelihood of increasingly more frequent glitches as the observatory gets older, it is doubtful that it will still be operational in 2011 when JWST enters space. Hubble's eventual demise will probably come as a shower of burning debris over the Indian Ocean and humanity will have lost a vital tool that will be difficult to replace.

On board Columbia for what would be her last fully successful space mission, the astronauts reported on the morning of 11 March that a latch holding one of the telescope's old solar arrays in place had unexpectedly moved slightly in the payload bay. After some analysis, it was determined that the array would pose no safety hazard during re-entry or landing. Fears of problems with the troublesome freon coolant loop as the Shuttle sped back to Earth the following day also proved unfounded and Altman brought Columbia through the pre-dawn darkness onto KSC's runway at 9:32 am. Little did he know that the veteran orbiter had just made her last successful landing.

"JUST A RESEARCH MISSION"

It was in the spring of 2001 that the STS-107 mission first sparked my interest: for two reasons. One was the fact that its crew was so comparatively inexperienced – only three veteran astronauts among the seven-strong team – and the other was that it was the first 'standalone', 16-day scientific research flight carried out by the Shuttle for almost three years. With most of the missions around it exclusively committed to hauling hardware into orbit for the International Space Station, Commander Rick Husband's flight seemed strangely out of place.

Four years later, I am glad that I took an interest in STS-107. I already intended to write about the kind of research that Husband and his crew would be doing during their fortnight aloft, but the reaction of one space magazine surprised me. "Why do you want to write about *that* one?" they asked. "It's *just* a research mission." It was true: Columbia's 28th flight would not feature any 'glamorous' spacewalks, but my curiosity as to *why* NASA still needed research missions now that the International Space Station was operational led me to arrange an interview with Husband himself.

We spoke for half an hour on 3 April 2001, over the telephone from his Houston office, and I could scarcely have imagined that in less than two years he would be dead; killed, along with his crew, in the second Shuttle disaster as they hypersonically knifed their way back through Earth's atmosphere after an otherwise highly successful mission. A decade and a half after the painful loss of Challenger, the reusable spacecraft had never achieved its 1970s goal of flying once every two weeks, but nonetheless its record as one of the world's most successful launch vehicles seemed assured.

Admittedly, the 'controlled explosion' of liftoff remained one of the most hazardous portions of any mission and even as recently as July 1999, seconds after STS-93 left the pad, everyone was reminded that it was still an experimental, phenomenally complex and temperamental ship, although one with such redundancy that it was able to push on to orbit regardless. The end of a mission, however, was considered quite the opposite, even though the thermal extremes of a high-speed atmospheric re-entry were well known and the consequences of a major failure of the protective heat-resistant tiles and blankets did not bear thinking about.

"I really don't enjoy launches," said Mike Anderson, one of Husband's crewmates on the STS-107 mission, before blasting off, "[but] entries are a little bit better. It's a little quieter ... not quite as violent [and] you can enjoy it a little bit. But still, for me on this flight's entry, I'm just going to sit down in my seat and hopefully reflect on the 16 days on orbit that we've had ... just anxious to get back to Earth and give the scientists all their research results. I'll be happy to have the flight behind us."

Both Anderson and Husband had flown once before STS-107; the former on a docking flight to the Mir space station, the latter on an assembly mission to the International Space Station. When Husband was named to the Commander's position in December 2000, it came as a surprise to many, as it would only be his second flight: to gain experience, most Commanders typically flew two or even three times as a Pilot before being promoted to the coveted left-hand seat in the Shuttle's cockpit.

Some observers speculate that Husband was simply one of NASA's best pilots, having turned in an admirable performance on his STS-96 flight in mid-1999, but combining it with a humble low profile. In Husband's mind, however, he owed his position exclusively to "God's blessing and provision", which had already gained him the place as an astronaut that he had dreamed about since boyhood. He was joined on STS-107 by one of the most culturally and academically diverse Shuttle crews ever selected: an Israeli Jew, an Indian woman, an African-American man, two medical doctors and a former US Navy test pilot.

The awesome challenge facing any Commander does not just encompass the mission, but also requires the ability to bring together a group of highly qualified, highly motivated individuals and knit them into a virtual 'mini-family'. One of Husband's methods was to take his crew, in August 2001, on a team-building excursion sponsored by the National Outdoor Leadership School (NOLS). "This was a course where ... the seven of us went out into the mountains in Wyoming with two instructors from [NOLS] and backpacked with 60-pound packs for nine nights and 10 days out there," he told an interviewer before lifting off on STS-107.

"We got to see some incredible scenery. We got to learn a lot about how each of us, as individuals, deal with the kind of situations that they put us into. It's a physical challenge with the backpacks and the walking up and down. It's also a challenge learning how to keep track of all your equipment personally [and] then learning to work together, pulling together and learning more about each other, so that when you come back ... you know each other's strengths and weaknesses and so you can maximise that during the rest of your training flow."

The trek took them through dense areas of the Shoshone and Bridger-Teton national forests, which feature treeless peaks in the 3,600-metre range and high mountain lakes. Their guides led the expedition for the first few days, then the seven astronauts elected a new leader each day and evaluated each other's performance at nightfall. Despite the serious, team-building nature of the excursion, the crew nevertheless treated it with aplomb and humour, even phoning Houston from the 3,960-metre Wind River Peak to inform their fellow astronauts that they had "landed".

Said John Kanengieter, one of the two NOLS guides: "They have a terrific sense of humour and they're really good fun. These guys are all business, for sure, but man, they have a good time doing it!"

Nevertheless, Husband's team *was* inexperienced for a Shuttle crew, making them the butt of good-natured jests, some of which they invented themselves. "We joke around sometimes, saying that before Jerry Ross flew his seventh mission, he had *six flights* to his credit," added Husband. "Our crew had only *half* that amount of flight experience, but after our flight we'll have caught up with him and then some!" Added STS-107 Pilot Willie McCool: "He's got us beat by a factor of two, [but] when we come back, we'll have 10 flights among all of us and we'll jump ahead of Jerry."

AN ISRAELI IN SPACE

McCool, together with Mission Specialists Dave Brown and Laurel Clark – both medical doctors, as well as active-duty US Navy officers – were making their first flights, as was Payload Specialist Ilan Ramon, the first Israeli spacefarer. Rounding out the seven-member crew was STS-87 veteran Kalpana Chawla.

"Personally, I think it's very peculiar to be the first Israeli up in space," Ramon said before the mission, "but my background is kind of a symbol of a lot of other Israelis' backgrounds. My mother is a Holocaust survivor; she was in Auschwitz [and] my father fought for the independence of Israel. I was born in Israel and I'm kind of the proof for them, and for the whole Israeli people, that whatever we fought for and we've been going through in the last century is coming true. [During training I talked] to a lot of Holocaust survivors ... and when you talk to these people, who are pretty old today, and tell them that you're going to [go into] space, they look at you as a dream that they could have never dreamed of. So it's very exciting for me to be able to fulfil their dream that they wouldn't dare to dream."

Yet Ramon's place on the crew was not a purely political stunt: he had a number of genuine scientific experiments to conduct during Columbia's 16-day mission. The seed of the idea to send an Israeli up on the Shuttle dates back to December 1995, when then-US President Bill Clinton and then-Israeli Prime Minister Shimon Peres agreed "to proceed with space-based experiments in sustainable water use and environmental protection". As part of the deal, an Israeli astronaut was selected to operate a multispectral camera to investigate the migration of airborne dust from the Sahara and its impact on global climatic change.

Other Israeli experiments on the flight would examine the metabolic processes responsible for weakening astronauts' bones during weightlessness, which could provide insights into new treatments for osteoporosis. Selected in April 1997, Ramon and his backup Yitzhak May flew to JSC to begin initial physical and psychological screening and commenced their official training in the summer of the following year. A Colonel in the Israeli Air Force and veteran fighter pilot of the 1973 Yom Kippur War, he reportedly also took part in an airstrike that destroyed the Osirak Iraqi nuclear reactor near Baghdad. In addition, he was a qualified electrical and computer engineer.

His primary focus was the Mediterranean Israeli Dust Experiment (MEIDEX), which was essentially a Xybion radiometric camera and wide-field-of-view video camera mounted on a pallet at the rear end of Columbia's payload bay. Operating at

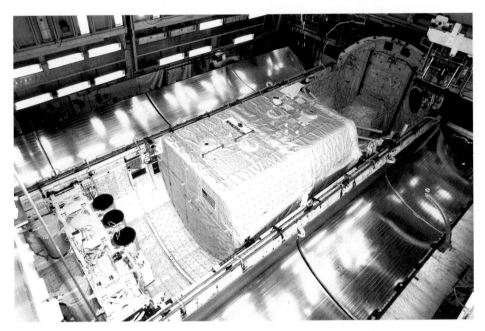

The FREESTAR pallet (far left) and Spacehab Research Double Module are pictured on Columbia in the OPF before the orbiter's payload bay doors were closed.

six spectral bands from ultraviolet to infrared, it would be used to investigate the geographical variation of the optical, physical and chemical properties of desert aerosol particles over the Mediterranean and Atlantic Saharan regions. "These particles", said Ramon, "have [an] amazing impact on things like global warming and precipitation, so this experiment is trying to monitor the sources and sinks of the dust.

"Secondly, [it analyses] all the characteristics of the dust particles to try to understand how it affects the global warming and precipitation. Taking these measurements, together with a measurement that we'll be taking from an airplane flying on the same footprint that we are looking at, and together with a lot of ground station measurements, scientists are going to analyse this data after flight and try to better understand the impact of the dust on our global climate." Additional targets included making sea-surface and desert reflectivity observations and monitoring strange, high-altitude electrical phenomena known as 'sprites'.

"[Sprites] are kind of 'ghostie lightning' that are going from above thunderstorms up to the [ionosphere], up to about 100 kilometres," explained Ramon. "These phenomena [are] pretty new for the scientists and our camera is actually the best way to monitor and try to catch these sprites." It would mark the first time that observations of this nature had been conducted simultaneously from spacecraft-based, aircraft-based and ground-based sensors. Typically, Ramon would operate the camera from a laptop computer, watching images on a television-type display on Columbia's aft flight deck.

Despite his multitude of tasks, Israel's first spacefarer had no intention of forgetting his heritage. Packed away with his personal belongings was a pencil drawing entitled 'Moonscape', sketched by a 14-year-old Czechoslovakian Jew named Petr Ginz, who perished at Auschwitz in 1944. "The drawing is as he imagined the Earth looking from the Moon," explained Ramon, who had requested it from the Holocaust-era Yad Vashem art museum. "This drawing was made long before anyone dreamed about actually going to the Moon. I'm taking [it] along to symbolise the winning spirit of this boy."

Although he described himself as a 'secular Jew', Ramon made a point of taking kosher food into orbit and intended to observe the three Sabbaths he would spend aloft. Before launch, it was uncertain how these could be precisely timed: the Jewish Sabbath runs from Friday's sundown until Saturday's sundown, but with 16 sunrises and sunsets during each 24-hour period posed a problem. Eventually, it was decided Ramon would stick to the way sundowns were measured on Earth. As events would transpire, he was so busy that he only had chance to partly observe a single Sabbath.

'Partly', that is, because *that* Sabbath was the last of the STS-107 mission: it began as Ramon and his crewmates were packing away their equipment for re-entry and ended just a few hours after Columbia disintegrated 60 kilometres above Texas ...

AN IMPORTANT MISSION FOR SCIENCE

Of course, MEIDEX and Ramon's presence were high-profile elements of the mission, but the main purpose of STS-107 was to fill a genuine need of the life and microgravity research communities to have a flight opportunity for their experiments while the fledgling International Space Station was being built. Since December 1998, when the STS-88 crew hauled the Unity node up to the Russian Zarya control module and kicked-off the multi-billion-dollar project, all but four of 20 Shuttle missions had been devoted to assembling and supplying the outpost.

Of those four 'outsiders', two had serviced the Hubble Space Telescope, one deployed Chandra and another extensively radar-mapped Earth's surface from orbit. Not since the autumn of 1998 had a Shuttle mission been exclusively dedicated to science and this caused concern not only to researchers, but also to Congress, who were aware that the United States could lose its global 'lead' in the microgravity arena. "We can't expect the scientific community to remain engaged if researchers do not see hope that there will be research flight opportunities on a regular basis," Congressman Dana Rohrabacher told a March 2000 hearing.

There were other benefits of carrying experiments on two-week Shuttle flights rather than six-month station expeditions. "In our case," said Rick Husband, "there are some experiments that can be designed for shorter duration so that [scientists] can send these up, they can get the results back, they can do some analysis and then they can turn them around and try to go fly again. On the space station, they may have some experiments that are designed for a longer duration, to take a look at a process over a longer period of time than ... you can achieve on the Shuttle."

That STS-107 was inextricably linked to the International Space Station had never been in doubt. In fact, John Charles, NASA's mission scientist for biological and physical research, had long referred to it as doing "simulated space station science ... although the science itself stands on its own right." He added that many of the life and physical science studies allocated to the mission would have an overwhelming emphasis on improving crew health and safety in readiness for extended stays in low-Earth orbit.

Congressman Dave Weldon, a Republican for the Shuttle's home state of Florida and colleague of Rohrabacher, agreed that research missions like STS-107 were essential for demonstrating scientific experiments before committing them, on a longer-term basis, to the station. Unfortunately, for a time, this government backing was crippled by a mission called 'Triana' which, bizarrely, had been conceived in a *dream* by then-US Vice President Al Gore. Named after Rodrigo de Triana – the lookout on Christopher Columbus' first voyage to the New World – it was billed as a 21st-century Earth-gazing Internet lookout with questionable scientific merits.

Gore's 'challenge' to NASA was to build a relatively inexpensive satellite for between $25 and $50 million that would broadcast real-time, 24-hour views of our home planet over the Internet. However, the simple, camera-equipped satellite quickly acquired several other instruments and cost *four times* as much as Gore had estimated. It received negative assessments from NASA's inspector-general and an angry Congressional review of its scientific worthiness. By the time it was being built, a place had already been earmarked for Triana on STS-107. This annoyed the microgravity community, who felt their experiments were being hampered by weight limitations imposed by the satellite.

As the National Academy of Sciences debated the merits of Triana, it soon became clear that it could not be finished in time for Columbia's launch – then targeted for the summer of 2001 – and it was pulled from the flight. In its place was a facility with a mouthful of an acronym: the Fast Reaction Experiments Enabling Science, Technology, Applications and Research (FREESTAR), which would house six high-priority instruments, including the Israeli dust-and-sprite-watching camera, on a pallet at the rear of the payload bay.

A PURPOSE-BUILT RESEARCH MODULE

The experiments that required to operate within a pressurised environment and be tended by the astronauts in a 'hands-on' capacity faced a problem. The two venerable Spacelab modules, which had flown numerous times and supported hundreds of experiments across a variety of disciplines since November 1983, had been officially retired following the 16-day Neurolab mission in April 1998. In its place, NASA had hatched a $47-million deal with a company known as Spacehab Inc., which had, since 1984, been developing its own pressurised research modules for the Shuttle.

In physical appearance, Spacehab was not dissimilar to Spacelab, with the exception that it had a flat roof and thus was not fully cylindrical. Connected to the

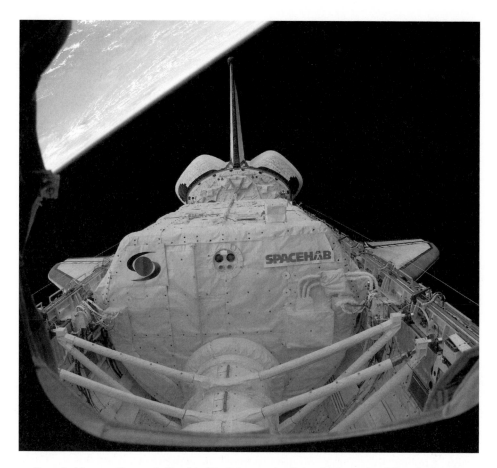

Spacehab's new Research Double Module seen in Columbia's payload bay during the 16-day mission. Note the dark-grey-coloured RCC panels lining the leading edges of the orbiter's wings.

Shuttle's crew cabin by a pressurised tunnel, it occupied the forward portion of the payload bay and virtually quadrupled the room available for conducting scientific experiments in the cramped middeck. The original module, which first flew in June 1993, was 3 m long and offered 28.3 m^3 of space, much of which was commercially 'sold' by Spacehab Inc. to experimenters who desired relatively easy access to microgravity.

Later adjustments to the company's contract with NASA led to the development of cargo-carrying Spacehab modules – including a 'double' version, in which one module was bolted back-to-back onto a structural-test article for trucking research hardware, equipment and consumables to and from the Mir space station and the International Space Station. In fact, both Rick Husband and Mike Anderson had flown with such cargo modules on their first missions. Inevitably, with the retirement of Spacelab and the need for a similarly large laboratory for standalone flights, the concept of a fully equipped Research Double Module (RDM) was explored in greater depth.

The new module would make its first flight on STS-107, but was much more than 'just' an experiment-carrying canister. "The [module]", said Husband before the mission, "is a double Spacehab that has had some additional equipment added to [its] aft portion. That primarily adds up to some environmental control systems that will remove moisture from the air back in the Spacehab. The other modules – the Logistics Double Modules that deliver supplies to [the] space station – don't have that humidity-removal system in there and so the orbiter is responsible for removing all the humidity. With this additional equipment back there, they figure that it will take some of the load off the orbiter and be able to better control the environment back there for people to work in. They've also added some additional power capability and data-handling so that it can accommodate more experiments and power more experiments as a result ... and then ship that data to the orbiter to then be sent to the ground or recorded onboard. It's a module that is much more specifically designed to be an on-orbit laboratory in the back of the orbiter['s] payload bay."

The end product, therefore, was a facility 6.1 m long, 4.3 m wide and 3.4 m high that provided a pressurised volume of 62 m^3 for over 4,000 kg of experiments. On its maiden flight, it would carry 3,400 kg of research equipment to support numerous scientific investigations. Its aft portion, said Husband, enabled a new interface with Columbia's uplink-and-downlink system to enable investigators to command their experiments from the ground and receive real-time data from them. For STS-107, Spacehab Inc. marketed 18% of the module's capacity – netting $22 million in revenue – while the remainder was reserved by NASA.

"It's roomy," Willie McCool said of the new module, "and it gives us, the operators, space to operate [and] run the payloads. So that's it in a nutshell. Room. Power. Cooling. Everything you need to make the experiments work."

Those experiments were both multidisciplinary and roughly divided into life and physical sciences. The former included studies of pulmonary and cardiovascular changes during rest and exercise and investigated bone cell activity. Dave Brown hoped the Canadian-provided bone cell experiments could yield insights into the future treatment of osteoporosis. "That's something that's a problem for everybody when we get older, particularly women," he said before launch. "It turned out [that] astronauts – when we go to space and no longer have the stress of gravity on our skeletons – lose calcium and so we're going to be studied for that. *We're* actually a very good model – a very accelerated model – for what happens to people over many years, so it's useful to study *us* to learn more about how to slow or prevent osteoporosis here on Earth. We also have quite a few [middeck-style] locker experiments that have bone cells in them, again, to study the same metabolism and why bone cells either gain or lose calcium." During the flight, the experiments required crew members to gulp down pills and inject fluids containing 'tracer' chemicals to indicate the rate at which calcium was being lost from their bones.

Elsewhere in the Spacehab module was an experiment provided by Michael Delp of Texas A&M University to investigate the impact of microgravity on blood vessels.

On Earth, the pull of gravity causes these vessels to constrict and thus prevent blood from collecting in the lower extremities; however, in space, they become less able to dilate or constrict and, over long periods of time, the circulatory system weakens. Upon returning to Earth, most astronauts feel dizzy when standing upright – a condition known as 'orthostatic intolerance' – and around 60% are unable to pass a 10-minute stand test without losing consciousness.

"Gravity pulls blood down to the feet normally [and] arteries resist that pull," explained Delp. "In microgravity, there's no weight bearing. The body responds to the lack of force by remodelling itself. Look at what happens if a weightlifter stops working out. If a muscle is no longer stressed, it loses mass." Delp's study involved a complement of rats, which were being flown on board Columbia to explore the extent to which their hind limbs grew thinner and weaker in response to pressure changes and chemical signals essential to their vascular health.

Earlier rodent experiments had shown that they react more rapidly than humans to space-induced physiological changes and, as a result, a 16-day mission for *them* was roughly equivalent to a six-month period in microgravity for *us*. They were housed in special enclosures in the Spacehab module and, it was planned, the blood vessels in their hind limb skeletal muscles would be analysed after Columbia's landing to track structural and genetic changes.

"There are similarities to what happens in microgravity and what happens in old age," said Delp. "When the elderly go to the emergency room, the reason is likely due to orthostatic intolerance, either directly or indirectly. They can't stay upright, and when they *do* go down, they injure themselves."

As well as volunteering for blood draws and urine tests in support of the life science experiments, the astronauts also put their names down as test subjects for Charles Czeisler's sleep study. Throughout the mission, they wore watch-sized 'actigraphs', which were essentially tiny accelerometers that measured wrist movements as part of efforts to track disturbances in their sleep–wake cycles. Previous experiments during the Neurolab mission had already highlighted the disruptive effect of 16 sunrises and sunsets during each 24-hour period on astronauts' ability to 'sleep' and 'wake' normally.

Known as 'circadian rhythms', these cycles are essentially daily repeating biological clocks, the disturbance of which can ultimately cause physical or mental impairment, interfere with an individual's concentration and potentially affecting the immune system. According to Laura Barger of Harvard Medical School and Brigham and Women's Hospital in Boston, Massachusetts, astronauts typically are unable to sleep well before a mission due to excitement and shifting their rest times to accommodate launch schedules, and in orbit generally sleep for no more than six hours per night.

Rick Linnehan, who flew on board Neurolab, once commented that, due to his immense workload with the life science experiments, most of his 16 days aloft seemed to roll into one and he slept far less. He was not alone.

"During Shuttle flights, the light-dark cycle is about 90 minutes long," said Barger. "Seeing a sunset and sunrise every 90 minutes can send potentially disruptive signals to the area of the brain that regulates circadian rhythmicity. Additionally, the lighting onboard the Shuttle might not be sufficiently intense to maintain circadian alignment. Consequently, sleep could be disturbed. If the astronaut sleeps one to two hours *less* per night, over a 16-day mission, that can add up to a 32-hour sleep deficit." Undoubtedly, such a deficit could have a detrimental effect on their performance.

Not only did STS-107 combine many scientific disciplines, it brought together many nations, with the European, Canadian, German and Japanese space agencies sponsoring a multitude of different experiments. Students from Australia, China, Israel, Japan, Liechtenstein and the United States also investigated the impact of spaceflight on fish, spiders, ants, silkworms, bees and inorganic crystals. "It's a humbling experience when you realise how many people work to put together a mission like this," Rick Husband said before Columbia's launch, "and to be able to go to various places and visit the people who have worked very hard on all these experiments."

Even with more than two weeks aloft, it would have been virtually impossible to complete the crew's overflowing plate of scientific objectives without a dual-shift system operating around-the-clock. The Red Team consisted of Husband, Clark, Ramon and Chawla, while their Blue counterparts were McCool, Anderson and Brown. "You might say, if we have seven people on one shift, they could just divvy up the experiments and, hence, you should be able to do the same number of things," Chawla said before the mission. "The issue is that, on our orbiter, there are lots of attitude requirements. The orbiter should be in a certain attitude to do the ozone measurements [and] in a different attitude to the dust measurements. In a free-drift attitude – meaning that no jets should be firing and it's just 'drifting' – to do some of our very microgravity-sensitive experiments. One of the [Combustion Module] experiments needs a very quiescent environment ... so you need to take advantage of the *whole day*. It really helps to use the crew much more efficiently by doing that."

Also, the fact that at least three crew members were asleep during most periods of the day or night, made Columbia herself and the Spacehab module – which, eitherway, could only comfortably accommodate four people – considerably more roomy. "It gives the waking crew members a little more space to work," said Mike Anderson, "[and] a little bit more elbow room to do their job." He was perhaps the person most intimately involved with the experiments – having worked on them for two years before STS-107 finally set off – and, as such, was named as its Payload Commander.

"When you have a complex mission like this," Anderson said, "you need a point of contact for answering all the crew questions ... to help make the decisions about things and try to find the best way to make this mission a success. Who do you train for which payload? How do you take advantage of each crew member's strengths to assign them to a payload that's most appropriate for them? How do you choreograph the operations on-orbit? My job ... was to pull this mission together and that we were going to get the best science we could out of this 16-day flight."

Like so many previous science missions, therefore, STS-107 had two Commanders: Rick Husband had overall responsibility for the safety of the flight and its crew, while Mike Anderson took care of the experiments crammed into Columbia's payload bay. After launch, the pace was expected to be hectic as both Commanders busied themselves with their respective duties. "Mike will be the 'post-insertion guru'," Husband said before the flight. "He's the one who's ... in charge of running everything on the middeck and making sure everything gets set up, whereas I'll be running things on the flight deck."

A LONG WAIT

Their launch, unfortunately, was a long time coming. When Husband and McCool were assigned in December 2000, eight weeks after their five colleagues, they were due to fly in August 2001. Columbia would then stage Scott Altman's Hubble servicing mission in November of the same year. A rejuggling of priorities eventually led to STS-109 leapfrogging Husband's flight in the launch pecking order; ultimately, neither mission got off the ground that year. In fact, when I spoke to Husband in April 2001, he did not expect STS-107 to fly until the following spring.

Further delays in getting Columbia ready for her first trip into space in more than two years, together with postponements caused by the need to further train the STS-109 crew for a reaction wheel replacement, pushed the Hubble mission into March 2002 and Husband's crew eventually received a firm launch target of 19 July. Then, only five weeks before they were scheduled to depart for KSC, and with the Shuttle herself almost ready to roll to the VAB for stacking, a potentially serious problem was found deep within the main propulsion systems of sister ships Discovery and Atlantis.

Six cracks, each measuring around 2.5 mm, had turned up on metal liners which fitted inside each main engine's plumbing and helped cryogenic propellants flow past accordion-shaped bellows. As the liners do not hold pressure, NASA was quick to stress that they did not imply leaking liquid oxygen or hydrogen, but spokesman James Hartsfield told a press conference of concern that debris shed by a crack could work its way into an engine and trigger an explosion. Repair work demanded the removal of all three of Columbia's main engines and suspended preparations for the already-long-delayed mission.

"These cracks may pose a safety concern and we have teams at work investigating all aspects of the situation," said Ron Dittemore as June drew to a close. Rick Husband and his crew, characteristically, took it in their stride. "We've had a fair number of slips through the course of our training," he said, "but we've made good use of those. This will be no different." As the summer wore on, NASA opted, as soon as this 'grounding' order was lifted, to fly two high-priority International Space Station assembly missions in late autumn and reschedule STS-107 for mid-January 2003.

"We tried to make sure we took advantage of the time, not only from the standpoint of training, but also from the standpoint of being able to take some time off," said Husband of the enforced delay. "We saw a tremendous opportunity there to take some summer vacation that we wouldn't have otherwise had, and then we were able to take some time at Thanksgiving and Christmas as well." By the end of the year, the liner cracks – which, engineers found, *did* affect Columbia, although it used a different material – were resolved, but around the New Year another problem surfaced.

This time, a crack was found in a 5-cm metal bearing in a propellant line tie rod assembly on board Discovery and triggered an assessment of potential damage. By this time, the STS-107 stack was on the pad and by 14 January 2003 managers were satisfied that Columbia was safe to fly. Like the six Shuttle missions that preceded it, these final preparations were surrounded by intense security – including patrols by F-15 fighters, reconnaissance aircraft and US Army attack helicopters – that NASA had put in place in the wake of the terrorist attacks on New York and Washington in September 2001.

Ilan Ramon's presence on the crew, coupled with an increasingly unstable political situation in the Middle East and tottering peace process, placed even more pressure on the space agency's security forces. "We adjust our security posture for every launch," said David Saleeba, who worked as a Secret Service agent before being drafted in as NASA's head of security. "There have been adjustments upwards since 9/11, things that will stay in place for the foreseeable future. We've also ratcheted some things up. We've also backed off in some areas of security in the last few launches."

"There is a great deal of security. Some you can see and some you can't," added Joan Heller, a spokeswoman for the Brevard County Department of Emergency Management. "It's unprecedented." Indeed, Saleeba told journalists just before Columbia's launch that NASA routinely monitored intelligence reports from the Department of Homeland Security, all of the armed forces, the FBI, Customs Service, Coast Guard and the Federal Aviation Administration to enhance security. The astronauts' departure and arrival times were no longer made public until the last minute, and precise liftoff times were not disclosed until 24 hours before launch.

Ramon, however, was not overly worried and seemed satisfied with the precautions laid on for him, his crewmates and several hundred Israeli dignitaries and family members attending Columbia's launch. Those precautions extended well beyond KSC itself and were particularly visible, as the countdown clock ticked towards a 16 January liftoff, at the beachfront hotel in Cocoa Beach where high-ranking Israeli VIPs were in residence. Cars were stopped and searched by armed officers and bomb-sniffing dogs and the families of the astronauts were treated to police escorts as a matter of course.

"If there ever was a time to use the phrase 'all good things come to people who wait', *this* is that one time," Launch Director Mike Leinbach radioed to Columbia's crew as the final minutes ticked away on the morning of 16 January. "From the many, many people who put this mission together, good luck and Godspeed."

"We appreciate it, Mike," replied Rick Husband from the cockpit. "The Lord has blessed us with a beautiful day here and we're going to have a great mission. We're ready to go." Earlier that same morning, his crew rode from the Operations and Checkout Building to the pad under an intensive armed guard. For Husband, Anderson and Chawla, what came next was not a total surprise, but despite having flown the Shuttle before they would still be gripped by the exhilaration and even terror of knowing that they were essentially riding a controlled explosion all the way into orbit.

"When you launch in a rocket, you're not really *flying* that rocket," said Anderson. "You're just *hanging on*! You're trying to harness that energy in a way that will propel you into space, and we're very successful in doing that, but there are a million things that can go wrong. We train and try to prepare for the things that may go wrong, but there's always that unknown. [However], the benefits for what we can do on-orbit, the science that we do and the benefits we gain from exploring space are well worth the risk."

Columbia roared into a cloudless azure-blue sky at 3:39 pm, precisely on the opening of the launch window, to the cheers and Hebrew prayers of the assembled onlookers. The picture-perfect liftoff, not surprisingly, made front-page headlines in Tel Aviv, providing a momentary distraction from the ongoing conflict with the Palestinians. Even Tiewfiek Khateeb, an Arab-Israeli member of parliament, lauded the historic nature of the flight. "It's a happy occasion for Israel and the world's scientific community," he told an interviewer, but stopped short on being drawn on whether the goodwill might extend to 'other areas'. "Let's not get carried away. We can't isolate ourselves from reality. This joyous day is little more than an aberration ... until the issue of the Israeli occupation [of Palestinian lands] is resolved and issues of social inequity are addressed." Watching Columbia's ascent from the viewing area at KSC was Daniel Ayalon, Israel's ambassador to the United States, who was moved by the spectacle. "Taking it only two generations after the Jewish people were at their lowest ebb," he said, "on the very demise ... here we are soaring up and making great achievements together with our best allies."

Ramon, meanwhile, securely strapped into his seat on the Shuttle's middeck for the rattling climb to orbit, later told an interviewer that he was most struck by "a lot of noise and shaking." He and his six crewmates, however, had little time to dwell on their launch and within minutes of reaching space immediately set to work converting their ship from a rocket into an orbiting laboratory for two weeks of intensive research. After opening up the Spacehab module, he and Clark set to work activating the first of the life science experiments.

"BINGO! WE NAILED ONE!"

Elsewhere, on the crowded flight deck, McCool and Brown – who, as members of the Blue Team, would shortly be bedding-down for their first sleep shift – busied themselves activating the FREESTAR experiments on a pallet at the rear end of Columbia's payload bay and setting up laptop computers to support the SOLSE instrument. An acronym for Shuttle Ozone Limb Sounding Experiment, this device consisted of a visible and ultraviolet spectrograph designed to measure the vertical distribution of ozone in Earth's atmosphere using a relatively new, limb-viewing geometry.

Its first calibration run was over Hilo in Hawaii, conducted to coincide with a ground-based balloon launch, and its performance highlighted potential improvements of a limb-viewing instrument rather than a 'downward-looking' one. Although the instrument itself operated more or less autonomously, it did require thruster firings to position it for observations and was primarily Willie McCool's responsibility. "I think the world as a whole is interested in global warming and the preservation of our ozone layer," he said, "and this will certainly help in understanding the ozone and its depletion."

Like SOLSE, the other FREESTAR experiments were essentially autonomous, but required Shuttle manoeuvres for 'pointing'. The Israeli-provided MEIDEX camera, obviously, required precision targeting to observe sprites and dust storms, as did the Solar Constant Experiment (SOLCON) which explored changes in the Sun's energy output during its 11-year 'cycle' of sunspot activity. Its data thus helped to ensure continuity of the 'solar constant' levels obtained by other instruments on board free-flying satellites over longer periods of time.

The other three experiments attached to the FREESTAR pallet were the Critical Viscosity of Xenon (CVX), Low Power Transceiver (LPT) and Prototype Synchrotron Radiation Detector (PSRD). Xenon, a very pure fluid with a simple structure and a 'critical temperature' just below room temperature, was employed to measure viscosity changes and better understand 'shear thinning' in complex fluids like paint, which need to flow easily during application and stand firm afterwards. The other experiments demonstated a low-power, lightweight S-band navigation and communications device for future spacecraft and measured cosmic-ray background radiation in support of a future International Space Station instrument.

According to FREESTAR Mission Manager Tom Dixon, all six experiments performed admirably – with a few minor hiccups – throughout the flight, as did the two Mission-Peculiar Equipment Support Structure (MPESS) pallets from which they operated. "This has been one of the smoothest missions I've ever had a payload on," he said as STS-107 wore on, "and the closest we've ever stuck to the pre-mission timeline." Dixon's praise was echoed by the principal investigators of many of the FREESTAR experiments, including Joachim Joseph of Tel Aviv University, whose team busily supervised their MEIDEX results.

"We are excited that we were able to achieve all mission objectives and more," Joseph told journalists. "We were able to record ... sprites and 'elves' and many scientists are asking us for daily bulletins about our studies." 'Elves', first identified in 1994, are luminous red, bagel-shaped electrical phenomena that materialise above thunderstorms in a tenth of a millisecond. Although it was hoped that MEIDEX would see one, Dave Brown – and atmospheric scientist Yoav Yair of the Open University of Israel in Tel Aviv – could not believe their luck when one was serendipitously photographed over a South Pacific thunderhead on 19 January.

"Bingo! We nailed one almost in the first data-take!" exulted Yair. Around the same time, less than a week into the 16-day mission, two sprites had already been identified over Australia and the South Pacific. Additionally, a pall of grey smoke was spotted hanging over the Amazon rain forest, illustrating how complex interactions between it and the atmosphere could influence weather and climatic conditions. Observations of dust storms in the Mediterranean region – one of

MEIDEX's key tasks – was hampered during this time by heavy clouds. In fact, this prevented Ramon from seeing his homeland until a week after launch.

Unfortunately, January is among the worst times to see Mediterranean dust storms, but the repeated delays of STS-107 had pushed it from an originally planned summertime launch to the least-optimum point of the year. Eventually, on 26 January, Ramon finally saw his first Mediterranean dust storm. "The observation today went pretty well, I think," he told Mission Control. "We had the first dust, I hope," adding that his first flyover of a crystal-clear Israel had been "marvellous".

HOT SCIENCE IN A WARM LAB

As the FREESTAR experiments commenced operations, the Spacehab module was also a hive of activity as two shifts of astronauts activated and began their extensive programme of around-the-clock research. Only the most minor of problems, it seemed, was troubling Columbia herself: one of two heaters in a cryogenic fluid storage tank in the payload bay refused to work and part of the intercom between the module and the crew cabin exhibited difficulties.

Then, on the afternoon of 20 January, managers noticed an electrical 'spike' in one of two dehumidifiers used to collect and distribute water produced from condensation buildup in Spacehab. An identical system had sprung a leak the night before and was shut down, prompting Willie McCool to "stop talking and start mopping". A valve was reconfigured to allow cool air from Columbia's crew cabin to flow into the module, which stabilised the temperature level at a balmy 29 Celsius. Later in the mission, an air duct was routed into Spacehab to bring this down to around 22 Celsius.

"Even though the air temperatures are a little bit warmer in the hab right now than they would normally be, it's not really a factor for the crew," said Mission Operations Manager Phil Engelauf. "It's really not even outside their comfort level." In fact, the temperature rise in the module had even drawn envious comments from scientists at the Goddard Space Flight Center, who told the astronauts that the overnight forecast in Maryland was for a 'low' of 8 degrees and a wind chill of minus 5 degrees. Even Cape Canaveral had seen snow flurries on 17 January.

Despite the slight rise in temperature, the research inside Spacehab continued at breakneck pace, although the astronauts had to move some of their life sciences gear over to Columbia's middeck to keep it sufficiently cool. One of the most important pieces of equipment on board STS-107 was ESA's Advanced Respiratory Monitoring System (ARMS), which required the crew to participate in an extensive series of breathing exercises to measure their cardiac output, with particular focus on the critical first few hours in microgravity. Its first measurements were taken just six hours after launch.

"They're looking specifically at the changes in respiratory or lung function when patients have to be on their backs for extended periods of time," said Laurel Clark. "In fact, there's some good evidence that patients in intensive care units would be better off on their *stomachs* as opposed to their backs. This entails a huge amount of

Demonstrating the 'roominess' of the new Spacehab Research Double Module, Kalpana Chawla tends experiments during her shift as part of STS-107's Red Team. Note the foot restraints covering the 'floor' of the laboratory.

overhead to take care of them that way, but certainly if there's much better oxygen exchange and lung function, then it's worth that overhead. They've been studying us in different positions, on our stomachs and on our backs, studying the air exchange."

All four 'science crew' members – Anderson, Brown, Clark and Ramon – worked with the ARMS device during their respective shifts throughout the first two-and-a-half days of the mission, acquiring data both at rest and during periods of moderate to medium-high exercise on a stationary bicycle ergometer. The only problems were a higher-than-anticipated noise from one of its two gas analysers, which was later resolved, and a temporary 'lock-up' of its computer. This initially led to the cancellation of one of the planned pulmonary tests, although a rescheduling of priorities succeeded in recovering it and performing it satisfactorily.

Even 10 days into the mission, with almost a week to go before Columbia's landing, ARMS investigators were lauding a 100% success rate for their device. They did, however, stress that most of the critical data would not be complete until a battery of tests could be conducted on the astronauts, beginning a few hours after their 1 February touchdown back in Florida ...

Meanwhile, in the field of biological research, cell cultures were being grown as part of efforts to better understand their genetic characteristics and ultimately combat prostate cancer and improve crop yields. Many of the life science and biological studies were sponsored by European and Canadian investigators, while the German space agency provided an experiment to observe the development of gravity-sensing organs in fish and students from across the globe explored the effects of microgravity on spiders, inorganic crystals, silkworms, bees and ants.

The Advanced Protein Crystallisation Facility (APCF), which had ridden several previous Spacelab missions, supported several hundred samples of key proteins, while the Biobox payload cultivated specimens of mammalian cells as part of four experiments provided by Belgian, French and Italian researchers. Both were supplied by ESA, as was Biopack, which explored the influences of acceleration and high-energy cosmic radiation on biological processes.

Flying commercially after several years of tests was Astroculture in Columbia's middeck, which was being used on STS-107 to investigate whether *roses* grown and nurtured in low-Earth orbit could produce new fragrances. As part of a study sponsored by perfume giant International Flavours and Fragrances (IFF), a tiny rose known as 'Overnight Scentsation' had flown on board STS-95 in October 1998 and yielded a fragrance which the company's Braja Mookherjee called "a very green, fresh rosy note". Fibres were used to 'collect' the scent for post-flight analysis and the fragrance was later incorporated into a new perfume known as 'Zen'.

On STS-107, two different plants – a rose and an Asian rice flower – were housed inside Astroculture and, as of 24 January, their growth was proceeding well. "It's truly fascinating to see two flower plants doing very well up there," said project manager Weijia Zhou of the Wisconsin Center for Space Automation and Robotics, which worked jointly with IFF on the study, adding that "the [space-grown] rose is a mild and pleasant kind of aroma; the other [Earth-grown] is strong".

Elsewhere, combustion scientists were continuing the highly successful research begun on Columbia's MSL-1 missions in 1997 by reflying their versatile

Combustion Module, which was primarily under Kalpana Chawla's supervision. She had proved to be a rising star in NASA's Astronaut Office since her first bittersweet journey into space more than five years earlier. She had been ultimately exonerated from blame for the botched deployment of the Spartan-201 satellite during STS-87, and perhaps to highlight his own confidence in her abilities, Rick Husband had nominated her as the flight engineer for STS-107. Her performance was nothing short of exemplary.

"We've been busier than I ever imagined," she told Mission Control on 18 January, "since things do take longer up here." Her duties with the Combustion Module encompassed three separate experiments: Paul Ronney's SOFBALL study of flame balls in microgravity, Gerard Faeth's Laminar Soot Processes (LSP) analysis of the formation and behaviour of soot and a new investigation known as 'Mist'. The latter was intended to investigate the use of fine water mists in firefighting, although its first run on 27 January encountered problems when it sprung a leak.

After repairing a faulty seal, then struggling to introduce mist into the experiment's chamber, Chawla began more than a dozen Mist experiments. During one of the tests, the pace of which had been deliberately increased to make up for lost time, video footage revealed an impressive slowing-down, elongation and 'wiggling' of one particularly weak flame. "[Mist studies] how you can use water [vapour] to better extinguish fires," said Brown, adding that the combustion experiments on STS-107 were the second step towards a fully fledged Fluids and Combustion Facility for the space station.

Ronney and Faeth were also more than happy with their studies. "It's been a great experiment to date," Ronney told journalists on 22 January as the facility's cameras detected and videotaped faint amounts of light and heat emitted by flames just 5 to 10 mm in diameter. Even taking account of the earlier MSL-1 research, he added STS-107's SOFBALL studies had set at least three new records: the weakest flame ever burned (in orbit *or* on Earth), the least amount of fuel mixed with air and – at more than 81 minutes – the longest-lived flame ever burned in space.

"I guess you could also say maybe the most excitement by a combustion researcher," Ronney joked. "Is that a record, too?" Flame balls had first been theoretically predicted by the Russian physicist Yakov Zeldovich in 1944, but were not seen under experimental conditions until Ronney's drop-tower tests four decades later. During STS-107, no fewer than 39 tests were performed, utilising 15 different fuel mixtures and triggering 55 flame balls which burned for a total of six-and-a-quarter hours. To minimise disturbances, Columbia's thrusters were disabled and the orbiter kept in 'free-drift' while the OARE monitor carefully measured acceleration levels.

It was hoped that SOFBALL data could lead to cleaner and more efficient car engines, as well as improving fire safety. It also gave the crew the opportunity to 'name' a few of the hardier flame balls. "[I'm calling this one] Howard," deadpanned Dave Brown as the mission's final week drew to a close. "After that, *everyone* started naming them," added Ronney. "It was fun. It also helped us to keep track of some of the strange things we saw." Two balls which flew around in a spiral, DNA-like pattern were dubbed 'Crick and Watson' after the two Nobel Prize-winning biophysicists.

One of the largest flame balls was named 'Zeldovich' and another honoured Paul Ronney himself, although, as the principal investigator pointed out with a chuckle, "it turned out to be small and short-lived ... a wimp!" Overall, the flame ball studies were an outstanding success. "We didn't think flame balls could last for more than a few minutes," said Ronney of the MSL-1 studies, "but we were wrong. Many of them were still burning when SOFBALL's computer automatically ended the test. We needed *more time*." That additional time was provided during the STS-107 experiment runs.

"Before the mission began, I said I wanted to send a flame ball around the world. 'Kelly' almost made it," Ronney said of the longest-lasting of all flame balls, which burned for 81 of the 90 minutes needed for Columbia to circle the globe. "Kelly's experience is a fascinating example of group dynamics among flame balls. She was created, one of nine flame balls, in a gaseous mixture of hydrogen, oxygen and sulphur hexafluoride. All the others began drifting around the chamber ... competing with one another, while Kelly remained motionless at the centre. Before long, the others were exhausted; they had drifted too close to the walls and winked out. Kelly was left alone with a chamber full of fuel."

Other important experiments included Buddy Guynes' Mechanics of Granular Materials (MGM), which sought to test 'sand columns' under microgravity conditions that could not be mimicked on Earth. "The possibility of retirement is appealing to me, but I want to work for a good while yet," said Guynes, a researcher from NASA's Marshall Space Flight Center. "I'm expecting exciting results from the mission and would like to have a hand in getting the good news out to the public." The experiment looked at how environmental changes could drastically change the properties of a bulk granular material.

The obvious household example of this phenomenon is a packet of coffee, which, originally brick-solid when packed, becomes much softer and easily shifted once it is opened and air is admitted. "Before that air comes in," explained Guynes, "you can almost use that coffee [package] like a hammer. As soon as you let air in, it gets real loose. The [thixotropic] soil [effects] you see in an earthquake can be similar, especially if there's water around. Water is a lubricant between the grains. When they're shaken, they also get loose."

The strength of soils comes from the friction and geometric 'interlocking' between the faces of individual grains; but this can also introduce weaknesses as their craggy surfaces form small voids and make them behave like liquids when moisture and air are trapped. As external pressure increases, intergranular pressures drop and soften the material, eventually liquefying it. During this process of liquefaction, buildings can sink and tilt, bridge piers can move and buried structures can float. On STS-107, three columns of sand were 'squeezed' between tungsten plates in a water-filled Lexan jacket.

Previous MGM flights in 1996 and 1998 had already yielded a number of insights into soil mechanics and behaviour and it was hoped that the experiments on Columbia's mission would enable scientists to design better models of soil movement under stress. This could eventually lead to new ways of strengthening building foundations, managing undeveloped land and handling powdered and granular materials in chemical, agricultural and other industries. "We might end up changing our building techniques," Guynes speculated. "This is one of the things NASA is doing that has a very strong application for the guy on the street."

SMALL WONDERS AND JOYS

Despite their hectic work schedule, the crew – and particularly the four rookie spacefarers on board Columbia – were nevertheless able to take time out to fully

From a roll of undeveloped film miraculously recovered from among the wreckage of Columbia, the crew 'flies' through the Spacehab module for their in-flight portrait. Top row (left to right) are Dave Brown, Willie McCool and Mike Anderson and bottom row are Kalpana Chawla, Rick Husband, Laurel Clark and Ilan Ramon.

appreciate their unique surroundings. For Laurel Clark, the noises came as a surprise. "Obviously, everything floats," she told an interviewer from orbit. "The zippers and all the belts that have D-rings that we hold things down with are always floating and hitting each other and jingling. It makes this beautiful tinkling music in the background all the time. It just caught me off-guard. It was beautiful."

Willie McCool, who spent most of his shifts up on the flight deck orchestrating manoeuvres in support of the FREESTAR experiments, found the sunrises and sunsets, coupled with sweeping panoramic views of the Himalayas and Australia, as some of the most awe-inspiring views of the mission. Yet both he and Ilan Ramon stressed the apparent delicacy of our world when seen from orbit. "The atmosphere is so thin and fragile and I think everybody, all of us ... have to keep it clean and good," said Ramon. "It *saves* our life and *gives* our life."

Kalpana Chawla had flown before, but when she floated onto the flight deck on 29 January she beheld an unusual sight. As orbital sunset overtook her view, she saw her *own reflection* in Columbia's overhead windows, as well as the sunlit and darkened faces of Earth. "In the retina of my eye, the whole Earth and the sky could be seen reflected," she said, "so I called all the crew members, one by one, and they saw it and everybody said 'Oh, wow!'." Not surprisingly for a former gymnast, Dave Brown spent his spare time doing backflips in the Spacehab ...

"ABSOLUTELY NO CONCERN FOR ENTRY"

A week and a half into the mission, the Red Team members – Husband, Chawla, Clark and Ramon – had the opportunity to speak to the three-man crew of the International Space Station. By that time, Expedition Six Commander Ken Bowersox and Flight Engineers Nikolai Budarin and Don Pettit were entering their tenth week in orbit and, as far as they knew, would be returning to Earth in mid-March on STS-114. At the time of the ship-to-ship radio call, the station was flying over eastern Ukraine as Columbia soared high over northern Brazil.

"Columbia, this is Alpha," radioed Bowersox, identifying himself by the station's official callsign.

"Hey Alpha, this is Columbia, how you doing over there?" replied Husband.

"We're doing great. We're so glad to see you guys made it into orbit," said Bowersox.

"We're glad to be here too," came Husband's answer. "We're really excited to be able to talk to you guys, one space lab to another big old space lab on that beautiful station of yours."

"I tell you, it's certainly an amazing place to live and to work," Pettit added. The conversation lasted a matter of minutes before the crews bade each other farewell and drifted out of radio contact. The following morning – 28 January – they both united with thousands of people on Earth in remembering the victims of the Challenger accident 17 years earlier. As Husband spoke of his crew's profound sadness at the loss of those seven brave lives, he could hardly have imagined that he,

McCool, Brown, Chawla, Anderson, Clark and Ramon would join their ranks in barely four days' time ...

Two years later, everyone knows the root cause for Columbia's loss on 1 February 2003 and although the STS-107 crew were unaware that their ship was irreparably damaged, concerns had already been raised on the ground. Video footage taken during their climb to orbit on 16 January had clearly shown a briefcase-sized chunk of foam falling from the External Tank about 81 seconds after launch and impacting somewhere on Columbia's left wing. Debris of this kind had, however, been falling from tanks on virtually every mission since 1981 and few NASA managers were overly worried.

Nevertheless, the media had already latched on to the foam strike, to such an extent that an email was sent to Husband a week into the mission to advise him of what had occurred, lest he be caught off-guard if asked about it during a space-to-ground press conference. The email informed him of the strike and assured him that, after painstaking analysis by experts on the ground, any damage to the wing was expected to be superficial and there was "absolutely no concern for entry".

Rick Husband thanked the sender of the email, Flight Director Steve Stich, and pressed on with his mission.

LAST DAY

On 30 January, as the science crew wrapped up the final few experiments in the Spacehab module, a quick peek was taken under its floor to make sure that the balky dehumidifier would not leak during re-entry. No moisture was found. The following day, the entire crew shut down FREESTAR and the module and Husband, McCool and Chawla tested Columbia's steering thrusters, hydraulics and other systems that would be needed during their descent towards a Florida touchdown.

After lauding the success of the mission, Entry Flight Director Leroy Cain – whose words from Mission Control would become almost as famous as Apollo 13's "Houston, we have a problem" over the following days – announced that the STS-107 crew would return to Earth on 1 February. Weather conditions at KSC were predicted to be excellent and "this [entry] will be a very good visual sighting for folks, particularly on the West Coast, as well as in [the] mid-Arizona, New Mexico area. It should be a pretty spectacular event for folks that have never seen a Shuttle sighting, particularly at night."

It *was*, indeed, going to be a spectacular sight, although not as Cain had intended, and certainly not a sight that anyone watching would want to see again. For the seven astronauts, however, as they readied their ship for landing and packed away their equipment, the mood was one of jubilation at a job well done. "Science-wise, this flight's been absolutely fantastic," said Anderson. "I think a lot of our experiments have exceeded our expectations by one hundred percent. We've seen things we never expected to see."

Early on 1 February, the crew donned their orange pressure suits and took their seats for the hour-long glide to Earth. On the flight deck were Husband, McCool,

Chawla and Clark; the others would occupy seats on Columbia's locker-studded middeck. Although Chawla was the flight engineer, responsible for helping the Commander and Pilot monitor the systems, it was Clark's job to assist her. Ironically, Clark's words to an interviewer shortly before launch give an unsettling insight into the increasingly desperate situation in the cockpit as the crew realised they were in deep trouble.

"[Kalpana] helps the Commander back up his systems if anything is unusual with their systems, and [I] back up the Pilot," she said. "I work with Willie, monitoring the electrical power systems and hydraulic systems; I have a computer screen near my seat where I can monitor the overall health of the vehicle and pick up any problems that might be occurring early on or once we see any kind of a malfunction or anything unusual that's happening . . . [although] most of the time, you don't have to do much, other than monitor the normal entry profile."

At 1:15 pm GMT on that fateful Saturday, as the ship flew upside down and backwards 270 km above the Indian Ocean at a speed of more than 28,000 km/h, Husband fired the OMS engines in an irreversible burn to drop Columbia out of orbit and place her on course to land in Florida exactly an hour later. For the first 30 minutes of re-entry, the ship simply fell like a stone through orbital darkness towards its predetermined runway on the other side of the planet. So far, true to Clark's prediction, the crew was following a 'normal' re-entry profile.

The second half of re-entry, however, was far more interesting as compression of the steadily thickening air at hypersonic speeds produced a brilliant light show outside the flight deck windows.

"That might be some plasma now," McCool remarked as he glimpsed the peculiar, salmon-coloured glow that gradually replaced the pitch blackness he had become used to over the last 16 days.

"Think so, already?" asked Clark from her seat directly behind the Pilot, as she aimed a handheld video camera through Columbia's overhead windows.

"That's some plasma," confirmed Husband.

"Copy, and there's some good stuff outside," replied Clark. "I'm filming overhead right now."

"It's kinda dull," added McCool.

"Oh, it'll be obvious when the time comes," Husband told him. Minutes later, the glow steadily brightened and McCool told Brown, Anderson and Ramon down on the middeck that he could now see vivid orange and yellow flashes across Columbia's nose. In the chatter that followed, Husband likened what he was seeing to "a blast furnace". The time was 1:44 pm and the Shuttle had reached an altitude of around 120 km as it hurtled over the eastern Pacific, nine minutes from the California coastline and a continent away from its landing site in Florida.

Flying with her nose angled upwards to subject her reinforced carbon–carbon nosecap and the leading edges of her wings to the most extreme re-entry temperatures of close to 3,000 Celsius, Columbia was still dropping at more than 24 times the speed of sound as aerodynamic pressures on her steadily doubled, then tripled, then quadrupled. Aware of the intensely dangerous situation they were in, but unaware of the fact that their ship was fatally damaged, Husband's crew calmly donned their gloves, pressurised their suits and conducted communications checks. Armed with her camera, Clark planned to videotape the entire re-entry.

At 1:50 pm, with the computers still flying the vehicle, Columbia's right-hand RCS jets automatically fired to adjust the position of her nose. This was one of several manoeuvres designed to bleed off speed. Three minutes later, precisely on time, she crossed the California coastline and ground-based observers were able to watch her streaking, meteor-like, across the night sky, "at incredible speed", according to freelance photographer Gene Blevins. It was around this time, however, that he and colleague Bill Hartenstein saw something strange: "a big red flare [came] from underneath the Shuttle and [was] forced downward . . . something *came off the Shuttle!*"

A continent away, at the KSC viewing site, the crew's families, Israeli dignitaries and high-level NASA managers – including Administrator Sean O'Keefe and his new Associate Administrator for Space Flight, Bill Readdy, a former Shuttle Commander – had begun to gather on a beautiful Florida morning for the landing. STS-107 was the first of six missions planned for 2003: the others would be exclusively dedicated to assembling the International Space Station. In fact, assuming everything went as planned on those missions, the station's football-field-sized truss structure and electricity-generating solar arrays should be in place by the end of the year.

That would allow research on board the outpost to commence in earnest: the European lab, Columbus, was due to be installed in October 2004 by the STS-123 crew and the Japanese lab, Kibo, would follow over three missions a year later. Columbia was expected to have a minor, though important, role in the continuing construction work: she would carry a piece of starboard-side truss known as the 'S-5' segment into orbit in November 2003, fly the final Hubble servicing mission in April 2004 and then deliver logistics and a new crew to the station in January 2005.

All that changed abruptly a few minutes before 2:00 pm GMT on 1 February 2003 as Columbia's otherwise-normal re-entry profile began to go horribly wrong.

8

"In my book, they always landed ..."

PREMONITIONS

For Andy Cline, the guide for the National Outdoor Leadership School (NOLS) who both worked and played with Rick Husband's crew, the morning of Saturday 1 February 2003 brought a peculiar and inexplicable sense of dread. Alone in his cabin in Wyoming, he got up early, aware that his seven friends were returning to Earth and eager to check if they had begun their descent. However, in his head, he instinctively knew that something was not quite right.

"My wife had gotten up early to go into the mountains for some work," he told BBC journalist Leo Enright a few months later, the emotion of that terrible day still raw in his voice, "and I was lying in bed thinking that the Columbia crew was about to land. It was one of those uncanny moments when I realised that something was ... desperately wrong. I put on my clothes and started running down the path to the road. My wife was just in the process of driving back into the driveway and she told me about the [accident]."

For Evelyn Husband and her two children, on the other hand, that Saturday started with great joy as they awaited the return of their husband and father from his 16-day flight. Like the other STS-107 family members, they were at the KSC viewing site waiting to be reunited with their loved ones. At 2:05 pm GMT, with 11 minutes to go before Columbia was scheduled to touchdown on the 5-km-long runway, a photographer snapped Evelyn and her children in front of the countdown clock as it ticked away the final seconds, little realising that the Shuttle had already disintegrated ...

For at least two other people, the awareness that something was wrong had already appeared days earlier. One was Laurel Clark's 8-year-old son Iain, who repeatedly begged his mother not to fly: he had cried on the morning of Columbia's launch and complained to her about leaving at each space-to-ground video conference. It could have been due to the near-fatal air crash that he and his parents suffered in December 2002, but to his father it was "something *more* than typical separation anxiety. What do kids know that we *don't* know? What do they see that we *don't* see?"

In the following months, Jon Clark would wonder if it would have been better for all three of them to have died in that air crash; perhaps then, he rationalised, his family could have remained together, the STS-107 launch would have been delayed, and the lives of the other six astronauts might have been saved. Perhaps. On the other hand, the root cause of the Columbia disaster might have remained undetected, uncorrected and would have lain in wait for another unlucky Shuttle crew ...

Speculation had arisen immediately after launch that a briefcase-sized chunk of foam insulation had fallen from the External Tank and hit the Shuttle's left wing; film footage showed a spectacular shower of particles, although it remained unclear if these were from the foam itself or fragments of shattered tiles from Columbia's wing. If it was the latter, it did not bode well: for the Reinforced Carbon Carbon (RCC) protected her from the brunt of near-3,000-Celsius temperatures during re-entry. Senior NASA managers, however, doubted that the foam could cause such severe damage to the panels and concluded it was not a 'safety-of-flight' issue.

Nonetheless, lingering doubts remained. On 31 January, in a chilling preview of what would unfold, an engineer named Kevin McCluney gave a hypothetical description to colleagues in JSC's flight control team of the kind of 'signature' they might see if a large hole *had* been punched through one of Columbia's tiles, allowing super-heated plasma to enter. "Let's surmise," he began, "what sort of signature we'd see if a limited stream of plasma did get into the [main landing gear] wheel well, roughly from entry interface until about 200,000 feet; in other words, a 10- or 15-minute window.

"First would be a temperature rise for the tyres, brakes, strut actuator and the uplock actuator return ... "

At 1:52:17 pm GMT on 1 February, nine minutes after entry interface – the point at which the Shuttle begins to encounter the uppermost traces of the atmosphere – Flight Director Leroy Cain and his team saw unusual data on their monitors. Cain had begun his shift that morning with "Let's go get 'em, guys," before giving Husband the green light to fire Columbia's OMS engines. Most of the re-entry profile was conducted under GPC control, as was the normal procedure, but with 23 minutes to go before touchdown, Maintenance, Mechanical, Arm and Crew Systems (MMACS) officer Jeff Kling spotted something strange.

In what flight controllers refer to as an 'off-nominal' event, Kling noted that two downward-pointing arrows had appeared next to readings from a pair of sensors deep inside Columbia's left wing. They were designed to measure hydraulic fluid temperatures in lines leading back to the elevons; a few seconds later, two more sensors also failed. The attention of Kling and two other engineers was instantly riveted on the quartet, which showed all the signs of having had their wiring cut. They tried vainly to identify some common 'thread' to explain the fault.

"FYI, I've just lost four separate temperature transducers on the left side of the vehicle," he told the flight director. "Hydraulic return temperatures. Two of them on system one and one in each of systems two and three."

"Four hyd return temps?" asked Cain.

"To the left outboard and left inboard elevon." Cain's thoughts were following a similar pattern to Kling's own: he wondered if there was a common root cause for the four failures to have occurred in such close proximity. When the mechanical

Taken at 2:15 pm GMT on 1 February 2003, precisely when Columbia should have been landing in Florida, this photograph shows a tense Mission Control in Houston.

systems officer told him that there was "no commonality" between them, he was perplexed. Immediately, his thoughts returned to his own words to journalists the previous afternoon, in which he confirmed that the foam impact to Columbia's left wing during launch posed absolutely no concern for re-entry. Were these four sensor failures linked, in fact, to the foam strike?

Cain was nervous and admitted later that he thought at that moment that hot gas may have found its way through a hole in the wing and was gradually wreaking havoc on the internal sensors, cables and systems. However, upon checking with Guidance, Navigation and Control (GNC) officer Mike Sarafin, he was assured that Columbia's performance as she passed high above the California–Nevada state line at 22.5 times the speed of sound was normal. Cain checked with Kling that all other hydraulic systems were functioning normally and was assured that there were no other problems.

"... *Tyre pressure would rise, given enough time, and assuming the tyre[s] don't get holed,*" continued Kevin McCluney's chilling hypothesis from a day earlier, "*the data would start dropping out as the electrical wiring is severed ...*"

Suddenly, at 1:58 pm, Rick Husband made his first radio call since entry interface a quarter of an hour earlier. He started to call Houston, but his transmission was abruptly cut off. A few seconds later came a loss of temperature and pressure data for both the inboard and outboard tyres of the landing gear in the left well. If Columbia's tyres were holed or losing pressure, it would be a bad day: she was already heavier-than-normal with the Spacehab module and FREESTAR pallet in her payload bay and many doubted a 'wheels-up' belly landing would be survivable.

" ... *Data loss would include that for tyre pressures and temperatures, brake pressures and temperatures,"* McCluney had concluded.

After hearing Kling's report, Capcom Charlie Hobaugh – who was scheduled to fly Columbia on her next mission, STS-118 – called Husband to inform him of the tyre pressure messages and ask him to repeat whatever he had tried to say. There was no reply. Meanwhile, Cain was pressing Kling for answers on whether the messages were due to faulty instrumentation, but was advised all of the associated sensors were reading 'off-scale low'; in other words, they had simply stopped working. Seconds later, at 1:59:32 pm, Husband tried again to contact Mission Control. They were the last words ever spoken from Columbia.

"Roger," he said, presumably in acknowledgement of Hobaugh's earlier tyre pressure call, "uh ... buh ... " At that point, his voice was cut off in mid-sentence, together with all data flowing from the ship. Communications were never restored. Thirty-two seconds after Husband's partial transmission, a ground-based observer equipped with a camcorder shot video footage of multiple debris contrails streaking across the Texan sky ...

With the flow of telemetry broken, the situation in Mission Control was becoming increasingly uncomfortable, with Kling telling Cain that there was no common thread between the tyre pressure messages and the earlier hydraulic sensor failures and adding that *other* instrumentation for monitoring the positions of the nose and main landing gear had also been lost. As the seconds of radio silence drew longer, Cain asked Instrumentation and Communications Officer (INCO) Laura Hoppe how long she expected intermittent communications to last. She admitted that she expected some 'ratty' communications, but was puzzled by how protracted and 'solid' the silence was.

"Columbia, Houston, comm check," radioed Hobaugh at 2:03 pm, checking to see if the crew was experiencing communication difficulties with Mission Control. His words were greeted only by static.

A minute later, he repeated the call. Again, there was no reply.

"LOCK THE DOORS"

Half a continent away, in Florida, astronauts Jerry Ross and Bob Cabana were chatting outside the convoy commander's van at the Shuttle runway when they heard communications had been lost. At first, they were unconcerned – at least, that is, until they were informed that powerful radars near KSC, which were supposed to lock onto the incoming spacecraft at 2:04 pm and track its final approach, saw nothing coming over the horizon. In Leroy Cain's words, "that was the absolute black-and-white end. If the radar is looking and there is nothing coming over the horizon, the vehicle is *not there*."

Unlike an aircraft, which can adjust its flight profile to make a second approach to a runway, the Shuttle falls from orbit with the aerodynamic grace of a brick and has only one shot at landing. Throughout re-entry, its enormous velocity of 25 times the speed of sound is gradually slowed by a series of sweeping turns to such a point that it touches down at 220 mph. Its trajectory through the atmosphere can thus be plotted very precisely and reliably and, as soon as the OMS engines have fired, its touchdown time can be predicted to the *second*.

"We're a 220,000-pound glider coming in," said Digger Carey, who flew as Pilot on Columbia's previous mission in March 2002. "We actually enter the Earth's atmosphere over the Indian Ocean and we have enough speed to glide halfway around the world and land in Florida."

The assembled crowds at KSC were following the countdown clock as it ticked towards a scheduled 2:16 pm landing. They confidently expected to hear the trademark double sonic boom, announcing Columbia's arrival in Florida airspace, and the sight of a small, rapidly descending black-and-white speck on the horizon. It never came. Astronaut Steve Lindsey, who had the important job of taking care of the families of the STS-107 crew, was on hand and recalled hearing people innocently remark "They're always late" when the countdown clock reached zero and began ticking back upwards.

Lindsey, by now a three-flight veteran, knew better: returning Shuttles were *never* late. They were always *exactly* on time. When the sonic booms did not reverberate across the marshy Florida landscape, when Columbia did not appear over the horizon, and when she did not settle onto the runway at 2:16 pm, he instinctively knew that something was terribly wrong.

So too did Bill Readdy, another veteran astronaut who had been NASA's Associate Administrator for Space Flight since the summer of 2002. He was standing next to Administrator Sean O'Keefe, who later described the former fighter pilot and three-time Shuttle flier as ashen-faced and visibly trembling. Jerry Ross, meanwhile, was a few months into his new job as the agency's chief of vehicle integration and his first act upon learning that tracking had been lost was to say a brief prayer. Then he set to work instructing his staff to secure KSC and round up the crew's families.

By this time, Texan police were being inundated with 911 calls, reporting bright flashes in the sky, loud explosions and falling debris; CNN quickly picked up on the stories and began reporting them. In Mission Control, however, televisions were not tuned in to outside broadcasts. Many flight controllers, including Leroy Cain, suspected the worst but prayed for the best. It was an off-duty NASA engineer named Ed Garske, who was watching Columbia pass overhead from the roadside south of Houston, who phoned colleague Don McCormack in Mission Control with the devastating news.

"Don, Don, I saw it!" Garske yelled, clearly out of breath. "It broke up!"

"Slow down," McCormack replied. "What are you telling me?"

"I saw the orbiter. It *broke up!*"

Sitting behind Cain, at around the same time, veteran flight director Phil Engelauf received a telephone call from off-duty flight director Bryan Austin – who had supervised Columbia's previous mission, STS-109 – and heard first-hand testimony of the Shuttle disintegrating in the skies above Texas. By this time, although no one in Mission Control had actually *seen* the ship's destruction, they had resigned themselves to it. Already, at 9:05 am, Cain had asked Flight Dynamics Officer (FDO) Richard Jones when he expected tracking data from the radars in Florida – and was told it should have happened at least a *minute earlier* ...

Now, as Engelauf relayed Austin's emotional report to Cain, the flight director slowly shook his head, composed himself and turned to the now-silent control room to declare an emergency. At 2:12 pm, he instructed Ground Control (GC) officer Bill Foster to "lock the doors" – a *de facto* admission that all hope was gone – and ordered flight controllers not to leave the building, but to begin preserving their data and writing up their logbook notes for use in the subsequent investigation. After checking with Jones that no further tracking data had been acquired, Cain referred his team to their emergency plans.

"Okay," he told them, "all flight controllers on the flight loop, we need to kick off the FCOH [Flight Control Operations Handbook] contingency plan procedure, FCOH checklist page 2.8-5." He then proceeded to talk them through the required actions: preserving their logbook notes and display printouts, communicating only on the flight loop and restricting outside telephone calls and transmissions.

Fifteen hundred kilometres to the east, at KSC, the families of Columbia's astronauts had been shepherded from the viewing site to the crew quarters by 2:30 pm. It was left to Bob Cabana to tell them the dreadful news. Mission Control, he said, had not picked up any radio beacon signals that would have been activated had the crew managed to bail out of their disintegrating ship and that, at an altitude of more than 60 km and a velocity of 23,600 km/h, it was extremely unlikely that anyone could possibly have survived.

Later that morning, Clark Barnett spotted something while driving through Sabine County in eastern Texas; he gave it no more thought until he received a call from his friend Mike Gibbs, who told him about Columbia. The two men decided to investigate and found an astronaut's charred torso, thigh bone and skull with front teeth intact. "I wouldn't want anybody seeing what I saw," Gibbs said later. "It was pretty gruesome." Elsewhere in Texas, near Hemphill, Roger Coday found some human remains, said a quiet prayer and built a small wooden cross by the roadside.

THE GOLDEN EGG

On the evening of 18 March 2003, Florida firefighter Art Baker prayed for success. One of thousands of volunteers searching a vast area of Texas for debris from Columbia, he had just heard that a single find could shed light on exactly why the spacecraft disintegrated during its descent six weeks earlier. Suspicions that 'the foam did it' were in most people's minds, although Shuttle manager Ron Dittemore and others doubted that such an "inconsequential" debris strike could have brought down a $2-billion national asset. Even Sean O'Keefe feared the root cause might never been found.

"Was it something that happened after launch? Was it something that happened

The recovered Orbiter Experiments (OEX) recorder: battered, but in one piece. It was to provide key evidence for what had happened to Columbia in her final minutes.

during the re-entry? Or was it something that happened during ascent and we didn't see it? Those are all possibilities," Dittemore had told a press conference on 5 February. "It just does not make sense to us that a piece of debris would be the root cause for the loss of Columbia and its crew. There's *got* to be another reason."

The definitive answer might come, however, from a square, breadbox-sized metal device called the Orbiter Experiments (OEX) recorder – part of Columbia's Modular Auxiliary Data System (MADS) and essentially the ship's 'black box' – which stored readings from more than 570 sensors scattered throughout the fuselage. In fact, as NASA's oldest orbiter, Columbia was the *only* member of the fleet to have such a device; it had been fitted for her first four test flights and, although some sensors had been removed, most were still operational by the time STS-107 lifted off.

One of the cruellest ironies was that, had any of the *other* Shuttles been lost in the same manner as Columbia, investigators might never have found the technical cause with absolute certainty. Only with the discovery of the OEX recorder by Art Baker as he trudged hilly terrain near Hemphill on 19 March could they begin piecing together the ship's final moments. The recorder's location was pinpointed by carefully plotting the discovery of other debris – including boxes originally mounted either side of the recorder, which had been mangled or torn to shreds – and it was found, miraculously, in near-pristine condition.

"We're obviously excited that it has been found," said Tyrone Woodyard, a spokesman for the newly formed Columbia Accident Investigation Board (CAIB), which was under presidential directive to find the cause of the disaster. "We're cautious, because we don't know what the data is going to be like. In other words, we don't know how intact that information is going to be." Late on 21 March, NASA shipped the recorder to data-storage-tape specialist Imation Corporation in Minnesota, which began the painstaking task of cleaning and stabilising it. Next, it returned to KSC for copying and then to JSC for engineering analysis.

Although the tape had broken between its 'supply' and 'take-up' reels inside the device and a portion had been stretched or 'wrinkled', Imation specialists described it as being in surprisingly good condition. It was hand-cleaned – repeatedly immersed in a filtered, de-ionised water bath, dried with lint-free cloth and nitrogen and wound back onto its original hub with new flanges – and returned to KSC on 25 March. Analysis revealed a strong signal and valid data lasting until 2:00:18 pm GMT, almost a full minute after Rick Husband's final radio transmission and 14 seconds after Texas skywatchers videotaped the first debris contrails.

After that, when Columbia's fuselage broke apart and electrical power to the OEX was severed, the tape had stopped running.

TIMELINE TO TRAGEDY

Nevertheless, it was enough to piece together Columbia's horrifying last minutes. Nor was the recorder the only remarkable find. A week after the disaster, Carl Vita – an engineer with United Space Alliance, the company responsible for Shuttle operations at KSC – and NASA engineer Marty Pontecorvo found a video cassette lying on a road near Palestine in Texas. They put it in a greasy Wal-Mart fried chicken bag and sent it to JSC for analysis. "It'll probably turn out to be Waylon Jennings or a Merle Haggard tape," Vita joked, "and the lab guys will get a kick out of it."

Not until the end of February did they realise how significant it was.

It was, in fact, *the videotape* filmed by Laurel Clark during the first part of Columbia's descent and contained 13 minutes of footage showing herself, Chawla, Husband and McCool in jubilant spirits putting on their gloves, telling jokes and admiring the spectacular light show outside the flight deck windows. Little did they know that even as they chatted and looked forward to their Florida homecoming, ionised atoms from the gradually thickening air were entering a hole in their ship's left wing and soon would begin destroying it from the inside outwards.

"Some might view it as a miracle," said Charles Figley of Florida State University's Traumatology Institute. "Suddenly, here is a postcard of these men and women." He added that it might provide additional peace of mind for the crew's families by assuring them that their loved ones were happy, blissfully unaware of the situation that would engulf them in the next few minutes and died doing what they loved. "Each time I watch it", Laurel Clark's husband Jon said later, "I get a sense of joy to know they were just having such a great time."

The surviving videotape ended at around 1:47:30 pm; Clark obviously continued recording after that time, but what had happened next – stored on the outermost edges of the cassette – had evidently burned away during its fall to Earth. About a minute after the surviving tape finished, a 'strain gauge' attached to an aluminium 'spar' just behind one of the RCC panels which wrap themselves in a 'U' shape around the leading edge of the wings measured an unusual increase of structural stress. It seemed that the aluminium was expanding under the steadily increasing heat and beginning to soften.

Each wing consists of upper and lower surfaces connected by an aluminium framework with a total of 22 individually shaped RCC panels attached to the leading edge. The panels are numbered, with '1' being closest to the fuselage and '22' furthest away. Detailed image analysis of Columbia's ascent suggested a piece of foam from the External Tank probably hit Panel 8, which was not visible to the astronauts from the crew cabin or the Spacehab module while in orbit, their view being blocked by the open payload bay door. Even if they had known the panel was severely damaged, they could have done nothing to fix it.

The strain gauge that started measuring unusual levels of stress was situated directly behind Panel 9, ironically in a region of the leading edge that was subjected to the most extreme temperatures during re-entry. Its data was among that recovered from the OEX recorder, and clearly pointed to trouble brewing within a couple of minutes of entry interface and progressively worsening as each second passed. When the CAIB published its final report in August 2003, its authors were convinced beyond any doubt by the strain gauge data that Columbia began her re-entry with a breached RCC panel.

That conclusion was later refined to a *particular* panel – Panel 8 – by the strength of the gauge's readings and investigators judged the gauge to have been within 40 cm of the point where hot gas was impinging on the aluminium spar. *That* ultimately allowed them to zero-in on Panel 8. Twenty seconds after the first indication of trouble from the strain gauge, a MADS sensor in the hollow cavity behind Panels 9 and 10 and just in front of the aluminium spars began measuring unusual temperature increases. Moreover, the sensor was both *heavily insulated* and some distance from the breach.

This led investigators to conclude that the hole in the lower part of Panel 8 must have been at least 15–25 *centimetres* in diameter. Anything smaller would probably not generate the observed readings from such a heavily insulated sensor so many centimetres away from the actual breach site. It was around five-and-a-half minutes after entry interface, at about 1:50 pm, that Columbia's computers began the delicate process of actively guiding the spacecraft towards Florida by smoothly swinging the nose 80 degrees to the right. Seconds later, sensors attached to the left-hand OMS pod registered an unusual change in temperature.

Instead of steadily climbing, temperatures around the pod rose very slowly; wind tunnel tests would later confirm some of the hot air entering Panel 8 was blowing metallic vapour from melted insulation through air vents in the top side of the left wing. This interfered with the normal airflow around the Shuttle and slowed anticipated temperature increases on the left side of the fuselage. As the re-entry heating worsened, melted Inconel – a heat-resistant alloy used to seal the RCC panels – started spraying Columbia's metallic skin. Then, seconds after 1:52 pm, the aluminium spar behind RCC Panel 8 finally burned through.

The plume now had access to the interior of the wing and immediately cut through sensor wiring attached to the spar, and started heating the aluminium trusses which supported its upper and lower surfaces. All of this sensor data was recorded and stored only on the OEX recorder; none of it was available to the crew or to Mission Control, who still believed they were following a normal re-entry profile. Judging from the amount of spar and other debris recovered, CAIB investigator Pat Goodman estimated that a hole of at least 20 cm across was burned through the aluminium.

By this stage, temperatures imposed on the RCC exceeded 8,000 Celsius – more than twice as much as they were built to withstand. Under normal re-entry conditions, the Shuttle compresses the thin air to generate two shockwaves. This forms a 'boundary layer', just a few centimetres thick, which resists further compression and provides a natural insulator, keeping temperatures on the RCC panels around 3,000 Celsius. However, a smooth surface is essential for the boundary layer to form. With a ragged hole in Panel 8, the protective properties of Columbia's boundary layer were severely disrupted.

"You have this massive thing ... slamming into the molecules as you get farther into the atmosphere," explained the CAIB's Jim Hallock, "and that collision basically disassociates these molecules into their constituents. *That's* where all this heating comes from. [The Shuttle] uses aerodynamic braking ... so you've got to get down to where the molecules are to slow it down. That's really what's slowing it down – running into these molecules and dissipating a lot of the energy of speed into heat. Once this boundary layer forms, it really protects you pretty well." But the hole disrupted the boundary layer, and let the hot plasma penetrate the wing like a blowtorch.

When the aluminium spar burned through, Columbia was still flying high over the Pacific Ocean, 300 miles from the California coastline, and her crew was oblivious to any danger. However, their sense of security would not last. Within 14 seconds, the hot plume started destroying three bundles of wiring running along the outboard wall of the main landing gear wheel well and gas began flowing into the box-like well itself through vents in its door hinges. It was at this point that the first unusual sensor reading popped up on Jeff Kling's monitor, highlighting increased hydraulic fluid temperatures.

By now, however, Columbia's left wing was literally being destroyed from the inside outwards, and the computers would shortly encounter problems with the ship's handling characteristics as it fell ever deeper into the atmosphere. Husband and McCool may have noticed an almost imperceptible tug as the nose pulled to the left under the effects of increasing aerodynamic drag, but this was corrected automatically by the computers which commanded the elevons at the back of the left wing to balance out the discrepancy to the right. By 1:53:28 pm, however, when Columbia crossed the California coastline, she was in severe distress.

Not only were the trusses inside her left wing softening and melting, but so too was its aluminium skin and – critically – the adhesive needed to hold heat-resistant tiles on the bottom and insulation blankets on the top in place. It was almost certainly fragments of these tiles and blankets that ground-based observers saw falling from the spacecraft as they watched its glowing plasma trail traverse the sky from west to east. Some observers saw a bright flash at around this time, which could have represented the point when the plume finally burned through the upper surface of the left wing.

The CAIB's Roger Tetrault later showed journalists a map pinpointing *where* each piece of debris was found. Interestingly, the most 'westerly' tiles – those that fell away early in Columbia's breakup – were those associated with RCC Panels 8 and 9. "What that indicates is there was probably a breach somewhere in the 8 and 9 area," Tetrault said. "The hot gas flowed into that breach, heated up the inside of the wing, the [adhesive] which holds the tiles to the outside skin heated up and basically lost its adhesion capability at 400 Fahrenheit and fell off to the west."

At about 1:54:11 pm, Columbia's computers responded to further disruption of the normal flight path, although on this occasion the forces trying to pull the nose to the left and roll the spacecraft to the left suddenly *reversed* and the left wing seemingly gained additional 'lift'. The computers readjusted the elevons to counteract the unwanted motions. Again, Husband and McCool may have noticed the adjustments on their displays, but made no attempt to contact Mission Control. Investigators would later blame the perceived lift on the now-weakened lower surface of the left wing, which was bowing inwards under increasing air pressure.

Seconds earlier, ground-based observers had videotaped a large piece of debris fall away from Columbia and vanish in her plasma trail; subsequent analysis concluded that it may have been a large portion of the left wing's lower surface. Additionally, Pat Goodman speculated that conditions might have been so bad at this stage that the melted aluminium spars and trusses could have been lying "in pools on the bottom of the wing." In all likelihood, the only structural support keeping the upper and lower surfaces of the wing together were the RCC panels and the leading-edge spar behind them.

By 1:56:16 pm, the plume of superheated gas had burned through the left-hand wheel well box and began scorching one of Columbia's landing gear struts; shortly thereafter, as the Shuttle swept over New Mexico, a pair of sensors recorded slowly increasing pressures in the outboard tyre. Meanwhile, on the bottom surface of the wing, the concave depression was growing in size and the ship's computers again commanded the elevons to counteract the unwanted deviations from the flight profile. Columbia was still, it seemed, barely in control. That was about to change abruptly.

Sometime after 1:58 pm, her flying characteristics shifted; possibly the left wing began to collapse. "The vehicle was in control and responding to commands up to that point and, after that point, something changed," said United Space Alliance manager Doug White. "[The flight system] continued to be in control and respond to commands, but the rates and the amount of muscle it needed to continue flying the vehicle the way it should be flown were continuing to increase. Something definitely happened to cause the flight control system to need more muscle and start having to fight harder."

It was not until the loss of tyre pressure data triggered an alarm in the cockpit that Husband was prompted to radio Mission Control, but his transmission was cut off in mid-sentence. In fact, STS-1 veteran John Young later remarked that the astronauts inside Columbia's cabin might not notice or 'feel' even major changes in the shape of the left wing, even if it *fell off* entirely. Presumably, Laurel Clark was still videotaping as the first of what would undoubtedly have been many master alarms over the next few minutes began blaring. We shall never know.

By this time the elevon adjustments were proving hopeless and no longer capable of preventing the ship's nose from yawing to the left. At 1:59 pm, two thruster firings, each lasting about 1.5 seconds, were automatically executed by the left-hand RCS jets in an unsuccessful bid to regain control. A second master alarm sounded in the cabin as the elevons' control circuitry failed and seconds later ground-based observers in western Texas saw a large piece of debris falling away from the fast-moving Columbia. In all likelihood, it was most of the left wing.

This was followed, in the first couple of seconds after 2:00 pm, by two other pieces of debris, which could have been the vertical stabiliser fin and part of one of the OMS pods. Despite the dire predicament in which she and her crew now found themselves – out of control, falling nose-first with her left wing missing and probably several other major structural components also gone – one last, reasonably reliable snippet of data hinted that her hydraulic power units and electrical generators were running, her main engine compartment was intact and her communications and navigation equipment were operating normally.

However, the astronauts knew they were in deep trouble. Master alarms would have sounded as Columbia's computers advised Husband and McCool of fuel leaks from the left-hand OMS pod and a multitude of other problems: the cooling system had failed, extreme temperatures were being measured by sensors on the ship's belly and along the left side of her fuselage and the electrical system was experiencing intermittent shorts. Columbia's orientation was changing rapidly, at more than 20 degrees *per second*, according to recovered data from the OEX recorder, subjecting its remaining structure to aerodynamic stresses for which it was not designed.

"Based on the debris and the track, I think the [left] wing disintegrated in pieces and we probably ended up with very little of the left wing actually attached before the entire orbiter broke up," said the CAIB's Scott Hubbard. The computers "lost control and the wing started to come apart, the sense from the data is that the orbiter went into a flat spin, possibly even going *backwards* for some period of time."

At around this time, Columbia's backup flight system computer recorded that one of the flight deck hand controllers – normally used by the pilots to manually fly the vehicle during its final approach to the runway – was moved beyond its normal position. There was some speculation that Husband or McCool, realising the danger, attempted to take manual control, but CAIB investigators would later conclude that the joystick-like device was probably just inadvertently bumped out of position during those horrifying final moments. In fact, until the loss of telemetry, Columbia's digital autopilot remained in control of the ship's descent.

Veteran astronaut Rick Searfoss had followed an almost identical re-entry profile during his stint in command of STS-90 five years before, conducting the first of several roll reversals to bleed off speed while high over west Texas. He wondered if Husband had had a chance, seconds before the alarms began sounding in the cockpit, to glance out of his window to see the area around the Red River where he grew up. "We'll never know," Searfoss told an interviewer sadly in February 2003, "at least not in this lifetime. I hope perhaps this was one of his very last thoughts."

Nonetheless, a shudder runs down the spine when one imagines the situation in the cabin during those last seconds. We do not know how the astronauts reacted upon realising that their ship was disintegrating around them. Perhaps, mercifully, they might have be so intently focused on their flight deck instruments, trying to comprehend and respond to master alarm calls and multiple malfunctions, that the end came blissfully quickly. All that is known with certainty is that the OEX recorder, found that March day in Hemphill by Art Baker, stopped working at 2:00:18 pm.

Sometime within the next minute or so – as Charlie Hobaugh tried to regain radio contact, as Leroy Cain and Jeff Kling pondered a problem with Columbia's tyres, as Laura Hoppe wondered how long intermittent communications would last, as Ed Garske and Bryan Austin saw multiple contrails over their homes in Texas, as Andy Cline woke with a sense of foreboding in the woods of Wyoming and as thousands of spectators in Florida awaited Columbia's triumphant homecoming – seven astronauts perished in the United States' second Shuttle disaster.

DEBRIS SEARCHING AND SOUL SEARCHING

"At this point, we're just trying to get it out to the public to not touch or tamper with this debris in any manner," said police spokesman Greg Sowell of the east Texan city of Nacogdoches on the afternoon of 1 February, "due to the possibility of toxic substances being on the debris." A city of some 30,000 inhabitants, located a few hundred kilometres northeast of Houston, Nacogdoches suffered the brunt of falling fragments from Columbia that day. Dentist Jeff Hancock, for example, was shocked when a metal bracket crashed through the ceiling of his office ...

Elsewhere in Texas, Gary Hunziker in Plano, just north of Dallas, described his astonishment as he saw "two bright objects" – obviously fragments of debris – flying separately on either side of the fast-moving Columbia. Still other eyewitnesses heard loud explosions that were variously likened to rolling thunder, sonic booms or "a car crashing into my house" while others imagined that the United States was under imminent attack from terrorists or UFOs. It is perhaps one of the greatest miracles of that horrific day, however, that no one on the ground was killed or seriously injured.

Circumstances could have been very different. During the subsequent investigation, it was determined that had Columbia begun to disintegrate barely a minute earlier, *three* times as many people could have been exposed to falling debris. Unfortunately, NASA's options for avoiding populated areas in the event of a catastrophe during re-entry were, and still are, limited. To accomplish a landing in Florida requires one of two flightpaths: one carries the Shuttle over California and across the United States' heavily populated heartland, the other traverses Central America, close to the Yucatan Peninsula, and over the Gulf of Mexico.

As debris from Columbia flooded in from search parties across several states, a floor grid was set up in a hangar at KSC to 'reassemble' what was found for analysis.

Even diverting Columbia to Edwards Air Force Base in California would not have been inherently 'safer' in terms of people on the ground. Although most of the re-entry flightpath would have been conducted over the Pacific Ocean, an Edwards touchdown would have still required Husband to risk crossing heavily populated Los Angeles. In addition to concerns of public safety, questions also arose in the weeks after the disaster of what – if anything – could have been done to lessen the immense atmospheric heating on the Shuttle itself during re-entry. The answer: virtually nothing.

Had NASA known at an early stage in the 16-day mission that Columbia's left wing was fatally crippled and she would not survive the return to Earth, the only remotely viable option was to send Anderson and Brown on emergency spacewalks to remove *all* unnecessary hardware from the payload bay – including the Spacehab module and FREESTAR pallet – and dump them overboard. This would reduce the re-entry weight by around 15,000 kg, which would slightly lower heating stress during re-entry, but the vehicle itself weighed 90,000 kg and releasing the payload would not be nearly enough to prevent a catastrophic breakup.

Moreover, that work would require at least two lengthy excursions. Several NASA managers also doubted that Anderson and Brown could have safely got into a position over Columbia's payload bay wall to 'see' the wing damage, let alone do anything about it. "Just the nature of them trying to position themselves in space underneath the vehicle", said Dittemore, "could cause more damage than we were trying to fix." However, when interviewed by the CAIB during the investigation, veteran spacewalker and six-time Shuttle flier Story Musgrave disagreed. "It's *not* difficult and *not* dangerous. There is zero risk involved. It's a 15-minute walk. Another thing that's important is you've *involved* the people whose lives matter. You've *involved* them in the process of how you come home. There's a particular moral element to that."

A 'rescue' flight by another Shuttle was considered a risky option, although many within NASA deemed it irresponsible to send *another* crew aloft when the malfunction that had doomed STS-107 was still not understood. Even as late as March 2003, Ron Dittemore and others doubted that a mere chunk of foam could possibly have been responsible for the disaster. Columbia's sister ship Atlantis was, on 1 February, almost ready to move to the launch pad for a scheduled International Space Station visit four weeks later. Could her flight have been radically changed at the last minute to rescue Husband's crew?

Technically, yes, but only on the assumption that the implications of the damaged left wing had been identified as early as the fourth day of the STS-107 mission. If that had happened, Atlantis might have been hurriedly readied for launch as early as 9 February with a skeleton crew of four – Commander Eileen Collins, Pilot Jim Kelly and Mission Specialists Soichi Noguchi and Steve Robinson – to rendezvous with Columbia and transfer spacesuits and carbon dioxide-scrubbing lithium hydroxide canisters from ship to ship in what would likely be an operation taking several days.

Fortunately, the extended-duration nature of STS-107 meant that the EDO pallet was housed in Columbia's payload bay; this would have enabled Husband's crew to remain comfortably aloft until 5 February, but if they conserved their consumables, shut down the science payload and reduced their movements they could have survived another 10 days. "You just power down to give yourself more time," former astronaut Blaine Hammond has observed. "You power down the orbiter to minimum levels and conserve as much as you can and you can stay up there 30 days with the minimum systems needed to keep the orbiter alive."

Hammond also addressed fears that, without fully understanding what caused the initial damage to Columbia, the 'rescue' mission might itself suffer a similar fate. "Then you've got to ask yourself how the shortcuts affect the safety of that flight," he said. "You'd have to get a crew, but you'd have no shortage of volunteers to go up, but you're putting *those* guys at risk and putting another orbiter at risk by short-cutting all of the normal checks and inspections. To save seven lives, I think that's certainly a good trade-off, [but] you're putting a lot on the line."

Sadly, such a rescue flight never came to pass, which former astronaut Mark Brown saw as a lost opportunity of Apollo 13-style proportions. "I think it potentially could have been NASA's finest hour," the STS-28 veteran said. "We could have made a heroic effort to try to bring the crew back." Nonetheless, some satisfaction arose from Laurel Clark's unfinished videotape that showed a crew living their dream and not realising until the final seconds that they were in mortal danger. It could have been a much more protracted and lingering end, according to space analyst Jim Oberg.

Had the crew known what was happening, their ambitious science mission would have been lost in its entirety, the joy of being in space would have been tarnished and if Atlantis' rescue flight had failed or been unable to reach them in time, they would have simply *existed* in space until their oxygen finally ran out. Within a couple of months, the eerily silent Columbia would have been gradually dragged into the atmosphere in a fireball. "It would be visible at dawn and dusk and *that* would be pretty creepy," said Oberg, likening it to a Viking funeral.

As debris rained down from the Texan skies that February day, it might be supposed that all of the painstaking scientific research done during the STS-107 mission would have been lost forever. In most cases, it *was* lost, and its loss after such a highly successful flight proved devastating to many scientists whose experiments had taken years to develop. "I hope they can salvage something," said biochemist Hideaki Moriyama of the University of Nebraska, who had supplied various protein samples as part of ongoing studies of HIV-AIDS, Huntington's and Parkinson's diseases. "It took more than four years to prepare these experiments."

Other scientists, including David Warmflash of JSC, who had been investigating bacteria cultures, regarded the loss of experiments as "not in the same category as losing a crew". However, in a final status report on the STS-107 research, released by NASA on 19 February 2003, it was revealed that experiments whose results were downlinked to Mission Control – primarily the combustion, materials science and fluid physics investigations – had returned between 50% and 90% of their data. Others, including the vast majority of the medical and biological experiments, whose specimens were to be analysed after Columbia's landing, were destroyed.

Except, that is, for one quite remarkable discovery.

In April 2003, searchers in a Texan field unexpectedly found a canister filled with a set of *live* moss and thousands of roundworms from one of the biological experiments inside the Spacehab module. The tiny, pencil-tip-sized worms, known as *C. elegans*, have typical lifespans of about a week and were actually the fourth or fifth generation of their forebears that were in orbit during Columbia's mission. Designed by researchers at Ohio State University and NASA's Ames Research Center in California, the experiment had been looking at new synthetic nutrients for the worms.

Another experiment of sorts that was recovered was *Columbia* – or, at least, debris from around 38% of her, weighing some 41,000 kg – according to Mike Leinbach, the chairman of the ship's reconstruction team at KSC. Unlike Challenger, whose shattered remains had been buried in a disused missile silo, it was decided that Columbia's fragments would be 'lent out' to qualified researchers seeking to design new hypersonic aircraft and next-generation spaceplanes. "We're going to learn from Columbia," Leinbach said. "This is the legacy we're going to leave the STS-107 crew and their families."

The first set of debris loaned to researchers was shipped to The Aerospace Corporation of El Segundo, California, in June 2004, who conducted non-destructive tests to develop better analytical models for predicting the behaviour of composite materials during hypersonic flight. Of particular interest to the corporation's materials scientists were graphite/epoxy honeycomb skin fragments from Columbia's OMS pods and debris from her main engine and RCS helium pressurant tanks. According to researcher Gary Steckel, such information could provide insights into the kinds of wreckage most likely to survive re-entry and possibly present hazards to people or property.

"All the pieces that we have, we know where they were found," explained Steckel, whose group were lent the pieces for a year. "Based on the mass of those articles and their geometry, our people will do a trajectory analysis to try to figure out exactly where it was when it broke free. Once they've got that, the thermal history of the debris can be predicted. I'll be looking at what peak temperatures were seen, how much heating the debris saw and how that jibes with what other people predict."

The first educational institution given permission to use debris from the spacecraft was Lehigh University of Pennsylvania which, in March 2005, received a loan of 50 fragments from NASA – including shards of windshield and wing pieces – as part of efforts to understand why certain components failed and how they responded under extreme hypersonic forces. Employing powerful electron and light microscopy, materials scientist Arnold Marder and his team hoped to identify "tell-tale signs of the mode of failure" and possibly gain new insights into the design of future hypersonic spaceplanes.

For the three-man team on board the International Space Station – including former Columbia flier Ken Bowersox – the impact of losing their friends was profound. "My first reaction was pure shock," Bowersox told a space-to-ground news conference shortly after the disaster. "I was numb and it was hard to believe that what we were experiencing was really happening. As that reality wore on, we were able to feel some sadness." He added that Mission Control had reduced his schedule, along with those of his comrades Nikolai Budarin and Don Pettit, to give them time for reflection. "We've had time to grieve for our friends and that was very important. When you're up here for this long, you can't just bottle up your emotions and focus all of the time. It's important for us to acknowledge that the people on STS-107 were our friends, that we had a connection with them, and that we feel their loss, and each of us had a chance to shed some tears." The station crew was entering their eleventh week in orbit at the time of the disaster and was due to return to Earth on board Atlantis in mid-March 2003.

That changed abruptly on 1 February. Contingency plans had been laid before the Columbia disaster to run the station unmanned if necessary, but it was decided that bringing Bowersox's crew home and leaving it unoccupied would make it harder to cope with unexpected problems. Nonetheless, without frequent Shuttle visits, three-member crews could no longer be effectively supported and the option was taken to fly two-man 'caretaker' teams on six-month tours of duty to keep the station operational in the interim. The first caretaker crew – Russian cosmonaut Yuri Malenchenko and American astronaut Ed Lu – were sent aloft in April 2003.

In the temporary absence of the Shuttle, launching these crews to and from the orbital outpost was only possible using the Russian Soyuz spacecraft. As a result, on 6 May 2003 Bowersox and Pettit became the first NASA astronauts to return to Earth on a Russian spacecraft and land on foreign soil. They and Budarin also became the first crew to return safely to Earth in the wake of the Columbia disaster.

MORE STABLE THAN LAUNCH?

Shortly after the top-secret STS-28 mission in August 1989, veteran astronaut Robert 'Hoot' Gibson was assigned to investigate an unusual aerodynamic 'shift' experienced by Columbia during her return to Earth. Gibson, who had commanded Columbia on STS-61C three-and-a-half years earlier and later served as chief of the astronaut corps, found the surfaces of Columbia's wings were two to four times 'rougher' than those of her sister ships Discovery and Atlantis. Moreover, he discovered that her *left* wing was considerably rougher than her right wing and suspected that this had been responsible for a slight change in her flying characteristics during re-entry.

Another, similar, incident cropped up during the STS-73 re-entry in November 1995. Although the cause of the aerodynamic shift – an unusual and earlier-than-predicted increase in heating and atmospheric drag associated with the left wing – was explained by NASA, Gibson suspected that the surface roughness could have been a factor. On both previous occasions, Columbia's crews survived re-entry, but according to aerodynamicist John Anderson of the National Air and Space Museum, the combination of surface roughness *and* severe RCC damage from a chunk of foam could have conspired to pull the ship into a fatal 'sideways' flight angle.

Anderson speculated that aerodynamic shifts alone probably would be insufficient to cause catastrophic damage, but that the damaged RCC panel – in the particularly critical region of Panels 8 and 9, which are subjected to the most extreme re-entry heating – "just might have been enough to throw things over the edge." Documents released by United Space Alliance on 3 February 2003 revealed that more than 70% of Columbia's heat-resistant tiles were *original* ones installed before STS-1, leading Anderson to suggest that drag effects could have pulled the ship into the 'sideways' orientation that would have caused her to rapidly break apart.

He pointed to the two – and possibly as many as four – yaw thruster firings executed in Columbia's final seconds. "Anything to cause increased drag on that left wing would certainly have caused it to yaw," Anderson said. "The Shuttle is designed to fly straight. It's *not* designed to fly sideways. That would have been absolute disaster if something had yawed it so much that it was basically trying to fly sideways." In his STS-28 study, Gibson found that temperature increases normally associated with the formation of the insulating boundary layer had essentially begun *too early*.

Normally, during re-entry, the Shuttle's wings experience a smooth 'laminar' airflow in the uppermost reaches of the atmosphere. As the spacecraft dips deeper into the steadily thickening air, this becomes more turbulent and imparts greater heating and aerodynamic drag on the wings. During STS-28, the transition from laminar to turbulent airflows occurred just 15 minutes after entry interface – five minutes earlier than it should have done – said Gibson, adding that "the left wing saw a significantly higher heating environment than the rest of the orbiter." A similar situation developed on STS-73.

Although NASA engineers explained the early transition as being due to protruding gap-fillers between the tiles, which Gibson partly accepted, he also believed surface roughness to have been the primary problem. Still, he never expected a Shuttle to be lost during re-entry; both he and his wife, former astronaut Rhea Seddon, had flown Columbia and had eight missions between them. "We are the state-of-the-art in aerospace technology in bringing those things down," he said after the disaster. "It surprises me that something happened at all during re-entry, even though we've known that re-entry could be stressful."

Having said that, Gibson admitted that, during his stint as chief of the Astronaut Office, he never lost sight of the risks involved. "I was continually fighting my astronauts who didn't want to wear pressure suits for re-entry. I said 'No, we're going to wear the suits'. Their argument was that re-entry is more stable than launch, and it is, but that's *relative* stability. It's still an extremely difficult task," he said, but added quietly "I never thought we'd lose one during re-entry."

Many veteran flight controllers agreed. "Any arrogance I may have had went out the window on 1 February," said flight director Wayne Hale. "In my personal life, I thought we had it pretty much knocked ... I would have told you we understood what we were doing and we had mature processes and good hardware. I think all those assumptions have been shattered." Over-confidence, and the general opinion that the Shuttle was an 'operational' vehicle rather than an 'experimental' one, would be one of the criticisms levelled at the space agency by the CAIB's final report.

Series of images of a damaged RCC panel following a foam-strike test at the Southwest Research Institute (SWRI) on 7 July 2003.

"I think there's been a little bit of denial that NASA, at least in the Shuttle programme, has modified its organisational structure over the years into one that no longer contains the attributes that they built their reputations on," said CAIB chairman Hal Gehman, a retired US Navy admiral who had also led the Pentagon's inquiry into the USS Cole terrorist bombing in Yemen in 2000. "There may be some people who deny that, but the Board is absolutely convinced [that] the management team they have right now is *not* capable of safely operating the Shuttle over the long term."

However, the report also blamed Congress and the White House for operating NASA under an unrealistic set of rules and guidelines. "Exploring space on a fixed-cost basis is not realistic," Gehman said. "Launching Shuttles on a calendar basis, instead of an event-driven basis, is not realistic. Demanding that you save money and run this thing in an efficient and effective way and that you get graded on schedule and things like that is not realistic. The whole nation and Congress and the White House has an unrealistic view of how we do space exploration."

The Board shied away, however, from blaming individuals for mistakes made during the mission. Former flight director Linda Ham, for example, had become a lightning rod for criticism in her role as chairwoman of the Mission Management Team: some observers condemned her for 'ignoring' requests to acquire spy-satellite imagery of Columbia while in orbit, overlooking engineers' concerns about the severity of the foam strike and failing to question a hurried, Boeing-led analysis of the wing damage. The CAIB report, by its own admission, saw no value in blaming individuals but instead encouraged broader 'cultural' change within NASA.

It is perhaps this cultural change that will be the biggest challenge to face the agency, for the technical 'root cause' of Columbia's destruction was relatively simple and had been in most people's minds since the day of the disaster: the foam *did* do it. That suspicion went from hypothesis to leading theory to absolute certainty on 7 July 2003 when CAIB member Scott Hubbard watched a chunk of foam being fired at an RCC panel. The test took place at the Southwest Research Institute in San Antonio, Texas, using the facility's unique nitrogen-gas cannon.

Not only was the chunk of similar size and weight to the piece that fell from STS-107's External Tank, but its velocity was also comparable at 850 km/h and it was fired at an RCC Panel 8 'borrowed' from another Shuttle. To precisely mimic aerodynamic conditions 81.7 seconds into Columbia's ascent – at which point the vehicle was racing towards orbit at twice the speed of sound – a slight rotational spin was imparted on the chunk of foam and these factors produced a gaping 40 × 43 cm hole in the panel!

The crowd of spectators watching the test gasped with both horror and amazement.

In fact, the impact was so violent that it damaged some instrumentation, imparted 1,000 kg of force and blew two RCC fragments back into the hole. "We've found the smoking gun!" exulted Hubbard. Additionally, the fragments blown 'back' into the hole helped to set another mystery to rest: an Air Force ground radar had detected a strange object trailing Columbia on 17 January, flying slightly 'slower' than the Shuttle but travelling in the same orbit. It was without doubt a fragment of RCC debris from the impact and burned up in the atmosphere a couple of days later.

Several long-distance camera shots of the foam impact were acquired, but one was out of focus and another lacked the resolution to show the impact site in sufficient detail. Boeing's analysis of the impact, using a computer-modelling program known as 'Crater', predicted regions of localised damage to the underside of Columbia's left wing, but did not anticipate a catastrophic RCC breach. The CAIB later heard that several mid-level engineers had serious misgivings about the Boeing analysis, but that their concerns were not effectively passed up through the chain of command to Dittemore or Ham.

"Communication did not flow effectively up to or down from managers," the Board's report read. "After the accident, managers stated privately and publicly that if engineers had a safety concern, they were obligated to communicate their concerns to management. Managers did not seem to understand that, as leaders, they had a corresponding and perhaps greater obligation to create viable routes for the engineering community to express their views and receive information. This barrier to communications not only blocked the flow of information to managers but it also prevented the downstream flow of information from managers to engineers."

Ultimately, the source of the debris was pinpointed at the 'bipod ramp': an area at which the Shuttle's nose is connected to the External Tank by a spindly, two-legged strut. Before launch, insulating foam was poured around the strut and carefully shaped to make it aerodynamically 'smooth'. Next, a wire brush was used to create airholes, allowing trapped gases to escape and preventing the formation of 'voids' in the foam. Exactly *why* the foam fell from Columbia's tank may never be known but it is possible that a void formed and caused a piece of debris to break away during ascent.

"As a physicist conducting a test," Hubbard said later, "I feel gratified that after months of work, we were able to demonstrate this connection between the foam and the damage. But I know it was a source of tragedy, so that makes me feel very sad. This whole six months, we've constantly been reminded by pictures of the seven lost astronauts what this all means."

The bipod ramp, it was decided, would be eliminated from all future External Tanks. However, Gehman went further. "We view this as a *system*," he said. "NASA has to cut down on the amount of debris that comes off. They have to toughen the orbiter, they have to be able to inspect and repair the orbiter and then they've also got to give the crew a better chance to survive. All four of these contribute to safer operations and not any one of them, in my view, is a fix."

Techniques for repairing the orbiter had been considered before STS-1. In fact, as early as January 1980, Martin Marietta had been awarded a $2.1-million contract to build a repair kit which, coupled with a 'jet backpack' called the Manned Maneuvering Unit (MMU), might be used to fix damaged tiles in orbit. It would, however, be tricky, according to STS-61C's George 'Pinky' Nelson, who worked on developing the kit and flew the MMU in April 1984. "Imagine standing on perfectly slick ice and trying to work on a wall," he said. "Every time you touch it, you slide away."

Despite testing problems, development of the kit went well until NASA cancelled it in mid-1980. Later, as each Shuttle crew returned safely and the patchwork of thermal tiles held up to the aerodynamic and thermal stresses of a hypersonic re-entry regime, such kits seemed increasingly unnecessary. In the wake of STS-107, all future crews will carry as many as five possible repair kits, including a cure-in-place caulk-like substance and gun; furthermore, in all likelihood they will be directed exclusively to the International Space Station, from where other astronauts will use high-resolution, long-distance cameras to photograph the orbiters' wings and undersides.

"This is going to be the most-photographed Space Shuttle mission that's ever been launched," STS-114 Commander Eileen Collins told an interviewer in the spring of 2004, during a break in her training to lead the first post-Columbia mission. "We're going to have cameras on the ground, on the External Tank, on the Shuttle, and cameras on airplanes that will be flying during the ascent itself. I have

Astronauts Jim Halsell (left, in white shirt) and Scott Parazynski (right, wearing EVA 'glove') discuss the pinkish, cure-in-place caulking material under consideration as a repair measure for patching future RCC or tile damage.

no doubt we're going to have plenty of pictures of this launch. Our crew will be taking pictures of the External Tank after it separates. On Flight Day Two, we're going to be doing an exterior inspection with cameras and lasers of the wing leading edge and the underside of the Shuttle. On Flight Day Three, we're going to do a pitch-around manoeuvre as we approach the space station so [that its] crew can look at us. I have no doubt that if there's any damage, we're going to know it." The key issue, Collins said, was eliminating the source of the debris in the first place, but having additional repair techniques in place as a 'just in case' measure.

Among the management decisions attacked in the CAIB's report was the fact that virtually *every one* of the 113 Shuttle missions sent aloft since April 1981 had been hit by External Tank debris. During the first 15 flights, until January 1985, the orbiters returned home with an average of 123 dents and scrapes to their heat-resistant tiles; by this time, NASA had already accepted that the primary cause was falling foam insulation and ice from the tanks. The level of damage rose significantly, however, during the remainder of 1985, with Shuttles suffering an average of 227 damaged tiles per flight.

Most serious were the STS-51G and STS-51F missions, which suffered 315 and

553 impacts respectively. The cause was attributed to a faulty batch of resin used to bond foam onto their External Tanks. Concerns were also raised by high-level managers from prime contractor Rockwell International, including then-vice-president for Shuttle integration Martin Cioffoletti. However, as each mission returned safely, he had no firm evidence to declare that the tank debris posed a threat. "We couldn't stand up and say this is going to kill anyone," he said, "because there was no evidence this was causing that kind of significant damage."

On the second flight after the Challenger disaster, STS-27 in December 1988, Atlantis suffered a debris strike during ascent of such severity that Mission Control asked the crew to take a look at it with their RMS cameras. "I shiver with fear as I imagine what would happen to Atlantis if there is major damage," Mission Specialist Mike Mullane later wrote. When the astronauts saw the damage, he added, "We gasped. Hundreds of tiles are scraped and gouged. At least one tile is completely missing. What's going to happen to us on re-entry?"

Fortunately, Atlantis returned safely to Earth, doubtless with five extremely anxious astronauts on board. Damage occurred on most flights, but three other missions fared particularly badly. On STS-2 in November 1981, hot gas blasted past Columbia's RCC panels near a spot where they joined to heat-resistant tiles on the ship's belly; then, on STS-5 in November of the following year, a small hole was melted in a tile-covered metal bar. More recently, in May 2000, Atlantis came home from STS-101 with a dislodged seal between two RCC panels in her left wing.

Each incident was, to be fair, addressed and corrective action taken, but NASA's general attitude towards tile damage was that it was an irksome maintenance issue rather than a disaster waiting to happen. Tile repairs "seemed to be a necessary pain-in-the-butt kind of thing," said Seymour Himmel, a former member of the Aerospace Safety Advisory Panel set up by Congress in the wake of the January 1967 Apollo 1 fire. This comes as quite a surprise when one considers that the tiles can be damaged by oil from an uncovered finger, dropped tools or even a *coin*.

Ironically, throughout KSC, signs have been present for years instructing workers to ensure that all 'foreign objects' – coins, glasses, jewellery – are kept secured or zipped safely into pockets to avoid falling a few feet and cracking the orbiters' fragile skins and yet damage from foam falling at several hundred kilometres per hour during ascent had never been properly addressed or fully prevented ...

A VERY 'HUMAN' MACHINE

Eight precious individuals were lost on 1 February 2003.

It is this fact that is so often overlooked, for in addition to Husband, McCool, Brown, Chawla, Anderson, Clark and Ramon, the world lost its pioneering Space Shuttle. Columbia was near the end of her 28th flight and, despite her distress, fought with valiant and remarkably 'human-like' characteristics to save her crew. *She* had repeatedly ordered elevon adjustments to cope with increasing aerodynamic drag on her left wing, *she* had commanded RCS firings in a desperate bid to remain on course and *she* had kept the astronauts alive for at least the first few seconds after her fuselage broke apart.

She "was doing well ... but losing the battle," said Shuttle manager Ron Dittemore at an emotional press conference on 5 February. Added Rick Searfoss: "It was beyond her control and she was *trying*. It's a poignant thought for me to think about that." For his STS-90 crewmate Jay Buckey, however, justifiable pride in Columbia's achievements was tinged with heartbroken sadness at being robbed of the chance to take his grandchildren to see her someday in an aerospace museum. "Columbia is a part of us," Buckey said. "It was a privilege to have been on the Columbia."

STS-109 veteran Duane 'Digger' Carey, who left the astronaut corps in 2004 to pursue around-the-world motorcycling, also regarded flying NASA's oldest Shuttle as a "humbling experience", adding that, due to its sheer complexity, "there were an awful lot of smart folks who came before us and more or less wrote the book on how to operate the machine." He and Mike Massimino visited KSC in April 2003 to view debris from the ship they had flown a year earlier. "It was really moving to see all those pieces laid out," Massimino said later. "It was a day I'll never forget."

In spite of her achievements, she earned a reputation for being notoriously difficult to prepare for flight, but that as soon as she lifted off she performed beautifully. John Young, one of those who 'wrote the book' on how to operate her, was in Mission Control when she broke up. "I knew the instant we lost the trajectory [data] that we had lost the vehicle," he said. "It was a good vehicle." Nonetheless, he stressed it was "not an operational vehicle and [NASA] needs to stop treating it like one. This is still an *experimental* vehicle. STS-107 proved that pretty well."

Columbia was the best orbiter to fly through the atmosphere, according to Ken Cockrell, who rode all four Shuttles and commanded STS-80 in late 1996. "Endeavour and Discovery and Columbia are solid as a rock after the boosters come off during ascent. They're just *smooth*. It's like you're in an electrically powered vehicle," he said in the spring of 2002, adding that "Columbia is the best glider and probably the easiest one to land because it's a little heavier, a little smoother aerodynamically and it glides just a tiny bit better ... [which] makes it easier to make a sweet landing."

"Occasionally, good ships can run aground," said STS-87's Winston Scott. "It depends on the winds, the weather, the hands of fate and that's exactly what happened to Columbia." Congressman Bill Nelson, who flew STS-61C, called her "a beautiful old girl. When [we came] back in that searing heat of re-entry, we were on the night side of the Earth and I remember being stunned when I looked out the window and it was daylight ... then I realised it was that hot plasma that had enveloped the orbiter and it was this beautiful tangerine pink, almost like it was at sunrise ... "

"You rely on those machines for your life," remarked Rick Searfoss. "It felt like a home, being protected with no fear. It's somewhat akin to the way old sailors in clipper ships were brought home safely from a journey. When [Columbia] was lost, it affected me in that way."

"She was often bad-mouthed for being a little heavy on the rear end," remembered STS-1 Pilot Bob Crippen at a KSC memorial on 7 February 2003, "but many of us can relate to *that*. Many said she was old and past her prime, [but she] had a great many missions ahead of her. She, along with her crew, had their life snuffed out in her prime." Columbia's next mission would have been a trip to the International Space Station in November 2003, led by astronaut Scott Kelly. He provided some insights into Rick Husband's final actions on board the venerable ship.

"He was probably focusing on what he was trained for," Kelly told a CBS journalist in March 2003, "and that is to respond to whatever malfunctions they had in the best way he could." No one doubted Husband's capabilities. "You really only need to know two things," said Jim Halsell, one of few people to have flown Columbia three times. "First: we recruited him into the Astronaut Office because of his wide-known reputation within the Air Force. Second: Rick was offered his own Shuttle command after only one flight as a Pilot, instead of the standard two. He was *that good.*"

He was not reckless or unaware of the risks involved. "We all knew the risks because of Challenger, but we all have a passion for it," said Kevin Kregel, who flew Columbia twice and inspected her debris as part of his new job helping to design next-generation hypersonic spaceplanes. "I flew [on STS-99] with a Japanese and a German. Just to think that 60 years ago, our parents were fighting each other and

Chief astronaut Kent Rominger, who commanded Rick Husband's STS-96 mission in mid-1999, speaks about the crew at a JSC memorial on 5 February 2003. Seated behind him, NASA Administrator Sean O'Keefe looks on.

now we're working together in space ..." Fellow astronaut Jack Lousma agreed, adding that the risk of being killed "is something that comes with the turf."

Franklin Chang-Diaz, who holds a joint record with Jerry Ross for having flown into space seven times, went further. Speaking to children at Priceville Elementary School in Alabama in March 2003, he admitted that his profession *was* dangerous, but pointed out that it "ensures our survival as a species. Our planet has six billion people and our resources like food and water are getting scarce. We need a presence in space to develop new ways and find new sources." Added Rick Searfoss: "Humans aren't meant to grub around on the surface of the planet. Our destiny is to move beyond."

Four-flight Shuttle veteran Dan Brandenstein was in Mission Control on the day Columbia disintegrated – not as an astronaut, but on behalf of aerospace contractor Lockheed, which had developed the workstations and equipment used by the flight controllers that day. "You hope you don't have to go through it again in a lifetime," he said, remembering the Challenger disaster 17 years before. "My role is somewhat different now, but anyone who has flown in space ... recognises what it takes and the risks. It's very sad and it hits the families hard."

Nor was it exclusively the families of the crew who were shattered by the tragedy; their colleagues in the Astronaut Office at JSC, who lived with the risks on a day-to-day basis, expressed their own feelings of devastation at the loss of people they regarded as family members. "It's just a body blow to your psyche," STS-80 veteran Tom Jones told a *Baltimore Sun* journalist on 2 February. "These are all my friends." He admitted, however, that there was no way Husband or his crewmates could have foreseen the catastrophe about to engulf them or do anything to prevent it.

For two other former Columbia fliers, the pain was even closer. Mike Lopez-Alegria was on board Endeavour for STS-113 in November 2002, the last fully successful mission before the disaster. He recalled with fondness NASA's oldest orbiter. "She was a flagship of our fleet and she represented the Space Shuttle programme ... our nation's human spaceflight efforts and, in part, our national pride," he said at the opening ceremony of a memorial fountain erected to the STS-107 crew at Clear Lake in Houston in March 2003. "We are wounded, but we'll recover and return to flight, just as Columbia's crew would ask of us."

Dave Wolf, a medical doctor who flew Columbia in 1993 and later spent four months on the Russian Mir space station, narrowly missed being selected to fly on Husband's mission. He was assigned instead to an International Space Station flight in October 2002. This mission also experienced foam falling from *its* External Tank, which hit an electronics box on one of the SRBs. Fortunately, Wolf's crew made it into orbit – and back home – safely. "It's emotionally trying," he said. "When you lose seven close people in your office, it has a large impact. This was a great group of folks."

Jeff Ashby, another Columbia veteran, spoke at a memorial service in Lufkin, Texas, on 8 February. He recalled the warmth, humanity – and humour – of the STS-107 crew. "They actually baked cakes for their training instructors on their birthdays," he told a congregation at the First Baptist Church. "The crew mascot was a small furry hamster ... that sang the 'Kung-Fu Fighting' song. They referred to their crew secretary as 'The Great and Powerful Roz' and they successfully convinced her that if she put candy on her desk, she'd see them more often. I remember the Astronaut [Office] Christmas party just a couple of months ago, and, soon after it began, it was evident that there was one area in the party – one table – that was a little livelier than most, and of course it was their crew. About an hour into the party, I went over to their table to kind of share in the joy that they had and was very quickly helped into a seat by Ilan Ramon, at which point his wife tattooed a small 'STS-107' emblem on my cheek! They were *all* wearing them!"

Their secretary, Roz Hobgood, was virtually an eighth crew member. "I was greeted with hugs in the morning," she told an *Amarillo Globe* journalist. "Every day I had candy or flowers." She recalled helping Rick Husband clear boxes when he moved into his new office, talking to Mike Anderson about his beloved Porsche, taking her first trip in a hot-air balloon thanks to Dave Brown, tackling Kalpana Chawla's brain-teasing questions, enjoying the childlike enthusiasm of Willie McCool and the warm friendship of Laurel Clark ... and even mistakenly thinking Ilan Ramon was *dyslexic* when he wrote, it seemed, 'backwards' in Hebrew!

Two Japanese astronauts – one of whom flew with Chawla, the other who served as Deputy Mission Scientist for STS-107 – expressed their sorrow and stories of the lost crew members. "I can readily imagine Mike [Anderson] sitting comfortably in discussion with Aristotle on issues of scientific knowledge," said Chiaki Mukai, a veteran of two Shuttle flights, including one trip on board Columbia, "for Mike was at his heart a philosopher ... a most kind and gracious philosopher." Takao Doi agreed, describing Anderson as "calm and composed ... [with] a heart filled with warmth" and his fallen crewmate Chawla as "a person overflowing with gentleness and charm".

Despite the horrific tragedy, very few astronauts expressed doubts about the value of their work and barely a handful have resigned from the corps. Columbia's loss did, however, force some soul-searching, particularly from the point of view of the astronauts' families. Mike Massimino remarked that, for his children, the accident was "kind of right in your face. This is what Daddy does for a living", but after discussing his future with his wife Carola he decided to remain on active flight status. "I like the job day to day," he said. "Flying in space is the icing on the cake."

Scott Altman had no reservations about the levels of safety put in place before each flight. "Every time that I strapped on the Shuttle", he said, "I felt confident everybody had done what I thought was everything possible to make sure we were ready to go." However, he added, "maybe we just haven't imagined enough what could go wrong. As you try to make safety improvements, you go at what you perceive are the biggest risks. Was it a bad decision to go after things that were more of a threat before? I'm not sure I could say it was."

"Safety and reliability have to be *designed* into systems," pointed out Tom Henricks, who flew Columbia twice, once in the Pilot's seat and later as Commander. "If that's done properly, it reduces the requirement of inspections. Every time you touch a system like that you are opening yourself up [to] breaking the components. You're better off just leaving it alone. One reason you saw the number of [Shuttle] inspections decrease over time was because things were designed better." However, Henricks felt that safety was compromised and NASA began accepting too many risks for his liking.

After returning from his fourth flight, STS-78, he had been offered the chance to command the final Shuttle docking mission to the Russian Mir space station, but turned it down due to safety concerns. "The pendulum [after the Challenger disaster] had swung to as conservative as they could make it," he said, "but then that pendulum started swinging back almost immediately and it was *very* prevalent by the time we were going to Mir. We were still sending Americans to Mir after a fire and a collision. Near the post-Challenger timeframe, that wouldn't have happened."

Much has been written about Columbia and her achievements. She flew 28 times, making her the second most-flown orbiter after Discovery, and spent over 300 days in orbit. One hundred and twenty-six men and women have gazed Earthward from her windows, of whom 26 flew her twice and four astronauts – Tammy Jernigan, Jim Halsell, Don Thomas and Rick Linnehan – have served on board her three times. She holds, and in all likelihood will continue to hold, the record for the longest Shuttle mission and even her last tragic flight established itself as the fourth-longest in the programme's history.

In the shadow of these and other awesome achievements, it is difficult to produce an epitaph that sums up both the ship herself and her final, fallen crew. Then, a few months after the Columbia disaster, NOLS guides John Kanengieter and Andy Cline spoke of the time they had spent with the STS-107 crew in the summer of 2001. Eager to learn exactly what a launch and re-entry 'felt' like, the two men asked the astronauts to describe it for them. Rick Husband and his crewmates, never ones to shy away from a challenge, did more than that.

Dressed in their backpacks, camping and walking gear, high in the mountains of Wyoming, they described for Kanengieter and Cline their activities on launch day, from suiting-up, to strapping into their seats on board Columbia, to the vigorous shaking as their spacecraft thundered skyward. Then, in a particularly poignant memory for Kanengieter and Cline, they described their positions and responsibilities during re-entry. Just as Laurel Clark's videotape showed, she, Chawla, Husband and McCool took their 'seats' in two lines on the flight deck, while Anderson, Brown and Ramon settled in another line into the locker-studded middeck for a smooth ride home.

"One of the things that struck me on 1 February," Cline said later, "was when I was watching the Shuttle break up, one of the first thoughts [I had] was of them sitting in that line and knowing exactly where they were sitting – because they had *let us experience that*, beforehand. It was a magical moment and I'll always cherish that because, in my book, they always landed."

Columbia's missions

Designation   STS-1
Sequence   1st Shuttle flight and 1st flight of Columbia
Milestones   1st flight of winged orbital spacecraft with a crew on board and
                     1st flight of a manned spacecraft which had not first been tested
                     in an unmanned capacity
Launch   12 April 1981 at 12:00:04 pm GMT from Pad 39A at the
                     Kennedy Space Center in Florida
Payload   Development Flight Instrumentation (DFI) and Aerodynamic
                     Coefficient Identification Package (ACIP)
Max. altitude   270 km
Max. inclination   40.3 degrees
Landing   14 April 1981 at 6:20:57 pm GMT on Runway 23 at Edwards Air
                     Force Base in California
Distance   1,730,000 km
Orbits   36
Duration   2 days, 6 hours, 20 minutes and 53 seconds
Crew   John Watts Young Jr, 50, commander
                     Capt Robert Laurel Crippen, 43, US Navy, pilot

Designation   STS-2
Sequence   2nd Shuttle flight and 2nd flight of Columbia
Milestones   1st reuse of a winged orbital spacecraft with a crew on board, 1st
                     in-space testing of Canadian-built robot arm, 1st scientific
                     payload to fly on board the Shuttle, 1st flight of European-built
                     Spacelab pallet and 1st Shuttle mission to return home early due
                     to mechanical problems
Launch   12 November 1981 at 3:09:59 pm GMT from Pad 39A at
                     Kennedy Space Center in Florida
Payload   Development Flight Instrumentation (DFI), Aerodynamic
                     Coefficient Identification Package (ACIP), Office of Space and
                     Terrestrial Applications (OSTA)-1 and Interim Environmental
                     Contamination Monitor (IECM)
Max. altitude   250 km

Max. inclination	38 degrees
Landing	14 November 1981 at 9:23:12 pm GMT on Runway 23 at Edwards Air Force Base in California
Distance	1,730,000 km
Orbits	36
Duration	2 days, 6 hours, 13 minutes and 13 seconds
Crew	Col Joseph Henry Engle, 49, US Air Force, commander
	Capt Richard Harrison Truly, 44, US Navy, pilot

Designation	STS-3
Sequence	3rd Shuttle flight and 3rd flight of Columbia
Milestones	Longest Shuttle mission to date, 1st flight of an External Tank without Fire Retardant Latex primer, 1st test of a deployable payload with the robot arm and 1st Shuttle mission kept in orbit an additional day due to unacceptable weather conditions at the landing site
Launch	22 March 1982 at 4:00:00 pm GMT from Pad 39A at Kennedy Space Center in Florida
Payload	Development Flight Instrumentation (DFI), Interim Environmental Contamination Monitor (IECM), Office of Space Science (OSS)-1, Monodisperse Latex Reactor (MLR), Electrophoresis Equipment Verification Test (EEVT), Heflex Bioengineering Test (HBT), Shuttle Student Involvement Project (SSIP) and one Getaway Special (GAS) 'practice' canister
Max. altitude	235 km
Max. inclination	38 degrees
Landing	30 March 1982 at 4:04:46 pm GMT on Runway 17 at White Sands Missile Range in New Mexico
Distance	5,367,000 km
Orbits	129
Duration	8 days, 0 hours, 4 minutes and 46 seconds
Crew	Col Jack Robert Lousma, 46, US Marine Corps, commander
	Col Charles Gordon Fullerton, 45, US Air Force, pilot

Designation	STS-4
Sequence	4th Shuttle flight and 4th flight of Columbia
Milestones	1st classified Department of Defense payload to fly on board the Shuttle, last two-man Shuttle crew, 1st middeck test of Shuttle spacesuit and last Orbital Flight Test of Columbia
Launch	27 June 1982 at 3:00:00 pm GMT from Pad 39A at Kennedy Space Center in Florida
Payload	Development Flight Instrumentation (DFI), Interim Environmental Contamination Monitor (IECM), Department of Defense Payload (DoD-82-1), Continuous Flow Electrophoresis System (CFES), Monodisperse Latex Reactor (MLR), Shuttle

	Student Involvement Project (SSIP), Vapour Phase Compression Freezer (VPCF) and one Getaway Special (GAS) canister
Max. altitude	315 km
Max. inclination	28.5 degrees
Landing	4 July 1982 at 4:09:31 pm GMT on Runway 22 at Edwards Air Force Base in California
Distance	4,668,000 km
Orbits	112
Duration	7 days, 1 hour, 9 minutes and 31 seconds
Crew	Capt Thomas Kenneth Mattingly II, 46, US Navy, commander Henry Warren Hartsfield Jr, 48, pilot

Designation	**STS-5**
Sequence	5th Shuttle flight and 5th flight of Columbia
Milestones	1st 'commercial' Shuttle mission, 1st Shuttle flight to have a spacewalk cancelled due to space sickness and spacesuit problems, 1st deployment of communications satellites from the Shuttle and 1st time four astronauts had flown into orbit in the same spacecraft
Launch	11 November 1982 at 12:19:00 pm GMT from Pad 39A at Kennedy Space Center in Florida
Payload	Anik-C3 and Satellite Business Systems (SBS)-3, both with Payload Assist Module (PAM)-Ds, Development Flight Instrumentation (DFI), Shuttle Student Involvement Project (SSIP) and one Getaway Special (GAS) canister
Max. altitude	295 km
Max. inclination	28.5 degrees
Landing	16 November 1982 at 2:33:26 pm GMT on Runway 22 at Edwards Air Force Base in California
Distance	3,398,000 km
Orbits	81
Duration	5 days, 2 hours, 14 minutes and 26 seconds
Crew	Vance DeVoe Brand, 51, commander Col Robert Franklyn Overmyer, 46, US Marine Corps, pilot Dr Joseph Percival Allen IV, 45, mission specialist 1 Dr William Benjamin Lenoir, 43, mission specialist 2

Designation	**STS-9**
Sequence	9th Shuttle flight and 6th flight of Columbia
Milestones	1st flight of European-built Spacelab module and pallet in fully outfitted capacity, longest Shuttle mission to date, 1st time six astronauts had flown into orbit in the same spacecraft, 1st flight of a European (and non-US) astronaut on an American spacecraft, 1st man to fly into space for a sixth time and 1st time a crew had been divided into teams to operate experiments around-the-clock

Launch	28 November 1983 at 4:00:00 pm GMT from Pad 39A at Kennedy Space Center in Florida
Payload	Spacelab-1
Max. altitude	250 km
Max. inclination	57 degrees
Landing	8 December 1983 at 11:47:24 pm GMT on Runway 17 at Edwards Air Force Base in California
Distance	6,913,500 km
Orbits	166
Duration	10 days, 7 hours, 47 minutes and 24 seconds
Crew	John Watts Young Jr, 53, commander (Red Team)
	Maj Brewster Hopkinson Shaw Jr, 38, US Air Force, pilot (Blue Team)
	Dr Owen Kay Garriott, 53, mission specialist 1 (Blue Team)
	Dr Robert Allan Ridley Parker, 46, mission specialist 2 (Red Team)
	Dr Byron Kurt Lichtenberg, 35, payload specialist 1 (Blue Team)
	Dr Ulf Dietrich Merbold, 42, payload specialist 2 (Red Team)

Designation	**STS-61C**
Sequence	24th Shuttle flight and 7th flight of Columbia
Milestones	1st flight of Columbia following two-year refit, 1st flight of two astronauts with the same surname on the same spacecraft, 1st congressman to fly on board Columbia and 1st flight of Columbia to have landing opportunities waved off on two successive days due to unacceptable weather conditions at the landing site
Launch	12 January 1986 at 11:55:00 am GMT from Pad 39A at Kennedy Space Center in Florida
Payload	Satcom Ku-1 with Payload Assist Module (PAM)-D2, Materials Science Laboratory (MSL)-2, Comet Halley Active Monitoring Program (CHAMP), Infrared Imaging Experiment (IR-IE), Initial Blood Storage Experiment (IBSE), Hand-held Protein Crystal Growth (HPCG), Shuttle Student Involvement Project (SSIP), Hitchhiker-G1 and a Getaway Special (GAS) bridge with 12 canisters, plus one additional canister attached to the payload bay wall
Max. altitude	340 km
Max. inclination	28.5 degrees
Landing	18 January 1986 at 1:58:51 pm GMT on Runway 22 at Edwards Air Force Base in California
Distance	4,070,000 km
Orbits	97
Duration	6 days, 2 hours, 3 minutes and 51 seconds
Crew	Cdr Robert Lee 'Hoot' Gibson, 39, US Navy, commander
	Lt-Col Charles Frank Bolden Jr, 39, US Marine Corps, pilot

Dr George Driver 'Pinky' Nelson, 35, mission specialist 1
Dr Steven Alan Hawley, 34, mission specialist 2
Dr Franklin Ramon Chang-Diaz, 35, mission specialist 3
Robert Joseph Cenker Jr, 37, payload specialist 1
Congressman Clarens William Nelson, 43, payload specialist 2

Designation	**STS-28**
Sequence	30th Shuttle flight and 8th flight of Columbia
Milestones	1st flight of Columbia in the wake of the Challenger disaster and 1st fully classified Department of Defense mission on board Columbia
Launch	8 August 1989 at 12:37:00 pm GMT from Pad 39B at Kennedy Space Center in Florida
Payload	Classified Department of Defense payload, probably a second-generation Satellite Data Systems (SDS)-B military communications satellite, plus a lighter 'ferret' satellite, Cosmic Ray Upset Experiment (CRUX), Heavy Ion Environment at Low Altitudes (HIEN-LO) and Cloud Logic to Optimise the Use of Defence Systems (CLOUDS)
Max. altitude	305 km
Max. inclination	57 degrees
Landing	13 August 1989 at 1:37:08 pm GMT on Runway 17 at Edwards Air Force Base in California
Distance	3,380,000 km
Orbits	80
Duration	5 days, 1 hour, 0 minutes and 8 seconds
Crew	Col Brewster Hopkinson Shaw Jr, 44, US Air Force, commander Cdr Richard Noel Richards, 42, US Navy, pilot Lt-Col James Craig Adamson, 43, US Army, mission specialist 1 Cdr David Cornell Leestma, 40, US Navy, mission specialist 2 Maj Mark Neil Brown, 37, US Air Force, mission specialist 3

Designation	**STS-32**
Sequence	33rd Shuttle flight and 9th flight of Columbia
Milestones	Longest Shuttle mission to date, 1st women astronauts to fly on board Columbia and 1st satellite retrieval by Columbia
Launch	9 January 1990 at 12:35:00 pm GMT from Pad 39A at Kennedy Space Center in Florida
Payload	Leasat-5, Long Duration Exposure Facility (LDEF) retrieval, Interim Operational Contamination Monitor (IOCM), IMAX payload bay camera, Characterisation of Neurospora Circadian Rhythms (CNCR), Protein Crystal Growth (PCG), Fluid Experiment Apparatus (FEA), American Flight Echocardiograph (AFE), Latitude/Longitude Locator (L3), Mesoscale Lightning Experiment (MLE) and Air Force Maui Optical Site (AMOS)

Max. altitude	330 km
Max. inclination	28.5 degrees
Landing	20 January 1990 at 9:36:39 am GMT on Runway 22 at Edwards Air Force Base in California
Distance	7,260,000 km
Orbits	171
Duration	10 days, 21 hours, 1 minute and 39 seconds
Crew	Capt Daniel Charles Brandenstein, 46, US Navy, commander
	Lt-Cdr James Donald Wetherbee, 37, US Navy, pilot
	Dr Bonnie Jeanne Dunbar, 40, mission specialist 1
	Marsha Sue Ivins, 38, mission specialist 2
	George David Low, 33, mission specialist 3

Designation	**STS-35**
Sequence	38th Shuttle flight and 10th flight of Columbia
Milestones	1st Shuttle mission totally dedicated to astronomical research, 1st fully outfitted pallet-and-igloo combination to fly on board Columbia, oldest man to fly into space to date and 1st seven-man crew to fly on board a Shuttle mission since before the Challenger disaster
Launch	2 December 1990 at 6:49:01 am GMT from Pad 39B at Kennedy Space Center in Florida
Payload	ASTRO-1, Broad-Band X-Ray Telescope (BBXRT), Shuttle Amateur Radio Experiment (SAREX) and Air Force Maui Optical Site (AMOS)
Max. altitude	300 km
Max. inclination	28.5 degrees
Landing	11 December 1990 at 5:54:08 am GMT on Runway 22 at Edwards Air Force Base in California
Distance	6,000,650 km
Orbits	143
Duration	8 days, 23 hours, 5 minutes and 7 seconds
Crew	Vance DeVoe Brand, 59, commander (Red/Blue Team)
	Col Guy Spence Gardner, 42, US Air Force, pilot (Red Team)
	Dr Jeffrey Alan Hoffman, 46, mission specialist 1 (Blue Team)
	John Michael Lounge, 44, mission specialist 2 (Blue Team)
	Dr Robert Allan Ridley Parker, 53, mission specialist 3 (Red Team)
	Dr Samuel Thornton Durrance, 47, payload specialist 1 (Blue Team)
	Dr Ronald Anthony Parise, 39, payload specialist 2 (Red Team)

Designation	**STS-40**
Sequence	41st Shuttle flight and 11th flight of Columbia

Milestones	1st Shuttle mission totally dedicated to life sciences research and 1st time three women had flown into orbit in the same spacecraft
Launch	5 June 1991 at 1:24:51 pm GMT from Pad 39B at Kennedy Space Center in Florida
Payload	Spacelab Life Sciences (SLS)-1, Getaway Special (GAS) bridge with 12 canisters, Orbital Acceleration Research Experiment (OARE) and Middeck Zero-Gravity Dynamics Experiment (MODE)
Max. altitude	260 km
Max. inclination	39 degrees
Landing	14 June 1991 at 3:39:11 pm GMT on Runway 22 at Edwards Air Force Base in California
Distance	6,083,000 km
Orbits	145
Duration	9 days, 2 hours, 14 minutes and 20 seconds
Crew	Col Bryan Daniel O'Connor, 44, US Marine Corps, commander
	Lt-Col Sidney Michael Gutierrez, 39, US Air Force, pilot
	Dr James Philip Bagian, 39, mission specialist 1
	Dr Tamara Elizabeth Jernigan, 32, mission specialist 2
	Dr Margaret Rhea Seddon, 43, mission specialist 3
	Dr Francis Andrew Gaffney, 44, payload specialist 1
	Dr Millie Elizabeth Hughes-Fulford, 45, payload specialist 2

Designation	**STS-50**
Sequence	48th Shuttle flight and 12th flight of Columbia
Milestones	Longest Shuttle mission to date, 1st flight of Extended Duration Orbiter pallet, 1st landing of Columbia in Florida and 1st use of drag chute by Columbia on the runway
Launch	25 June 1992 at 4:12:23 pm GMT from Pad 39A at Kennedy Space Center in Florida
Payload	United States Microgravity Laboratory (USML)-1, Orbital Acceleration Research Experiment (OARE), Extended Duration Orbiter (EDO) pallet, Investigations into Polymer Membrane Processing (IPMP), Shuttle Amateur Radio Experiment (SAREX) and Ultraviolet Plume Instrument (UVPI)
Max. altitude	257 km
Max. inclination	28.5 degrees
Landing	9 July 1992 at 11:42:27 am GMT on Runway 33 at Kennedy Space Center in Florida
Distance	9,266,600 km
Orbits	220
Duration	13 days, 19 hours, 30 minutes and 4 seconds
Crew	Capt Richard Noel Richards, 45, US Navy, commander (Red/Blue Team)
	Lt-Cdr Kenneth Duane Bowersox, 35, US Navy, pilot (Red Team)

Dr Bonnie Jeanne Dunbar, 43, mission specialist 1 (Red Team)
Dr Ellen Louise Baker, 39, mission specialist 2 (Blue Team)
Lt-Col Carl Joseph Meade, 41, US Air Force, mission specialist 3 (Red Team)
Dr Lawrence James DeLucas, 41, payload specialist 1 (Red Team)
Dr Eugene Huu-Chau Trinh, 41, payload specialist 2 (Blue Team)

Designation	**STS-52**
Sequence	51st Shuttle flight and 13th flight of Columbia
Milestones	1st Canadian astronaut to fly on board Columbia
Launch	22 October 1992 at 5:09:39 pm GMT from Pad 39B at Kennedy Space Center in Florida
Payload	Laser Geodynamics Satellite (LAGEOS)-2 with Italian Research Interim Stage (IRIS) and LAGEOS Apogee Stage (LAS), United States Microgravity Payload (USMP)-1, Canadian Experiments (CANEX)-2, Ultraviolet Plume Instrument (UVPI), Commercial Protein Crystal Growth (CPCG), Commercial Materials Dispersion Apparatus Instrument Technology Associates Experiments (CMIX), Chemical Vapour Transport Experiment/Heat Pipe Performance Experiment (CVTE/HPPE), Physiological Systems Experiment (PSE), Shuttle Plume Impingement Experiment (SPIE) and Tank Pressure Control Experiment/Thermal Phenomena (TPCE/TP)
Max. altitude	300 km
Max. inclination	28.5 degrees
Landing	1 November 1992 at 2:05:52 pm GMT on Runway 33 at Kennedy Space Center in Florida
Distance	6,632,000 km
Orbits	158
Duration	9 days, 20 hours, 56 minutes and 13 seconds
Crew	Cdr James Donald Wetherbee, 39, US Navy, commander
	Capt Michael Allen Baker, 38, US Navy, pilot
	Charles Lacy Veach, 48, mission specialist 1
	Capt William McMichael Shepherd, 43, US Navy, mission specialist 2
	Dr Tamara Elizabeth Jernigan, 33, mission specialist 3
	Dr Steven Glenwood MacLean, 37, payload specialist 1
Designation	**STS-55**
Sequence	55th Shuttle flight and 14th flight of Columbia
Milestones	1st on-the-pad main engine shutdown in Columbia's career and STS-55 surpassed a cumulative 365 days-in-space mark by all five Shuttles since April 1981

Launch	26 April 1993 at 2:50:00 pm GMT from Pad 39A at Kennedy Space Center in Florida
Payload	Spacelab-D2 and Shuttle Amateur Radio Experiment (SAREX)
Max. altitude	300 km
Max. inclination	28.5 degrees
Landing	6 May 1993 at 2:29:59 pm GMT on Runway 22 at Edwards Air Force Base in California
Distance	6,673,900 km
Orbits	159
Duration	9 days, 23 hours, 39 minutes and 59 seconds
Crew	Col Steven Ray Nagel, 46, US Air Force, commander (Blue Team)
	Col Terence Thomas Henricks, 40, US Air Force, pilot (Blue Team)
	Col Jerry Lynn Ross, 45, US Air Force, mission specialist 1 (Blue Team)
	Lt-Col Charles Joseph Precourt, 37, US Air Force, mission specialist 2 (Red Team)
	Dr Bernard Andrew Harris Jr, 36, mission specialist 3 (Red Team)
	Dr Ulrich Walter, 39, payload specialist 1 (Blue Team)
	Hans Wilhelm Schlegel, 41, payload specialist 2 (Red Team)

Designation	**STS-58**
Sequence	58th Shuttle flight and 15th flight of Columbia
Milestones	1st dissection of rats in space, longest Shuttle mission to date and 1st professional veterinarian to fly into orbit
Launch	18 October 1993 at 2:53:10 pm GMT from Pad 39B at Kennedy Space Center in Florida
Payload	Spacelab Life Sciences (SLS)-2, Orbital Acceleration Research Experiment (OARE), Extended Duration Orbiter (EDO) pallet, Shuttle Amateur Radio Experiment (SAREX) and Pilot In-Flight Landing Operations Trainer (PILOT)
Max. altitude	286 km
Max. inclination	39 degrees
Landing	1 November 1993 at 3:05:42 pm GMT on Runway 22 at Edwards Air Force Base in California
Distance	9,350,100 km
Orbits	224
Duration	14 days, 0 hours, 12 minutes and 32 seconds
Crew	John Elmer Blaha, 51, commander
	Lt-Col Richard Alan Searfoss, 37, US Air Force, pilot
	Dr Margaret Rhea Seddon, 45, mission specialist 1
	Lt-Col William Surles McArthur Jr, 42, US Army, mission specialist 2

Dr David Alan Wolf, 37, mission specialist 3
Dr Shannon Wells Lucid, 50, mission specialist 4
Dr Martin Joseph Fettman, 36, payload specialist 1

Designation	**STS-62**
Sequence	61st Shuttle flight and 16th flight of Columbia
Milestones	1st all-veteran astronaut crew to fly on board Columbia
Launch	4 March 1994 at 1:53:00 pm GMT from Pad 39B at Kennedy Space Center in Florida
Payload	United States Microgravity Payload (USMP)-2, Office of Aeronautics and Space Technology (OAST)-2, Dexterous End Effector (DEE), Shuttle Solar Backscatter Ultraviolet Instrument (SSBUV), Advanced Protein Crystal Growth (APCG), Physiological Systems Experiment (PSE), Commercial Protein Crystal Growth (CPCG), Middeck Zero-Gravity Dynamics Experiment (MODE), Air Force Maui Optical Site (AMOS), Commercial Generic Bioprocessing Apparatus (CGBA), Bioreactor Demonstration System (BDS), Limited Duration Space Candidate Materials Exposure (LDCE), Biospecimen Specimen Temperature Controller (BSTC) and Extended Duration Orbiter (EDO) pallet
Max. altitude	300 km (temporarily lowered to 265-170 km later in the mission)
Max. inclination	39 degrees
Landing	18 March 1994 at 1:09:41 pm GMT on Runway 33 at Kennedy Space Center in Florida
Distance	9,330,900 km
Orbits	223
Duration	13 days, 23 hours, 16 minutes and 41 seconds
Crew	Col John Howard Casper, 50, US Air Force, commander
	Lt-Col Andrew Michael Allen, 38, US Marine Corps, pilot
	Cdr Pierre Joseph Thuot, 38, US Navy, mission specialist 1
	Lt-Col Charles Donald 'Sam' Gemar, 38, US Army, mission specialist 2
	Marsha Sue Ivins, 42, mission specialist 3

Designation	**STS-65**
Sequence	63rd Shuttle flight and 17th flight of Columbia
Milestones	1st Japanese astronaut to fly on board Columbia and longest Shuttle mission to date
Launch	8 July 1994 at 4:43:00 pm GMT from Pad 39A at Kennedy Space Center in Florida
Payload	International Microgravity Laboratory (IML)-2, Advanced Protein Crystallisation Facility (APCF), Commercial Protein Crystal Growth (CPCG), Air Force Maui Optical Site (AMOS), Orbital Acceleration Research Experiment (OARE), Shuttle Amateur Radio Experiment (SAREX), Military Applications of

	Ship Tracks (MAST) and Extended Duration Orbiter (EDO) pallet
Max. altitude	300 km
Max. inclination	28.5 degrees
Landing	23 July 1994 at 10:38:01 am GMT on Runway 33 at Kennedy Space Center in Florida
Distance	9,781,700 km
Orbits	234
Duration	14 days, 17 hours, 55 minutes and 1 second
Crew	Col Robert Donald Cabana, 45, US Marine Corps, commander (Red Team)
	Lt-Col James Donald Halsell Jr, 37, US Air Force, pilot (Red Team)
	Richard James Hieb, 38, mission specialist 1 (Red Team)
	Lt-Col Carl Erwin Walz, 38, US Air Force, mission specialist 2 (Blue Team)
	Dr Leroy Chiao, 33, mission specialist 3 (Blue Team)
	Dr Donald Alan Thomas, 39, mission specialist 4 (Blue Team)
	Dr Chiaki Mukai, 42, payload specialist 1 (Red Team)

Designation	**STS-73**
Sequence	72nd Shuttle flight and 18th flight of Columbia
Milestones	Youngest Shuttle Commander to date
Launch	20 October 1995 at 1:53:00 pm GMT from Pad 39B at Kennedy Space Center in Florida
Payload	United States Microgravity Laboratory (USML)-2, Orbital Acceleration Research Experiment (OARE), Three-Dimensional Microgravity Accelerometer (3DMA), Suppression of Transient Accelerations by Levitation Evaluation (STABLE) and Extended Duration Orbiter (EDO) pallet
Max. altitude	277 km
Max. inclination	39 degrees
Landing	5 November 1995 at 11:45:22 am GMT on Runway 33 at Kennedy Space Center in Florida
Distance	10,621,670 km
Orbits	255
Duration	15 days, 21 hours, 52 minutes and 22 seconds
Crew	Cdr Kenneth Duane Bowersox, 38, US Navy, commander (Red Team)
	Cdr Kent Vernon Rominger, 39, US Navy, pilot (Red Team)
	Capt Dr Catherine Grace Coleman, 34, US Air Force, mission specialist 1 (Blue Team)
	Lt-Cdr Michael Eladio Lopez-Alegria, 37, US Navy, mission specialist 2 (Blue Team)
	Dr Kathryn Cordell Thornton, 43, mission specialist 3 (Red Team)

Dr Fred Weldon Leslie, 43, payload specialist 1 (Blue Team)
Dr Albert Sacco Jr, 46, payload specialist 2 (Red Team)

Designation	**STS-75**
Sequence	75th Shuttle flight and 19th flight of Columbia
Milestones	1st mission of Columbia to carry astronauts of three nationalities, 1st Italian to fly on board Columbia and 1st Swiss to fly on board Columbia
Launch	22 February 1996 at 8:18:00 pm GMT from Pad 39B at Kennedy Space Center in Florida
Payload	Tethered Satellite System (TSS)-1 Reflight, United States Microgravity Payload (USMP)-3, Orbital Acceleration Research Experiment (OARE), Commercial Protein Crystal Growth (CPCG), Middeck Glovebox (MGBX) and Extended Duration Orbiter (EDO) pallet
Max. altitude	300 km
Max. inclination	28.5 degrees
Landing	9 March 1996 at 1:58:21 pm GMT on Runway 33 at Kennedy Space Center in Florida
Distance	10,460,740 km
Orbits	251
Duration	15 days, 17 hours, 40 minutes and 21 seconds
Crew	Lt-Col Andrew Michael Allen, 40, US Marine Corps, commander (White/Blue Team)
	Maj Dr Scott Jay 'Doc' Horowitz, 38, US Air Force, pilot (Red Team)
	Dr Jeffrey Alan Hoffman, 51, mission specialist 1 (White/Red Team)
	Lt-Col Maurizio Cheli, 36, Italian Air Force, mission specialist 2 (Red Team)
	Claude Nicollier, 51, mission specialist 3 (Blue Team)
	Dr Franklin Ramon Chang-Diaz, 45, mission specialist 4 (Blue Team)
	Umberto Guidoni, 41, payload specialist 1 (Red Team)
Designation	**STS-78**
Sequence	78th Shuttle flight and 20th flight of Columbia
Milestones	Longest Shuttle mission to date and 1st French astronaut to fly on board Columbia
Launch	20 June 1996 at 2:49:00 pm GMT from Pad 39B at Kennedy Space Center in Florida
Payload	Life and Microgravity Spacelab (LMS), Space Acceleration Measurement System (SAMS), Orbital Acceleration Research Experiment (OARE) and Extended Duration Orbiter (EDO) pallet

Max. altitude	280 km
Max. inclination	39 degrees
Landing	7 July 1996 at 12:37:30 pm GMT on Runway 33 at Kennedy Space Center in Florida
Distance	11,265,000 km
Orbits	271
Duration	16 days, 21 hours, 48 minutes and 30 seconds
Crew	Col Terence Thomas Henricks, 43, US Air Force, commander
	Kevin Richard Kregel, 39, pilot
	Dr Richard Michael Linnehan, 38, mission specialist 1
	Lt-Col Susan Jane Helms, 38, US Air Force, mission specialist 2
	Cdr Dr Charles Eldon Brady Jr, 44, US Navy, mission specialist 3
	Dr Jean-Jacques Favier, 47, payload specialist 1
	Dr Robert Brent Thirsk, 43, payload specialist 2

Designation	**STS-80**
Sequence	80th Shuttle flight and 21st flight of Columbia
Milestones	Longest Shuttle mission to date, oldest man to fly into space to date, 1st double satellite deployment and retrieval conducted by Columbia and 1st spacewalk cancelled due to airlock hatch problems
Launch	19 November 1996 at 7:55:47 pm GMT from Pad 39B at Kennedy Space Center in Florida
Payload	Orbiting and Retrievable Far and Extreme Ultraviolet Spectrometer-Shuttle Pallet Satellite (ORFEUS-SPAS)-2, Wake Shield Facility (WSF)-3, Extravehicular Activity Development Flight Test (EDFT), Space Experiment Module (SEM), National Institutes of Health (NIH)-R4, Commercial Materials Dispersion Apparatus Instrument Technology Associates Experiments (CMIX), Biological Research In Canisters (BRIC), Cell Culture Module (CCM) and Extended Duration Orbiter (EDO) pallet
Max. altitude	350 km
Max. inclination	28.5 degrees
Landing	7 December 1996 at 11:49:05 am GMT on Runway 33 at Kennedy Space Center in Florida
Distance	12,000,000 km
Orbits	277
Duration	17 days, 15 hours, 53 minutes and 18 seconds
Crew	Kenneth Dale Cockrell, 46, commander
	Cdr Kent Vernon Rominger, 40, US Navy, pilot
	Dr Tamara Elizabeth Jernigan, 37, mission specialist 1
	Dr Thomas David Jones, 41, mission specialist 2
	Dr Franklin Story Musgrave, 61, mission specialist 3

Designation   **STS-83**
Sequence   83rd Shuttle flight and 22nd flight of Columbia
Milestones   1st female Pilot to fly on board Columbia
Launch   4 April 1997 at 7:20:32 pm GMT from Pad 39A at Kennedy
                       Space Center in Florida
Payload   Microgravity Science Laboratory (MSL)-1, Shuttle Amateur
                       Radio Experiment (SAREX) and Extended Duration Orbiter
                       (EDO) pallet
Max. altitude   300 km
Max. inclination   28.5 degrees
Landing   8 April 1997 at 6:33:11 pm GMT on Runway 33 at Kennedy
                       Space Center in Florida
Distance   2,414,000 km
Orbits   63
Duration   3 days, 23 hours, 12 minutes and 39 seconds
Crew   Lt-Col James Donald Halsell Jr, 40, US Air Force, commander
                       (Red Team)
                       Lt-Cdr Susan Leigh Still, 35, US Navy, pilot (Red Team)
                       Dr Janice Elaine Voss, 40, mission specialist 1 (Blue Team)
                       Dr Michael Landon Gernhardt, 40, mission specialist 2 (Blue
                       Team)
                       Dr Donald Alan Thomas, 41, mission specialist 3 (Red Team)
                       Dr Roger Keith Crouch, 56, payload specialist 1 (Blue Team)
                       Dr Gregory Thomas Linteris, 39, payload specialist 2 (Red
                       Team)

Designation   **STS-94**
Sequence   85th Shuttle flight and 23rd flight of Columbia
Milestones   1st astronaut crew launched into orbit on a second occasion to
                       complete the same mission, post-Challenger record turnaround
                       time for Columbia and shortest interval between two flights for
                       an astronaut crew
Launch   1 July 1997 at 6:02:00 pm GMT from Pad 39A at Kennedy Space
                       Center in Florida
Payload   MSL-1 Reflight, Shuttle Amateur Radio Experiment (SAREX)
                       and Extended Duration Orbiter (EDO) pallet
Max. altitude   300 km
Max. inclination   28.5 degrees
Landing   17 July 1997 at 10:46:36 am GMT on Runway 33 at Kennedy
                       Space Center in Florida
Distance   9,977,930 km
Orbits   250
Duration   15 days, 16 hours, 44 minutes and 36 seconds
Crew   Lt-Col James Donald Halsell Jr, 40, US Air Force, commander
                       (Red Team)

Lt-Cdr Susan Leigh Still, 35, US Navy, pilot (Red Team)
Dr Janice Elaine Voss, 40, mission specialist 1 (Blue Team)
Dr Michael Landon Gernhardt, 41, mission specialist 2 (Blue Team)
Dr Donald Alan Thomas, 42, mission specialist 3 (Red Team)
Dr Roger Keith Crouch, 56, payload specialist 1 (Blue Team)
Dr Gregory Thomas Linteris, 39, payload specialist 2 (Red Team)

Designation	**STS-87**
Sequence	88th Shuttle flight and 24th flight of Columbia
Milestones	1st flight of Columbia to carry representatives of four nations, 1st Ukrainian cosmonaut to fly on board the Shuttle, 1st spacewalk from Columbia and 1st Indian woman in space
Launch	19 November 1997 at 7:46:00 pm GMT from Pad 39B at Kennedy Space Center in Florida
Payload	United States Microgravity Payload (USMP)-4, Shuttle Pointed Autonomous Research Tool for Astronomy (SPARTAN)-201-4, Loop Heat Pipe/Sodium Sulphur Battery Experiment (LHP/NaSBE), Turbulent Gas Jet Diffusion Flames (TGDF), Shuttle Ozone Limb Sounding Experiment (SOLSE), Extravehicular Activity Development Flight Test (EDFT), Orbital Acceleration Research Experiment (OARE), one Getaway Special (GAS) canister, Collaborative Ukrainian Experiment (CUE), Middeck Glovebox (MGBX), Autonomous Extravehicular Activity Robotic Camera/Sprint (AERCam/Sprint), Shuttle Ionospheric Modification with Pulsed Local Exhaust (SIMPLEX) and Extended Duration Orbiter (EDO) pallet
Max. altitude	280 km
Max. inclination	28.5 degrees
Landing	5 December 1997 at 12:20:02 pm GMT on Runway 33 at Kennedy Space Center in Florida
Distance	10,460,730 km
Orbits	251
Duration	15 days, 16 hours, 34 minutes and 2 seconds
Crew	Kevin Richard Kregel, 41, commander
	Maj Steven Wayne Lindsey, 37, US Air Force, pilot
	Dr Kalpana Chawla, 36, mission specialist 1
	Capt Winston Elliott Scott, 47, US Navy, mission specialist 2
	Dr Takao Doi, 43, mission specialist 3
	Col Leonid Konstantinovich Kadenyuk, 46, Russian Air Force, payload specialist 1

Designation	**STS-90**
Sequence	90th Shuttle flight and 25th flight of Columbia

Milestones	Last dedicated Spacelab mission using the pressurised research module and Columbia became the 1st Shuttle to make a 25th flight
Launch	17 April 1998 at 6:19:00 pm GMT from Pad 39B at Kennedy Space Center in Florida
Payload	Neurolab, three Getaway Special (GAS) canisters and Extended Duration Orbiter (EDO) pallet
Max. altitude	275 km
Max. inclination	39 degrees
Landing	3 May 1998 at 4:08:58 pm GMT on Runway 33 at Kennedy Space Center in Florida
Distance	10,260,000 km
Orbits	255
Duration	15 days, 21 hours, 49 minutes and 58 seconds
Crew	Col Richard Alan Searfoss, 41, US Air Force, commander
	Cdr Scott Douglas 'Scooter' Altman, 38, US Navy, pilot
	Dr Richard Michael Linnehan, 40, mission specialist 1
	Kathryn Patricia Hire, 38, mission specialist 2
	Dr Dafydd Rhys Williams, 43, mission specialist 3
	Dr Jay Clark Buckey Jr, 41, payload specialist 1
	Dr James Anthony Pawelczyk, 37, payload specialist 2

Designation	**STS-93**
Sequence	95th Shuttle flight and 26th flight of Columbia
Milestones	1st space mission under the command of a woman
Launch	23 July 1999 at 4:31:00 am GMT from Pad 39B at Kennedy Space Center in Florida
Payload	Chandra X-ray Observatory (CXO) with Inertial Upper Stage (IUS), Midcourse Space Experiment (MSX), Shuttle Ionospheric Modification with Pulsed Local Exhaust (SIMPLEX), Gelation of Sols: Applied Materials Research (GOSAMR), Space Tissue Loss (STL), Cell Culture Module (CCM), Lightweight Flexible Solar Array Hinge (LFSAH), Shuttle Amateur Radio Experiment (SAREX), EarthKAM, Plant Growth Investigations in Microgravity (PGIM), Commercial Generic Bioprocessing Apparatus (CGBA), Micro-Electrical Mechanical System (MEMS), Biological Research in Canisters (BRIC) and Southwest Ultraviolet Imaging System (SWUIS)
Max. altitude	270 km
Max. inclination	28.5 degrees
Landing	28 July 1999 at 3:20:36 am GMT on Runway 33 at Kennedy Space Center in Florida
Distance	2,900,000 km
Orbits	79
Duration	4 days, 22 hours, 49 minutes and 36 seconds

Crew	Col Eileen Marie Collins, 42, US Air Force, commander
	Capt Jeffrey Shears Ashby, 45, US Navy, pilot
	Lt-Col Dr Catherine Grace Coleman, 38, US Air Force, mission specialist 1
	Dr Steven Alan Hawley, 47, mission specialist 2
	Col Michel Ange-Charles Tognini, 49, French Air Force, mission specialist 3

Designation	**STS-109**
Sequence	108th Shuttle flight and 27th flight of Columbia
Milestones	Last fully successful mission by Columbia
Launch	1 March 2002 at 11:22:02 am GMT from Pad 39A at Kennedy Space Center in Florida
Payload	Hubble Space Telescope Servicing Mission (HST SM)-3B
Max. altitude	580 km
Max. inclination	28.5 degrees
Landing	12 March 2002 at 9:32:00 am GMT on Runway 33 at Kennedy Space Center in Florida
Distance	8,700,000 km
Orbits	165
Duration	10 days, 22 hours, 9 minutes and 58 seconds
Crew	Cdr Scott Douglas 'Scooter' Altman, 42, US Navy, commander
	Lt-Col Duane Gene 'Digger' Carey, 44, US Air Force, pilot
	Dr John Mace Grunsfeld, 43, mission specialist 1
	Lt-Col Dr Nancy Jane Currie, 43, US Army, mission specialist 2
	Dr Richard Michael Linnehan, 44, mission specialist 3
	Dr James Hansen Newman, 45, mission specialist 4
	Dr Michael James Massimino, 39, mission specialist 5

Designation	**STS-107**
Sequence	113th Shuttle flight and 28th flight of Columbia
Milestones	1st US manned space mission to be lost during re-entry, 4th-longest flight in Shuttle history, 1st flight of Spacehab on board Columbia and 1st Israeli astronaut to fly into space
Launch	16 January 2003 at 3:39:00 pm GMT from Pad 39A at Kennedy Space Center in Florida
Payload	Spacehab Research Double Module (RDM), Fast Reaction Experiments Enabling Science, Technology, Applications and Research (FREESTAR) and Extended Duration Orbiter (EDO) pallet
Max. altitude	280 km
Max. inclination	39 degrees
Landing	Scheduled to land at 2:16 pm on 1 February 2003 on Runway 33 at Kennedy Space Center in Florida, Columbia disintegrated

	during re-entry at approximately 2:00 pm, while flying 61 km over north-central Texas
Distance	Approximately 10,700,000 km
Orbits	255
Duration	15 days, 22 hours, 20 minutes and 22 seconds (to loss of tracking)
Crew	Col Rick Douglas Husband, 45, US Air Force, commander (Red Team)
	Cdr William Cameron McCool, 41, US Navy, pilot (Blue Team)
	Capt Dr David McDowell Brown, 46, US Navy, mission specialist 1 (Blue Team)
	Dr Kalpana Chawla, 41, mission specialist 2 (Red Team)
	Lt-Col Michael Phillip Anderson, 43, US Air Force, mission specialist 3 (Blue Team)
	Capt Dr Laurel Blair Clark, 41, US Navy, mission specialist 4 (Red Team)
	Col Ilan Ramon, 48, Israeli Air Force, payload specialist 1 (Red Team)

Index

Boldface figures indicate an illustration.

Printing: Mercedes-Druck, Berlin
Binding: Stein+Lehmann, Berlin